Race, Class, and Gender

An Anthology

Race, Class, and Gender

An Anthology

NINTH EDITION

MARGARET L. ANDERSEN
University of Delaware

PATRICIA HILL COLLINS
University of Maryland

Australia • Brazil • Japan • Korea • Mexico • Singapore • Spain • United Kingdom • United States

***Race, Class, and Gender: An Anthology*, Ninth Edition**
Margaret L. Andersen and Patricia Hill Collins

Product Director: Marta Lee-Perriard

Product Manager: Jennifer Harrison

Content Developer: Lori Bradshaw

Product Assistant: Chelsea Meredith

Media Developer: John Chell

Marketing Manager: Kara Kindstrom

Content Project Manager: Cheri Palmer

Art Director: Michelle Kunkler

Manufacturing Planner: Judy Inouye

Production, Composition, and Illustration: MPS Limited

Text Researcher: Lumina Datamatics

Copy Editor: Heather McElwain

Cover Designer: Tin Box Studio

Cover Image: Otto Krause/iStock

WCN: 01-100-101

Library of Congress Control Number: 2014957667

ISBN: 978-1-305-09361-4

Cengage Learning
20 Channel Center Street
Boston, MA 02210
USA

Printed in the United States of America
Print Number: 01 Print Year: 2015

Contents

Preface

We write this preface at a time when the social dynamics of race, class, and gender are changing. Many believe that race no longer matters as a predictor of one's chances in life. After all, the United States elected a Black president, something that would have been unimaginable not that many years ago. Some would argue that social class matters more than race, especially given the presence of an African American and Latino middle class. Women are now CEOs of major technology companies. Yet, these visible signs of change may cloak the simultaneous presence of highly segregated Black and Latino communities, many of them struggling in the face of poverty. The middle class, long the hallmark of our democratic society, struggles to keep a firm foothold in the American economy. Although women at the top are doing very well, the majority of women still struggle with low wages and the challenge of working and supporting their families, often on their own.

The U.S. economy, though recovering somewhat from the financial crisis that began in 2009, has also left many people in financial distress. People who had worked their entire lives have seen their life's savings disappear as their houses dropped in value and their savings accounts for sending their kids to college or for their own retirement shrank. Homelessness continues to mark our city streets. Even college students, whom you might think of as immune to global economic trends, are seeing rising tuition costs as states reduce their support for public education. Debt and economic uncertainty are the result.

Not everyone experiences these changes in the same way. To explain people's life situations requires understanding how race, class, and gender shape the American opportunity structure. How are race, class, and gender systematically interrelated, and what is their relationship to other social factors?

That is the theme of this book: how race, class, and gender simultaneously shape social issues and experiences in the United States. Central to the book is the idea that race, class, and gender are interconnected and that they must be understood as operating together if you want to understand the experiences of diverse groups and particular issues and events in society. We want this book to help students see how the lives of different groups develop in the context of their race, class, and gender location in society.

Since the publication of the first edition of this book, the study of race, class, and gender has become much more present in people's thinking. Over the years that this book has been published, there has also been an enormous growth in the research scholarship that is anchored in an intersectional framework. Still, people continue to treat race, class, and gender in isolation from the others; some also treat race, class, and gender as if they were equivalent experiences. Although we see them as interrelated—and sometimes similar in how they work—we also understand that each has its own dynamic, but a dynamic that can only be truly understood in relationship to the others. With the growth of race, class, and gender studies, we can also now better understand how other social factors, such as sexuality, nationality, age, and disability, are connected to the social structures of race, class, and gender. We hope that this book helps students understand how these structural phenomena—that is, the social forces of race, class, and gender and their connection with other social variables—are deeply embedded in the social structure of society.

This anthology is, thus, more than a collection of readings. Our book is strongly centered in an analytical framework about the interconnections among race, class, and gender. In this edition we continue our efforts to further develop a framework of the *intersectionality* of race, class, and gender, or as Patricia Hill Collins has labeled it, the *matrix of domination*. The organization of the book features this framework. Our introductory essay distinguishes an intersectional framework from other models of studying "difference." The four parts of the book are intended to help students see the importance of this intersectional framework, to engage critically the core concepts on which the framework is based, and to analyze different social institutions and current social issues using this framework, including being able to apply it to understanding social change.

ORGANIZATION OF THE BOOK

The four major parts of the book reflect these goals. We introduce each of the four parts with an essay we wrote to analyze the issues raised by the reading selections. These essays are an important part of this book because they establish the conceptual foundation that we use to think about race, class, and gender.

As in past editions, we include essays in Part I ("Why Race, Class, and Gender Still Matter") that engage students in personal narratives, as a way of helping them step beyond their own social location and to see how race, class, gender, sexuality, and other social factors shape people's lives differently. In this edition, we have also included some of the now classic pieces in intersectional studies, and we include two pieces that provide some historical foundation for how race, class, and gender have evolved in American society. We want this section to show students the very different experiences that anchor the study of race, class, and gender. We therefore begin our book with essays that show their continuing, even if changing, significance.

Part II, "Systems of Power and Inequality," provides the conceptual foundation for understanding how race, class, and gender are linked together and how they link with other systems of power and inequality, especially ethnicity and sexuality. Here we want students to understand that individual identities and experiences are structured by intersecting systems of power. The essays in Part II link ethnicity, nationality, and sexuality to the study of race, class, and gender. We treat each of these separately here, not because we think they stand alone, but to show students how each operates so they can better see their interlocking nature. The introductory essay provides working definitions for these major concepts and presents some of the contemporary data that will help students see how race, class, and gender stratify contemporary society.

Part III, "The Structure of Social Institutions," examines how intersecting systems of race, class, and gender shape the organization of social institutions and how, as a result, these institutions affect group experience. Social scientists routinely document how Latinos, African Americans, women, workers, and other distinctive groups are affected by institutional structures. We know this is true but want to go beyond these analyses to scrutinize how institutions are themselves constructed through race, class, and gender relations. As categories of social experience, race, class, and gender shape all social institutions and systems of meaning. In this edition we have added a new section on "Bodies, Beauty, and Sports" to analyze the institutions that shape the bodily experiences of people, especially younger people.

We have revised Part IV, "An Intersectional Framework for Change: From the Local to the Global," to show students the very different contexts in which an intersectional perspective can inform social practices. Many anthologies use their final section to show how students can make a difference in society, once they understand the importance of race, class, and gender. We think this is a tall order for students who may have had only a few weeks to begin understanding how race, class, and gender matter—and matter together. By showing the different contexts for social change—ranging from group-based education in counseling to global transformation—we hope to show students how an intersectional framework can shape one's action in both local, national, and global contexts.

This book is grounded in a sociological perspective, although the articles come from different perspectives, disciplines, and experiences. Several articles provide a historical foundation for understanding how race, class, and gender have emerged. We also include materials that bring a global dimension to the study of race, class, and gender—not just by looking comparatively at other cultures but also by analyzing how globalization is shaping life in the United States.

Not all articles accomplish this as much as we would like, but we try not to select articles that focus exclusively on one issue while ignoring the others. In this regard, our book differs significantly from other anthologies on race, class, and gender that include many articles on each factor, but do less to show how they are connected. We also distinguish our book from those that are centered in a multicultural perspective. Although multiculturalism is important, we think that race, class, and gender go beyond the appreciation of cultural differences. Rather, we see race, class, and gender as embedded in the structure of society and significantly influencing group cultures and opportunities. Race, class, and gender are structures of group opportunity, power, and privilege, not just cultural differences. We search for articles that are conceptually and theoretically informed and at the same time accessible to undergraduate readers. Although it is important to think of race, class, and gender as analytical categories, we do not want to lose sight of how they affect human experiences and feelings; thus, we include personal narratives that are reflective and analytical. We think that personal accounts generate empathy and also help students connect personal experiences to social structural conditions.

We know that developing a complex understanding of the interrelationships between race, class, and gender is not easy and involves a long-term process engaging personal, intellectual, and political change. We do not claim to be models of perfection in this regard. We have been pleased by the strong response to the first eight editions of this book, and we are fascinated by how race, class, and gender studies have developed since the publication of our first edition. We know further work is needed. Our own teaching and thinking has been transformed by developing this book. We imagine many changes still to come.

NEW TO THE NINTH EDITION

In the years since race/class/gender studies first evolved, a large and growing body of scholarship and activism has emerged utilizing this inclusive perspective. Such work makes the job of assembling this collection both easier and harder: It is easier because there is more intersectional work to choose from, but harder because of having to make difficult decisions about what to include. As in earlier editions, we have selected articles based primarily on two criteria: (1) accessibility to undergraduate readers and the general public, not just highly trained specialists; and, (2) articles that are grounded in race *and* class *and* gender—in other words, intersectionality.

We have made several changes in the ninth edition, including the following:

- 31 new readings;
- a new section on bodies, beauty, and sports;
- a completely revised final section focusing on intersectional change in different contexts;
- more readings with a global framework;
- more empirical research framed by an intersectional analysis;
- continued focus on the media and popular culture, but new readings on youth, social networking, and the Internet;
- four revised introductions, one of the noted strengths of our book compared to others; and,

- new material on race, class, and gender based on important current issues, including immigration, social media, police in poor, urban communities, growing inequality, white privilege, sexuality, jobs, family leave, school segregation, disability studies, and food sustainability.

PEDAGOGICAL FEATURES

We realize that the context in which you teach matters. If you teach in an institution where students are more likely to be working class, perhaps how the class system works will be more obvious to them than it is for students in a more privileged college environment. Many of those who use this book will be teaching in segregated environments, given the high degree of segregation in education. Thus, how one teaches this book should reflect the different environments where faculty work. Ideally, the material in this book should be discussed in a multiracial, multicultural atmosphere, but we realize that is not always the case. We hope that the content of the book and the pedagogical features that enhance it will help bring a more inclusive analysis to educational settings than might be there to start with.

We see this book as more than just a collection of readings. The book has an analytical logic to its organization and content, and we think it can be used to format a course. Of course, some faculty will use the articles in an order different from how we present them, but we hope the four parts will help people develop the framework for their course. We also provide pedagogical tools to help people expand their teaching and learning beyond the pages of the book.

We have included features with this edition that provide faculty with additional teaching tools. They include the following:

- *Instructor's manual.* This edition includes an instructor's manual with suggestions for classroom exercises, discussion and examination questions, and course assignments.
- *Index.* The index will help students and faculty locate particular topics in the book quickly and easily.
- *Cengage Learning Testing, powered by Cognero Instant Access.* This is a flexible, online system that allows you to author, edit, and manage test bank content from multiple Cengage Learning solutions; create multiple test versions in an instant; and deliver tests from your LMS, your classroom, or wherever you want.

A NOTE ON LANGUAGE

Reconstructing existing ways of thinking to be more inclusive requires many transformations. One transformation needed involves the language we use when referring to different groups. Language reflects many assumptions about race, class, and gender; and for that reason, language changes and evolves as knowledge changes. The term *minority,* for example, marginalizes groups, making them seem somehow outside the mainstream or dominant culture. Even worse, the phrase *non-White,* routinely used by social scientists, defines groups in terms of what they are not and assumes that Whites have the universal experiences against which the experiences of all other groups are measured. We have consciously avoided using both of these terms throughout this book, although this is sometimes unavoidable.

We have capitalized Black in our writing because of the specific historical experience, varied as it is, of African Americans in the United States. We also capitalize White when referring to a particular group experience; however, we recognize that White American is no more a uniform experience than is African American. We use *Hispanic* and *Latina/o* interchangeably, though we recognize that is not how groups necessarily define themselves. When citing data from other sources (typically government documents), we use *Hispanic* because that is usually how such data are reported.

Language becomes especially problematic when we want to talk about features of experience that different groups share. Using shortcut terms like Hispanic, Latina/o, Native American, and women of color homogenizes distinct historical experiences. Even the term *White* falsely unifies experiences across such factors as ethnicity, religion, class, and gender, to name a few. At times, though, we want to talk of common experiences across different groups, so we have used labels such as Latina/o, Asian American, Native American, and women of color to do so. Unfortunately, describing groups in this way reinforces basic categories of oppression. We do not know how to resolve this problem but want readers to be aware of the limitations and significance of language as they try to think more inclusively about diverse group experiences.

ACKNOWLEDGMENTS

An anthology rests on the efforts of more people than the editors alone. This book has been inspired by our work with scholars and teachers from around the country who are working to make their teaching and writing more inclusive and sensitive to the experiences of all groups. Over the years of our own collaboration, we have each been enriched by the work of those trying to make higher education a more equitable and fair institution. In that time, our work has grown from many networks that have generated new race, class, and gender scholars. These associations continue to sustain us. Many people contributed to the development of this book. We especially thank D. Stanley Eitzen, Maxine Baca Zinn, Elizabeth Higginbotham, Valerie Hans, and the Boston Area Feminist Scholars Group for the inspiration, ideas, suggestions, and support.

We appreciate the support given by our institutions, with special thanks to President Patrick Harker, Vice President Patricia Wilson, and Executive Vice President Scott Douglass of the University of Delaware for providing the financial support that supported Dana Alvare to assist with this edition. Many thanks go to Dana for helping so much with this edition. Thank you as well to Delaware's Provost Domenico Grasso for supporting the scholarship of his senior administration. A very special thanks go to Judy Allarey, Dana Brittingham, Sandy Buchanan, Sarah Hedrick, Rita Scott, Joan Stock, and Susan Williams for the help and good cheer that they provide every day; their efforts make the task of balancing an administrative job and writing much more possible and enjoyable.

We thank the team at Cengage for their encouragement and support for this project. Most particularly, we thank Lori Bradshaw for expertly overseeing all of the production details. We also thank the anonymous reviewers who provided valuable commentary on the prior edition and thus helped enormously in the development of the ninth edition.

This book has evolved over many years and through it all we have been lucky to have the love and support of Richard, Roger, Valerie, and Lauren. We thank them for the love and support that anchors our lives. And, with this edition, we welcome Aubrey Emma Hanerfeld and Harrison Collins Pruitt with hopes that the worlds they encounter will be just, and inclusive, helping them thrive in whatever paths they take.

About the Editors

Margaret L. Andersen (B.A. Georgia State University; M.A., Ph.D., University of Massachusetts, Amherst) is the Edward F. and Elizabeth Goodman Rosenberg Professor of Sociology at the University of Delaware where she also holds joint appointments in Black American Studies and Women's Studies; she has also served as the Vice Provost for Faculty Affairs and Diversity. She has received two teaching awards at the University of Delaware. She has published numerous books and articles, including *Thinking about Women: Sociological Perspectives on Sex and Gender* (10th ed., Pearson, 2015); *Race and Ethnicity in Society: The Changing Landscape* (edited with Elizabeth Higginbotham, 3rd ed., Cengage, 2012); *On Land and On Sea: A Century of Women in the Rosenfeld Collection* (Mystic Seaport Museum, 2007); *Living Art: The Life of Paul R. Jones, African American Art Collector* (University of Delaware Press, 2009); and *Sociology: The Essentials* (co-authored with Howard F. Taylor and Kim Logio; Cengage, 2014). She received the American Sociological Association's Jessie Bernard Award for expanding the horizons of sociology to include the study of women and the Eastern Sociological Society's Merit Award and Robin Williams Lecturer Award. She is a past vice president of the American Sociological Association and past president of the Eastern Sociological Society.

Patricia Hill Collins (B.A., Brandeis University; M.A.T., Harvard University; Ph.D., Brandeis University) is distinguished university professor of sociology at the University of Maryland, College Park, and Charles Phelps Taft Emeritus Professor of African American Studies and Sociology at the University of Cincinnati. She is the author of numerous articles and books including *On Intellectual Activism* (Temple University, 2013), *Another Kind of Public Education: Race, Schools, the Media and Democratic Possibilities* (Beacon, 2009), *From Black Power to Hip Hop: Racism, Nationalism and Feminism* (Temple University, 2006); *Black Sexual Politics: African Americans, Gender and the New Racism* (Routledge, 2004), which won the Distinguished Publication Award from the American Sociological Association; *Fighting Words* (University of Minnesota, 1998); and *Black Feminist Thought: Knowledge, Consciousness, and the Politics of Empowerment* (Routledge, 1990, 2000), which won the Jessie Bernard Award of the American Sociological Association and the C. Wright Mills Award of the Society for the Study of Social Problems. In 2008–2009, she served as the 100th president of the American Sociological Association.

About the Contributors

Joan Acker is professor emerita of sociology at the University of Oregon. She founded and directed the University of Oregon's Center for the Study of Women in Society and is the recipient of the American Sociological Association's Career of Distinguished Scholarship Award as well as the Jessie Bernard Award for feminist scholarship. She is the author of *Class Questions, Feminist Answers* as well as many other works in the areas of gender, institutions, and class.

Michelle Alexander is associate professor of law at the Moritz College of Law at Ohio State University. She has served as director of the Racial Justice Project for the ACLU of Northern California, and has clerked for Justice Harry A. Blackmun on the United States Supreme Court.

Maxine Baca Zinn is professor emerita of sociology, Michigan State University. Widely known for her work on Latina families and family diversity, she is the co-author (with D. Stanley Eitzen) of *Diversity in Families; In Conflict and Order; Globalization: The Transformation of Social Worlds;* and *Social Problems,* and the co-editor of *Gender through the Prism of Difference: A Sex and Gender Reader* (with Pierrette Hondagneu-Sotelo and Michael A. Messner). She is a recipient of the American Sociological Association Jessie Bernard Award for opening the horizons of sociology for women.

Janani Balasubramanian is a graduate of Stanford University where she double majored in feminist studies (queer studies) and engineering (atmosphere and energy). She is a South Asian performance and literary artist.

Marianne Bertrand is Chris P. Dialynas Professor of Economics and Neubauer Family Faculty Fellow at Chicago Booth University School of Business. Her work has been published in the *Quarterly Journal of Economics,* the *Journal of Political Economy,* the *American Economic Review,* and the *Journal of Finance,* among others.

Hanne Blank is a historian, writer, editor, and public speaker about the intersection of bodies, self, and culture. Her books include *The Surprisingly Short History of Heterosexuality,* and *The Unapologetic Fat Girl's Guide to Exercise and Other Incendiary Acts,* among others.

Denise Brennan is associate professor and chair of anthropology in the Department of Anthropology at Georgetown University. She is the author of several books and articles about the global sex trade, human trafficking, and women's labor, primarily in Latin America and the Caribbean.

Rod K. Brunson is associate professor in the School of Criminal Justice at Rutgers University where he is also the Vice Dean for Academic Affairs. His research focuses on youth experiences in neighborhood contexts, focusing on the dynamics of race, class, and gender.

Erica Chito Childs is associate professor of sociology at Hunter College, City University of New York. She is a leading qualitative researcher of multiracism, families, media, and popular culture. Her books include *Navigating Interracial Borders: Black-White Couples and Their Social Worlds* and *Fade to Black and White: Interracial Images in Popular Culture.*

Judith Ortiz Cofer is the Regents' and Franklin Professor of English and Creative Writing, Emerita at the University of Georgia. She is the author of numerous books of poetry and essays, and a novel, *The Line of the Sun.* She received the 2003 Américas Award for her book *The Meaning of Consuelo.*

Bethany M. Coston is a doctoral candidate in sociology at Stony Brook University. She has authored numerous publications in her areas of research interest including gender, sexualities, and intimate partner violence.

Tressie McMillan Cottom is a doctoral candidate in sociology at Emory University. Her research focuses on the socioeconomic conditions and social mobility associated with for-profit higher education. She writes a regular column in *Slate* titled "Counter Narrative."

Jessie Daniels is a professor at the Graduate Center, CUNY and Hunter College. Her areas of study are racism, new media, and incarceration and health. She is the author of *Cyber Racism: White Supremacy Online and the New Attack on Civil Rights* and *White Lies.*

Bonnie Thornton Dill is professor of women's studies and dean of the College of Arts and Humanities at the University of Maryland, College Park. Her books include *Women of Color in U.S. Society,* co-edited with Maxine Baca Zinn, and *Across the Boundaries of Race and Class: An Exploration of Work and Family among Black Female Domestic Servants.*

Marlese Durr is professor of sociology and anthropology at Wright State University. Among other works, she has published *The New Politics of Race: From Du Bois to the 21st Century* and *Race, Work, and Families in the Lives of African Americans.*

Nirmala Erevelles is professor of social and cultural studies in education at the University of Alabama. She is the author of *Disability and Difference in Global Contexts: Enabling a Transformative Body Politic.*

Abby L. Ferber is professor of sociology, director of the Matrix Center, and co-director of Women's and Ethnic Studies at the University of Colorado at Colorado Springs. She is the author of numerous books, including *White Man Falling: Race, Gender and White Supremacy,* and *Hate Crime in America: What Do We Know?* She is co-author of *Sex, Gender, and Sexuality: The New Basics,* and co-editor of *Privilege: A Reader* with Michael S. Kimmel.

Charles A. Gallagher is the chair of the Department of Sociology at LaSalle University with research specialties in race and ethnic relations, urban sociology, and inequality. He has published several articles on subjects such as color-blind political narratives, racial categories within the context of interracial marriages, and perceptions of privilege based on ethnicity.

Herbert J. Gans has been a prolific and influential sociologist for more than fifty years. His published works on urban renewal and suburbanization are intertwined with his personal advocacy and participant observation, including a stint as consultant to the National Advisory Commission on Civil Disorder. He is the author of the classic *The Urban Villagers* as well as the more recent *Democracy and the News.*

Amy Hanser is associate professor of sociology at the University of British Columbia. She is author of *Service Encounters: Class, Gender, and the Market for Social Distinction in Urban China.*

Rebecca Hayes-Smith is a professor in the Department of Sociology, Anthropology and Social Work at Central Michigan University. Her research focuses on gender and racial inequalities in the criminal justice system and the "CSI effect."

Debra Henderson is associate professor and director of graduate studies for the Department of Sociology and Anthropology at Ohio University. She is author of numerous publications on intersectional welfare and poverty.

Pierrette Hondagneu-Sotelo is professor of sociology, American studies, and ethnicity at the University of Southern California. Her books include *God's Heart Has No Borders: How Religious Activists Are Working for Immigrant Rights; Gendered Transitions: Mexican Experiences of Immigration;* and *Doméstica: Immigrant Workers Cleaning and Caring in the Shadows of Affluence,* which won the Society for Social Problems' C. Wright Mills Award.

Lawrence R. Jacobs is the Walter F. and Joan Mondale Chair for Political Studies and director of the Center for the Study of Politics and Governance at the University of Minnesota. His numerous works include *Politicians Don't Pander: Political Manipulation and the Loss of Democratic Responsiveness,* for which he has won major prices. He is a frequent commentator in the national media.

Sujatha Jesudason is the executive director and founder of Generations Ahead, a social justice organization that brings diverse communities together to promote policies on genetic technologies that protect human rights.

Miliann Kang is professor of Women, Gender, and Sexuality at University of Massachusetts, Amherst. She has won numerous national awards for her book, *The Managed Hand: Race, Gender and the Body in Beauty Service Work.*

Jonathan Ned Katz was the first tenured professor of lesbian and gay studies in the United States (Department of Lesbian and Gay Studies, City College of San Francisco). He is the founder of the Queer Caucus of the College Art Association. He is also the co-founder of the activist group Queer Nation.

Michael S. Kimmel is Distinguished Professor of Sociology at Stony Brook University. He is a leading researcher and writer on men and masculinity, authoring and editing over twenty volumes in the field, including *Guyland: The Perilous World Where Boys Become Men; Men's Lives;* and *The Gendered Society.*

Meghan Kuebler is a Ph.D. candidate in sociology at CUNY University of Albany. She is the author of numerous publications about racial and ethnic disparities in home ownership.

Gloria Ladson-Billings is the Kellner Family Professor of Urban Education in the Department of Curriculum and Instruction and the Assistant Vice Chancellor of Academic Affairs at the University of Wisconsin–Madison. She is the author of *The Dreamkeepers: Successful Teachers of African-American Children* and former president of the American Educational Research Association.

Audre Lorde was a poet, essayist, teacher, activist, and writer dedicated to confronting and addressing the injustices of racism, sexism, and homophobia. Her numerous writings include, among others: *Sister Outsider; The Cancer Journals; From a Land Where Other People Live;* and *The Black Unicorn.*

Gregory Mantsios is the director of Worker Education at Queens College, the City University of New York.

Tiffany Manuel is vice president of Knowledge, Impact, and Strategy at Enterprise Community Partners, a nonprofit organization that creates affordable housing opportunities for low- and moderate-income people in the United States.

Marie Friedmann Marquardt is a scholar-in-residence at the Candler School of Theology at Emory University. She is co-author of *Living "Illegal": The Human Face of Unauthorized Immigration.*

Peggy McIntosh is associate director of the Wellesley College Centers for Women. She is the founder and co-director of the National SEED Project on Inclusive Curriculum—a project that helps teachers make school climates fair and equitable. She is the co-founder of the Rocky Mountain Women's Institute.

Michael Messner is professor of sociology and gender studies at the University of Southern California. He is the author of *It's All For the Kids: Gender, Families, and Youth Sports; Taking the Field: Women, Men and Sports;* and *Politics of Masculinities,* among others. He is a recipient of the American Sociological Association's Jessie Bernard Award.

Doug Meyer is visiting assistant professor of sociology and anthropology at the College of Wooster. His work focuses on the race, class, and gender dynamics of anti-queer violence.

Jody Miller is professor in the School of Criminal Justice at Rutgers University. Among other publications, she is the author of *Getting Played: African American Girls, Urban Inequality, and Gendered Violence.*

Andrea Minear is assistant professor of elementary education at the University of West Alabama, Livingston. Her academic and research interests include social justice and equity in education.

Alfonso Morales is associate professor in the Department of Urban and Regional Planning at the University of Wisconsin–Madison. His work examines how urban agriculture, food distribution, and community and economic development.

James A. Morone is professor of political science at Brown University and the past president of the Politics and History Section of the American Political Science Association. He has published numerous books and essays, including *Hellfire Nation: The Politics of Sin in American History* (2003), which was nominated for a Pulitzer Prize.

Sendhil Mullainathan is professor of economics at Harvard University. He is the co-founder of The Abdul Latif Jameel Poverty Action Lab at MIT that uses randomized evaluations to study poverty alleviation. He is the co-author of *Scarcity: Why Having Too Little Means So Much.*

Carolina Bank Muñoz is associate professor of sociology at Brooklyn College. She is the author of *Transnational Tortillas: Race, Gender and Shop-Floor Politics in Mexico and the United States,* as well as many other publications.

Timothy Noah is a journalist who writes twice weekly for MSNBC.com. He is the author of *The Great Divergence: America's Growing Inequality Crisis and What We Can Do about It.*

C. J. Pascoe is assistant professor of sociology at the University of Oregon. She is the author of *Dude, You're a Fag: Masculinity and Sexuality in High School,* which won the American Educational Research Association's 2007 Book of the Year Award.

Gina M. Pérez is an associate professor of comparative American studies at Oberlin College. She studies Latinas/os, migration, and transnationalism and is the author of *The Near Northwest Side Story: Migration, Displacement, and Puerto Rican Families.*

Bandana Purkayastha is professor of sociology and Asian American studies at the University of Connecticut. She is the author of *Negotiating Ethnicity: Second-Generation South Asian Americans Traverse a Transnational World.*

Dorothy Roberts is the George A. Weiss University Professor of Law and Sociology and the Raymond Pace and Sadie Tanner Mossell Alexander Professor of Civil Rights at University of Pennsylvania Law School. She is an acclaimed scholar of race, gender, and the law, and author of *Killing the Black Body: Race, Reproduction and the Meaning of Liberty;* and *Fatal Intervention: How Science, Politics and Big Business Re-Create Race in the 21st Century,* among others.

Lillian B. Rubin lives and works in San Francisco. She is an internationally known lecturer and writer. Some of her books include *The Man with the Beautiful Voice; Tangled Lives; The Transcendent Child;* and *Intimate Strangers.*

C. Matthew Snipp is the Burnet C. and Mildred Finley Wohlford Professor of Humanities and Sciences in the Department of Sociology at Stanford University where he founded the Center for Native American Excellence. He is the author of *American Indians: The First of This Land;* and *Public Policy Impacts on American Indian Economic Development.* His tribal heritage is Oklahoma Cherokee and Choctaw.

Natalie J. Sokoloff is professor emerita of sociology at the John Jay College of Criminal Justice, SUNY. She is author of numerous books and publications about intimate partner violence, and women and men's incarceration. She is the editor of *Domestic Violence at the Margins: Readings on Race, Class, Gender, and Culture.*

Timothy J. Steigenga is professor of political science and chair of the social sciences and humanities at the Wilkes Honors College of Florida Atlantic University. He is the author of numerous books and publications, including *A Place to Be: Brazilian, Guatemalan, and Mexican Immigrants in Florida's New Destinations.*

Jesse A. Steinfeldt is associate professor in the Department of Counseling and Educational Psychology at Indiana University, Bloomington.

Matthew Clint Steinfeldt is a lecturer in exercise science at Fort Lewis College in Durango, Colorado.

Ronald T. Takaki was professor of ethnic studies at the University of California, Berkeley, and distinguished historian whose scholarship provided much of the foundation for inclusive historical studies. He authored several books, including *Iron Cages: Race and Culture in 19th Century America; Strangers from a Different Shore: A History of Asian Americans;* and *A Different Mirror: A History of Multicultural America.*

Beverly Tatum is president of Spelman College. She is a clinical psychologist, author, and teacher whose areas of research interest include Black families in

White communities, racial identity in teens, and the role of race in the classroom. She is the author of *Can We Talk about Race? And Other Conversations in an Era of School Resegregation,* and *Why Are All the Black Kids Sitting Together in the Cafeteria? And Other Conversations about Race.*

Bhoomi K. Thakore is a research associate at Northwestern University. She is the author of numerous papers on racial representations in the popular media.

Ann Tickamyer is professor and head of the Department of Agricultural Economics, Sociology, and Education at Pennsylvania State University. Her work focuses on poverty, livelihood practices, and welfare provision in rural Appalachia and Indonesia.

Jeremiah Torres is a graduate of Stanford University, where he studied symbolic systems. His article "Label Us Angry" appeared in the book *Asian American X,* a collection of essays about the experiences of contemporary Asian Americans.

Haunani-Kay Trask is a Hawaiian scholar and poet and has been an indigenous rights activist for the Native Hawaiian community for over 25 years. She is a former professor of Hawaiian Studies at the University of Hawaii at Manoa and is the author of several books of poetry and nonfiction.

Deborah R. Vargas is associate professor of ethnic studies and director of graduate studies in the Department of Ethnic Studies at University of California, Riverside. She is author of *Dissonant Divas in Chicana Music: The Limits of La Onda.*

Manuel A. Vásquez is professor and chair of University of Florida's Religion Department. He has authored numerous books, including *More than Belief: A Materialist Theory of Religion.* He is a co-author of *Living "Illegal": The Human Face of Unauthorized Immigration* (2nd edition, 2013).

Matt Vidal is senior lecturer in work and organizations at King's College, London. His areas of expertise include the sociology of work, organizations, labor markets, and comparative political economy.

Mary C. Waters is M.E. Zukerman Professor of Sociology and Harvard College Professor at Harvard University. She is the author of *Black Identities: West Indian Immigrant Dreams and American Realities; Ethnic Options: Choosing Identities in America;* and numerous articles on race, ethnicity, and immigration.

Sandra E. Weissinger is assistant professor of sociology at the Southern Illinois University Edwardsville. Her work focuses on intragroup marginalization, inequalities, community activism, and African American communities and institutions.

Kath Weston is a sociological anthropologist who has written several books, including *Families We Choose: Lesbians, Gays, and Kinship; Render Me, Gender Me; Long Slow Burn;* and *Gender in Real Times.*

Christine L. Williams is a professor of sociology at the University of Texas at Austin. Her research interests include gender and sexuality; work, occupations, and organizations; qualitative methodology; and sociological theory. She is the author of *Inside Toyland: Working, Shopping, and Social Inequality; Still a Man's World: Men Who Do Women's Work;* and *Gender Differences at Work: Women and Men in Nontraditional Occupations.*

Dana M. Williams is assistant professor of sociology at Valdosta State University. His work focuses on social inequalities, social movements, political sociology, and complex organizations.

Philip J. Williams is professor of political science and Latin American studies at University of Florida where he is director of the Center for Latin American Studies. He is the author of *The Catholic Church and Politics in Nicaragua and Costa Rica* and *Militarization and Demilitarization in El Salvador's Transition to Democracy.*

Adia M. Harvey Wingfield is an associate professor of sociology at Georgia State University, specializing in race, class, and gender, work and occupations, and social theory. She is the author of *Yes We Can? White Racial Framing and the 2008 Presidential Campaign* (with Joe Feagin) and *Doing Business with Beauty: Black Women, Hair Salons, and the Racial Enclave Economy.*

Jennifer Wriggins is Sumner T. Bernstein Professor of Law at University of Maine School of Law. She specializes in torts, insurance, and family law with a focus on race and gender. She is co-author of *The Measure of Injury: Race, Gender, and Tort Law.*

Ruth Enid Zambrana is professor of women's studies and director of the Consortium on Race, Gender, and Ethnicity at the University of Maryland, College Park. She is the author of *Latinos in American Society: Families and Communities in Transition* and the co-editor of *Emerging Intersections: Race, Class, and Gender in Theory, Policy, and Practice.*

Min Zhou is professor of sociology and Asian American studies, Walter and Shirley Wang Endowed Chair in U.S.–China Relations and Communications, and the founding chair of the Asian American Studies Department at UCLA. She is the author of *Chinatown: The Socioeconomic Potential of an Urban Enclave; The Transformation of Chinese America;* and *Contemporary Chinese America: Immigration, Ethnicity, and Community Transformation.*

PART I

Why Race, Class, and Gender Still Matter

MARGARET L. ANDERSEN
AND PATRICIA HILL COLLINS

The United States is a nation where people are supposed to be able to rise above their origins. Those who want to succeed, it is believed, can do so through hard work and solid effort. Although equality has historically been denied to many, there is now a legal framework in place that guarantees protection from discrimination and equal treatment for all citizens.

Historic social movements, such as the civil rights movement and the feminist movement, raised people's consciousness about the rights of African Americans and women. Moreover, these movements have generated new opportunities for multiple groups—African Americans, Latinos, white women, disabled people, lesbian, gay, bisexual, transgendered (LGBT) people, and older people, to name some of the groups that have been beneficiaries of civil rights action and legislation.

We have also now had an African American president; gays and lesbians increasingly have the rights to same-sex marriage; women sit in very high places—as Supreme Court justices and CEOs of major companies; disabled people have rights of access to work and schools and are protected under federal laws. The vast majority of Americans, when asked, say that support equal rights and nondiscrimination policies; indeed, over 90 percent say they would vote for a woman as president of the United States (Streb et al. 2008). Why, then, do race, class, and gender still matter?

Race, class, and gender still matter because they continue to structure society in ways that value some lives more than others. Currently, some groups have more opportunities and resources, while other groups struggle. Race, class, and gender matter because they remain the foundations for systems of power and inequality that, despite our nation's diversity, continue to be among the most significant social facts of people's lives. Despite having removed the formal barriers to opportunity, the United States is still highly unequal along lines of race, class, and gender.

In this book, we ask students to think about race, class, and gender as *systems of power.* We want to encourage readers to imagine ways to transform, rather than reproduce, existing social arrangements. This starts with shifting one's thinking so that groups who are so often silenced or ignored become heard. All social groups are located in a system of power relationships wherein your social location can shape what you know—and what others know about you. As a result, dominant forms of knowledge have been constructed largely from the experiences of the most powerful—that is, those who have the most access to systems of education and communication. To acquire a more inclusive view—one that pays attention to group experiences that may differ from your own—requires that you form a new frame of vision.

You can think of this as if you were taking a photograph. For years, poor people, women, and people of color—and especially poor women of color—were totally outside the frame of vision of more powerful groups or distorted by the views of the powerful. If you move your angle of sight to include those who have been overlooked, however, you may be surprised by how incomplete or just plain wrong your earlier view was. Completely new subjects can also appear. This is more than a matter of sharpening one's focus, although that is required for clarity. Instead, this new angle of vision means actually seeing things differently, perhaps even changing the lens you look through—thereby removing the filters (or stereotypes and misconceptions) that you bring to what you see and think.

DEVELOPING A RACE, CLASS, AND GENDER PERSPECTIVE

In this book, we ask you to think about how race, class, and gender matter in shaping everyone's lived experiences. We focus on the United States, but increasingly the inclusive vision we present here matters on a global scale as well. Thinking from a perspective that engages race, class, and gender is not just about illuminating the experiences of oppressed groups. It changes how we understand groups who are on both sides of power and privilege. For example, the development of women's studies has changed what we know and how we think about women. At the same time, it has changed what we know and how we think about men. This does not mean that women's studies is about "male bashing." It means taking the experiences of women and men seriously and analyzing how race, class, and gender shape the experiences of both men and women—in different, but interrelated, ways. Likewise, the study of racial and ethnic groups begins by learning the diverse histories and experiences of these groups. In doing so, we also transform our understanding of White people's experiences. Rethinking class means seeing the vastly different experiences of both wealthy, middle-class, working-class, and poor people in the United States, and learning to think differently about privilege and opportunity. The exclusionary thinking that comes from past frames of vision simply does not reveal the intricate interconnections that exist among the different groups that comprise the U.S. society.

It is important to stress that thinking about race, class, and gender is not just a matter of studying victims. Relying too heavily on the experiences of poor people, women, and people of color can erase our ability to see race, class, and gender as an integral part of everyone's experiences. We remind students that race, class, and gender have affected the experiences of all individuals and groups. For example, gender is not just about women and class is not only about the poor. Therefore, it is important to study White people when analyzing race, the experiences of the affluent when analyzing class, and all people when analyzing gender. Such a perspective focuses your attention on the dynamics of privilege, not just oppression.

So you might ask, how does reconstructing knowledge about excluded groups matter? To begin with, knowledge is not just some abstract thing—good to have, but not all that important. There are real consequences to having partial or distorted knowledge. First, knowledge is not just about content and information; it provides an orientation to the world. What you know frames how you behave and how you think about yourself and others. If what you know is wrong because it is based on exclusionary thought, you are likely to act in exclusionary ways, thereby reproducing the racism, anti-Semitism, sexism, class oppression, and homophobia of society. This may not be because you are intentionally racist, anti-Semitic, sexist, elitist, or homophobic. It can simply be because you do not know any better. Challenging oppressive race, class, and gender relations in society requires reconstructing what we know so that we have some basis from which to change these damaging and dehumanizing systems of oppression.

Second, learning about other groups helps you realize the partiality of your own perspective; this is true for both dominant and subordinate groups. Knowing only the history of Puerto Rican women, for example, or seeing their history only in single-minded terms will not reveal the historical linkages between the oppression of Puerto Rican women and the exclusionary and exploitative treatment of African Americans, working-class Whites, Asian American men, and similar groups. Ronald T. Takaki discusses this in his essay on the multicultural history of American society ("A Different Mirror").

Finally, having misleading and incorrect knowledge leads to the formation of bad social policy—policy that then reproduces, rather than solves, social problems. As an example, U.S. immigration policy has often taken a one-size-fits-all approach, failing to recognize that vast differences among groups coming to the United States privilege some and disadvantage others. Taking a broader view of social issues fosters more effective social policy.

RACE, CLASS, AND GENDER AS A MATRIX OF DOMINATION

Race, class, and gender shape the experiences of all people in the United States. This fact has been widely documented in research and, to some extent, is commonly understood. For years, social scientists have studied the consequences of race, class, and gender inequality for different groups in society. The framework of race, class, and gender studies presented here, however, explores how race, class, and gender operate *together* in people's lives. Fundamentally, race, class, and gender are *intersecting* categories of experience that affect all aspects of human life; they *simultaneously* structure the experiences of all people in this society. At any moment, race, class, or gender may feel more salient or meaningful in a given person's life, but they are overlapping and cumulative in their effects.

In this volume, we focus on several core features of this intersectional framework for studying race, class, and gender. First, we emphasize *social structure* in our efforts to conceptualize intersections of race, class, and gender. We use the approach of a *matrix of domination* to analyze race, class, and gender. A matrix of domination sees social structure as having multiple, interlocking levels of domination that stem from the societal configuration of race, class, and gender relations. This structural pattern affects individual consciousness, group interaction, and group access to institutional power and privileges (Collins 2000). Within this structural framework, we focus less on comparing race, class, and gender as separate systems of power than on investigating the structural patterns that join them. Because of the simultaneity of race, class, and gender in people's lives, intersections of race, class, and gender can be seen in individual stories and personal experience. In fact, much exciting work on the intersections of race, class, and gender appears in autobiographies, fiction, and personal essays. We do recognize the significance of these individual narratives and include many here, but we also emphasize social structures that provide the context for individual experiences.

Second, studying interconnections among race, class, and gender within a context of social structures helps us understand how race, class, and gender are manifested differently, depending on their configuration with the others. Thus, one might say African American men are privileged *as men,* but this may not be true when their race and class are also taken into account. Otherwise, how can we possibly explain the particular disadvantages African American men experience in the criminal justice system, in education, and in the labor market? For that matter, how can we explain the experiences that Native American women undergo—disadvantaged by the unique experiences that they have based on race, class, *and* gender—none of which is isolated from the effects of the others? Studying the connections among race, class, and gender reveals that divisions by race and by class and by gender are not as clear-cut as they may seem. White women, for example, may be disadvantaged because of gender but privileged by race and perhaps (but not necessarily) by class. Increasing class differentiation within racial-ethnic groups also reminds us that race is not a monolithic category, as can be seen in the fact that poverty among White people is increasing more than poverty among other groups, even while some Whites are the most powerful members of society.

Third, the matrix of domination approach to race, class, and gender studies is historically grounded. We have chosen to emphasize the intersections of race, class, and gender as institutional systems that have had a special impact in the United States. Yet race, class, and gender intersect with other categories of experience, such as sexuality, ethnicity, age, ability, religion, and nationality. Historically, these intersections have taken varying forms from one society to the next; within any given society, the connections among them also shift. Thus, race is not inherently more important than gender, just as sexuality is not inherently more significant than class and ethnicity.

Given the complex and changing relationships among these categories of analysis, we ground our analysis in the historical, institutional context of the United States. Doing so means that race, class, and gender emerge as fundamental categories of analysis in the U.S. setting, so significant that in many ways they influence all of the other categories. Systems of race, class, and gender have been so consistently and deeply codified in U.S. laws that they have had intergenerational effects on economic, political, and social institutions. For example, the capitalist class relations that have characterized all phases of U.S. history have routinely privileged or penalized groups organized by gender and by race. U.S. social institutions have reproduced economic equalities for poor people, women, and people of color from one generation to the next. Thus, in the United States, race, class, and gender demonstrate visible, long-standing, material effects that in many ways foreshadow more recently visible categories of ethnicity, religion, age, ability, and/or sexuality.

DIFFERENCE, DIVERSITY, AND MULTICULTURALISM

How does the matrix of domination framework differ from other ways of conceptualizing race, class, and gender relationships? We think this can be best understood by contrasting the *matrix of domination framework* to what might be called a *difference framework* of race, class, and gender studies. A difference framework, though viewing some of the common processes in race, class, and gender relations, tends to focus on unique group experiences. Emphasizing diversity and multiculturalism, the difference framework will likely examine the different experiences of various groups in society. Valuable as such studies may be, they tend to treat each group separately, or perhaps they focus mostly on the culture of particular groups, seldom looking at the systems of power that link groups together. Such studies are valuable because of how they document unique histories and cultural contributions, but we distinguish our work by looking at the *interrelationships* among race, class, and gender, not just their unique ways of being experienced.

You might think of the distinction between the two approaches as one of thinking comparatively, which is an example of one of the core features of a difference framework, versus thinking relationally, which is the hallmark of the matrix of domination approach. For example, in the difference framework individuals are encouraged to compare their experiences with those supposedly unlike them. When you think comparatively, you might look at how different groups have, for example, encountered prejudice and discrimination or you might compare laws prohibiting interracial marriage to current debates about same-sex marriage. These are important and interesting questions, but they are taken a step further when you think beyond comparison to the structural relationships between different group experiences. In contrast, when you think relationally, you see the social structures that *simultaneously* generate unique group histories and link them together in society. You then untangle the workings of social systems that shape the experiences of different people and groups, and you move beyond just comparing (for example) gender oppression with race oppression or the oppression of gays and lesbians with that of racial groups. Recognizing how intersecting systems of power shape different groups' experiences positions you to think about changing the system, not just documenting the effects of such systems on different people.

The language of difference encourages comparative thinking. People think comparatively when they learn about experiences other than their own and begin comparing and contrasting the experiences of different groups. This is a step beyond centering one's thinking in a single group (typically one's own), but it is nonetheless limited. For example, when students encounter studies of race, class, and gender for the first time, they often ask, "How is this group's experience like or not like my own?" This is an important question and a necessary first step, but it is not enough. For one thing, this question frames one's understanding of different groups only within the context of other groups' experiences. It can assume an artificial norm against which different groups are judged. Furthermore, such a question tends to promote ranking the oppression of one group compared to another, as if the important thing were to determine who is most victimized. Thinking comparatively tends to assume that race, class, and gender constitute separate and independent components of human experience that can be compared for their similarities and differences.

We should point out that comparative thinking can foster greater understanding and tolerance, but comparative thinking alone can also leave intact the power relations that create race, class, and gender relations. Because the concept of difference contains the unspoken question "different from what?," the difference framework can privilege those who are deemed to be "normal" and stigmatize people who are labeled as "different." Because it is based on comparison, the very concept of difference fosters dichotomous (either/or) thinking. Some approaches to difference place people in either/or categories, as if one is either Black or White, oppressed or oppressor, powerful or powerless, normal or different when few of us fit neatly into any of these restrictive categories.

Some difference frameworks try to move beyond comparing systems of race, class, and gender by thinking in terms of an *additive* approach. The additive approach is reflected in terms such as *double* and *triple jeopardy*. Within this logic, poor African American women seemingly experience the triple oppression of race, gender, and class, whereas poor Latina lesbians encounter quadruple oppression, and so on. But social inequality cannot necessarily be quantified in this fashion. Adding together "differences" (thought to lie in one's difference from the norm) produces a hierarchy of difference that ironically reinstalls those who are additively privileged at the top while relegating those who are additively oppressed to the bottom. We do not think of race and gender oppression in the simple additive terms implied by phrases such as double and triple jeopardy. The effects of race, class, and gender do "add up," both over time and in intensity of impact, but seeing race, class, and gender only in additive terms misses the social structural connections among them and the particular ways in which different configurations of race, class, and gender affect group experiences.

Within difference frameworks, additive thinking can foster another troubling outcome. One can begin with the concepts of race, class, and gender and continue to "add on" additional types of difference. Ethnicity, sexuality, religion, age, and ability all can be added on to race, class, and gender in ways that suggest that any of these forms of difference can substitute for others. This use of difference fosters a view of oppressions as equivalent and as being the same. Recognizing that difference encompasses more than race, class, and gender is a step in the right direction. But continuing to add on many distinctive forms of difference can be a never-ending process. After all, there are as many forms of difference as there are individuals. Ironically, this form of recognizing difference can erase the workings of power just as effectively as diversity initiatives.

When it comes to conceptualizing race, class, and gender relations, the matrix of domination approach also differs from another version of the focus on difference, namely, thinking about diversity. *Diversity* has become a catchword for trying to understand the complexities of race, class, and gender in the United States. What does *diversity* mean? Because the American public has become a more heterogeneous population, *diversity* has become a buzzword—popularly used, but loosely defined. People use *diversity* to mean cultural variety, numerical representation, changing social norms, and the inequalities that characterize the status of different groups. In thinking about diversity, people have recognized that race, class, gender, age, sexual orientation, and ethnicity matter; thus, groups who have previously been invisible, including people of color, gays, lesbians, transgender and bisexual people, older people, and immigrants, are now in some ways more visible. At the same time that diversity is more commonly recognized, however, these same groups continue to be defined as "other"; that is, they are perceived through dominant group values, treated in exclusionary ways, and subjected to social injustice and economic inequality.

The movement to "understand diversity" has made many people more sensitive and aware of the intersections of race, class, and gender. Thinking about diversity has also encouraged students and social activists to see linkages to other categories of analysis, including sexuality, age, religion, physical disability, national identity, and ethnicity. But appreciating diversity is not the only point. The very term *diversity* implies that understanding race, class, and gender is simply a matter of recognizing the plurality of views and experiences in society—as if race, class, and gender were benign categories that foster diverse experiences instead of systems of power that produce social inequalities.

Diversity initiatives hold that the diversity created by race, class, and gender differences are pleasing and important, both to individuals and to society as a whole—so important, in fact, that diversity should be celebrated. Under diversity initiatives, ethnic foods, costumes, customs, and festivals are celebrated, and students and employees receive diversity training to heighten their multicultural awareness. Diversity initiatives also advance a notion that, despite their differences, "people are really the same." Under this view, the diversity created by race, class, and gender constitutes cosmetic differences of style, not structural opportunities.

Certainly, opening our awareness of distinct group experiences is important, but some approaches to diversity can erase the very real differences in power that race, class, and gender create. For example, diversity initiatives have asked people to challenge the silence that has surrounded many group experiences. In this framework, people think about diversity as "listening to the voices" of a multitude of previously silenced groups. This is an important part of coming to understand race, class, and gender, but it is not enough. One problem is that people may begin hearing the voices as if they were disembodied from particular historical and social conditions. This perspective can make experience seem to be just a matter of competing discourses, personifying "voice" as if the voice or discourse itself constituted lived experience. Second, the "voices" approach suggests that any analysis is incomplete unless every voice is heard. In a sense, of course, this is true, because inclusion of silenced people is one of the goals of race/class/gender work. But in situations where it is impossible to hear every voice, how does one decide which voices are more important than others? One might ask, who are the privileged listeners within these voice metaphors?

We think that the matrix of domination model is more analytical than either the difference or diversity frameworks *because of its focus on structural systems of power and inequality.* This means that race, class, and gender involve more than either comparing and adding up oppressions or privileges or appreciating cultural diversity. The matrix of domination model requires analysis and criticism of existing systems of power and privilege; otherwise, understanding diversity becomes just one more privilege for those with the greatest access to education—something that has always been a mark of the elite class. Therefore, race, class, and gender studies mean more than just knowing the cultures of an array of human groups. Instead, studying race, class, and gender means recognizing and analyzing the hierarchies and systems of domination that permeate society and limit our ability to achieve true democracy and social justice.

Finally, the matrix of domination framework challenges the idea that race, class, and gender are important only at the level of culture—an implication of the catchword *multiculturalism. Culture* is traditionally defined as the "total way of life" of a group of people. It encompasses both material and symbolic components and is an important dimension of understanding human life. Analysis of culture per se, however, tends to look at the group itself rather than at the broader conditions within which the group lives. Of course, as anthropologists know, a sound analysis of culture situates group experience within these social structural conditions. Nonetheless, a narrow focus on culture tends to ignore social conditions of power, privilege, and prestige. The result is that multicultural studies often seem tangled up with notions of cultural pluralism—as if knowing a culture other than one's own is the only goal of a multicultural education.

Because we approach the study of race, class, and gender with an eye toward transforming thinking, we see our work as differing somewhat from the concepts implicit in the language of difference, diversity, or multiculturalism. Although we think it is important to see the diversity and plurality of different cultural forms, in our view this perspective, taken in and of itself, misses the broader point of understanding how racism, class relations, and sex and gender privilege have shaped the experience of groups. Imagine, for example, looking for the causes of poverty solely within the culture of currently poor people, as if patterns of unemployment, unexpected health care costs, rising gas prices, and home mortgage foreclosures had no effect on people's opportunities and life decisions. Or, imagine trying to study the oppression of LGBT people in terms of gay culture only. Obviously, doing this turns attention to the group itself and away from the dominant society. Likewise, studying race only in terms of Latino culture or Asian American culture or African American culture, or studying gender only by looking at women's culture, encourages thinking that blames the victims for their own oppression. For all of these reasons, the focus of this book is on the institutional, or structural, bases for race, class, and gender relations.

This book is not just about comparing differences, understanding diversity, or describing multicultural societies. Instead, we attempt to develop a structural perspective on the relationships among race, class, and gender as systems of power, the hallmark of the matrix of domination framework. We recognize that reaching these goals will require rejecting the kind of exclusionary thinking that has virtually erased some groups' experiences, and instead embracing an inclusive perspective that incorporates neglected groups and themes. Inclusive perspectives begin with the recognition that the United States is a multicultural and diverse society. Population data and even casual observations reveal that obvious truth, but developing an inclusive perspective requires more than recognizing the plurality of experiences in this society. Understanding race, class, and gender means coming to see the systematic exclusion and exploitation of some groups as well as the intergenerational privileges of others. This is more than just adding in different group experiences to already established frameworks of thought. It means constructing new analyses that are focused on the centrality of race, class, and gender in the experiences of us all.

DEVELOPING AN INCLUSIVE PERSPECTIVE

We want readers to understand that race, class, and gender are linked experiences, no one of which is more important than the others; the three are interrelated and together configure the structure of U.S. society. You can begin to develop a more inclusive perspective by asking: How does the world look different if we put the experiences of those who have been excluded at the center of our thinking?

Inclusive perspectives see the interconnections between these experiences and do not reduce a given person's or group's life to a single factor. In addition, developing an inclusive perspective entails more than just summing up the experiences of individual groups, as in the additive model discussed previously. Race, class, and gender are social structural categories. This means that they are embedded in the institutional structure of society. Understanding them requires a social structural analysis—by which we mean revealing the race, class, and gender patterns and processes that form the very framework of society.

Once you understand that race, class, and gender are *simultaneous* and *intersecting* social structural systems of relationship and meaning, you also can see the distinctive ways that other categories of experience intersect in society. Age, religion, sexual orientation, nationality, physical ability, region, and ethnicity also shape systems of privilege and inequality. To emphasize this, we have opened this anthology with the classic essay by noted feminist Audre Lorde, now deceased. Her essay, "Age, Race, Class and Sex: Women Redefining Difference" remains a critical and inspiring statement about the importance of thinking inclusively and across the differences that too often divide us.

The opening section of this book also includes two important historical essays, intended to anchor the book in an admittedly brief, but nonetheless essential, presentation of how our nation's institutions have evolved as a result of the exclusionary treatment—indeed, even annihilation—of certain groups. This is essential to understanding who we are as a society and a culture. We open this section with Ronald T. Takaki's "A Different Mirror." Takaki shows common connections in the histories of African Americans, Chicanos, Irish Americans, Jews, and Native Americans. He argues that only when we understand a multidimensional history that encompasses race, class, and gender will we see ourselves in the full complexity of our humanity. C. Matthew Snipp's review of how American Indians have been removed, restricted, relocated, and robbed of resources recounts the extent to which "American" settlement and seizing of lands—along with cultural annihilation—have marked our nation's history.

Understanding the diverse histories, cultures, and experiences of groups who have been defined as marginal in society has been vital to the formation of our social institutions. Beyond this important fact is the reality that groups outside of the dominant culture have been silenced over the course of history, leading to distortions and incomplete knowledge of how society has been organized. Ignoring the experiences of those who have been on the margins results in a distorted view of how the nation itself has developed.

You might ask yourself how much you learned about the history of group oppression in your formal education. Maybe you touched briefly on topics such as the labor movement, slavery, women's suffrage, perhaps even the Holocaust, but most likely these were brief excursions from an otherwise dominant narrative that largely ignored the perspectives of working-class people, women, and people of color, along with others. How much of what you study now is centered in the experiences of the most dominant groups in society? Have studies that have been defined as classic been generated from research samples that have only studied certain populations (such as middle-class college students in psychology studies or mostly White people in other social science research)? Is the literature you are assigned or artistic creations that you study mostly the work of Europeans? Men? White people? How much has the creative work of Native Americans, Muslim Americans, new immigrant populations, Asian Americans, Latinos/as, African Americans, gays, lesbians, or women informed the study of "American" art and culture?

By minimizing the experiences and creations of these different groups, we communicate that their work and creativity is less important and less central to the development of culture than is the history of White American men. What false or incomplete conclusions does this exclusionary thinking generate? When you learn, for example, that democracy and egalitarianism were central cultural beliefs in the early history of the United States, how do you explain the enslavement of millions of African Americans, the genocide of Native Americans, the absence of laws against child labor, the presence of laws forbidding intermarriage between Asian Americans and White Americans?

Haunani-Kay Trask shows ("From a Native Daughter") that there can also be a gap between dominant cultural narratives and people's actual experiences. As she—a native Hawaiian—tells it, the official history she learned in schools was not what she was taught in her family and community. Dominant narratives can try to justify the oppression of different groups, but the unwritten, untold, subordinated truth can be a source for knowledge in pursuit of social justice. This book asks you to think more inclusively. Without doing so, you are prone to understand society, your own life within it, and the experiences of others through stereotypes and the misleading information that is all around you. What new experiences, understandings, theories, histories, and analyses do these readings inspire? What does it take for a member of one group (say a Latino male) to be willing to learn from and value the experiences of another (for example, an Indian Muslim woman)? These essays show that, although we are caught in multiple systems, we can learn to see our connection to others.

Engaging oneself at the personal level is critical to this process of thinking differently about race, class, and gender. Changing one's mind is not just a matter of assessing facts and data, although that is important; it also requires examining one's feelings. We incorporate personal narratives that reflect the diverse experiences of race, class, gender, and/or sexual orientation into the opening section of this book to encourage you to think about your personal story. We intend for the personal nature of these accounts—especially those that describe what exclusion means and how it feels—to build empathy among groups. We think that empathy encourages an emotional stance that is critical to relational thinking and developing an inclusive perspective. As an example, Jeremiah Torres's essay ("Label Us Angry") describes how his seemingly trouble-free childhood changed overnight when he became the victim of a hate crime in a community known for its acceptance of multiculturalism. His narrative reminds us that learning about race, class, and gender is not just an intellectual exercise; experiences with these forms of oppression are personal and often painful.

We hope that understanding the significance of race, class, and gender will encourage readers to put the experiences of the United States itself into a broader context. Knowing how race, class, and gender operate within U.S. national borders should help you see beyond those borders. We hope that developing an awareness of how the increasingly global basis of society influences the configuration of race, class, and gender relationships in the United States will encourage readers to cast an increasingly inclusive perspective on the world itself.

REFERENCES

Collins, Patricia Hill. 2000. *Black Feminist Thought: Knowledge, Consciousness, and Empowerment*. New York: Routledge.

Streb, Michael J., Barbara Burrell, Brian Frederick, and Michael A. Genovese. 2008. "Social Desirability Effects and Support for a Female American President." *Public Opinion Quarterly* 72 (Spring): 76–89.

1

Age, Race, Class and Sex

Women Redefining Difference

AUDRE LORDE

Much of Western European history conditions us to see human differences in simplistic opposition to each other: dominant/subordinate, good/bad, up/down, superior/inferior. In a society where the good is defined in terms of profit rather than in terms of human need, there must always be some group of people who, through systematized oppression, can be made to feel surplus, to occupy the place of the dehumanized inferior. Within this society, that group is made up of Black and Third World people, working-class people, older people, and women.

As a forty-nine-year-old Black lesbian feminist socialist mother of two, including one boy, and a member of an interracial couple, I usually find myself a part of some group defined as other, deviant, inferior, or just plain wrong. Traditionally, in American society, it is the members of oppressed, objectified groups who are expected to stretch out and bridge the gap between the actualities of our lives and the consciousness of our oppressor. For in order to survive, those of us for whom oppression is as American as apple pie have always had to be watchers, to become familiar with the language and manners of the oppressor, even sometimes adopting them for some illusion of protection. Whenever the need for some pretense of communication arises, those who profit from our oppression call upon us to share our knowledge with them. In other words, it is the responsibility of the oppressed to teach the oppressors their mistakes. I am responsible for educating teachers who dismiss my children's culture in school. Black and Third World people are expected to educate white people as to our humanity. Women are expected to educate men. Lesbians and gay men are expected to educate the heterosexual world. The oppressors maintain their position and evade responsibility for their own actions. There is a constant drain of energy which might be better used in redefining ourselves and devising realistic scenarios for altering the present and constructing the future.

Institutionalized rejection of difference is an absolute necessity in a profit economy which needs outsiders as surplus people. As members of such an economy, we have *all* been programmed to respond to the human differences between us with fear and loathing and to handle that difference in one of three

SOURCE: Lorde, Audre. "Age, Race, Class, and Sex." *Sister Outsider*, published by Crossing Press, Random House Inc. Copyright © 1984, 2007 by Audre Lorde. Used herein by permission of the Charlotte Sheedy Literary Agency, Inc.

ways: ignore it, and if that is not possible, copy it if we think it is dominant, or destroy it if we think it is subordinate. But we have no patterns for relating across our human differences as equals. As a result, those differences have been misnamed and misused in the service of separation and confusion.

Certainly there are very real differences between us of race, age, and sex. But it is not those differences between us that are separating us. It is rather our refusal to recognize those differences, and to examine the distortions which result from our misnaming them and their effects upon human behavior and expectation.

Racism, the belief in the inherent superiority of one race over all others and thereby the right to dominance. Sexism, the belief in the inherent superiority of one sex over the other and thereby the right to dominance. Ageism. Heterosexism. Elitism. Classism.

It is a lifetime pursuit for each one of us to extract these distortions from our living at the same time as we recognize, reclaim, and define those differences upon which they are imposed. For we have all been raised in a society where those distortions were endemic within our living. Too often, we pour the energy needed for recognizing and exploring difference into pretending those differences are insurmountable barriers, or that they do not exist at all. This results in a voluntary isolation, or false and treacherous connections. Either way, we do not develop tools for using human difference as a springboard for creative change within our lives. We speak not of human difference, but of human deviance.

Somewhere, on the edge of consciousness, there is what I call a *mythical* norm, which each one of us within our hearts knows "that is not me." In America, this norm is usually defined as white, thin, male, young, heterosexual, Christian, and financially secure. It is with this mythical norm that the trappings of power reside within this society. Those of us who stand outside that power often identify one way in which we are different, and we assume that to be the primary cause of all oppression, forgetting other distortions around difference, some of which we ourselves may be practising. By and large within the women's movement today, white women focus upon their oppression as women and ignore differences of race, sexual preference, class, and age. There is a pretense to a homogeneity of experience covered by the word *sisterhood* that does not in fact exist.

Unacknowledged class differences rob women of each others' energy and creative insight. Recently a women's magazine collective made the decision for one issue to print only prose, saying poetry was a less "rigorous" or "serious" art form. Yet even the form our creativity takes is often a class issue. Of all the art forms, poetry is the most economical. It is the one which is the most secret, which requires the least physical labor, the least material, and the one which can be done between shifts, in the hospital pantry, on the subway, and on scraps of surplus paper. Over the last few years, writing a novel on tight finances, I came to appreciate the enormous differences in the material demands between poetry and prose. As we reclaim our literature, poetry has been the major voice of poor, working-class, and Colored women. A room of one's own may be a necessity for writing prose, but so are reams of paper, a typewriter, and plenty of time. The actual requirements to produce the visual arts also help determine, along class lines, whose art is whose. In this day of inflated prices for material, who are our sculptors, our painters, our photographers? When we speak of a broadly based women's culture, we need to be aware of the effect of class and economic differences on the supplies available for producing art.

As we move toward creating a society within which we can each flourish, ageism is another distortion of relationship which interferes without vision. By ignoring the past, we are encouraged to repeat its mistakes. The "generation gap" is an important social tool for any repressive society. If the younger members of a community view the older members as contemptible or suspect or excess, they will never be able to join hands and examine the living memories of the community, nor ask the all-important question, "Why?" This gives rise to a historical amnesia that keeps us working to invent the wheel every time we have to go to the store for bread.

We find ourselves having to repeat and relearn the same old lessons over and over that our mothers did because we do not pass on what we have learned, or because we are unable to listen. For instance, how many times has this all been said before? For another, who would have believed that once again our daughters are allowing their bodies to be hampered and purgatoried by girdles and high heels and hobble skirts?

Ignoring the differences of race between women and the implications of those differences presents the most serious threat to the mobilization of women's joint power.

As white women ignore their built-in privilege of whiteness and define woman in terms of their own experience alone, then women of Color become "other," the outsider whose experience and tradition is too "alien" to comprehend. An example of this is the signal absence of the experience of women of Color as a resource for women's studies courses. The literature of women of Color is seldom included in women's literature courses and almost never in other literature courses, nor in women's studies as a whole. All too often, the excuse given is that the literatures of women of Color can only be taught by Colored women, or that they are too difficult to understand, or that classes cannot "get into" them because they come out of experiences that are "too different." I have heard this argument presented by white women of otherwise quite clear intelligence, women who seem to have no trouble at all teaching and reviewing work that comes out of the vastly different experiences of Shakespeare, Moliere, Dostoyevsky, and Aristophanes. Surely there must be some other explanation.

This is a very complex question, but I believe one of the reasons white women have such difficulty reading Black women's work is because of their reluctance to see Black women as women and different from themselves. To examine Black women's literature effectively requires that we be seen as whole people in our actual complexities—as individuals, as women, as human—rather than as one of those problematic but familiar stereotypes provided in this society in place of genuine images of Black women. And I believe this holds true for the literatures of other women of Color who are not Black.

The literatures of all women of Color re-create the textures of our lives, and many white women are heavily invested in ignoring the real differences. For as long as any difference between us means one of us must be inferior, then the recognition of any difference must be fraught with guilt. To allow women of Color to step out of stereotypes is too guilt provoking, for it threatens the complacency of those women who view oppression only in terms of sex.

Refusing to recognize difference makes it impossible to see the different problems and pitfalls facing us as women.

Thus, in a patriarchal power system where white-skin privilege is a major prop, the entrapments used to neutralize Black women and white women are not the same. For example, it is easy for Black women to be used by the power structure against Black men, not because they are men, but because they are Black. Therefore, for Black women, it is necessary at all times to separate the needs of the oppressor from our own legitimate conflicts within our communities. This same problem does not exist for white women. Black women and men have shared racist oppression and still share it, although in different ways. Out of that shared oppression we have developed joint defenses and joint vulnerabilities to each other that are not duplicated in the white community, with the exception of the relationship between Jewish women and Jewish men.

On the other hand, white women face the pitfall of being seduced into joining the oppressor under the pretense of sharing power. This possibility does not exist in the same way for women of Color. The tokenism that is sometimes extended to us is not an invitation to join power; our racial "otherness" is a visible reality that makes that quite clear. For white women there is a wider range of pretended choices and rewards for identifying with patriarchal power and its tools.

Today, with the defeat of ERA, the tightening economy, and increased conservatism, it is easier once again for white women to believe the dangerous fantasy that if you are good enough, pretty enough, sweet enough, quiet enough, teach the children to behave, hate the right people, and marry the right men, then you will be allowed to co-exist with patriarchy in relative peace, at least until a man needs your job or the neighborhood rapist happens along. And true, unless one lives and loves in the trenches it is difficult to remember that the war against dehumanization is ceaseless.

But Black women and our children know the fabric of our lives is stitched with violence and with hatred, that there is no rest. We do not deal with it only on the picket lines, or in dark midnight alleys, or in the places where we dare to verbalize our resistance. For us, increasingly, violence weaves through the daily tissues of our living—in the supermarket, in the classroom, in the elevator, in the clinic and the schoolyard, from the plumber, the baker, the saleswoman, the bus driver, the bank teller, the waitress who does not serve us.

Some problems we share as women, some we do not. You fear your children will grow up to join the patriarchy and testify against you, we fear our children will be dragged from a car and shot down in the street, and you will turn your backs upon the reasons they are dying.

The threat of difference has been no less blinding to people of Color. Those of us who are Black must see that the reality of our lives and our struggle does not make us immune to the errors of ignoring and misnaming difference. Within Black communities where racism is a living reality, differences among us often seem dangerous and suspect. The need for unity is often misnamed as a need for homogeneity, and a Black feminist vision mistaken for betrayal of our common interests as a people. Because of the continuous battle against racial erasure that Black women and Black men share, some Black women still refuse to recognize that we are also oppressed as women, and that sexual hostility against Black women is practiced not only by the white racist society, but implemented within our Black communities as well. It is a disease striking the heart of Black nationhood, and silence will not make it disappear. Exacerbated by racism and the pressures of powerlessness, violence against Black women and children often becomes a standard within our communities, one by which manliness can be measured. But these woman-hating acts are rarely discussed as crimes against Black women.

As a group, women of Color are the lowest paid wage earners in America. We are the primary targets of abortion and sterilization abuse, here and abroad. In certain parts of Africa, small girls are still being sewed shut between their legs to keep them docile and for men's pleasure. This is known as female circumcision, and it is not a cultural affair as the late Jomo Kenyatta insisted, it is a crime against Black women.

Black women's literature is full of the pain of frequent assault, not only by a racist patriarchy, but also by Black men. Yet the necessity for and history of shared battle have made us, Black women, particularly vulnerable to the false accusation that anti-sexist is anti-Black. Meanwhile, woman hating as a recourse of the powerless is sapping strength from Black communities, and our very lives. Rape is on the increase, reported and unreported, and rape is not aggressive sexuality, it is sexualized aggression. As Kalamu ya Salaam, a Black male writer points out, "As long as male domination exists, rape will exist. Only women revolting and men made conscious of their responsibility to fight sexism can collectively stop rape."

Differences between ourselves as Black women are also being misnamed and used to separate us from one another. As a Black lesbian feminist comfortable with the many different ingredients of my identity, and a woman committed to racial and sexual freedom from oppression, I find I am constantly being encouraged to pluck out some one aspect of myself and present this as the meaningful whole, eclipsing or denying the other parts of self. But this is a destructive and fragmenting way to live. My fullest concentration of energy is available to me only when I integrate all the parts of who I am, openly, allowing power from particular sources of my living to flow back and forth freely through all my different selves, without the restrictions of externally imposed definition. Only then can I bring myself and my energies as a whole to the service of those struggles which I embrace as part of my living.

A fear of lesbians, or of being accused of being a lesbian, has led many Black women into testifying against themselves. It has led some of us into destructive alliances, and others into despair and isolation. In the white women's communities, heterosexism is sometimes a result of identifying with the white patriarchy, a rejection of that interdependence between women-identified women which allows the self to be, rather than to be used in the service of men. Sometimes it reflects a die-hard belief in the protective coloration of heterosexual relationships, sometimes a self-hate which all women have to fight against, taught us from birth.

Although elements of these attitudes exist for all women, there are particular resonances of heterosexism and homophobia among Black women. Despite the fact that woman-bonding has a long and honorable history in the African and African American communities, and despite the knowledge and accomplishments of many strong and creative women-identified Black women in the political, social and cultural fields, heterosexual Black women often tend to ignore or discount the existence and work of Black lesbians. Part of this attitude has come from an understandable terror of Black male attack within the close confines of Black society, where the punishment for any female self-assertion is still to be accused of being a lesbian and therefore unworthy of the attention or support of the scarce Black male. But part of this need to misname and ignore Black lesbians comes from a very real fear that openly women-identified Black women who are no longer dependent upon men for their self-definition may well reorder our whole concept of social relationships.

Black women who once insisted that lesbianism was a white woman's problem now insist that Black lesbians are a threat to Black nationhood, are consorting with the enemy, are basically un-Black. These accusations, coming from the very women to whom we look for deep and real understanding, have served to keep many Black lesbians in hiding, caught between the racism of white women and the homophobia of their sisters. Often, their work has been ignored, trivialized, or misnamed, as with the work of Angelina Grimke, Alice Dunbar-Nelson, Lorraine Hansberry. Yet women-bonded women have always been some part of the power of Black communities, from our unmarried aunts to the amazons of Dahomey.

And it is certainly not Black lesbians who are assaulting women and raping children and grandmothers on the streets of our communities.

Across this country, as in Boston during the spring of 1979 following the unsolved murders of twelve Black women, Black lesbians are spearheading movements against violence against Black women.

What are the particular details within each of our lives that can be scrutinized and altered to help bring about change? How do we redefine difference for all women? It is not our differences which separate women, but our reluctance to recognize those differences and to deal effectively with the distortions which have resulted from the ignoring and misnaming of those differences.

As a tool of social control, women have been encouraged to recognize only one area of human difference as legitimate, those differences which exist between women and men. And we have learned to deal across those differences with the urgency of all oppressed subordinates. All of us have had to learn to live or work or coexist with men, from our fathers on. We have recognized and negotiated these differences, even when this recognition only continued the old dominant/subordinate mode of human relationship; where the oppressed must recognize the masters' difference in order to survive.

But our future survival is predicated upon our ability to relate within equality. As women, we must root out internalized patterns of oppression within ourselves if we are to move beyond the most superficial aspects of social change. Now we must recognize differences among women who are our equals, neither inferior nor superior, and devise ways to use each others' difference to enrich our visions and our joint struggles. The future of our earth may depend upon the ability of all women to identify and develop new definitions of power and new patterns of relating across difference. The old definitions have not served us, nor the earth that supports us. The old patterns, no matter how cleverly rearranged to imitate progress, still condemn us to cosmetically altered repetitions of the same old exchanges, the same old guilt, hatred, recrimination, lamentation, and suspicion.

For we have, built into all of us, old blueprints of expectation and response, old structures of oppression, and these must be altered at the same time as we alter the living conditions which are a result of those structures. For the master's tools will never dismantle the master's house.

As Paulo Freire shows so well in the *Pedagogy of the Oppressed*, the true focus of revolutionary change is never merely the oppressive situations which we seek to escape, but that piece of the oppressor which is planted deep within each of us, and which knows only the oppressors' tactics, the oppressors' relationships.

Change means growth, and growth can be painful. But we sharpen self-definition by exposing the self in work and struggle together with those whom we define as different from ourselves, although sharing the same goals. For Black and white, old and young, lesbian and heterosexual women alike, this can mean new paths to our survival.

We have chosen each other
and the edge of each other's battles
the war is the same
if we lose
someday women's blood will congeal
upon a dead planet
if we win
there is no telling
we seek beyond history
for a new and more possible meeting.

2

A Different Mirror

RONALD T. TAKAKI

I had flown from San Francisco to Norfolk and was riding in a taxi to my hotel to attend a conference on multiculturalism. Hundreds of educators from across the country were meeting to discuss the need for greater cultural diversity in the curriculum. My driver and I chatted about the weather and the tourists. The sky was cloudy, and Virginia Beach was twenty minutes away. The rearview mirror reflected a white man in his forties. "How long have you been in this country?" he asked. "All my life," I replied, wincing. "I was born in the United States." With a strong southern drawl, he remarked: "I was wondering because your English is excellent!" Then, as I had many times before, I explained: "My grandfather came here from Japan in the 1880s. My family has been here, in America, for over a hundred years." He glanced at me in the mirror. Somehow I did not look "American" to him; my eyes and complexion looked foreign.

Suddenly, we both became uncomfortably conscious of a racial divide separating us. An awkward silence turned my gaze from the mirror to the passing landscape, the shore where the English and the Powhatan Indians first encountered each other. Our highway was on land that Sir Walter Raleigh had renamed "Virginia" in honor of Elizabeth I, the Virgin Queen. In the English cultural appropriation of America, the indigenous peoples themselves would become outsiders in their native land. Here, at the eastern edge of the continent, I mused, was the site of the beginning of multicultural America. Jamestown, the English settlement founded in 1607, was nearby: the first twenty Africans were brought here a year before the Pilgrims arrived at Plymouth Rock. Several hundred miles offshore was Bermuda, the "Bermoothes" where William Shakespeare's Prospero had landed and met the native Caliban in *The Tempest.* Earlier, another voyager had made an Atlantic crossing and unexpectedly bumped into some islands to the south. Thinking he had reached Asia, Christopher Columbus mistakenly identified one of the islands as "Cipango" (Japan). In the wake of the admiral, many people would come to America from different shores, not only from Europe but also Africa and Asia. One of them would be my grandfather. My mental wandering across terrain and time ended abruptly as we arrived at my destination. I said good-bye to my driver and went into the hotel, carrying a vivid reminder of why I was attending this conference.

SOURCE: Takaki, Ronald T. *A Different Mirror: A History of Multicultural America.*

Questions like the one my taxi driver asked me are always jarring, but I can understand why he could not see me as American. He had a narrow but widely shared sense of the past—a history that has viewed American as European in ancestry. "Race," Toni Morrison explained, has functioned as a "metaphor" necessary to the "construction of Americanness": in the creation of our national identity, "American" has been defined as "white."[1]

But America has been racially diverse since our very beginning on the Virginia shore, and this reality is increasingly becoming visible and ubiquitous. Currently, one-third of the American people do not trace their origins to Europe; in California, minorities are fast becoming a majority. They already predominate in major cities across the country—New York, Chicago, Atlanta, Detroit, Philadelphia, San Francisco, and Los Angeles.

This emerging demographic diversity has raised fundamental questions about America's identity and culture. In 1990, *Time* published a cover story on "America's Changing Colors." "Someday soon," the magazine announced, "white Americans will become a minority group." How soon? By 2056, most Americans will trace their descent to "Africa, Asia, the Hispanic world, the Pacific Islands, Arabia—almost anywhere but white Europe." This dramatic change in our nation's ethnic composition is altering the way we think about ourselves. "The deeper significance of America's becoming a majority nonwhite society is what it means to the national psyche, to individuals' sense of themselves and their nation—their idea of what it is to be American."...[2]

What is fueling the debate over our national identity and the content of our curriculum is America's intensifying racial crisis. The alarming signs and symptoms seem to be everywhere—the killing of Vincent Chin in Detroit, the black boycott of a Korean grocery store in Flatbush, the hysteria in Boston over the Carol Stuart murder, the battle between white sportsmen and Indians over tribal fishing rights in Wisconsin, the Jewish-black clashes in Brooklyn's Crown Heights, the black-Hispanic competition for jobs and educational resources in Dallas, which *Newsweek* described as "a conflict of the have-nots," and the Willie Horton campaign commercials, which widened the divide between the suburbs and the inner cities.[3]

This reality of racial tension rudely woke America like a fire bell in the night on April 29, 1992. Immediately after four Los Angeles police officers were found not guilty of brutality against Rodney King, rage exploded in Los Angeles. Race relations reached a new nadir. During the nightmarish rampage, scores of people were killed, over two thousand injured, twelve thousand arrested, and almost a billion dollars' worth of property destroyed. The live televised images mesmerized America. The rioting and the murderous melee on the streets resembled the fighting in Beirut and the West Bank. The thousands of fires burning out of control and the dark smoke filling the skies brought back images of the burning oil fields of Kuwait during Desert Storm. Entire sections of Los Angeles looked like a bombed city. "Is this America?" many shocked viewers asked. "Please, can we get along here," pleaded Rodney King, calling for calm. "We all can get along. I mean, we're all stuck here for a while. Let's try to work it out."[4]

But how should "we" be defined? Who are the people "stuck here" in America? One of the lessons of the Los Angeles explosion is the recognition of the fact that we are a multiracial society and that race can no longer be defined in the binary terms of white and black. "We" will have to include Hispanics and Asians. While blacks currently constitute 13 percent of the Los Angeles population, Hispanics represent 40 percent. The 1990 census revealed that South Central Los Angeles, which was predominantly black in 1965 when the Watts rebellion occurred, is now 45 percent Hispanic. A majority of the first 5,438 people arrested were Hispanic, while 37 percent were black. Of the fifty-eight people who died in the riot, more than a third were Hispanic, and about 40 percent of the businesses destroyed were Hispanic-owned. Most of the other shops and stores were Korean-owned. The dreams of many Korean immigrants went up in smoke during the riot: two thousand Korean-owned businesses were damaged or demolished, totaling about $400 million in losses. There is evidence indicating they were targeted. "After all," explained a black gang member, "we didn't burn our community, just *their* stores."[5]

"I don't feel like I'm in America anymore," said Denisse Bustamente as she watched the police protecting the firefighters. "I feel like I am far away." Indeed, Americans have been witnessing ethnic strife erupting around the world—the rise of neo-Nazism and the murder of Turks in Germany, the ugly "ethnic cleansing" in Bosnia, the terrible and bloody clashes between Muslims and Hindus in India. Is the situation here different, we have been nervously wondering, or do ethnic conflicts elsewhere represent a prologue for America? What is the nature of malevolence? Is there a deep, perhaps primordial, need for group identity rooted in hatred for the other? Is ethnic pluralism possible for America? But answers have been limited. Television reports have been little more than thirty-second sound bites. Newspaper articles have been mostly superficial descriptions of racial antagonisms and the current urban malaise. What is lacking is historical context; consequently, we are left feeling bewildered.[6]

How did we get to this point, Americans everywhere are anxiously asking. What does our diversity mean, and where is it leading us? *How* do we work it out in the post–Rodney King era?

Certainly one crucial way is for our society's various ethnic groups to develop a greater understanding of each other. For example, how can African Americans and Korean Americans work it out unless they learn about each other's cultures, histories, and also economic situations? This need to share knowledge about our ethnic diversity has acquired new importance and has given new urgency to the pursuit for a more accurate history.…

While all of America's many groups cannot be covered [here], the English immigrants and their descendants require attention, for they possessed inordinate power to define American culture and make public policy. What men like John Winthrop, Thomas Jefferson, and Andrew Jackson thought as well as did mattered greatly to all of us and was consequential for everyone. A broad range of groups [is important]: African Americans, Asian Americans, Chicanos, Irish, Jews, and Indians. While together they help to explain general patterns in our society, each has contributed to the making of the United States.

African Americans have been the central minority throughout our country's history. They were initially brought here on a slave ship in 1619. Actually, these first twenty Africans might not have been slaves; rather, like most of the white laborers, they were probably indentured servants. The transformation of Africans into slaves is the story of the "hidden" origins of slavery. How and when was it decided to institute a system of bonded black labor? What happened, while freighted with racial significance, was actually conditioned by class conflicts within white society. Once established, the "peculiar institution" would have consequences for centuries to come. During the nineteenth century, the political storm over slavery almost destroyed the nation. Since the Civil War and emancipation, race has continued to be largely defined in relation to African Americans—segregation, civil rights, the underclass, and affirmative action. Constituting the largest minority group in our society, they have been at the cutting edge of the Civil Rights Movement. Indeed, their struggle has been a constant reminder of America's moral vision as a country committed to the principle of liberty. Martin Luther King clearly understood this truth when he wrote from a jail cell: "We will reach the goal of freedom in Birmingham and all over the nation, because the goal of America is freedom. Abused and scorned though we may be, our destiny is tied up with America's destiny."[7]

Asian Americans have been here for over one hundred and fifty years, before many European immigrant groups. But as "strangers" coming from a "different shore," they have been stereotyped as "heathen," exotic, and unassimilable. Seeking "Gold Mountain," the Chinese arrived first, and what happened to them influenced the reception of the Japanese, Koreans, Filipinos, and Asian Indians as well as the Southeast Asian refugees like the Vietnamese and the Hmong. The 1882 Chinese Exclusion Act was the first law that prohibited the entry of immigrants on the basis of nationality. The Chinese condemned this restriction as racist and tyrannical. "They call us 'Chink,'" complained a Chinese immigrant, cursing the "white demons." "They think we no good! America cuts us off. No more come now, too bad!" This precedent later provided a basis for the restriction of European immigrant groups such as Italians, Russians, Poles, and Greeks. The Japanese painfully discovered that their accomplishments in America did not lead to acceptance, for during World War II, unlike Italian Americans and German Americans, they were placed in internment camps. Two-thirds of them were citizens by birth. "How could I as a 6-month-old child born in this country," asked Congressman Robert Matsui years later, "be declared by my own government to be an enemy alien?" Today, Asian Americans represent the fastest-growing ethnic group. They have also become the focus of much mass media attention as "the Model Minority" not only for blacks and Chicanos, but also for whites on welfare and even middle-class whites experiencing economic difficulties.[8]

Chicanos represent the largest group among the Hispanic population, which is projected to outnumber African Americans. They have been in the United States for a long time, initially incorporated by the war against Mexico. The treaty had moved the border between the two countries, and the people of "occupied" Mexico suddenly found themselves "foreigners" in their "native land." As historian Albert Camarillo pointed out, the Chicano past is an integral part of America's westward expansion, also known as "manifest destiny." But while the early Chicanos were a colonized people, most of them today have immigrant roots. Many began the trek to El Norte in the early twentieth century. "As I had heard a lot about the United States," Jesus Garza recalled, "it was my dream to come here." "We came to know families from Chihuahua, Sonora, Jalisco, and Durango," stated Ernesto Galarza. "Like ourselves, our Mexican neighbors had come this far moving step by step, working and waiting, as if they were feeling their way up a ladder." Nevertheless, the Chicano experience has been unique, for most of them have lived close to their homeland—a proximity that has helped reinforce their language, identity, and culture. This migration to El Norte has continued to the present. Los Angeles has more people of Mexican origin than any other city in the world, except Mexico City. A mostly mestizo people of Indian as well as African and Spanish ancestries, Chicanos currently represent the largest minority group in the Southwest, where they have been visibly transforming culture and society.[9]

The Irish came here in greater numbers than most immigrant groups. Their history has been tied to America's past from the very beginning. Ireland represented the earliest English frontier: the conquest of Ireland occurred before the colonization of America, and the Irish were the first group that the English called "savages." In this context, the Irish past foreshadowed the Indian future. During the nineteenth century, the Irish, like the Chinese, were victims of British colonialism. While the Chinese fled from the ravages of the Opium Wars, the Irish were pushed from their homeland by "English tyranny." Here they became construction workers and factory operatives as well as the "maids" of America. Representing a Catholic group seeking to settle in a fiercely Protestant society, the Irish immigrants were targets of American nativist hostility. They were also what historian Lawrence J. McCaffrey called "the pioneers of the American urban ghetto," "previewing" experiences that would later be shared by the Italians, Poles, and other groups from southern and eastern Europe. Furthermore, they offer contrast to the immigrants from Asia. The Irish came about the same time as the Chinese, but they had a distinct advantage: the Naturalization Law of 1790 had reserved citizenship for "whites" only. Their compatible complexion allowed them to assimilate by blending into American society. In making their journey successfully into the mainstream, however, these immigrants from Erin pursued an Irish "ethnic" strategy: they promoted "Irish" solidarity in order to gain political power and also to dominate the skilled blue-collar occupations, often at the expense of the Chinese and blacks.[10]

Fleeing pogroms and religious persecution in Russia, the Jews were driven from what John Cuddihy described as the "Middle Ages into the Anglo-American world of the *goyim* 'beyond the pale.'" To them, America represented the Promised Land. This vision led Jews to struggle not only for themselves but also for other oppressed groups, especially blacks. After the 1917 East St. Louis race riot, the Yiddish *Forward* of New York compared this antiblack violence to a 1903 pogrom in Russia: "Kishinev and St. Louis—the same soil, the same people." Jews cheered when Jackie Robinson broke into the Brooklyn Dodgers in 1947. "He was adopted as the surrogate hero by many of us growing up at the time," recalled Jack Greenberg of the NAACP Legal Defense Fund. "He was the way we saw ourselves triumphing against the forces of bigotry and ignorance." Jews stood shoulder to shoulder with blacks in the Civil Rights Movement: two-thirds of the white volunteers who went south during the 1964 Freedom Summer were Jewish. Today Jews are considered a highly successful "ethnic" group. How did they make such great socioeconomic strides? This question is often reframed by neoconservative intellectuals like Irving Kristol and Nathan Glazer to read: if Jewish immigrants were able to lift themselves from poverty into the mainstream through self-help and education without welfare and affirmative action, why can't blacks? But what this thinking overlooks is the unique history of Jewish immigrants, especially the initial advantages of many of them as literate and skilled. Moreover, it minimizes the virulence of racial prejudice rooted in American slavery.[11]

Indians represent a critical contrast, for theirs was not an immigrant experience. The Wampanoags were on the shore as the first English strangers arrived in what would be called "New England." The encounters between Indians and whites not only shaped the course of race relations, but also influenced the very culture and identity of the general society. The architect of Indian removal, President Andrew Jackson, told Congress: "Our conduct toward these people is deeply interesting to the national character." Frederick Jackson Turner understood the meaning of this observation when he identified the frontier as our transforming crucible. At first, the European newcomers had to wear Indian moccasins and shout the war cry. "Little by little," as they subdued the wilderness, the pioneers became "a new product" that was "American." But Indians have had a different view of this entire process. "The white man," Luther Standing Bear of the Sioux explained, "does not understand the Indian for the reason that he does not understand America." Continuing to be "troubled with primitive fears," he has "in his consciousness the perils of this frontier continent.... The man from Europe is still a foreigner and an alien. And he still hates the man who questioned his path across the continent." Indians questioned what Jackson and Turner trumpeted as "progress." For them, the frontier had a different "significance": their history was how the West was lost. But their story has also been one of resistance. As Vine Deloria declared, "Custer died for your sins."[12]

By looking at these groups from a multicultural perspective, we can comparatively analyze their experiences in order to develop an understanding of their differences and similarities. Race, we will see, has been a social construction that has historically set apart racial minorities from European immigrant groups. Contrary to the notions of scholars like Nathan Glazer and Thomas Sowell, race in America has not been the same as ethnicity. A broad comparative focus also allows us to see how the varied experiences of different racial and ethnic groups occurred within shared contexts.

During the nineteenth century, for example, the Market Revolution employed Irish immigrant laborers in New England factories as it expanded cotton fields worked by enslaved blacks across Indian lands toward Mexico.

Like blacks, the Irish newcomers were stereotyped as "savages," ruled by passions rather than "civilized" virtues such as self-control and hard work. The Irish saw themselves as the "slaves" of British oppressors, and during a visit to Ireland in the 1840s, Frederick Douglass found that the "wailing notes" of the Irish ballads reminded him of the "wild notes" of slave songs. The United States annexation of California, while incorporating Mexicans, led to trade with Asia and the migration of "strangers" from Pacific shores. In 1870, Chinese immigrant laborers were transported to Massachusetts as scabs to break an Irish immigrant strike; in response, the Irish recognized the need for interethnic working-class solidarity and tried to organize a Chinese lodge of the Knights of St. Crispin. After the Civil War, Mississippi planters recruited Chinese immigrants to discipline the newly freed blacks. During the debate over an immigration exclusion bill in 1882, a senator asked: If Indians could be located on reservations, why not the Chinese?[13]

Other instances of our connectedness abound. In 1903, Mexican and Japanese farm laborers went on strike together in California: their union officers had names like Yamaguchi and Lizarras, and strike meetings were conducted in Japanese and Spanish. The Mexican strikers declared that they were standing in solidarity with their "Japanese brothers" because the two groups had toiled together in the fields and were now fighting together for a fair wage. Speaking in impassioned Yiddish during the 1909 "uprising of twenty thousand" strikers in New York, the charismatic Clara Lemlich compared the abuse of Jewish female garment workers to the experience of blacks: "[The bosses] yell at the girls and 'call them down' even worse than I imagine the Negro slaves were in the South." During the 1920s, elite universities like Harvard worried about the increasing numbers of Jewish students, and new admissions criteria were instituted to curb their enrollment. Jewish students were scorned for their studiousness and criticized for their "clannishness." Recently, Asian American students have been the targets of similar complaints: they have been called "nerds" and told there are "too many" of them on campus.[14]

Indians were already here, while blacks were forcibly transported to America, and Mexicans were initially enclosed by America's expanding border. The other groups came here as immigrants: for them, America represented liminality—a new world where they could pursue extravagant urges and do things they had thought beyond their capabilities. Like the land itself, they found themselves "betwixt and between all fixed points of classification." No longer fastened as fiercely to their old countries, they felt a stirring to become new people in a society still being defined and formed.[15]

These immigrants made bold and dangerous crossings, pushed by political events and economic hardships in their homelands and pulled by America's demand for labor as well as by their own dreams for a better life. "By all means let me go to America," a young man in Japan begged his parents. He had calculated that in one year as a laborer here he could save almost a thousand yen—an amount equal to the income of a governor in Japan. "My dear Father," wrote an immigrant Irish girl living in New York, "Any man or woman without a family are fools that would not venture and come to this plentyful Country where no man or woman ever hungered." In the shtetls of Russia, the cry "To America!" roared like "wildfire." "America was in everybody's mouth," a Jewish immigrant recalled. "Businessmen talked [about] it over their accounts; the market women made up their quarrels that they might discuss it from stall to stall; people who had relatives in the famous land went around reading their letters." Similarly, for Mexican immigrants crossing the border in the early twentieth century, El Norte became the stuff of overblown hopes. "If only you could see how nice the United States is," they said, "that is why the Mexicans are crazy about it."[16]

The signs of America's ethnic diversity can be discerned across the continent—Ellis Island, Angel Island, Chinatown, Harlem, South Boston, the Lower East Side, places with Spanish names like Los Angeles and San Antonio or Indian names like Massachusetts and Iowa. Much of what is familiar in America's cultural landscape actually has ethnic origins. The Bing cherry was developed by an early Chinese immigrant named Ah Bing. American Indians were cultivating corn, tomatoes, and tobacco long before the arrival of Columbus. The term *okay* was derived from the Choctaw word *oke,* meaning "it is so." There is evidence indicating that the name *Yankee* came from Indian terms for the English—from *eankke* in Cherokee and *Yankwis* in Delaware. Jazz and blues as well as rock and roll have African American origins. The "Forty-Niners" of the Gold Rush learned mining techniques from the Mexicans; American cowboys acquired herding skills from Mexican *vaqueros* and adopted their range terms—such as *lariat* from *la reata, lasso* from *lazo,* and *stampede* from *estampida.* Songs like "God Bless America," "Easter Parade," and "White Christmas" were written by a Russian-Jewish immigrant named Israel Baline, more popularly known as Irving Berlin.[17]

Furthermore, many diverse ethnic groups have contributed to the building of the American economy, forming what Walt Whitman saluted as "a vast, surging, hopeful army of workers." They worked in the South's cotton fields, New England's textile mills, Hawaii's cane fields, New York's garment factories, California's orchards, Washington's salmon canneries, and Arizona's copper mines. They built the railroad, the great symbol of America's industrial triumph....

Moreover, our diversity was tied to America's most serious crisis: the Civil War was fought over a racial issue—slavery....

... The people in our study have been actors in history, not merely victims of discrimination and exploitation. They are entitled to be viewed as subjects—as men and women with minds, wills, and voices.

In the telling and retelling
of their stories,
They create communities
of memory.

They also re-vision history. "It is very natural that the history written by the victim," said a Mexican in 1874, "does not altogether chime with the story of the victor." Sometimes they are hesitant to speak, thinking they are only "little people." "I don't know why anybody wants to hear my history," an Irish maid said apologetically in 1900. "Nothing ever happened to me worth the tellin.'"[18]

But their stories are worthy. Through their stories, the people who have lived America's history can help all of us, including my taxi driver, understand that Americans originated from many shores, and that all of us are entitled to dignity. "I hope this survey do a lot of good for Chinese people," an immigrant told an interviewer from Stanford University in the 1920s. "Make American people realize that Chinese people are humans. I think very few American people really know anything about Chinese." But the remembering is also for the sake of the children. "This story is dedicated to the descendants of Lazar and Goldie Glauberman," Jewish immigrant Minnie Miller wrote in her autobiography. "My history is bound up in their history and the generations that follow should know where they came from to know better who they are." Similarly, Tomo Shoji, an elderly Nisei woman, urged Asian Americans to learn more about their roots: "We got such good, fantastic stories to tell. All our stories are different." Seeking to know how they fit into America, many young people have become listeners; they are eager to learn about the hardships and humiliations experienced by their parents and grandparents. They want to hear their stories, unwilling to remain ignorant or ashamed of their identity and past.[19]

The telling of stories liberates. By writing about the people on Mango Street, Sandra Cisneros explained, "the ghost does not ache so much." The place no longer holds her with "both arms. She sets me free." Indeed, stories may not be as innocent or simple as they seem to be. Native American novelist Leslie Marmon Silko cautioned:

I will tell you something about stories …
They aren't just entertainment.
Don't be fooled.

Indeed, the accounts given by the people in this study vibrantly re-create moments, capturing the complexities of human emotions and thoughts. They also provide the authenticity of experience. After she escaped from slavery, Harriet Jacobs wrote in her autobiography. "[My purpose] is not to tell you what I have heard but what I have seen—and what I have suffered." In their sharing of memory, the people in this study offer us an opportunity to see ourselves reflected in a mirror called history.[20]

In his recent study of Spain and the New World, *The Buried Mirror,* Carlos Fuentes points out that mirrors have been found in the tombs of ancient Mexico, placed there to guide the dead through the underworld. He also tells us about the legend of Quetzalcoatl, the Plumed Serpent: when this god was given a mirror by the Toltec deity Tezcatlipoca, he saw a man's face in the mirror and realized his own humanity. For us, the "mirror" of history can guide the living and also help us recognize who we have been and hence are. In *A Distant Mirror,* Barbara W. Tuchman finds "phenomenal parallels" between the "calamitous fourteenth century" of European society and our own era. We can, she observes, have "greater fellow-feeling for a distraught age" as we painfully recognize the "similar disarray," "collapsing assumptions," and "unusual discomfort."[21]

But what is needed in our own perplexing times is not so much a "distant" mirror, as one that is "different." While the study of the past can provide collective self-knowledge, it often reflects the scholar's particular perspective or view of the world. What happens when historians leave out many of America's peoples? What happens, to borrow the words of Adrienne Rich, "when someone with the authority of a teacher" describes our society, and "you are not in it"? Such an experience can be disorienting—"a moment of psychic disequilibrium, as if you looked into a mirror and saw nothing."[22]

Through their narratives about their lives and circumstances, the people of America's diverse groups are able to see themselves and each other in our common past. They celebrate what Ishmael Reed has described as a society "unique" in the world because "the world is here"—a place "where the cultures of the world crisscross." Much of America's past, they point out, has been riddled with racism. At the same time, these people offer hope, affirming the struggle for equality as a central theme in our country's history. At its conception, our nation was dedicated to the proposition of equality. What has given concreteness to this powerful national principle has been our coming together in the creation of a new society. "Stuck here" together, workers of different backgrounds have attempted to get along with each other.

People harvesting
Work together unaware
Of racial problems,

wrote a Japanese immigrant describing a lesson learned by Mexican and Asian farm laborers in California.[23]

Finally, how do we see our prospects for "working out" America's racial crisis? Do we see it as through a glass darkly? Do the televised images of racial hatred and violence that riveted us in 1992 during the days of rage in Los Angeles frame a future of divisive race relations—what Arthur Schlesinger Jr. has fearfully denounced as the "disuniting of America"? Or will Americans of diverse races and ethnicities be able to connect themselves to a larger narrative? Whatever happens, we can be certain that much of our society's future will be influenced by which "mirror" we choose to see ourselves. America does not belong to one race or one group.... Americans have been constantly redefining their national identity from the moment of first contact on the Virginia shore. By sharing their stories, they invite us to see ourselves in a different mirror.[24]

NOTES

1. Toni Morrison, *Playing in the Dark: Whiteness in the Literary Imagination* (Cambridge, Mass., 1992), p. 47.
2. William A. Henry III, "Beyond the Melting Pot," in "America's Changing Colors," *Time*, vol. 135, no. 15 (April 9, 1990), pp. 28–31.
3. "A Conflict of the Have-Nots," *Newsweek*, December 12, 1988, pp. 28–29.
4. Rodney King's statement to the press, *New York Times*, May 2, 1992, p. 6.
5. Tim Rutten, "A New Kind of Riot," *New York Times Review of Books*, June 11, 1992, pp. 52–53; Maria Newman "Riots Bring Attention to Growing Hispanic

Presence in South-Central Area," *New York Times*, May 11, 1992, p. A10; Mike Davis, "In L.A. Burning All Illusions," *The Nation*, June 1, 1992, pp. 744–745; Jack Viets and Peter Fimrite, "S.F. Mayor Visits Riot-Torn Area to Buoy Businesses," *San Francisco Chronicle*, May 6, 1992, p. A6.

6. Rick DelVecchio, Suzanne Espinosa, and Carl Nolte, "Bradley Ready to Lift Curfew," *San Francisco Chronicle*, May 4, 1992, p. A1.
7. Abraham Lincoln, "The Gettysburg Address," in *The Annals of America*, vol. 9, *1863–1865: The Crisis of the Union* (Chicago, 1968), pp. 462–463; Martin Luther King, *Why We Can't Wait* (New York, 1964), pp. 92–93.
8. Interview with old laundryman, in "Interviews with Two Chinese," circa 1924, Box 326, folder 325, Survey of Race Relations, Stanford University, Hoover Institution Archives; Congressman Robert Matsui, speech in the House of Representatives on the 442 bill for redress and reparations, September 17, 1987, *Congressional Record* (Washington, D.C., 1987), p. 7584.
9. Albert Camarillo, *Chicanos in a Changing Society: From Mexican Pueblos to American Barrios in Santa Barbara and Southern California, 1848–1930* (Cambridge, Mass., 1979), p. 2; Juan Nepornuceno Seguín, in David J. Weber (ed.), *Foreigners in Their Native Land: Historical Roots of the Mexican Americans* (Albuquerque, N. Mex., 1973), p. vi; Jesus Garza, in Manuel Garnio, *The Mexican Immigrant: His Life Story* (Chicago, 1931), p. 15; Ernesto Galarza, *Barrio Boy: The Story of a Boy's Acculturation* (Notre Dame, Ind., 1986), p. 200.
10. Lawrence J. McCaffrey, *The Irish Diaspora in America* (Washington, D.C., 1984), pp. 6, 62.
11. John Murray Cuddihy, *The Ordeal of Civility: Freud, Marx, Levi Strauss, and the Jewish Struggle with Modernity* (Boston, 1987), p. 165; Jonathan Kaufman, *Broken Alliance: The Turbulent Times between Blacks and Jews in America* (New York, 1989), pp. 28, 82, 83–84, 91, 93, 106.
12. Andrew Jackson, First Annual Message to Congress, December 8, 1829, in James D. Richardson (ed.), *A Compilation of the Messages and Papers of the Presidents, 1789–1897* (Washington, D.C., 1897), vol. 2, p. 457; Frederick Jackson Turner, "The Significance of the Frontier in American History," in *The Early Writings of Frederick Jackson Turner* (Madison, Wis., 1938), pp. 185ff.; Luther Standing Bear, "What the Indian Means to America," in Wayne Moquin (ed.), *Great Documents in American Indian History* (New York, 1973), p. 307; Vine Deloria, Jr., *Custer Died for Your Sins: An Indian Manifesto* (New York, 1969).
13. Nathan Glazer, *Affirmative Discrimination: Ethnic Inequality and Public Policy* (New York, 1978); Thomas Sowell, *Ethnic America: A History* (New York, 1981); David R. Roediger, *The Wages of Whiteness: Race and the Making of the American Working Class* (London, 1991), pp. 134–136; Dan Caldwell, "The Negroization of the Chinese Stereotype in California," *Southern California Quarterly*, vol. 33 (June 1971), pp. 123–131.
14. Thomas Almaguer, "Racial Domination and Class Conflict in Capitalist Agriculture: The Oxnard Sugar Beet Workers' Strike of 1903," *Labor History*, vol. 25, no. 3 (summer 1984), p. 347; Howard M. Sachar, *A History of the Jews in America* (New York, 1992), p. 183.
15. For the concept of liminality, see Victor Turner, *Dramas, Fields, and Metaphors: Symbolic Action in Human Society* (Ithaca, N.Y., 1974), pp. 232, 237; and Arnold Van

Gennep, *The Rites of Passage* (Chicago, 1960). What I try to do is to apply liminality to the land called America.

16. Kazuo Ito, *Issei: A History of Japanese Immigrants in North America* (Seattle, 1973), p. 33; Arnold Schrier, *Ireland and the American Emigration, 1850–1900* (New York, 1970), p. 24; Abraham Cahan, *The Rise of David Levinsky* (New York, 1960; originally published in 1917), pp. 59–61; Mary Antin, quoted in Howe, *World of Our Fathers* (New York, 1983), p. 27; Lawrence A. Cardoso, *Mexican Emigration to the United States, 1897–1931* (Tucson, Ariz., 1981), p. 80.
17. Ronald Takaki, *Strangers from a Different Shore: A History of Asian Americans* (Boston, 1989), pp. 88–89; Jack Weatherford, *Native Roots: How the Indians Enriched America* (New York, 1991), pp. 210, 212; Carey McWilliams, *North from Mexico: The Spanish-Speaking People of the United States* (New York, 1968), p. 154; Stephan Themstrom (ed.), *Harvard Encyclopedia of American Ethnic Groups* (Cambridge, Mass., 1980), p. 22; Sachar, *A History of the Jews in America*, p. 367.
18. Weber (ed.), *Foreigners in Their Native Land*, p. vi; Hamilton Holt (ed.), *The Life Stories of Undistinguished Americans as Told by Themselves* (New York, 1906), p. 143.
19. "Social Document of Pany Lowe, interviewed by C. H. Burnett, Seattle, July 5, 1924," p. 6, Survey of Race Relations, Stanford University, Hoover Institution Archives; Minnie Miller, "Autobiography," private manuscript, copy from Richard Balkin; Tomo Shoji, presentation, Obana Cultural Center, Oakland, California, March 4, 1988.
20. Sandra Cisneros, *The House on Mango Street* (New York, 1991), pp. 109–110; Leslie Marmon Silko, *Ceremony* (New York, 1978), p. 2; Harriet A. Jacobs, *Incidents in the Life of a Slave Girl, written by herself* (Cambridge, Mass., 1987; originally published in 1857), p. xiii.
21. Carlos Fuentes, *The Buried Mirror: Reflections on Spain and the New World* (Boston, 1992), pp. 10, 11, 109; Barbara W. Tuchman, *A Distant Mirror: The Calamitous 14th Century* (New York, 1978), pp. xiii, xiv.
22. Adrienne Rich, *Blood, Bread, and Poetry: Selected Prose, 1979–1985* (New York, 1986), p. 199.
23. Ishmael Reed, "America: The Multinational Society," in Rick Simonson and Scott Walker (eds.), *Multi-cultural Literacy* (St. Paul, 1988), p. 160; Ito, *Issei*, p. 497.
24. Arthur M. Schlesinger, Jr., *The Disuniting of America: Reflections on a Multicultural Society* (Knoxville, Tenn., 1991); Carlos Bulosan, *America Is in the Heart: A Personal History* (Seattle, 1981), pp. 188–189.

3

The First Americans

American Indians

C. MATTHEW SNIPP

By the end of the nineteenth century, many observers predicted that American Indians were destined for extinction. Within a few generations, disease, warfare, famine, and outright genocide had reduced their numbers from millions to less than 250,000 in 1890. Once a self-governing, self-sufficient people, American Indians were forced to give up their homes and their land, and to subordinate themselves to an alien culture. The forced resettlement to reservation lands or the Indian Territory (now Oklahoma) frequently meant a life of destitution, hunger, and complete dependency on the federal government for material needs.

Today, American Indians are more numerous than they have been for several centuries. While still one of the most destitute groups in American society, tribes have more autonomy and are now more self-sufficient than at any time since the last century. In cities, modern pan-Indian organizations have been successful in making the presence of American Indians known to the larger community, and have mobilized to meet the needs of their people (Cornell 1988; Nagel 1986; Weibel-Orlando 1991). In many rural areas, American Indians and especially tribal governments have become increasingly more important and increasingly more visible by virtue of their growing political and economic power. The balance of this [reading] is devoted to explaining their unique place in American society.

THE INCORPORATION OF AMERICAN INDIANS

The current political and economic status of American Indians is the result of the process by which they were incorporated into Euro-American society (Hall 1989). This amounts to a long history of efforts aimed at subordinating an otherwise self-governing and self-sufficient people that eventually culminated in widespread economic dependency. The role of the U.S. government in this process can be seen in the five major historical periods of federal Indian relations:

SOURCE: Snipp, C. Matthew. 1996. "American Indians." In *Origins and Destinies*, 1e, edited by S. Pedraza and R. Rumbaut.

removal, assimilation, the Indian New Deal, termination and relocation, and self-determination.

Removal

In the early nineteenth century, the population of the United States expanded rapidly at the same time that the federal government increased its political and military capabilities. The character of Indian-American relations changed after the War of 1812. The federal government increasingly pressured tribes settled east of the Appalachian Mountains to move west to the territory acquired in the Louisiana Purchase. Numerous treaties were negotiated by which the tribes relinquished most of their land and eventually were forced to move west.

Initially the federal government used bargaining and negotiation to accomplish removal, but many tribes resisted (Prucha 1984). However, the election of Andrew Jackson by a frontier constituency signaled the beginning of more forceful measures to accomplish removal. In 1830 Congress passed the Indian Removal Act, which mandated the eventual removal of the eastern tribes to points west of the Mississippi River, in an area which was to become the Indian Territory and is now the state of Oklahoma. Dozens of tribes were forcibly removed from the eastern half of the United States to the Indian Territory and newly created reservations in the west, a long process ridden with conflict and bloodshed.

As the nation expanded beyond the Mississippi River, tribes of the plains, southwest, and west coast were forcibly settled and quarantined on isolated reservations. This was accompanied by the so-called Indian Wars—a bloody chapter in the history of Indian-White relations (Prucha 1984; Utley 1963). This period in American history is especially remarkable because the U.S. government was responsible for what is unquestionably one of the largest forced migrations in history.

The actual process of removal spanned more than a half-century and affected nearly every tribe east of the Mississippi River. Removal often meant extreme hardships for American Indians, and in some cases this hardship reached legendary proportions. For example, the Cherokee removal has become known as the "Trail of Tears." In 1838, nearly 17,000 Cherokees were ordered to leave their homes and assemble in military stockades (Thornton 1987, p. 117). The march to the Indian Territory began in October and continued through the winter months. As many as 8,000 Cherokees died from cold weather and diseases such as influenza (Thornton 1987, p. 118).

According to William Hagan (1979), removal also caused the Creeks to suffer dearly as their society underwent a profound disintegration. The contractors who forcibly removed them from their homes refused to do anything for "the large number who had nothing but a cotton garment to protect them from the sleet storms and no shoes between them and the frozen ground of the last stages of their hegira. About half of the Creek nation did not survive the migration and the difficult early years in the West" (Hagan 1979, pp. 77–81). In the West, a band of Nez Perce men, women, and children, under the leadership of Chief

Joseph, resisted resettlement in 1877. Heavily outnumbered, they were pursued by cavalry troops from the Wallowa valley in eastern Oregon and finally captured in Montana near the Canadian border. Although the Nez Perce were eventually captured and moved to the Indian Territory, and later to Idaho, their resistance to resettlement has been described by one historian as "one of the great military movements in history" (Prucha 1984, p. 541).

Assimilation

Near the end of the nineteenth century, the goal of isolating American Indians on reservations and the Indian Territory was finally achieved. The Indian population also was near extinction. Their numbers had declined steadily throughout the nineteenth century, leading most observers to predict their disappearance (Hoxie 1984). Reformers urged the federal government to adopt measures that would humanely ease American Indians into extinction. The federal government responded by creating boarding schools and the allotment acts—both were intended to "civilize" and assimilate American Indians into American society by Christianizing them, educating them, introducing them to private property, and making them into farmers. American Indian boarding schools sought to accomplish this task by indoctrinating Indian children with the belief that tribal culture was an inferior relic of the past and that Euro-American culture was vastly superior and preferable. Indian children were forbidden to wear their native attire, to eat their native foods, to speak their native language, or to practice their traditional religion. Instead, they were issued Euro-American clothes, and expected to speak English and become Christians. Indian children who did not relinquish their culture were punished by school authorities. The curriculum of these schools taught vocational arts along with "civilization" courses.

The impact of allotment policies is still evident today. The 1887 General Allotment Act (the Dawes Severalty Act) and subsequent legislation mandated that tribal lands were to be allotted to individual American Indians … and the surplus lands left over from allotment were to be sold on the open market. Indians who received allotted tribal lands also received citizenship, farm implements, and encouragement from Indian agents to adopt farming as a livelihood (Hoxie 1984, Prucha 1984).

For a variety of reasons, Indian lands were not completely liquidated by allotment, many Indians did not receive allotments, and relatively few changed their lifestyles to become farmers. Nonetheless, the allotment era was a disaster because a significant number of allottees eventually lost their land. Through tax foreclosures, real estate fraud, and their own need for cash, many American Indians lost what for most of them was their last remaining asset (Hoxie 1984).

Allotment took a heavy toll on Indian lands. It caused about 90 million acres of Indian land to be lost, approximately two-thirds of the land that had belonged to tribes in 1887 (O'Brien 1990). This created another problem that continues to vex many reservations: "checkerboarding." Reservations that were subjected to allotment are typically a crazy quilt composed of tribal lands, privately owned "fee" land, and trust land belonging to individual Indian families.

Checkerboarding presents reservation officials with enormous administrative problems when trying to develop land use management plans, zoning ordinances, or economic development projects that require the construction of physical infrastructure such as roads or bridges.

The Indian New Deal

The Indian New Deal was short-lived but profoundly important. Implemented in the early 1930s along with the other New Deal programs of the Roosevelt administration, the Indian New Deal was important for at least three reasons. First, signaling the end of the disastrous allotment era as well as a new respect for American Indian tribal culture, the Indian New Deal repudiated allotment as a policy. Instead of continuing its futile efforts to detribalize American Indians, the federal government acknowledged that tribal culture was worthy of respect. Much of this change was due to John Collier, a long-time Indian rights advocate appointed by Franklin Roosevelt to serve as Commissioner of Indian Affairs (Prucha 1984).

Like other New Deal policies, the Indian New Deal also offered some relief from the Great Depression and brought essential infrastructure development to many reservations, such as projects to control soil erosion and to build hydroelectric dams, roads, and other public facilities. These projects created jobs in New Deal programs such as the Civilian Conservation Corps and the Works Progress Administration.

An especially important and enduring legacy of the Indian New Deal was the passage of the Indian Reorganization Act (IRA) of 1934. Until then, Indian self-government had been forbidden by law. This act allowed tribal governments, for the first time in decades, to reconstitute themselves for the purpose of overseeing their own affairs on the reservation. Critics charge that this law imposed an alien form of government, representative democracy, on traditional tribal authority. On some reservations, this has been an on-going source of conflict (O'Brien 1990). Some reservations rejected the IRA for this reason, but now have tribal governments authorized under different legislation.

Termination and Relocation

After World War II, the federal government moved to terminate its long-standing relationship with Indian tribes by settling the tribes' outstanding legal claims, by terminating the special status of reservations, and by helping reservation Indians relocate to urban areas (Fixico 1986). The Indian Claims Commission was a special tribunal created in 1946 to hasten the settlement of legal claims that tribes had brought against the federal government. In fact, the Indian Claims Commission became bogged down with prolonged cases, and in 1978 the commission was dissolved by Congress. At that time, there were 133 claims still unresolved out of an original 617 that were first heard by the commission three decades earlier (Fixico 1986, p. 186). The unresolved claims that were still pending were transferred to the Federal Court of Claims.

Congress also moved to terminate the federal government's relationship with Indian tribes. House Concurrent Resolution (HCR) 108, passed in 1953, called for steps that eventually would abolish all reservations and abolish all special programs serving American Indians. It also established a priority list of reservations slated for immediate termination. However, this bill and subsequent attempts to abolish reservations were vigorously opposed by Indian advocacy groups such as the National Congress of American Indians. Only two reservations were actually terminated, the Klamath in Oregon and the Menominee in Wisconsin. The Menominee reservation regained its trust status in 1975 and the Klamath reservation was restored in 1986.

The Bureau of Indian Affairs (BIA) also encouraged reservation Indians to relocate and seek work in urban job markets. This was prompted partly by the desperate economic prospects on most reservations, and partly because of the federal government's desire to "get out of the Indian business." The BIA's relocation programs aided reservation Indians in moving to designated cities, such as Los Angeles and Chicago, where they also assisted them in finding housing and employment. Between 1952 and 1972, the BIA relocated more than 100,000 American Indians (Sorkin 1978). However, many Indians returned to their reservations (Fixico 1986). For some American Indians, the return to the reservation was only temporary; for example, during periods when seasonal employment such as construction work was hard to find.

Self-Determination

Many of the policies enacted during the termination and relocation era were steadfastly opposed by American Indian leaders and their supporters. As these programs became stalled, critics attacked them for being harmful, ineffective, or both. By the mid-1960s, these policies had very little serious support. Perhaps inspired by the gains of the Civil Rights movement, American Indian leaders and their supporters made "self-determination" the first priority on their political agendas. For these activists, self-determination meant that Indian people would have the autonomy to control their own affairs, free from the paternalism of the federal government.

The idea of self-determination was well received by members of Congress sympathetic to American Indians. It also was consistent with the "New Federalism" of the Nixon administration. Thus, the policies of termination and relocation were repudiated in a process that culminated in 1975 with the passage of the American Indian Self-Determination and Education Assistance Act, a profound shift in federal Indian policy. For the first time since this nation's founding, American Indians were authorized to oversee the affairs of their own communities, free of federal intervention. In practice, the Self-Determination Act established measures that would allow tribal governments to assume a larger role in reservation administration of programs for welfare assistance, housing, job training, education, natural resource conservation, and the maintenance of reservation roads and bridges (Snipp and Summers 1991). Some reservations also have their own police forces and game wardens, and can issue licenses and levy taxes. The Onondaga tribe in upstate New York has taken their sovereignty one step further by issuing passports that are internationally recognized. Yet there is a great deal of variability in terms of how much autonomy tribes have over reservation affairs. Some tribes, especially those on large and well-organized reservations, have nearly complete control over their reservations, while smaller reservations with limited resources often depend heavily on BIA services....

CONCLUSION

Though small in number, American Indians have an enduring place in American society. Growing numbers of American Indians occupy reservation and other trust lands, and equally important has been the revitalization of tribal governments. Tribal governments now have a larger role in reservation affairs than ever in the past. Another significant development has been the urbanization of American Indians. Since 1950, the proportion of American Indians in cities has grown rapidly. These American Indians have in common with reservation Indians many of the same problems and disadvantages, but they also face other challenges unique to city life.

The challenges facing tribal governments are daunting. American Indians are among the poorest groups in the nation. Reservation Indians have substantial needs for improved housing, adequate health care, educational opportunities, and employment, as well as developing and maintaining reservation infrastructure. In the face of declining federal assistance, tribal governments are assuming an ever-larger burden. On a handful of reservations, tribal governments have assumed completely the tasks once performed by the BIA.

As tribes have taken greater responsibility for their communities, they also have struggled with the problems of raising revenues and providing economic opportunities for their people. Reservation land bases provide many reservations with resources for development. However, these resources are not always abundant, much less unlimited, and they have not always been well managed. It will be yet another challenge for tribes to explore ways of efficiently managing their existing resources. Legal challenges also face tribes seeking to exploit unconventional resources such as gambling revenues. Their success depends on many complicated legal and political contingencies.

Urban American Indians have few of the resources found on reservations, and they face other difficult problems. Preserving their culture and identity is an especially pressing concern. However, urban Indians have successfully adapted to city environments in ways that preserve valued customs and activities—powwows, for example, are an important event in all cities where there is a large Indian community. In addition, pan-Indianism has helped urban Indians set aside tribal differences and forge alliances for the betterment of urban Indian communities.

These alliances are essential, because unlike reservation Indians, urban American Indians do not have their own form of self-government. Tribal governments do not have jurisdiction over urban Indians. For this reason, urban Indians must depend on other strategies for ensuring that the needs of their community are met, especially for those new to city life. Coping with the transition to urban life poses a multitude of difficult challenges for many American Indians. Some succumb to these problems, especially the hardships of unemployment, economic deprivation, and related maladies such as substance abuse, crime, and violence. But most successfully overcome these difficulties, often with help from other members of the urban Indian community.

Perhaps the greatest strength of American Indians has been their ability to find creative ways for dealing with adversity, whether in cities or on reservations. In the past, this quality enabled them to survive centuries of oppression and persecution. Today this is reflected in the practice of cultural traditions that Indian people are proud to embrace. The resilience of American Indians is an abiding quality that will no doubt ensure that they will remain part of the ethnic mosaic of American society throughout the twenty-first century and beyond.

REFERENCES

Cornell, Stephen. 1988. *The Return of the Native: American Indian Political Resurgence.* New York: Oxford University Press.

Fixico, Donald L. 1986. *Termination and Relocation: Federal Indian Policy, 1945–1960.* Albuquerque, NM: University of New Mexico Press.

Hagan, William T. 1979. *American Indians.* Chicago, IL: University of Chicago Press.

Hall, Thomas D. 1989. *Social Change in the Southwest, 1350–1880.* Lawrence, KS: University Press of Kansas.

Hoxie, Frederick E. 1984. *A Final Promise: The Campaign to Assimilate the Indians, 1880–1920.* Lincoln, NE: University of Nebraska Press.

Nagel, Joanne. 1986. "American Indian Repertoires of Contention." Paper presented at the annual meeting of the American Sociological Association, San Francisco, CA.

O'Brien, Sharon. 1990. *American Indian Tribal Governments.* Norman, OK: University of Oklahoma Press.

Prucha, Francis Paul. 1984. *The Great Father.* Lincoln, NE: University of Nebraska Press.

Snipp, C. Matthew, and Gene F. Summers. 1991. "American Indian Development Policies," pp. 166–180 in *Rural Policies for the 1990s*, edited by Cornelia Flora and James A. Christenson. Boulder, CO: Westview Press.

Sorkin, Alan L. 1978. *The Urban American Indian.* Lexington, MA: Lexington Books.

Thornton, Russell. 1987. *American Indian Holocaust and Survival: A Population History since 1942.* Norman, OK: University of Oklahoma Press.

Utley, Robert M. 1963. *The Last Days of the Sioux Nation.* New Haven: Yale University Press.

Weibel-Orlando, Joan. 1991. *Indian Country, L.A.* Urbana, IL: University of Illinois Press.

4

From a Native Daughter

HAUNANI-KAY TRASK

E noi'i wale mai nō ka haole, a,
'a'ole e pau nō hana a Hawai'i 'imi loa /
Let the haole freely research us in detail
But the doings of deep delving Hawai'i
will not be exhausted.

—*Kepelino*
19th-century Hawaiian historian

When I was young, the story of my people was told twice: once by my parents, then again by my school teachers. From my *'ohana* (family), I learned about the life of the old ones: how they fished and planted by the moon; shared all the fruits of their labors, especially their children; danced in great numbers for long hours; and honored the unity of their world in intricate genealogical chants. My mother said Hawaiians had sailed over thousands of miles to make their home in these sacred islands. And they had flourished, until the coming of the *haole* (whites).

At school, I learned that the "pagan Hawaiians" did not read or write, were lustful cannibals, traded in slaves, and could not sing. Captain Cook had "discovered" Hawai'i and the ungrateful Hawaiians had killed him. In revenge, the Christian god had cursed the Hawaiians with disease and death.

I learned the first of these stories from speaking with my mother and father. I learned the second from books. By the time I left for college, the books had won out over my parents, especially since I spent four long years in a missionary boarding school for Hawaiian children.

When I went away I understood the world as a place and a feeling divided in two: one *haole* (white), and the other *kānaka* (Native). When I returned ten years later with a Ph.D., the division was sharper, the lack of connection more painful. There was the world that we lived in—my ancestors, my family, and my people—and then there was the world historians described. This world, they had written, was the truth. A primitive group, Hawaiians had been ruled by bloodthirsty priests and despotic kings who owned all the land and kept our people in feudal subjugation. The chiefs were cruel, the people poor.

SOURCE: Trask, Haunani-Kay. 1993. *From a Native Daughter.* Monroe, ME: Common Courage Press. Reprinted by permission.

But this was not the story my mother told me. No one had owned the land before the *haole* came; everyone could fish and plant, except during sacred periods. And the chiefs were good and loved their people.

Was my mother confused? What did our *kūpuna* (elders) say? They replied: Did these historians (all *haole)* know the language? Did they understand the chants? How long had they lived among our people? Whose stories had they heard?

None of the historians had ever learned our mother tongue. They had all been content to read what Europeans and Americans had written. But why did scholars, presumably well-trained and thoughtful, neglect our language? Not merely a passageway to knowledge, language is a form of knowing by itself; a people's way of thinking and feeling is revealed through its music.

I sensed the answer without needing to answer. From years of living in a divided world, I knew the historian's judgment: *There is no value in things Hawaiian; all value comes from things haole.*

Historians, I realized, were very much like missionaries. They were a part of the colonizing horde. One group colonized the spirit; the other, the mind. Frantz Fanon had been right, but not just about Africans. He had been right about the bondage of my own people: "By a kind of perverted logic, [colonialism] turns to the past of the oppressed people, and distorts, disfigures, and destroys it" (1963:210). The first step in the colonizing process, Fanon had written, was the deculturation of a people. What better way to take our culture than to remake our image? A rich historical past became small and ignorant in the hands of Westerners. And we suffered a damaged sense of people and culture because of this distortion.

Burdened by a linear, progressive conception of history and by an assumption that Euro-American culture flourishes at the upper end of that progression, Westerners have told the history of Hawai'i as an inevitable if occasionally bittersweet triumph of Western ways over "primitive" Hawaiian ways. A few authors—the most sympathetic—have recorded with deep-felt sorrow the passing of our people. But in the end, we are repeatedly told, such an eclipse was for the best.

Obviously it was best for Westerners, not for our dying multitudes. This is why the historian's mission has been to justify our passing by celebrating Western dominance. Fanon would have called this missionizing, intellectual colonization. And it is clearest in the historian's insistence that *pre-haole* Hawaiian land tenure was "feudal"—a term that is now applied, without question, in every monograph, in every schoolbook, and in every tour guide description of my people's history.

From the earliest days of Western contact my people told their guests that *no one* owned the land. The land—like the air and the sea—was for all to use and share as their birthright. Our chiefs were *stewards* of the land; they could not own or privately possess the land any more than they could sell it.

But the *haole* insisted on characterizing our chiefs as feudal landlords and our people as serfs. Thus, a European term which described a European practice founded on the European concept of private property—feudalism—was imposed upon the people halfway around the world from Europe and vastly different from her in every conceivable way. More than betraying an ignorance of Hawaiian culture and history, however, this misrepresentation was malevolent in design.

By inventing feudalism in ancient Hawai'i, Western scholars quickly transformed a spiritually based, self-sufficient economic system of land use and occupancy into an oppressive, medieval European practice of divine right ownership, with the common people tied like serfs to the land. By claiming that the Pacific people lived under a European system—that the Hawaiians lived under feudalism—Westerners could then degrade a successful system of shared land use with a pejorative and inaccurate Western term. Land tenure changes instituted by Americans and in line with current Western notions of private property were then made to appear beneficial to the Hawaiians. But in practice, such changes benefited the *haole,* who alienated the people from the land, taking it for themselves.

The prelude to this land alienation was the great dying of the people. Barely half a century after contact with the West, our people had declined in number by eighty percent. Disease and death were rampant. The sandalwood forests had been stripped bare for international commerce between England and China. The missionaries had insinuated themselves everywhere. And a debt-ridden Hawaiian king (there had been no king before Western contact) succumbed to enormous pressure from the Americans and followed their schemes for dividing the land.

This is how private property land tenure entered Hawai'i. The common people, driven from their birthright, received less than one percent of the land. They starved while huge haole-owned sugar plantations thrived.

And what had the historians said? They had said that the Americans "liberated" the Hawaiians from an oppressive "feudal" system. By inventing a false feudal past, the historians justify—and become complicitous in—massive American theft.

Is there "evidence"—as historians call it—for traditional Hawaiian concepts of land use? The evidence is in the sayings of my people and in the words they wrote more than a century ago, much of which has been translated. However, historians have chosen to ignore any references here to shared land use. But there is incontrovertible evidence in the very structure of the Hawaiian language. If the historians had bothered to learn our language (as any American historian of France would learn French) they would have discovered that we show possession in two ways: through the use of an "a" possessive, which reveals acquired status, and through the use of an "o" possessive, which denotes inherent status. My body (*ko'u kino*) and my parents (*ko'u mãkua*), for example, take the "o" form; most material objects, such as food *(ka'u mea'ai)* take the "a" form. But land, like one's body and one's parents, takes the "o" possessive (*ko'u 'ãina*). Thus, in our way of speaking, land is inherent to the people; it is like our bodies and our parents. The people cannot exist without the land, and the land cannot exist without the people.

Every major historian of Hawai'i has been mistaken about Hawaiian land tenure. The chiefs did not own the land: they *could not* own the land. My mother was right and the *haole* historians were wrong. If they had studied our language they would have known that no one owned the land. But was their failing merely ignorance, or simple ethnocentric bias?

No, I did not believe them to be so benign. As I read on, a pattern emerged in their writing. Our ways were inferior to those of the West, to those of the historians' own culture. We were "less developed," or "immature," or "authoritarian." In some tellings we were much worse. Thus, Gavan Daws (1968), the most famed modern historian of Hawai'i, had continued a tradition established earlier by missionaries Hiram Bingham (1848; reprinted, 1981) and Sheldon Dibble (1909), by referring to the old ones as "thieves" and "savages" who regularly practiced infanticide and who, in contrast to "civilized" whites, preferred "lewd dancing" to work. Ralph Kuykendall (1938), long considered the most thorough if also the most boring of historians of Hawai'i, sustained another fiction—that my ancestors owned slaves, the outcast *kauwā*. This opinion, as well as the description of Hawaiian land tenure as feudal, had been supported by respected sociologist Andrew Lind.… Finally, nearly all historians had refused to accept our genealogical dating of A.D. 400 or earlier for our arrival from the South Pacific. They had, instead, claimed that our earliest appearance in Hawai'i could only be traced to A.D. 1100. Thus at least seven hundred years of our history were repudiated by "superior" Western scholarship. Only recently have archaeological data confirmed what Hawaiians had said these many centuries (Tuggle 1979).[1]

Suddenly the entire sweep of our written history was clear to me. I was reading the West's view of itself through the degradation of my own past. When historians wrote that the king owned the land and the common people were bound to it, they were saying that ownership was the only way human beings in their world could relate to the land, and in that relationship, some one person had to control both the land and the interaction between humans.

And when they said that our chiefs were despotic, they were telling of their own society, where hierarchy always results in domination. Thus any authority or elder is automatically suspected of tyranny.

And when they wrote that Hawaiians were lazy, they meant that work must be continuous and ever a burden.

And when they wrote that we were promiscuous, they meant that lovemaking in the Christian West is a sin.

And when they wrote that we were racist because we preferred our own ways to theirs, they meant that their culture needed to dominate other cultures.

And when they wrote that we were superstitious, believing in the *mana* of nature and people, they meant that the West has long since lost a deep spiritual and cultural relationship to the earth.

And when they wrote that Hawaiians were "primitive" in their grief over the passing of loved ones, they meant that the West grieves for the living who do not walk among their ancestors.

For so long, more than half my life, I had misunderstood this written record, thinking it described my own people. But my history was nowhere present. For we had not written. We had chanted and sailed and fished and built and prayed. And we had told stories through the great blood lines of memory: genealogy.

To know my history, I had to put away my books and return to the land. I had to plant *taro* in the earth before I could understand the inseparable bond between people and *'āina*. I had to feel again the spirits of nature and take gifts of plants and fish to the ancient altars. I had to begin to speak my language with our elders and leave long silences for wisdom to grow. But before anything else, I needed to learn the language like a lover so that I could rock within her and lie at night in her dreaming arms.

There was nothing in my schooling that had told me of this, or hinted that somewhere there was a longer, older story of origins, of the flowing of songs out to a great but distant sea. Only my parents' voices, over and over, spoke to me of a Hawaiian world. While the books spoke from a different world, a Western world.

And yet, Hawaiians are not of the West. We are of *Hawai'i Nei,* this world where I live, this place, this culture, this *'āina.*

What can I say, then, to Western historians of my place and people? Let me answer with a story.

A while ago I was asked to appear on a panel on the American overthrow of our government in 1893. The other panelists were all *haole.* But one was a *haole* historian from the American continent who had just published a book on what he called the American anti-imperialists. He and I met briefly in preparation for the panel. I asked him if he knew the language. He said no. I asked him if he knew the record of opposition to our annexation to America. He said there was no real evidence for it, just comments here and there. I told him that he didn't understand and that at the panel I would share the evidence. When we met in public and spoke, I said this:

There is a song much loved by our people. It was written after Hawai'i had been invaded and occupied by American marines. Addressed to our dethroned Queen, it was written in 1893, and tells of Hawaiian feelings for our land and against annexation to the United States. Listen to our lament:

Kaulana nā pua a'o Hawai'i	Famous are the children of Hawai'i
Kupa'a ma hope o ka 'āina	Who cling steadfastly to the land
Hiki mai ka 'elele o ka loko 'ino	Comes the evil-hearted with
Palapala 'ānunu me ka pākaha	A document greedy for plunder
Pane mai Hawai'i moku o Keawe	Hawai'i, island of Keawe, answers
Kokua nā hono a'o Pi'ilani	The bays of Pi'ilani [of Maui, Moloka'i, and Lana'i] help
Kako'o mai Kaua'i Mano	Kaua'i of Mano assists
Pau pu me ke one o Kakuhihewa	Firmly together with the sands of Kakuhihewa
'A'ole a'e kau i ka pūlima	Do not put the signature
Maluna o ka pepa o ka 'enemi	On the paper of the enemy
Ho'ohui 'āina kū'ai hewa	Annexation is wicked sale
I ka pono sivila a'o ke kānaka	Of the civil rights of the Hawaiian people
Mahope mākou o Lili'ūlani	We support Lili'uokalani
A loa'a 'e ka pono o ka 'āina	Who has earned the right to the land
Ha'ina 'ia mai ana ka puana	The story is told
'O ka po'e i aloha i ka 'āina	Of the people who love the land

This song, I said, continues to be sung with great dignity at Hawaiian political gatherings, for our people still share the feelings of anger and protest that it conveys.

But our guest, the *haole* historian, answered that this song, although beautiful, was not evidence of either opposition or of imperialism from the Hawaiian perspective.

Many Hawaiians in the audience were shocked at his remarks, but, in hindsight, I think they were predictable. They are the standard response of the historian who does not know the language and has no respect for its memory.

Finally, I proceeded to relate a personal story, thinking that surely such a tale could not want for authenticity since I myself was relating it. My *tūtū* (grandmother) had told my mother who had told me that at the time of the overthrow a great wailing went up throughout the islands, a wailing of weeks, a wailing of impenetrable grief, a wailing of death. But he remarked again, this too is not evidence.

And so, history goes on, written in long volumes by foreign people. Whole libraries begin to form, book upon book, shelf upon shelf. At the same time, the stories go on, generation to generation, family to family.

Which history do Western historians desire to know? Is it to be a tale of writings by their own countrymen, individuals convinced of their "unique" capacity for analysis, looking at us with Western eyes, thinking about us within Western philosophical contexts, categorizing us by Western indices, judging us by Judeo-Christian morals, exhorting us to capitalist achievements, and finally, leaving us an authoritative-because-Western record of their complete misunderstanding?

All this has been done already. Not merely a few times, but many times. And still, every year, there appear new and eager faces to take up the same telling, as if the West must continue, implacably, with the din of its own disbelief. But there is, as there has been always, another possibility. If it is truly our history Western historians desire to know, they must put down their books, and take up our practices. First, of course, the language. But later, the people, the *ãina*, the stories. Above all, in the end, the stories. Historians must listen, they must hear the generational connections, the reservoir of sounds and meanings.

They must come, as American Indians suggested long ago, to understand the land. Not in the Western way, but in the indigenous way, the way of living within and protecting the bond between people and *'ãina*. This bond is cultural, and it can be understood only culturally. But because the West has lost any cultural understanding of the bond between people and land, it is not possible to know this connection through Western culture. This means that the history of indigenous people cannot be written from within Western culture. Such a story is merely the West's story of itself.

Our story remains unwritten. It rests within the culture, which is inseparable from the land. To know this is to know our history. To write this is to write of the land and the people who are born from her.

NOTES

1. See also Fornander (1878–85; reprinted, 1981). Lest one think these sources antiquated, it should be noted that there exist only a handful of modern scholarly works on the history of Hawai'i. The most respected are those by Kuykendall (1938) and

Daws (1968), and a social history of the 20th century by Lawrence Fuchs (1961). Of these, only Kuykendall and Daws claim any knowledge of pre-*haole* history, while concentrating on the 19th century. However, countless popular works have relied on these two studies which, in turn, are themselves based on primary sources written in English by extremely biased, anti-Hawaiian Westerners such as explorers, traders, missionaries (e.g., Bingham [1848; reprinted, 1981] and Dibble [1909]), and sugar planters. Indeed, a favorite technique of Daws's—whose *Shoal of Time* was once the most acclaimed and recent general history—is the lengthy quotation without comment of the most racist remarks by missionaries and planters. Thus, at one point, half a page is consumed with a "white man's burden" quotation from an 1886 *Planters Monthly* article ("It is better here that the white man should rule ...," etc., p. 213). Daws's only comment is, "The conclusion was inescapable." To get a sense of such characteristic contempt for Hawaiians, one has but to read the first few pages, where Daws refers several times to the Hawaiians as "savages" and "thieves" and where he approvingly has Captain Cook thinking, "It was a sensible primitive who bowed before a superior civilization" (p. 2). See also—among examples too numerous to cite—his glib description of sacred *hula* as a "frivolous diversion," which, instead of work, the Hawaiians "would practice energetically in the hot sun for days on end ... their bare brown flesh glistening with sweat" (pp. 65–66). Daws, who repeatedly displays an affection for descriptions of Hawaiian skin color, taught Hawaiian history for some years at the University of Hawai'i. He once held the Chair of Pacific History at the Australian National University's Institute of Advanced Studies.

Postscript: Since this article was written, the first scholarly history by a Native Hawaiian was published in English: *Native Land and Foreign Desires* by Lilikalà Kame'eleihiwa (Honolulu: Bishop Museum Press, 1992).

REFERENCES

Bingham, Hiram. 1981. *A Residence of Twenty-one Years in the Sandwich Islands*. Tokyo: Charles E. Tuttle.

Daws, Gavan. 1968. *Shoal of Time: A History of the Hawaiian Islands*. Honolulu: University of Hawai'i Press.

Dibble, Sheldon. 1909. *A History of the Sandwich Islands*. Honolulu: Thrum Publishing.

Fanon, Frantz. 1963. *The Wretched of the Earth*. New York: Grove Press.

Fornander, Abraham. 1981. *An Account of the Polynesian Race, Its Origins, and Migrations and the Ancient History of the Hawaiian People to the Times of Kamehameha I*. Routledge, Vermont: Charles E. Tuttle.

Fuchs, Lawrence H. 1961. *Hawaii Pono: A Social History*. New York: Harcourt Brace & World.

Kuykendall, Ralph. 1938. *The Hawaiian Kingdom, 1778–1854: Foundation and Transformation*. Honolulu: University of Hawai'i Press.

Tuggle, H. David. 1979. "Hawai'i," in *The Prehistory of Polynesia*, ed. Jessie D. Jennings. Cambridge: Harvard University Press.

5

Label Us Angry

JEREMIAH TORRES

It hurts to know that the most painful and shocking event of my life happened in part because of my race—something I can never change. On October 23, 1998, my friend and I experienced what would forever change our perceptions of our hometown and society in general.

We both attended elementary, middle, and high school in the quiet, prosperous, seemingly sophisticated college town of Palo Alto. In the third grade, we happily sang "It's a Small World," holding hands with the children of professors, graduate students, and professionals of the area, oblivious to our diversity in race, culture, or experience. Our small world grew larger as we progressed through the school system, each year learning more about what made us different from each other. But on that October evening, the world grew too large for us to handle.

Carlos and I were ready for a night out with the boys. It was his seventeenth birthday, and we were about to celebrate at the pool hall. I pulled out of the Safeway driveway as a speeding driver delivered a jolting honk. I followed him out, speeding to catch up with him, my immediate anger getting the better of me.

We lined up at the stoplight, and the passenger, a young white man dressed for the evening, rolled down his window; I followed. He looked irritated.

"He wasn't honking at you, you stupid fuck!"

His words slapped me across the face. I opened my stunned mouth, only to deliver an empty breath, so I gave him my middle finger until I could return some angry words. He grimaced and reached under his seat to pull out a bottle of mace, spraying it directly in my face, barely missing Carlos, who witnessed the bizarre scene in shock. It burned.

"Take that you fucking lowlifes! Stupid chinks!"

Carlos instinctively bolted out the door at those words. He started pounding the white guy without a second thought, with a new anger he had never known or felt before. Pssssht! The white guy hit Carlos point blank in the face with the mace. He screamed; tires squealed; "fuck you's" were exchanged.

We spent the next ten minutes half-blind, clutching our eyes in the burning pain, cursing in raging anger that made us forget for moments the intense, throbbing fire on our faces. I crawled out of my car to follow Carlos's screams and curses, opening my eyes to the still, spectating traffic surrounding us. I stumbled

SOURCE: Torres, Jeremiah. 2004. "Label Us Angry." In *Asian American X: An Intersection of 21st Century Asian American Voices*, edited by A. Har and J. Hsu. Ann Arbor, MI: University of Michigan Press. Reprinted by permission.

to the sidewalk, where Carlos pounded the ground and recalled the words of the white guy. We needed water.

I stumbled further to a nearby house that had lights in the living room. I doorbelled frantically, but nobody answered. I appealed to the traffic for help. They just watched, forming a new route around my car to continue about their evening. The mucous membranes in our sinuses cut loose, and we spit every few seconds to sustain our gasping breaths. After nearly five minutes of appeals, a kind woman stopped to call the cops and give us water to quench the burning.

The cops came within minutes with advice for dealing with the mace. We tried to identify the car and the white guy who had sprayed us, and they sent out the obligatory all points bulletin. They questioned us soon after, asking if we were in a gang. I returned a blank stare with a silent "no." Apparently, two Filipino teenagers finding trouble on a Friday evening raised suspicions of a new Filipino gang in Palo Alto—yeah, all five of us.

I often ask myself if it would have been different had I been driving a BMW and dressed in an ironed polo shirt and slacks, like a typical Palo Alto kid. Maybe then the white guy would not have been afraid and called us lowlifes and chinks. I don't think so. He wasn't afraid of us; he initiated the curses and maced us from a safe distance. He reached out to hurt us because he was having a bad day and we looked different.

That night was our first encounter with overt racism that stems from a hatred of difference. We hadn't seen it through the smiles and happy songs of elementary school or the isolated cliques of middle and high school, but now we knew it was there. We hadn't seen it through the clean-cut, sophisticated facade of the Palo Alto white guy, but now we knew it was there. The "lowlife," "chink," and "gangster" labels made us different, marginalizing us from the town we called home.

Those labels made us angry, but we hesitated to project that anger. At first, we didn't tell anyone except our closest friends, afraid our parents would find out and react irrationally by locking us in our rooms to keep us away from trouble. But then we realized that the trouble had found us, and we decided to voice our anger.

We wrote an anonymous article in the school newspaper narrating the incident and the underlying racism that had come to surface. We noted that the incident wasn't purely racial, or a hate crime, but proof that racist tendencies still exist, even in open-minded suburban towns like Palo Alto. Parents, students, and teachers were shocked, maybe because they knew the truth in what we were saying. Many asked if it was Carlos and me who had been maced, but I responded, "Does it matter? What matters is that some people in this town still can't accept diversity. It's sad." We confronted the community with an issue previously reserved for hypothetical classroom discussions and brought it into the open. It was the least we could do to release our anger and expose its roots, hoping for a change in those who chose to label us.

After the article, Carlos and I took different routes. I continued with my studies, complying with my regimen of high school classes and activities as my anger subsided. I tried to lay the incident aside, having exposed it and promoted self-inspection and possible change in others through writing. Carlos remained angry. Why not? He got a face full of mace and racist labels for his seventeenth birthday. He alienated himself from the white majority and returned the mean gestures of the white guy to the yuppie congregation of Palo Alto. He became an outsider. Whenever someone would look at him funny, he would stare back, sometimes too harshly.

On the day after finals, he was making his way through the front parking lot of school when a parent looked at him funny. He stared back. The parent called him a punk. Carlos exploded. He cursed and gestured all he could at the father, and when he sped away in his Suburban, Carlos followed. Carlos couldn't keep up with the Suburban, so he took a quarter from his pocket and threw it at the back window, shattering it to pieces. Carlos ran away when the cops came to school.

Within two days, students had identified Carlos as the perpetrator, and he was suspended from school as the father called his lawyer, indicting Carlos of "assault with the intent to hurt." Weeks passed until a court hearing, and Carlos attended anger management counseling, but he was still angry—angry that he was being tried over throwing a quarter and that once again "the white guys were winning." His mother scraped up the little money she had to spare to afford him a lawyer for the trial, but there was no contesting the father's accusations. Carlos was sentenced to a night in juvenile hall and two hundred hours of community service over some angry words and throwing a quarter. He became a convicted felon.

He had learned once again that he couldn't win against the labels thrown at him, the labels that hurt him more than the mace or the night in juvy, and so he became more of an outsider. In both cases, the labels distanced us from the "normal" Palo Altans: white, clean-cut, wealthy. That division didn't always exist, however; it was created by the generalizations "normal" Palo Altans made through labels. To them, we looked like lowlifes, chinks, gangsters, and punks. In truth, we were two Filipino Americans headed toward Stanford and Berkeley, living in a town that swiftly disowned us with four reckless labels after raising us for ten years. Label us angry.

PART II

Systems of Power and Inequality

MARGARET L. ANDERSEN AND
PATRICIA HILL COLLINS

One of the most important things to learn about race, class, and gender is that they are *systemic forms of inequality*. Although most people tend to think of them as individual characteristics (or identities), they are built into the very structure of society. It is this social fact that drives our analysis of race, class, and gender as *intersectional systems of inequality*. This does not make them irrelevant as individual or group characteristics but points you to the analysis of *social structure* to understand how race, class, and gender influence people's lives.

Using a social structural analysis of race, class, and gender turns your attention to how they work as *systems of power*—systems that advantage and disadvantage groups differently depending on their social location. No one of these social facts singularly determines where you will be situated within this system. To illustrate, not all men are equally powerful nor are all women equally disadvantaged. Some women, indeed, have more power and money than many men as well as other women. When you focus on the intermingling structural relationships of race, class, *and* gender, you see a more complex, ever-changing, and multidimensional social order.

Race, class, and gender operate, not alone, but within a system of simultaneous, interrelated social relationships—what we have earlier called the matrix of domination (Collins 2000). This means that they also engage other social realities—ethnicity, sexuality, age, disability, the region where you live, and so forth. These various social facts about a given life are intertwined within a system of inequality and unequal power. How they operate together depends on their particular configuration and changes at different points in time. Each shapes the others. Sexuality, for example, has long served to buttress the beliefs that support racial and gender subordination (Collins 2000, 2004). Likewise, a system of racial subordination has historically been a primary way that class structure was created. That is, some White people accumulated property through the appropriation of slave labor, even while Black slaves were denied basic rights of citizenship, including the right to vote, the right to marry, the right to own property, and the right to be considered a citizen.

To put it simply, *race, class, and gender are social structures, not just individual identities or experiences.* Furthermore, these structures are supported by ideological beliefs that make things appear "normal" and "acceptable," clouding our awareness of how the structure operates. Philosopher Marilyn Frye likens this to a birdcage. If you only look myopically at a single wire, you might not understand why the bird does not just fly away, but when you see the complete structure of the cage, you realize there is a whole structure of containment enclosing the bird (Frye 1983). Liken this to thinking today that sees racism as largely a thing of the past because formal barriers to racial discrimination have been removed. As Charles Gallagher ("Color-Blind Privilege: The Social and Political Functions of Erasing the Color Line in Post-Race America") points out in his formulation of *color-blind racism,* racism persists, even when it takes on new forms. Perhaps one "wire" from the cage has been removed—laws endorsing racial discrimination—but a whole structure of racism, including new belief systems, is in place.

Understanding the intersections among race, class, gender, ethnicity, and sexuality requires knowing how to conceptualize each. Although we would rather not treat them separately, we do so here to learn first what each means and how each is manifested in different group experiences. As you read the articles in this section, you will notice several common themes.

First, *each is a socially constructed category.* That is, their significance stems not from some "natural" state, but from what they have become as the result of social and historical processes.

Second, *each constructs groups in binary (or opposite; "either/or") terms:* Man/woman, Black/White, rich/poor, gay/straight, or citizen/alien. These binary constructions create a notion of "otherness" for those who are subordinated by this constellation of inequalities.

Third, *these are categories of individual and group identity, but note—and this is important—they are also social structures.* That is, they are not just about individuals or groups, but are also about one's group location in a system of power and inequality.

Finally, *neither race, class, nor gender is a fixed category*. Because they are social constructions, their form—and their interrelationship—changes over time. This also means that social change is possible.

As you learn about race, class, and gender, you should keep the intersectional model in mind. One way to think about their interrelationship in a social system is to imagine a typical college basketball game. This will probably seem familiar: The players on the court, the cheerleaders moving about on the side, the band playing, fans cheering, boosters watching from the best seats, and—if the team is ranked—perhaps a television crew. Everybody seems to have a place in the game. Everybody seems to be following the rules. But what explains the patterns that we see and we do not see?

Race clearly matters. Why do so many young Black men play basketball? Some people argue that African Americans are simply better in areas requiring physical skills such as sports, but there is another reality: For many young Black men, sports may seem the only hope for a good job, so sports, like the military, can seem like an attractive mobility route. The odds of actually doing so are extremely slim. Of the forty thousand African American boys playing high school basketball, only thirty-five will make it to the NBA (National Basketball Association) and only seven of those will be starters. This makes the odds of success 0.000175 (Eitzen 2012)!

But does a racial analysis alone fully explain the "rules" of college basketball? Not really. You also need a class analysis. Who benefits from college basketball? Players get scholarships and a chance to earn college degrees, though graduation rates for Black male basketball players, especially in the Division I schools, are abysmally low. Players reap the rewards, but who really benefits?

Winning teams also benefit educational institutions by increasing admissions applications, raising alumni giving and corporate support, and, sometimes, bringing in television revenues. Simply put, athletics—including college athletics—is big business. But there is a hierarchy here too—only a few universities actually reap big rewards from athletics; for most, it is a cost, not a financial boon. Corporate sponsors are the main beneficiaries. Corporations create and market products, branding the team and campus. Athletic shoes, workout clothing, cars, and beer all target the consumer dollars of those who enjoy watching basketball.

Many jobs are created by the enterprise of college basketball. Sports reporters, team physicians, trainers, and coaches are generally well remunerated as professionals. But there are also the service workers—preparing and selling food, cleaning the toilets and stands, and maintaining the stadium. A class system defines one's place in this system of inequality, even though service workers may be largely invisible to the fans. The stakes in this system of inequality get even higher when you move beyond college into the world of professional sports.

Do race and class fully explain the "rules" of basketball? A gender analysis is needed as well. Only in a few schools does women's basketball draw as large an audience as men's. Certainly in the media, men's basketball is generally the public's focal point, even though women's sports are increasingly popular. Where are the women? A few are coaches, rarely paid what the men receive—even on the most winning teams. Those closest to the action on the court may be cheerleaders—tumbling, dancing, and being thrown into the air in support of the exploits of the athletes. Others may be in the band. Some women are in the stands, cheering the team—many of them accompanied by their husbands, partners, boyfriends, parents, and children. Many work in the concession stands, fulfilling women's roles of serving others. Still others are even more invisible, left to clean the restrooms, locker rooms, and stands after the crowd goes home.

Men's behavior reveals a gendered dimension to basketball, as well. Where else are men able to put their arms around one another, slap one another's buttocks, hug one another, or cry in public without having their "masculinity" questioned? Sportscasters, too, bring gender into the play of sports, such as when they talk about men's athletic achievements as heroic but talk about women athletes' looks or their connection to children. For that matter, look at the prominence given to men's teams in sports pages of the daily newspaper compared with sports news about women, who are typically relegated to the back pages—if their athletic accomplishments are reported at all. Sometimes what we don't see can be just as revealing as what we do see. Gender, as a feature of the game on the court, is so familiar that it may go unquestioned.

This discussion of college basketball demonstrates how each factor—race, class, and gender—provides an important, yet partial, perspective on social action. Using an intersectional model, we not only see each of them in turn but also the connections among them. In fact, race, class, and gender are so inextricably intertwined that gaining a comprehensive understanding of a basketball game requires thinking about all of them and how they work together. New questions then emerge: Why are most of those serving the food in concession stands likely to be women and men of color? How are norms of masculinity and sexuality played out through sport? What class and racial ideologies are promoted through assuming that sports are a mobility route for those who try hard enough? If race, class, and gender relations are embedded in something as familiar and widespread as college basketball, to what extent are other social practices, institutions, relations, and social issues similarly structured?

Race, gender, and class divisions are deeply embedded in the structure of social institutions such as work, family, education, the media, and the state. They shape human relationships, identities, social institutions, and the social issues that emerge from within institutions. You can see the intersections of race, class, and gender in three realms of society: The representational realm, the social interaction realm, and the social structural realm. The representational realm includes the symbols, language, and images that convey racial meanings in society; the social interaction realm refers to the norms and behaviors observable in human relationships; and, the social structural realm involves the institutional sites where power and resources are distributed in society (Glenn 2002: 12).

This means that race, class, and gender affect all levels of our experience—our consciousness and ideas, our interaction with others, and the social institutions we live within. Because race, class, and gender are interconnected, no one can be subsumed under the other. In this section of the book, although we focus on each one to provide conceptual grounding, keep in mind that race, class, and gender are connected and overlapping—in all three realms of society: The realm of ideas, interaction, and institutions.

You might begin by considering a few facts:

- The United States is in the midst of a sizable redistribution of wealth, with a greater concentration of wealth and income in the hands of a few than at most previous periods of time. At the same time, a declining share of income is going to the middle class—a class that finds its position slipping relative to years past (Noah 2012).
- Women in the top 25 percent of income groups have seen the highest wage growth of any group over the last twenty years; the lowest earning groups of women, like men, have seen wages fall, while the middle has remained flat (Mishel et al. 2012). Class differences among women are hidden if you think of women as a monolithic group.
- Women of color, including Latinas, African American women, Native American women, and Asian American women, are concentrated in the bottom rungs of the labor market along with recent immigrant women (Bureau of Labor Statistics 2013).
- Poverty has been steadily increasing in the United States since 2000; it is particularly severe among women, especially among women of color and their children; in recent years, poverty has risen most among Hispanics (DeNavas-Walt, Proctor, and Smith 2013).
- At both ends of the economic spectrum, there is a growth of gated communities: Well-guarded, locked neighborhoods for the rich and prisons for the poor—particularly Latinos and African American men. At the same time, growth in the rate of imprisonment is highest among women (Glaze 2010).

None of these facts can be explained through an analysis that focuses only on race or class or gender. Clearly, race matters. Class matters. Gender matters. And they matter together. We turn now to defining some of the basic concepts involving the systems of power and inequality that we examine in this section.

RACE AND RACISM

Many think that we now live in a so-called postracial society—one where the nation has shed itself of the racially exclusionary practices of the past, moving toward a more tolerant, racially just and integrated society. After all, the thinking goes, we elected a Black president and multiracial people are becoming more numerous; open expression of racial prejudice is frowned upon. Yet, other signs of racial injustice are also visible—for those who look. Voting rights, so valiantly fought for in the civil rights movement, are being infringed upon. Incarceration of African American, Latino, and Native American men is shockingly high. Poverty keeps many of our urban zones and, increasingly, suburban places in serious trouble. Unemployment rates for young people of color are soaring. And, although overt prejudice may be lessened, people of color report very high levels of "microaggressions"—the kind of daily interactions with racism that remain invisible to many in more dominant groups.

These and other facts point to the ongoing structure of **racial stratification**—the inequality between and among racial groups. Racial stratification is an institutional phenomenon. Although racial inequality may be present in people's thinking, it is structured into social institutions. Although racial stratification changes over time, it is rooted in the practices of the past interacting with social forces of the present. It may be less visible than individual expressions of racial prejudice, but it is just as present and perhaps even more significant in shaping the experiences of diverse racial-ethnic groups.

Prejudice is defined as a hostile attitude toward people who are presumed to have alleged negative characteristics associated with a group to which they belong. *Prejudice* refers to people's attitudes, but racism is more than expressions of prejudice. **Racism** is a system of power and privilege; it can be manifested in people's attitudes but is rooted in society's structure. Racism is reflected in the different group advantages and disadvantages, based on their location in this societal system. Racism is structured into society, not just in people's minds. As such, it is built into the very fabric of dominant institutions in the United States and has been since the founding of the nation.

Understanding *racism as institutional* means that people may not be individually racist but can still benefit from a system that is organized to benefit some at the expense of others, as shown in Peggy McIntosh's now classic article, "Unpacking the Invisible Knapsack." She likens racism to an "invisible knapsack" of white privilege that typically goes unnoticed by Whites who benefit from a system of racism, *even when they do not intend to do so*. Similarly, in his essay on color-blind racism ("Color-Blind Privilege"), Charles Gallagher shows that the form of racism changes over time. Racial discrimination is no longer legal, but racism continues to structure relations among groups and to differentiate the power that different groups have. *Color-blind racism* is a new form of racism in which dominant groups assume that race no longer matters—even when society is highly racially segregated and when individual and group well-being is still strongly determined by race. Many people believe that being nonracist means being color-blind—that is, refusing to recognize or treat as significant a person's racial background and identity. But to ignore the significance of race in a society where racial groups have distinct historical and contemporary experiences is to deny the reality of their group experience. Being color blind in a society structured on racial privilege means assuming that everybody is "White," which is why people of color might be offended by friends who say, for example, "But I never think of you as Black." Blindness to the persistent realities of race also leads to an idea that there is nothing we should be doing about it—either individually or collectively—thus, racism is perpetuated.

Understanding racism also means having to understand the socially constructed meaning of race. Most people assume that race is biologically based, but the concept of race is more social than biological. Scientists working on the Human Genome Project have even found that there is no "race" gene, but you should not conclude from this that race is not "real." The reality of race stems from its social significance. That is, the meaning and significance of race stems from specific social, historical, and political contexts. These contexts make race meaningful, not just whatever physical differences may exist among groups.

Think about how racial categories are created, by whom, and for what purposes. Racial classification systems reflect prevailing views of race, thereby establishing groups that are presumed to be "natural." These constructed racial categories then serve as the basis for allocating resources; furthermore, once defined, the categories frame political issues and conflicts (Omi and Winant 1994). Omi and Winant define **racial formation** as "the sociohistorical process by which racial categories are created, inhabited, transformed, and destroyed" (1994: 55). In Nazi Germany, Jews were considered to be a race—a social construction that became the basis for the Holocaust. Anti-Semitism is, indeed, a form of racism. The history of anti-Semitism shows how understandings of who constitutes a race can change over time. Abby L. Ferber's essay ("What White Supremacists Taught a Jewish Scholar about Identity") shows the complexities that evolve in the social construction of race. As someone who studies White supremacist groups, she sees how White racism defines her as Jewish, even while she lives in society as White. Her reflections reveal, too, the interconnections between racism and anti-Semitism (the hatred of Jewish people), reminding us of the interplay between different systems of oppression.

Society constructs ideas, rules, and practices that define some groups in racial terms—and differently at different points in time. Racial meanings constantly change as institutions evolve and as different groups contest prevailing racial definitions. Some groups are "racialized"; others are not. Where, for example, did the term *Caucasian* come from? Although many take it to be "real" and don't think about its racist connotations, the term has racist origins. It was developed in the late eighteenth century by a German anthropologist, Johann Blumenbach. He developed a racial classification scheme that put people from the Russian Caucases at the top of the racial hierarchy because he thought Caucasians were the most beautiful and sophisticated people; darker people—Asians, Africans, Polynesians, and Native Americans—were put on the bottom of the list (Ferber 1999). It is amazing when you think about it that this term remains with us, with few questioning its racist origin and connotations.

In fact, the concept of race is relatively recent and it evolves only as social institutions develop that treat groups differently based on *presumed* racial differences (and note the emphasis on presumed here). Racism is linked to and develops from social and historical processes and thus shifts over time and with changing conditions. Now, for example, are Latinos being considered a race? Some would say they are in the process of becoming *racialized*—even though many Latino groups have lived within the borders of the United States since long before their lands were considered part of this nation. Chicanos originally held land in what is now the American Southwest, but it was taken following the Mexican American War. In 1848, Mexico ceded huge parts of what are now California, New Mexico, Nevada, Colorado, Arizona, and Utah to the United States for $15 million. Mexicans living there were one day Mexicans, the next day, "Americans" but without all the rights of citizens and defined as "other."

You can also see the social construction of race in the history of how people are counted and defined in the U.S. census. In 1860, only three "races" were presumed to exist—Whites, Blacks, and mulattoes. By 1890, however, these original three races had been joined by five others—quadroon, octoroon, Chinese, Japanese, and Indian. Ten short years later, this list shrank to five races—White, Black, Chinese, Japanese, and Indian—a situation reflecting the growth of strict segregation in the South (Rodriguez 2000). People of mixed racial heritage have long presented a challenge to census classifications. The U.S. government now allows people to check multiple boxes to identify themselves as more than one race. You can check "Hispanic" as a separate category. This change in the census (which began with the 2000 census) reflects the growing number of multiracial people in the United States. The census categories are not just a matter of accurate statistics; they have significant consequences for the apportionment of societal resources. Although some might argue that we should not "count" race at all, doing so is important because data on racial groups are used to enforce voting rights, to regulate equal employment opportunities, to determine various governmental supports, and to be able to track the socioeconomic status of various groups, among other things.

The concept of race is certainly changing. For many years, race and racism in the United States have been thought of in terms of Black/White relations. Changes in the racial landscape of the United States, however, now make such a framework inadequate for thinking about all group experiences. Black/White relations have defined racism in the United States for centuries, but a rapidly changing population that includes diverse Latino groups is forcing Americans to reconsider the nature of racism. Asian Americans, Native Americans, American Muslims, and others are demonstrating the ethnic and racial diversity in the United States. This reality has resulted in shifting understandings of race and ethnic relations. It is important to locate the study of race, racism, and ethnicity in the shifting relations of power that also characterize the treatment of different groups. This makes race and racism fundamentally about the character of U.S. social institutions.

For Asian Americans, changes in immigration laws have also shifted how race is constructed. Groups who at one time would have had national identities develop a pan-ethnic identity that stems from their experiences in the specific racial environment of the United States. As Min Zhou argues ("Are Asian Americans Becoming White?"), being defined as "white" is a social and historical process. As some groups achieve material success, they may become perceived as "white," even when—as in the case of Jewish and Irish immigrants, they have earlier been labeled as "nonwhite." Zhou shows how Asian Americans—an identity that merges different ethnic backgrounds—are simultaneously stereotyped as the "model minority" while still being perceived as outsiders—or alien to American society.

The overarching structure of racial power relations means that placement in this structure leads to differences in how people see racism—or don't—and what they are willing to do about it. Herbert J. Gans explores this idea in his essay, "Race as Class." Different groups are perceived within a racial hierarchy that distributes both material and symbolic resources in society. Some groups become less "racially marked" via upward class mobility, although, as Gans shows, this does not usually occur for African Americans—indicative of the stronghold that racial beliefs have on the general American public. As we will see, even with changes in attitudes and some optimism about racial progress, marked differences by race are still evident in employment, political representation, schooling, and other basic measures of group well-being.

CLASS

Class is a major force in American society. Class shapes the life chances of Americans even though there is a strong cultural belief that class mobility is possible. The consumer culture of America makes it seem that everyone is middle-class because the products associated with this class status are so ubiquitous.

What is social class? Most think of class as a rank held by an individual, but instead, think of class as a series of relations that pervade the entire society and shape our social institutions and relationships with one another. **Class** is a system that differentially structures group access to economic, political, cultural, and social resources.

The class system is currently undergoing profound changes, changes that are linked to patterns of economic transformation in the political economy—changes that are both global and domestic. Jobs are being exported overseas as vast multinational corporations seek to enhance their profits by promoting new markets and cutting the costs of labor. Within the United States, there is a shift from a manufacturing-based economy to a service economy, with corresponding changes in the types of jobs available and the wages attached to these jobs. Fewer skilled, decent-paying manufacturing jobs exist today than in the past. Fewer workers are covered by job benefits and unemployment insurance. This has become more evident to people, especially in light of the economic recession and downturn that began in 2009.

With unemployment high, jobs scarce, and a shrinking safety net to provide benefits to people in need, class divisions in the United States have perhaps become more obvious. But despite this perception, class distinctions are not fading. Quite the contrary. Inequality in the United States is not only growing and reaching historic gaps between the "haves" and everyone else, but inequality in the United States is also greater than in other Western, industrialized nations.

Income growth has been greatest for those at the top end of the population—both the upper 20 percent and the upper 5 percent of all income groups, regardless of race. For everyone else, income has remained flat. Indeed, as discussed by Timothy Noah in "The Great Divergence," class inequality has reached unprecedented levels. Such levels of inequality produce some of our most intransigent social problems, and threaten American democracy.

The inequality that characterizes U.S. society can be seen in multiple ways. Some of the basic ways of documenting inequality come from studying patterns of income distribution and wealth holdings. Both income and wealth are part of inequality. It is important to understand the difference. **Income** is the amount of money brought into a household in one year. Measures of income in the United States are based on annually reported census data drawn from a sample of the population. **Median income** is the income level above and below which half of the population lies. It is the best measure of group income standing. Thus, in 2012, median income for non-Hispanic White households was $57,009 (meaning half of such households earned more than this and half earned less); this is the "middle." Black households had a median income of $32,321; Hispanic households, $39,005; Asian Americans, $68,636 (DeNavas-Walt, Proctor, and Smith 2013). As you will see in Figure 2, family structure is also a factor in predicting income.

Income only tells part of the story. Differences in patterns of wealth are even more revealing and show how inequality is perpetuated in society. **Wealth** is determined by adding up all of one's financial assets and subtracting all of one's debt—that is, one's net worth. Income and wealth are related but they are not the same thing. As important as income can be in determining one's class status and measuring inequality, wealth is even more significant.

To understand this, imagine two recent college graduates. They graduate in the same year, from the same college, with the same major and the identical grade point average. Both get jobs with the same salary in the same company. One student's parents paid all college expenses and gave her a car upon graduation. The other student worked while in school and has graduated with substantial debt from student loans. This student's family has no money to help support the new worker. Who is better off? Same salary, same credentials, but one person has a clear advantage—one that will be played out many times over as the young worker buys a home, finances her own children's education, and possibly inherits additional assets. This shows you the significance of wealth—not just income—in structuring social class.

Thus, income data indicate quite dramatic differences in race, class, and gender standing. Wealth differences are also startling. The wealthiest one percent of

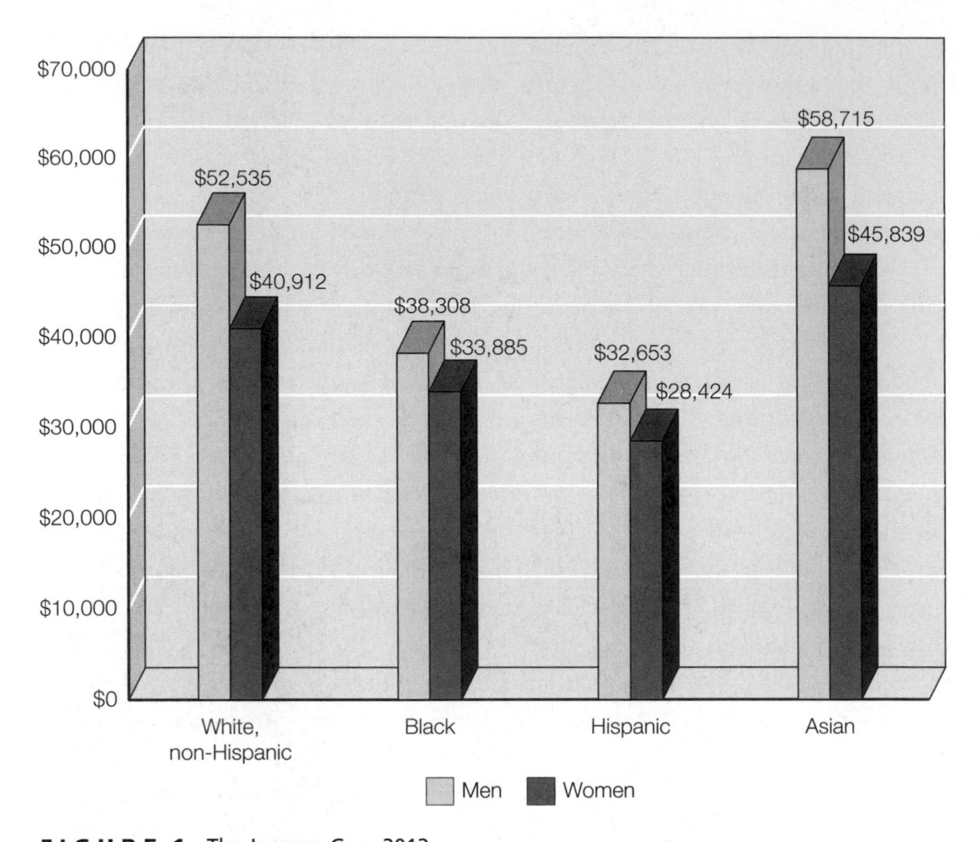

FIGURE 1 The Income Gap, 2012

NOTE: This reflects the income only of full-time, year-round workers.

SOURCE: Data from U.S. Census Bureau. 2013. *Detailed Income Tables,* Table PINC-05. www.census.gov

the population hold a larger share of the nation's total wealth than the bottom 90 percent combined—even while household debt has been soaring. Moreover, growth in wealth has become more skewed with those in the top one percent increasingly better off than the middle class (Bivens and Mishel 2011). For most Americans, debt, not wealth, is more common: 25 percent of White households, 61 percent of Black households, and 54 percent of Hispanic households have *no financial assets at all* (Oliver and Shapiro 2006). There are also vast differences in wealth holdings among different racial groups. The median net worth of White households is more than ten times that of African American and Latino households. This has meant that in the recent economic downturn, housing foreclosures have devastated whole communities. The impact has been greater for African Americans and other groups whose hold on such things as home mortgages has been more recent and more tenuous, as Meghan Kuebler shows in her analysis of race and the likelihood of home ownership ("Closing the Wealth

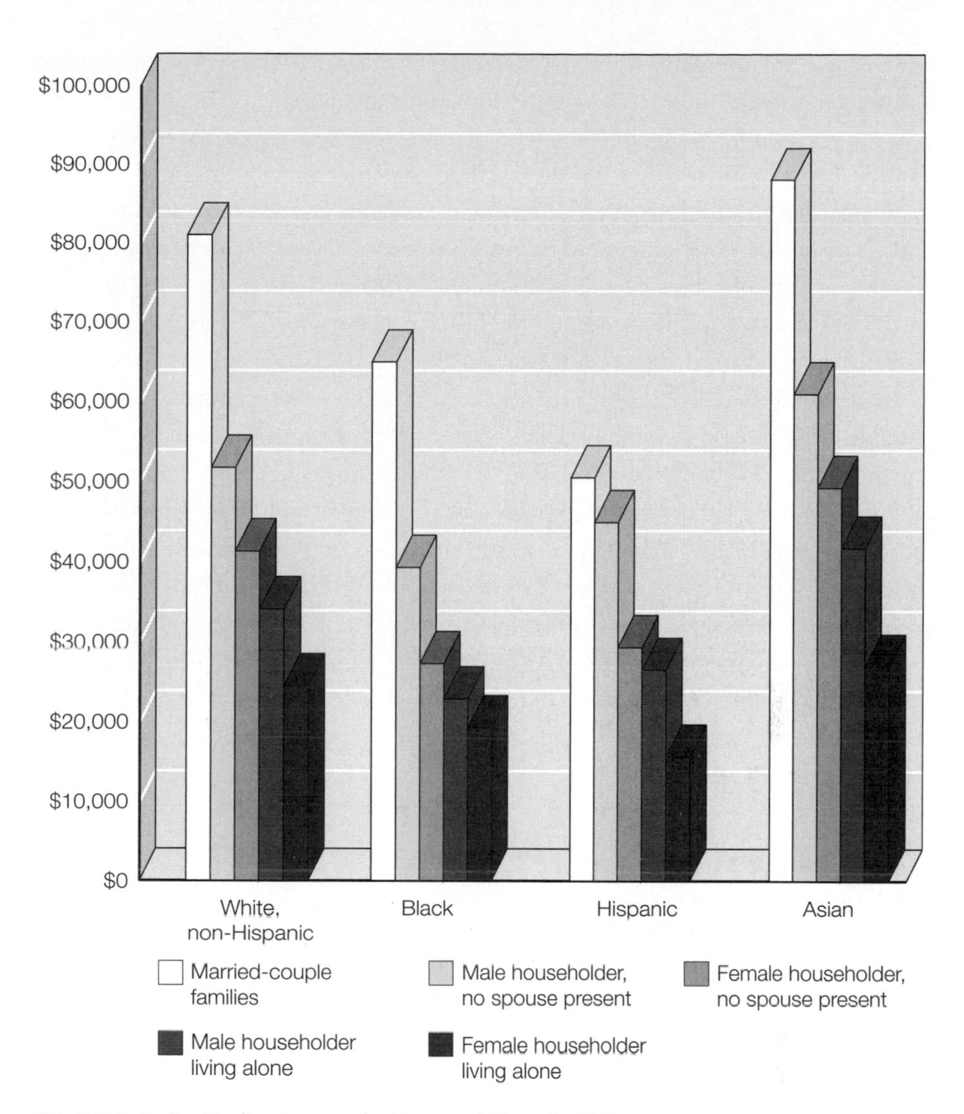

FIGURE 2 Median Income by Race and Household Type

SOURCE: Data from U.S. Census Bureau. 2013. *Detailed Poverty Tables,* Table HINC-01. www.census.gov

Gap: A Review of Racial and Ethnic Inequalities in Homeownership"). Home ownership has long been a dream for those seeking a stable foothold in U.S. society—a dream that, for many, is now fading.

Wealth, whether in the form of home ownership or other financial investments, provides a *cumulative* advantage to those who have it. Wealth produces more wealth because inheritance allows people to transmit economic status from one generation to the next. Wealth helps pay for college costs for children and down payments on houses; it can cushion the impact of emergencies, such as unexpected unemployment or sudden health problems. Even small amounts of wealth can provide the cushion that averts economic disaster for families. Buying a home, investing, being free of debt, sending one's children to college, and transferring economic assets to the next generation are all instances of class advantage that add up over time and produce advantage, even beyond one's current income level. Sociologists Melvin Oliver and Thomas Shapiro (2006) have found, for example, that even Black and White Americans at the same income level, with the same educational and occupational assets, have a substantial difference in their financial assets—an average difference of $43,143 per year! This means that, even when earning the same income, the two groups are in quite different class situations—although both may be considered middle class. When you study class in relationship to race, you will see that, overall, there are wide differences in the class status of Whites and people of color, but you should be careful not to see all Whites and Asian Americans as well off and all African Americans, Native Americans, and Latinos as poor. White households on average possess higher accumulated wealth and have higher incomes than Black, Hispanic, and Native American households, but large numbers of White households do not. White people also account for almost half (43 percent) of the nation's poor (DeNavas-Walt, Proctor, and Smith 2013). Class experiences even *within* racial groups can vary widely, as we have seen an expansion of the Black and Latino middle class in recent years.

These facts should caution us about conclusions based on *aggregate data* (that is, data that represent whole groups). Such data give you a broad picture of group differences, but they are not attentive to the more nuanced picture you see when taking into account race, class, and gender (along with other factors, such as age, level of education, occupation, and so forth). Aggregate data on Asian Americans, for example, show them as a group to be relatively well off. This portrayal, like the stereotypes of a model minority, however, obscures significant differences both when making comparisons between Asian Americans and other groups and between Asian American groups. So, for example, although Asian American median income is, in the aggregate, higher than for White Americans, this does not mean all Asian American families are better off than White families. Twelve percent of Asian Americans are poor, compared with 9.9 percent of White, non-Hispanic people (DeNavas-Walt, Proctor, and Smith 2013). Recent Asian immigrants have the highest rates of poverty, including the Hmong, Laotians, and Cambodians, whose rates of poverty match that of African Americans and Native Americans—about a 25 percent poverty rate (Le 2008).

Examining poverty is a revealing way to see the intersection of race, class, and gender. Women and their children are especially hard hit by poverty.

Forty-three percent of Hispanic families and 41 percent of Black families headed by women are poor, as are 21 percent of Asian American and 24 percent of White (not Hispanic) families headed by women (see Figure 2). Poverty rates among children (those under eighteen years of age) are especially disturbing: In 2012, 22 percent of all children in the United States lived below the *poverty line* ($23,283 for a family of four with two children). When adding race, the figures are even more disturbing: 37 percent of African American children, 33 percent of Hispanic children, 13 percent of Asian American children, and 10 percent of White (non-Hispanic) children are poor—astonishing figures for one of the most affluent nations in the world (DeNavas-Walt, Proctor, and Smith 2013).

Despite the evidence of poverty among different groups, as Debra Henderson and Ann Tickamyer point out ("The Intersection of Poverty Discourses"), poverty is associated in many people's minds only with African Americans. Such thinking has shaped social policies about poverty as if poor people are somehow undeserving. Were ideologies about poverty not so entangled with racism, would we be more generous in the social supports provided for the poor? As Henderson and Tickamyer document, stereotypes about different groups influence public policies about poverty. Poor people, whether in urban or rural environments, struggle with needs for economic resources, child care,

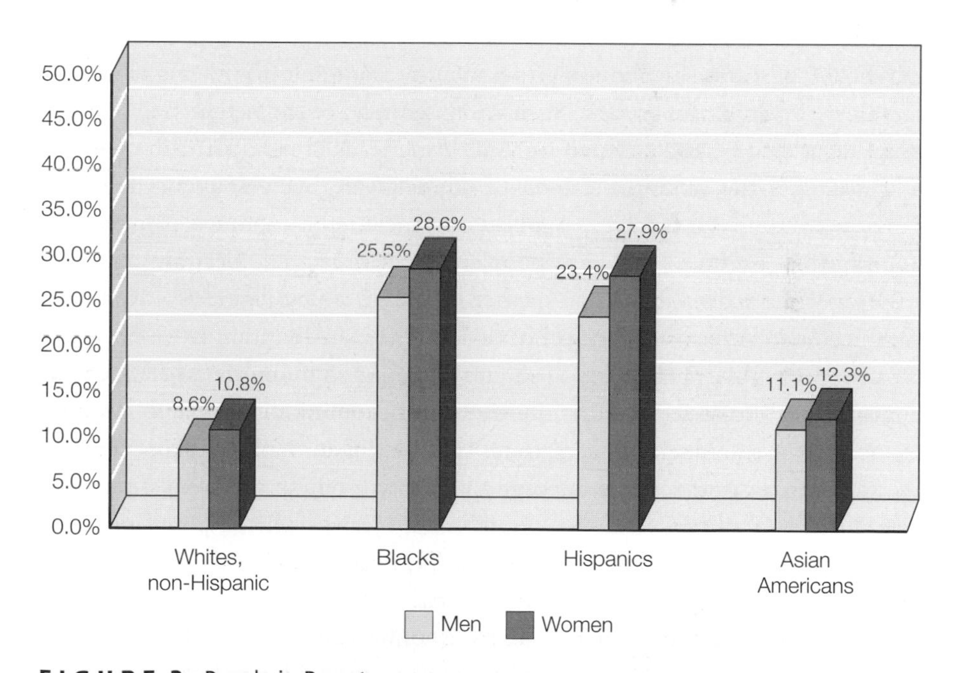

FIGURE 3 People in Poverty

SOURCE: Data from U.S. Census Bureau. 2013. *Detailed Poverty Tables,* Table POV01. www.census.gov

transportation, and health care—confronting a system that continues to blame the poor, not society, for their plight. In reality, most poverty stems from such *social factors* as unemployment, low wages, lack of child care, inequity in the delivery of health care. And, in many cases, family emergencies such as a medical emergency, a disabling accident, or the death of a family provider can throw a family with minimal resources to begin with into poverty.

All told, research consistently shows that class—and its relationship to gender and race—continue to shape the life chances of people in this nation, including the likelihood of one being poor. The consequences are real and are reflected in such basic measures of well-being as health, as explained by Lawrence R. Jacobs and James A. Morone ("Health and Wealth: Our Appalling Health Inequality Reflects and Reinforces Society's Other Gaps").

ETHNICITY, IMMIGRATION, AND MIGRATION

Sociologists traditionally define **ethnicity** to refer to groups who share a common culture. Like race, ethnicity develops within the context of systems of power. Thus, the meaning and significance of ethnicity can shift over time and in different social and political contexts. For example, groups may develop a sense of heightened ethnicity in the context of specific historical events; likewise, their feeling of sharing a common group identity can result from being labeled as "outsiders" by dominant groups. Think, for example, of the heightened sense of group identity that has developed for Arab Americans in the aftermath of 9/11.

Patterns in the shifting meaning of ethnicity are being influenced by the changing contours of the U.S. population. Racial-ethnic groups comprise an increasing proportion of the U.S. population, and their numbers are expected to rise over the years ahead. One-quarter of the U.S. population is now Black, Hispanic, Asian American, Pacific Islander, or Native American; by 2050, non-Hispanic Whites are predicted to make up only slightly more than half the total population. At the same time that the nation is becoming more diverse, changes are occurring regarding which groups predominate. Latinos have recently exceeded African Americans as the largest minority group in the population. Hispanic, Native American, Asian American, and African American populations are also growing more rapidly than White American populations, with the greatest growth among Hispanics and Asian Americans.

Immigration into the United States is only one facet of the increasing importance that ethnicity and migration play in contemporary social life. Changes in ethnicity and migration are being enacted on a world stage—a global context that increasingly links nations together in a global social order where diversity is more typical than not. You need only consult an older world atlas to get a sense of the widespread changes that characterize the current period. *World cities* have been formed that are linked through international systems of commerce; within these cities, migrants play an important role in providing the labor that is needed to keep pace with the global economy. Migrant (or "guest") workers may provide the labor for multinational corporations, or they may provide the service work that increasingly characterizes postindustrial society. Some workers may provide agricultural labor for multinational producers of food; others perform domestic labor for middle- and upper-class families whose own lives are being transformed by changes in the world economic system. Wherever you look, the world is being changed by the increasingly global basis of modern life.

Such changes indicate that globalization is not just about what is happening elsewhere in the world—as if global studies were just a matter of comparing societies. Globalization is everywhere. No nation, including the United States, can really be understood without seeing how globalization is affecting life even at home. Patterns of life in any one society are now increasingly shaped by the connections between societies, and this is evident when you look at how ethnicity and migration are changing life in the United States. Cultural features associated with different cultural traditions are increasingly evident, even in communities in the United States that have been thought of as "all-American" towns. Youth cultures embrace common features whether in Russia, Mexico, Japan, or the United States because of the penetration of world capitalist markets. Thus, hip-hop plays on the streets of Buenos Aires and world music is heard on radio stations in the United States.

Within the borders of the United States, changes associated with migration are raising new questions about national identity. How will the United States define itself as a nation-state in the changing global context? How will the increased visibility of people of color in the United States, in Latin America, in Africa, in Asia, and within the borders of former European colonial powers shape the future? How are cultural representations and institutions changed in a world context marked by such an ethnic mix—both within and between countries? In this context, which groups are defined as races, and how does that map onto concepts of citizenship and the division of labor in different nations? What will it mean to be American in a nation where racial-ethnic diversity is so much a part of the national fabric?

Such questions have taken on heightened meaning in this post-9/11 society. Following the terrorist attacks, most in the United States were catapulted into a strong sense of national identity. But accompanying this have been more restrictive immigration practices and policies and restrictions on many civil liberties. International and "ethnic" students are increasingly under surveillance. The nation's response to immigration has largely been one of criminalizing immigrants and questioning their motives for coming to the United States. Such tensions bring increased importance to addressing how the nation can maintain great diversity within a framework of social justice.

Several themes emerge in rethinking ethnicity and migration through the lenses of race, class, and gender. Despite the ideology of the "melting pot," national identity in the United States has been closely linked to a history of White privilege. Indeed, as Lillian Rubin points out in "Is This a White Country, or What?" the term *American* is usually assumed to mean White. Other types of Americans, such as African Americans and Asian Americans, become distinguished from the "real" Americans by virtue of their race. More importantly, certain benefits are reserved for those deemed to be "deserving Americans." You might think of her article as you listen to contemporary diatribes against so-called "illegal aliens." What different connotation emerges when you talk about immigrants as "illegal aliens" rather than simply as "undocumented workers?" Which nomenclature makes people seem to be "other," "distant," and "un-American?"

The very concept of a citizen in the United States has been implicitly defined by casting particular immigrant groups as "other." Different Asian American groups have contributed to American society even while being denied both the formal and informal rights of citizenship. As shown in Bhoomi K. Thakore's article ("Must-See TV: South Asian Characterization in American Popular Media"), stereotypes of Asian Americans also are rampant in the popular media, adding to the "otherness" with which Asian Americans are perceived by dominant groups.

Many people view ethnic and racial inequality as the failure of people of color to assimilate into the mainstream as White ethnic groups allegedly have done. But as Mary C. Waters points out in "Optional Ethnicities: For Whites Only?," this view seriously misreads the role of race and ethnicity in shaping American national identity. Waters suggests that White Americans of European ancestry have *symbolic ethnicity,* meaning that their ethnic identity does not influence their lives unless they want it to. Waters contrasts this symbolic ethnic identity among many Whites with the socially enforced and imposed racial identity among African Americans. Because race operates as a physical marker in the United States, intersections of race and ethnicity operate differently for Whites and for people of color. White ethnics can thus have "ethnicity" without cost, but people of color pay the price for their ethnic identity.

The politics of immigration have permeated American society in recent years. No longer are immigrants concentrated mostly in particular parts of the country. Economic restructuring has distributed immigrant groups throughout the nation and into new destinations, such as rural areas in the South and Midwest, as well as small towns throughout America. More people can now see the evidence of an increasingly diverse society. Such dispersion enriches the nation's culture, but also fans the flames of anti-immigrant prejudices. Marie Friedmann Marquardt and her co-authors show the consequences of anti-immigrant thinking in "Living 'Illegal': The Human Face of Unauthorized Immigration." With a sorely needed new immigration policy mired in the muck of politics, thousands of people who contribute to the U.S. economy still live in daily fear of deportation or other forms of hostile targeting. Marquardt et al. also show how the connotations of language, such as "legal"/"illegal" or "alien"/"undocumented" fuel this emotionally charged national debate.

Often such anti-immigration sentiments fall on those who are completely innocent of any wrong-doing. Carolina Bank Muñoz shows this to be the case for undocumented students—students who may have spent most or all of their lifetimes in U.S. schools, but then were denied the same rights to higher education as other U.S.-educated students. She articulates the need for social policies, such as the Dream Act (denied by Congress in 2011), that would extend certain rights of citizenship to such students. In this section, we see how ethnicity intertwines with race, class, and gender in ways that shape the life chances of different groups.

GENDER

Gender, like race, is a social construction, not a biological imperative. **Gender** is rooted in social institutions and results in patterns within society that structure the relationships between women and men and that give them differing positions of advantage and disadvantage within institutions. As an identity, gender is learned. That is, through gender socialization, people construct definitions of themselves and others that are marked by gender. Like race, however, gender cannot be understood at the individual level alone. Gender is structured in social institutions, including work, families, mass media, and education.

You can see this if you think about the concept of a gendered institution. **Gendered institution** is now used to define the total patterns of gender relations that are "present in the processes, practices, images, and ideologies, and distribution of power in the various sectors of social life" (Acker 1992: 567). This term brings a much more structural analysis of gender to the forefront. Rather than seeing gender only as a matter of interpersonal relationships and learned identities, this framework focuses the analysis of gender on relations of power—just as thinking about institutional racism focuses on power relations and economic and political subordination—not just interpersonal relations. Changing gender relations is not just a matter of changing individuals. As with race and class, change requires transformation of institutional structures.

Gender, however, is not a monolithic category. Maxine Baca Zinn, Pierrette Hondagneu-Sotelo, and Michael Messner argue ("Gender through the Prism of Difference") that, although gender is grounded in specific power relations, it is important to understand that gender is constructed differently depending on the specific social locations of diverse groups. Race, class, nationality, sexual orientation, and other factors produce varying social and economic consequences that cannot be understood by looking at gender differences alone. These authors ask us to move beyond studying differences and instead to use multiple "prisms" to see and comprehend the complexities of multiple systems of domination—each of which shapes and is shaped by gender.

Masculinity is also a fluid and social construction. Seen in this way, men appear as a less monolithic and unidimensional group as well, as Bethany M. Coston and Michael Kimmel (in "Seeing Privilege Where It Isn't: Marginalized Masculinities and the Intersectionality of Privilege"). Not all men benefit equally from gender privilege. The experiences of disabled men, gay men, and working-class men well capture the concept of a *prism of gender*—that is, how gender is refracted through the particular social locations of different groups in society.

Race, class, gender, and sexuality intersect in the construction of social stereotypes. Each gains meaning in relationship to the others (Glenn 2002). Judith Ortiz Cofer ("The Myth of the Latin Woman: I Just Met a Girl Named Maria") shows how the combination of her gender identity with her status as a Latina results in quite hurtful stereotypes that she continuously encounters in everyday life. You cannot understand her experience as a woman without also locating her in the ethnic, racial, national, and migration experiences that are also part of her daily life. In sum, gender oppression is maintained through multiple systems—systems that are reflected in stereotypes about different groups.

Stereotypes are also manifested in the everyday experiences of different groups. Marlese Durr and Adia M. Harvey Wingfield demonstrate this in their analysis of the stereotypes that African American professional women encounter in the workplace ("Keep Your 'N' in Check: African American Women and the Interactive Effects of Etiquette and Emotional Labor"). Durr and Wingfield show that, in the context of racial stereotyping, African American women have to manage the impression that others have of them, adding to the effort that professional women of color have to sustain in the workplace.

Amy Hanser's article ("The Gendered Rice Bowl") takes us in another direction—and beyond the United States. Looking at contemporary China and the work of women in the service industry, Hanser examines the merger of gender and sexual stereotypes, including how they interact with class, in shaping the expectations of women workers in service occupations. Although her research is based on Chinese women, you might ask how similar processes occur for women employed in service work in the United States—and other nations.

Altogether, the articles in this section show the social basis of gender. Although identities develop in the context of individual lives, they are shaped by the particular configurations of gender that emerge in a society also structured by race and class.

SEXUALITY

The linkage between race, class, and gender is revealed within studies of sexuality, just as sexuality is a dimension of each. For example, constructing images about Black sexuality is central to maintaining institutional racism. Similarly, beliefs about women's sexuality structure gender oppression. Thus, sexuality operates as a system of power and inequality comparable to and intersecting with the systems of race, class, and gender.

The connection between race, class, gender, and sexuality is explored in the opening essay here by Patricia Hill Collins ("Prisons for Our Bodies, Closets for Our Minds: Racism, Heterosexism, and Black Sexuality"). Collins shows how racism has historically been buttressed by beliefs about Black sexuality. Sexuality has been used as the vehicle to support racial fears and racial subordination. Racial subordination was built on the exploitation of Black bodies—both as labor and as sexual objects. Strictures against certain interracial, sexual relationships (but not those between White men and African American women) are a way to maintain White racism and patriarchy. Collins also shows how stigmatizing the sexuality of African Americans can in some ways be likened to the experiences of lesbian, gay, bisexual, and transgendered people. This is not to say that racial and sexual oppression are the same, but it is to say that these systems of oppression operate together, producing social beliefs and social actions that oppress and exploit African American men and women in particular ways, while also oppressing sexual minorities.

Underlying the analysis of sexuality are the concepts of **homophobia** (the fear and hatred of homosexuality) and **heterosexism** (the institutionalized power and privilege accorded to heterosexual behavior and identification). If only heterosexual forms of gender identity are labeled "normal," then gays, lesbians, and bisexuals become ostracized, oppressed, and defined as "socially deviant." Homophobia affects heterosexuals as well because it is part of the gender ideology used to distinguish "normal" men and women from those deemed deviant. Thus, young boys learn a rigid view of masculinity—one often associated with violence, bullying, and degrading others—to avoid being perceived as a "fag." And, as C. J. Pascoe shows ("Dude, You're a Fag"), the fag banter that boys use to insult each other is not necessarily directed just at gay boys or men. Rather, it is directed at men, especially adolescents, who are perceived as not conforming to the dominant norms of masculinity. In this way, *fag* is a mechanism of social control—controlling the dominant norms of gender in everyone's lives. The hatred directed toward lesbians, gays, and bisexuals is thus part of the system by which gender is created and maintained. In this regard, sexuality and gender are deeply linked.

The institutionalized structures and beliefs that define and enforce heterosexual behavior as the only natural and permissible form of sexual expression are what is meant by *heterosexism.* As Jonathan Ned Katz points out ("The Invention of Heterosexuality"), heterosexism is a specific historic construction; its meaning and presumed significance have evolved and changed at distinct historical times. Understanding this rests on understanding that sexuality, like race, class, gender, and ethnicity, is a social construction.

Ordinarily, sexual identity, like gender, is thought of in binary terms—as if one is either gay or straight, male or female. But Hanne Blank's very personal account of intersex identity ("Straight: The Surprisingly Short History of Heterosexuality") shows how heterosexuality, presumed to be the norm, has a specific history in its formulation and, thus, like gender, is a social and historical construction—not a fixed identity.

Denise Brennan ("Selling Sex for Visas: Sex Tourism as a Stepping-Stone to International Migration") reveals another dimension to the discussion of sexuality—sex as work. Sex workers sell their bodies or images of their bodies for money. Brennan emphasizes how sex workers are part of an international system of sex tourism, thus bringing a global perspective to the discussion of sexuality. Sex work on a global scale is also linked to world politics about race and class, given the specific position of women of color in international sex work. Women's sexuality is used to promote tourism and where images of the "exotic other" are used to attract more affluent classes to various regions of the world, thus linking sexuality to processes of migration and more global analyses of race, class, and gender.

REFERENCES

Acker, Joan. 1992. "Gendered Institutions: From Sex Roles to Gendered Institutions." *Contemporary Sociology* 21 (September): 565–569.

Bivens, Josh, and Lawrence Mishel. 2011. *Occupy Wall Streeters are Right about Skewed Economic Rewards in the United States*. Washington, DC: Economic Policy Institute. www.epi.org

Bureau of Labor Statistics. 2013. *Employment and Earnings*. Washington, DC: U.S. Department of Labor. www.bls.gov

Collins, Patricia Hill. 2004. *Black Sexual Politics: African Americans, Gender, and the New Racism*. New York: Routledge.

Collins, Patricia Hill. 2000. *Black Feminist Thought: Knowledge, Consciousness, and the Politics of Empowerment*. New York: Routledge.

DeNavas-Walt, Carmen, Bernadette D. Proctor, and Jessica C. Smith. 2013. *Income, Poverty, and Health Insurance: Coverage in the United States: 2012, P60-239*. Washington, DC: U.S. Census Bureau.

Eitzen, D. Stanley. 2012. *Fair and Foul: Beyond the Myths and Paradoxes of Sport*. Lanham, MD: Rowman and Littlefield.

Ferber, Abby. 1999. *White Man Falling: Race, Gender, and White Supremacy*. Lanham, MD: Rowman and Littlefield.

Frye, Marilyn. 1983. *The Politics of Reality*. Trumansburg, NY: The Crossing Press.

Glaze, Lauren E. 2010. *Correctional Populations in the United States 2009*. Washington, DC: Bureau of Justice Statistics.

Glenn, Evelyn Nakano. 2002. *Unequal Freedom: How Race and Gender Shaped American Citizenship and Labor*. Cambridge, MA: Harvard University Press.

Le, C. N. 2008 (April 11). "Socioeconomic Statistics & Demographics." *Asian-Nation: The Landscape of Asian America*. www.asian-nation.org

Mishel, Lawrence, Josh Bivens, Elise Gould, and Heidi Shierholz. 2012. *The State of Working America*, 12th ed. Ithaca, NY: Cornell University Press.

Noah, Timothy. 2012. *The Great Divergence: America's Growing Inequality Crisis and What We Can Do about It*. New York: Bloomsbury Press.

Oliver, Melvin L., and Thomas M. Shapiro. 2006. *Black Wealth/White Wealth: A New Perspective on Racial Inequality*, 2nd ed. New York: Routledge.

Omi, Michael, and Howard Winant. 1994. *Racial Formation in the United States: From the 1960s to the 1990s*, 2nd ed. New York: Routledge.

Rodriguez, Clara. 2000. *Changing Race: Latinos, the Census, and the History of Ethnicity*. New York: New York University Press.

6

White Privilege

Unpacking the Invisible Knapsack

PEGGY MCINTOSH

Through work to bring materials from Women's Studies into the rest of the curriculum, I have often noticed men's unwillingness to grant that they are overprivileged, even though they may grant that women are disadvantaged. They may say they will work to improve women's status, in the society, the university, or the curriculum, but they can't or won't support the idea of lessening men's. Denials which amount to taboos surround the subject of advantages which men gain from women's disadvantages. These denials protect male privilege from being fully acknowledged, lessened, or ended.

Thinking through unacknowledged male privilege as a phenomenon, I realized that since hierarchies in our society are interlocking, there was most likely a phenomenon of white privilege which was similarly denied and protected. As a white person, I realized I had been taught about racism as something which puts others at a disadvantage, but had been taught not to see one of its corollary aspects, white privilege, which puts me at an advantage.

I think whites are carefully taught not to recognize white privilege, as males are taught not to recognize male privilege. So I have begun in an untutored way to ask what it is like to have white privilege. I have come to see white privilege as an invisible package of unearned assets which I can count on cashing in each day, but about which I was "meant" to remain oblivious. White privilege is like an invisible weightless knapsack of special provisions, maps, passports, codebooks, visas, clothes, tools, and blank checks.

Describing white privilege makes one newly accountable. As we in Women's Studies work to reveal male privilege and ask men to give up some of their power, so one who writes about having white privilege must ask, "Having described it, what will I do to lessen or end it?"

After I realized the extent to which men work from a base of unacknowledged privilege, I understood that much of their oppressiveness was unconscious. Then I remembered the frequent charges from women of color that white women whom they encounter are oppressive. I began to understand why we are justly seen as oppressive, even when we don't see ourselves that way. I began to count

SOURCE: McIntosh, Peggy. 1989. "White Privilege: Unpacking the Invisible Knapsack." *Peace and Freedom Magazine* (July/August): 10–12. Women's International League for Peace and Freedom, Philadelphia. Reprinted by permission of the author.

the ways in which I enjoy unearned skin privilege and have been conditioned into oblivion about its existence.

My schooling gave me no training in seeing myself as an oppressor, as an unfairly advantaged person, or as a participant in a damaged culture. I was taught to see myself as an individual whose moral state depended on her individual moral will. My schooling followed the pattern my colleague Elizabeth Minnich has pointed out: whites are taught to think of their lives as morally neutral, normative, and average, and also ideal, so that when we work to benefit others, this is seen as work which will allow "them" to be more like "us."

I decided to try to work on myself at least by identifying some of the daily effects of white privilege in my life. I have chosen those conditions which I think in my case *attach somewhat more to skin-color privilege* than to class, religion, ethnic status, or geographical location, though of course all these other factors are intricately intertwined. As far as I can see, my African American co-workers, friends and acquaintances with whom I come into daily or frequent contact in this particular time, place, and line of work cannot count on most of these conditions.

1. I can if I wish arrange to be in the company of people of my race most of the time.
2. If I should need to move, I can be pretty sure of renting or purchasing housing in an area which I can afford and in which I would want to live.
3. I can be pretty sure that my neighbors in such a location will be neutral or pleasant to me.
4. I can go shopping alone most of the time, pretty well assured that I will not be followed or harassed.
5. I can turn on the television or open to the front page of the paper and see people of my race widely represented.
6. When I am told about our national heritage or about "civilization," I am shown that people of my color made it what it is.
7. I can be sure that my children will be given curricular materials that testify to the existence of their race.
8. If I want to, I can be pretty sure of finding a publisher for this piece on white privilege.
9. I can go into a music shop and count on finding the music of my race represented, into a supermarket and find the staple foods which fit with my cultural traditions, into a hairdresser's shop and find someone who can cut my hair.
10. Whether I use checks, credit cards, or cash, I can count on my skin color not to work against the appearance of financial reliability.
11. I can arrange to protect my children most of the time from people who might not like them.
12. I can swear, or dress in secondhand clothes, or not answer letters, without having people attribute these choices to the bad morals, the poverty, or the illiteracy of my race.

13. I can speak in public to a powerful male group without putting my race on trial.
14. I can do well in a challenging situation without being called a credit to my race.
15. I am never asked to speak for all the people of my racial group.
16. I can remain oblivious of the language and customs of persons of color who constitute the world's majority without feeling in my culture any penalty for such oblivion.
17. I can criticize our government and talk about how much I fear its policies and behavior without being seen as a cultural outsider.
18. I can be pretty sure that if I ask to talk to "the person in charge," I will be facing a person of my race.
19. If a traffic cop pulls me over or if the IRS audits my tax return, I can be sure I haven't been singled out because of my race.
20. I can easily buy posters, postcards, picture books, greeting cards, dolls, toys, and children's magazines featuring people of my race.
21. I can go home from most meetings of organizations I belong to feeling somewhat tied in, rather than isolated, out-of-place, outnumbered, unheard, held at a distance, or feared.
22. I can take a job with an affirmative action employer without having co-workers on the job suspect that I got it because of my race.
23. I can choose public accommodation without fearing that people of my race cannot get in or will be mistreated in the places I have chosen.
24. I can be sure that if I need legal or medical help, my race will not work against me.
25. If my day, week, or year is going badly, I need not ask of each negative episode or situation whether it has racial overtones.
26. I can choose blemish cover or bandages in "flesh" color and have them more or less match my skin.

I repeatedly forgot each of the realizations on this list until I wrote it down. For me white privilege has turned out to be an elusive and fugitive subject. The pressure to avoid it is great, for in facing it I must give up the myth of meritocracy. If these things are true, this is not such a free country; one's life is not what one makes it; many doors open for certain people through no virtues of their own.

In unpacking this invisible knapsack of white privilege, I have listed conditions of daily experience which I once took for granted. Nor did I think of any of these perquisites as bad for the holder. I now think that we need a more finely-differentiated taxonomy of privilege, for some of these varieties are only what one would want for everyone in a just society, and others give license to be ignorant, oblivious, arrogant and destructive.

I see a pattern running through the matrix of white privilege, a pattern of assumptions which were passed on to me as a white person. There was one main piece of cultural turf; it was my own turf, and I was among those who could control the turf. *My skin color was an asset for any move I was educated to want to make.* I could think of myself as belonging in major ways, and of making social systems work for me. I could freely disparage, fear, neglect, or be oblivious to anything outside of the dominant cultural forms. Being of the main culture, I could also criticize it fairly freely.

In proportion as my racial group was being made confident, comfortable, and oblivious, other groups were likely being made inconfident, uncomfortable, and alienated. Whiteness protected me from many kinds of hostility, distress, and violence, which I was being subtly trained to visit in turn upon people of color.

For this reason, the word "privilege" now seems to me misleading. We usually think of privilege as being a favored state, whether earned or conferred by birth or luck. Yet some of the conditions I have described here work to systematically overempower certain groups. Such privilege simply *confers dominance* because of one's race or sex.

I want, then, to distinguish between earned strength and unearned power conferred systemically. Power from unearned privilege can look like strength when it is in fact permission to escape or to dominate. But not all of the privileges on my list are inevitably damaging. Some, like the expectation that neighbors will be decent to you, or that your race will not count against you in court, should be the norm in a just society. Others, like the privilege to ignore less powerful people, distort the humanity of the holders as well as the ignored groups.

We might at least start by distinguishing between positive advantages which we can work to spread, and negative types of advantages which unless rejected will always reinforce our present hierarchies. For example, the feeling that one belongs within the human circle, as Native Americans say, should not be seen as privilege for a few. Ideally it is an *unearned entitlement*. At present, since only a few have it, it is an *unearned advantage* for them. This paper results from a process of coming to see that some of the power which I originally saw as attendant on being a human being in the U.S. consisted in *unearned advantage* and *conferred dominance*.

I have met very few men who are truly distressed about systemic, unearned male advantage and conferred dominance. And so one question for me and others like me is whether we will be like them, or whether we will get truly distressed, even outraged, about unearned race advantage and conferred dominance and if so, what we will do to lessen them. In any case, we need to do more work in identifying how they actually affect our daily lives. Many, perhaps most, of our white students in the U.S. think that racism doesn't affect them because they are not people of color; they do not see "whiteness" as a racial identity. In addition, since race and sex are not the only advantaging systems at work, we need similarly to examine the daily experience of having age advantage, or ethnic advantage, or physical ability, or advantage related to nationality, religion, or sexual orientation.

Difficulties and dangers surrounding the task of finding parallels are many. Since racism, sexism, and heterosexism are not the same, the advantaging associated with them should not be seen as the same. In addition, it is hard to disentangle aspects of unearned advantage which rest more on social class, economic class, race, religion, sex and ethnic identity than on other factors. Still, all of the oppressions are interlocking, as the Combahee River Collective Statement of 1977 continues to remind us eloquently.

One factor seems clear about all of the interlocking oppressions. They take both active forms which we can see and embedded forms which as a member of the dominant group one is taught not to see. In my class and place, I did not see myself as a racist because I was taught to recognize racism only in individual acts of meanness by members of my group, never in invisible systems conferring unsought racial dominance on my group from birth.

Disapproving of the systems won't be enough to change them. I was taught to think that racism could end if white individuals changed their attitudes. [But] a "white" skin in the United States opens many doors for whites whether or not we approve of the way dominance has been conferred on us. Individual acts can palliate, but cannot end, these problems.

To redesign social systems we need first to acknowledge their colossal unseen dimensions. The silences and denials surrounding privilege are the key political tool here. They keep the thinking about equality or equity incomplete, protecting unearned advantage and conferred dominance by making these taboo subjects. Most talk by whites about equal opportunity seems to me now to be about equal opportunity to try to get into a position of dominance while denying that *systems* of dominance exist.

It seems to me that obliviousness about white advantage, like obliviousness about male advantage, is kept strongly enculturated in the United States so as to maintain the myth of meritocracy, the myth that democratic choice is equally available to all. Keeping most people unaware that freedom of confident action is there for just a small number of people props up those in power, and serves to keep power in the hands of the same groups that have most of it already.

Though systemic change takes many decades, there are pressing questions for me and I imagine for some others like me if we raise our daily consciousness on the perquisites of being light-skinned. What will we do with such knowledge? As we know from watching men, it is an open question whether we will choose to use unearned advantage to weaken hidden systems of advantage, and whether we will use any of our arbitrarily-awarded power to try to reconstruct power systems on a broader base.

7

Color-Blind Privilege

The Social and Political Functions of Erasing the Color Line in Post-Race America

CHARLES A. GALLAGHER

The young white male sporting a FUBU (African-American owned apparel company "For Us By Us") shirt and his white friend with the tightly set, perfectly braided cornrows blended seamlessly into the festivities at an all white bar mitzvah celebration. A black model dressed in yachting attire peddles a New England, yuppie boating look in Nautica advertisements. It is quite unremarkable to observe white, Asian or African Americans with dyed purple, blond or red hair. White, black and Asian students decorate their bodies with tattoos of Chinese characters and symbols. In cities and suburbs young adults across the color line wear hip-hop clothing and listen to white rapper Eminem and black rapper 50-cent. It went almost unnoticed when a north Georgia branch of the NAACP installed a white biology professor as its president. Subversive musical talents like Jimi Hendrix, Bob Marley and The Who are now used to sell Apple Computers, designer shoes and SUVs. Du-Rag kits, complete with bandana headscarf and elastic headband, are on sale for $2.95 at hip-hop clothing stores and family centered theme parks like Six Flags. Salsa has replaced ketchup as the best selling condiment in the United States. Companies as diverse as Polo, McDonalds, Tommy Hilfiger, Walt Disney World, Master Card, Skechers sneakers, IBM, Giorgio Armani and Neosporin antibiotic ointment have each crafted advertisements that show an integrated, multiracial cast of characters interacting and consuming their products in [a] post-race, color-blind world.

Americans are constantly bombarded by depictions of race relations in the media, which suggests that discriminatory racial barriers have been dismantled. Social and cultural indicators suggest that America is on the verge, or has already become, a truly color-blind nation. National polling data indicate that a majority of whites now believe discrimination against racial minorities no longer exists. A majority of whites believe that blacks have "as good a chance as whites" in procuring housing and employment or achieving middle class status while a 1995 survey of white adults found that a majority of whites (58%) believed that African Americans were "better off" finding jobs than whites (Gallup, 1997; Shipler, 1998). Much of white America now see[s] a level playing field, while a majority of black

SOURCE: Gallagher, Charles A. 2003. *Race, Gender & Class* 10: 22–37. Reprinted by permission of the author.

Americans sees a field [that] is still quite uneven.... The colorblind or race neutral perspective holds that in an environment where institutional racism and discrimination have been replaced by equal opportunity, one's qualifications, not one's color or ethnicity, should be the mechanism by which upward mobility is achieved. Color as a cultural style may be expressed and consumed through music, dress or vernacular but race as a system which confers privileges and shapes life chances is viewed as an atavistic and inaccurate accounting of U.S. race relations.

Not surprisingly, this view of society blind to color is not equally shared. Whites and blacks differ significantly, however, on their support for affirmative action, the perceived fairness of the criminal justice system, the ability to acquire the "American Dream," and the extent to which whites have benefited from past discrimination (Moore, 1995; Moore & Saad, 1995; Kaiser Foundation, 1995). This article examines the social and political functions colorblindness serves for whites in the United States. Drawing on interviews and focus groups with whites from around the country I argue that color-blind depictions of U.S. race relations serves [*sic*] to maintain white privilege by negating racial inequality. Embracing a color-blind perspective reinforces whites' belief that being white or black or brown has no bearing on an individual's or a group's relative place in the socioeconomic hierarchy.

DATA AND METHOD

I use data from seventeen focus groups and thirty individual interviews with whites from around the country. Thirteen of the seventeen focus groups were conducted in a college or university setting, five in a liberal arts college in the Rocky Mountains and the remaining eight at a large urban university in the Northeast. Respondents in these focus groups were selected randomly from the student population. Each focus group averaged six respondents ... equally divided between males and females. An overwhelming majority of these respondents were between the ages of eighteen and twenty-two years of age. The remaining four focus groups took place in two rural counties in Georgia and were obtained through contacts from educational and social service providers in each county. One county was almost entirely white (99.54%) and in the other county whites constituted a racial minority. These four focus groups allowed me to tap rural attitudes about race relations in environments where whites had little or consistent contact with racial minorities....

COLORBLINDNESS AS NORMATIVE IDEOLOGY

The perception among a majority of white Americans that the socio-economic playing field is now level, along with whites' belief that they have purged themselves of overt racist attitudes and behaviors, has made colorblindness the dominant lens through which whites understand contemporary race relations.

Colorblindness allows whites to believe that segregation and discrimination are no longer [an] issue because it is now illegal for individuals to be denied access to housing, public accommodations or jobs because of their race. Indeed, lawsuits alleging institutional racism against companies like Texaco, Denny's, Coke, and Cracker Barrel validate what many whites know at a visceral level is true: firms which deviate from the color-blind norms embedded in classic liberalism will be punished. As a political ideology, the commodification and mass marketing of products that signify color but are intended for consumption across the color line further legitimate colorblindness. Almost every household in the United States has a television that, according to the U.S. Census, is on for seven hours every day (Nielsen 1997). Individuals from any racial background can wear hip-hop clothing, listen to rap music (both purchased at Wal-Mart) and root for their favorite, majority black, professional sports team. Within the context of racial symbols that are bought and sold in the market, colorblindness means that one's race has no bearing on who can purchase a Jaguar, live in an exclusive neighborhood, attend private schools or own a Rolex.

The passive interaction whites have with people of color through the media creates the impression that little, if any, socio-economic difference exists between the races....

Highly visible and successful racial minorities like [former] Secretary of State Colin Powell and ... [Secretary of State] Condelleeza Rice are further proof to white America that the state's efforts to enforce and promote racial equality have been accomplished.

The new color-blind ideology does not, however, ignore race; it acknowledges race while disregarding racial hierarchy by taking racially coded styles and products and reducing these symbols to commodities or experiences that whites and racial minorities can purchase and share. It is through such acts of shared consumption that race becomes nothing more than an innocuous cultural signifier. Large corporations have made American culture more homogenous through the ubiquitousness of fast food, television, and shopping malls but this trend has also created the illusion that we are all the same through consumption. Most adults eat at national fast food chains like McDonalds, shop at mall anchor stores like Sears and J.C. Penney's and watch major league sports, situation comedies or television drama. Defining race only as cultural symbols that are for sale allows whites to experience and view race as nothing more than a benign cultural marker that has been stripped of all forms of institutional, discriminatory or coercive power. The post-race, color-blind perspective allows whites to imagine that depictions of racial minorities working in high status jobs and consuming the same products, or at least appearing in commercials for products whites desire or consume, is the same as living in a society where color is no longer used to allocate resources or shape group outcomes. By constructing a picture of society where racial harmony is the norm, the color-blind perspective functions to make white privilege invisible while removing from public discussion the need to maintain any social programs that are race-based.

How then, is colorblindness linked to privilege? Starting with the deeply held belief that America is now a meritocracy, whites are able to imagine that the

socio-economic success they enjoy relative to racial minorities is a function of individual hard work, determination, thrift and investments in education. The color-blind perspective removes from personal thought and public discussion any taint or suggestion of white supremacy or white guilt while legitimating the existing social, political and economic arrangements which privilege whites. This perspective insinuates that class and culture, and not institutional racism, are responsible for social inequality. Colorblindness allows whites to define themselves as politically and racially tolerant as they proclaim their adherence to a belief system that does not see or judge individuals by the "color of their skin." This perspective ignores, as Ruth Frankenberg puts it, how whiteness is a "location of structural advantage" (2001, p. 1).... Colorblindness hides white privilege behind a mask of assumed meritocracy while rendering invisible the institutional arrangements that perpetuate racial inequality. The veneer of equality implied in colorblindness allows whites to present their place in the racialized social structure as one that was earned.

OPPORTUNITY HAS NO COLOR

Given this norm of colorblindness it was not surprising that respondents in this study believed that using race to promote group interests was a form of (reverse) racism....

Believing and acting as if America is now color-blind allows whites to imagine a society where institutional racism no longer exists and racial barriers to upward mobility have been removed. The use of group identity to challenge the existing racial order by making demands for the amelioration of racial inequities is viewed as racist because such claims violate the belief that we are a nation that recognizes the rights of individuals not rights demanded by groups....

The logic inherent in the color-blind approach is circular; since race no longer shapes life chances in a color-blind world there is no need to take race into account when discussing differences in outcomes between racial groups. This approach erases America's racial hierarchy by implying that social, economic and political power and mobility is [*sic*] equally shared among all racial groups. Ignoring the extent or ways in which race shapes life chances validates whites' social location in the existing racial hierarchy while legitimating the political and economic arrangements that perpetuate and reproduce racial inequality and privilege.

REFERENCES

Frankenberg, R. (2001). The mirage of an unmarked whiteness. In B. B. Rasmussen, E. Klineberg, I. J. Nexica & M. Wray (eds.) *The making and unmaking of whiteness.* Durham: Duke University Press.

Gallup Organization. (1997). *Black/white relations in the U.S.* June 10, pp. 1–5.

Kaiser Foundation. (1995). *The four Americas: Government and social policy through the eyes of America's multi-racial and multi-ethnic society*. Menlo Park, CA: Kaiser Family Foundation.

Moore, D. (1995). "Americans' most important sources of information: Local news." *The Gallup Poll Monthly*, September, pp. 2–5.

Moore, D. & Saad, L. (1995). No immediate signs that Simpson trial intensified racial animosity. *The Gallup Poll Monthly*, October, pp. 2–5.

Nielsen, A. C. (1997). *Information please almanac*. Boston: Houghton Mifflin.

Shipler, D. (1998). *A country of strangers: Blacks and whites in America*. New York: Vintage Books.

8

What White Supremacists Taught a Jewish Scholar about Identity

ABBY L. FERBER

A few years ago, my work on white supremacy led me to the neo-Nazi tract *The New Order,* which proclaims: "The single serious enemy facing the white man is the Jew." I must have read that statement a dozen times. Until then, I hadn't thought of myself as the enemy.

When I began my research for a book on race, gender, and white supremacy, I could not understand why white supremacists so feared and hated Jews. But after being immersed in newsletters and periodicals for months, I learned that white supremacists imagine Jews as the masterminds behind a great plot to mix races and, thereby, to wipe the white race out of existence.

The identity of white supremacists, and the white racial purity they espouse, requires the maintenance of secure boundaries. For that reason, the literature I read described interracial sex as "the ultimate abomination." White supremacists see Jews as threats to racial purity, the villains responsible for desegregation, integration, the civil-rights movement, the women's movement, and affirmative action—each depicted as eventually leading white women into the beds of black men. Jews are believed to be in control everywhere, staging a multi-pronged attack against the white race. For *WAR,* the newsletter of White Aryan Resistance, the Jew "promotes a thousand social ills ... [f]or which you'll have to foot the bills."

Reading white-supremacist literature is a profoundly disturbing experience, and even more difficult if you are one of those targeted for elimination. Yet, as a Jewish woman, I found my research to be unsettling in unexpected ways. I had not imagined that it would involve so much self-reflection. I knew white supremacists were vehemently anti-Semitic, but I was ambivalent about my Jewish identity and did not see it as essential to who I was. Having grown up in a large Jewish community, and then having attended a college with a large Jewish enrollment, my Jewishness was invisible to me—something I mostly ignored. As I soon learned, to white supremacists, that is irrelevant.

Contemporary white supremacists define Jews as non-white: "not a religion, they are an Asiatic *race,* locked in a mortal conflict with Aryan man," according to *The New Order.* In fact, throughout white-supremacist tracts, Jews are

SOURCE: Ferber, Abby L. From *The Chronicle of Higher Education,* May 7, 1999, pp. B6–B7. Reprinted with permission of the author.

described not merely as a separate race, but as an impure race, the product of mongrelization. Jews, who pose the ultimate threat to racial boundaries, are themselves imagined as the product of mixed-race unions.

Although self-examination was not my goal when I began, my research pushed me to explore the contradictions in my own racial identity. Intellectually, I knew that the meaning of race was not rooted in biology or genetics, but it was only through researching the white-supremacist movement that I gained a more personal understanding of the social construction of race. Reading white-supremacist literature, I moved between two worlds: one where I was white, another where I was the non-white seed of Satan; one where I was privileged, another where I was despised; one where I was safe and secure, the other where I was feared and thus marked for death.

According to white-supremacist ideology, I am so dangerous that I must be eliminated. Yet, when I put down the racist, anti-Semitic newsletters, leave my office, and walk outdoors, I am white.

Growing up white has meant growing up privileged. Sure, I learned about the historical persecutions of Jews, overheard the hushed references to distant relatives lost in the Holocaust. I knew of my grandmother's experiences with anti-Semitism as a child of the only Jewish family in a Catholic neighborhood. But those were just stories to me. Reading white supremacists finally made the history real.

While conducting my research, I was reminded of the first time I felt like an "other." Arriving in the late 1980s for the first day of graduate school in the Pacific Northwest, I was greeted by a senior graduate student with the welcome: "Oh, you're the Jewish one." It was a jarring remark, for it immediately set me apart. This must have been how my mother felt, I thought, when, a generation earlier, a college classmate had asked to see her horns. Having lived in predominantly Jewish communities, I had never experienced my Jewishness as "otherness." In fact, I did not even *feel* Jewish. Since moving out of my parents' home, I had not celebrated a Jewish holiday or set foot in a synagogue. So it felt particularly odd to be identified by this stranger as a Jew. At the time, I did not feel that the designation described who I was in any meaningful sense.

But whether or not I define myself as Jewish, I am constantly defined by others that way. Jewishness is not simply a religious designation that one may choose, as I once naïvely assumed. Whether or not I see myself as Jewish does not matter to white supremacists.

I've come to realize that my own experience with race reflects the larger historical picture for Jews. As whites, Jews today are certainly a privileged group in the United States. Yet the history of the Jewish experience demonstrates precisely what scholars mean when they say that race is a social construction.

At certain points in time, Jews have been defined as a non-white minority. Around the turn of the last century, they were considered a separate, inferior race, with a distinguishable biological identity justifying discrimination and even genocide. Today, Jews are generally considered white, and Jewishness is largely considered merely a religious or ethnic designation. Jews, along with other European ethnic groups, were welcomed into the category of "white" as beneficiaries of one of the largest affirmative-action programs in history—the 1944 GI Bill of Rights. Yet, when I read white-supremacist discourse, I am reminded that my ancestors were expelled from the dominant race, persecuted, and even killed.

Since conducting my research, having heard dozens of descriptions of the murders and mutilations of "race traitors" by white supremacists, I now carry with me the knowledge that there are many people out there who would still wish to see me dead. For a brief moment, I think that I can imagine what it must feel like to be a person of color in our society ... but then I realize that, as a white person, I cannot begin to imagine that.

Jewishness has become both clearer and more ambiguous for me. And the questions I have encountered in thinking about Jewish identity highlight the central issues involved in studying race today. I teach a class on race and ethnicity, and usually, about midway through the course, students complain of confusion. They enter my course seeking answers to the most troubling and divisive questions of our time, and are disappointed when they discover only more questions. If race is not biological or genetic, what is it? Why, in some states, does it take just one black ancestor out of 32 to make a person legally black, yet those 31 white ancestors are not enough to make that person white? And, always, are Jews a race?

I have no simple answers. As Jewish history demonstrates, what is and is not a racial designation, and who is included within it, is unstable and changes over time—and that designation is always tied to power. We do not have to look far to find other examples: The Irish were also once considered non-white in the United States, and U.S. racial categories change with almost every census.

My prolonged encounter with the white-supremacist movement forced me to question not only my assumptions about Jewish identity, but also my assumptions about whiteness. Growing up "white," I felt raceless. As it is for most white people, my race was invisible to me. Reflecting the assumption of most research on race at the time, I saw race as something that shaped the lives of people of color—the victims of racism. We are not used to thinking about whiteness when we think about race. Consequently, white people like myself have failed to recognize the ways in which our own lives are shaped by race. It was not until others began identifying me as the Jew, the "other," that I began to explore race in my own life.

Ironically, that is the same phenomenon shaping the consciousness of white supremacists: They embrace their racial identity at the precise moment when they feel their privilege and power under attack. Whiteness historically has equaled power, and when that equation is threatened, their own whiteness becomes visible to many white people for the first time. Hence, white supremacists seek to make racial identity, racial hierarchies, and white power part of the natural order again. The notion that race is a social construct threatens that order. While it has become an academic commonplace to assert that race is socially constructed, the revelation is profoundly unsettling to many, especially those who benefit most from the constructs.

My research on hate groups not only opened the way for me to explore my own racial identity, but also provided insight into the question with which I began this essay: Why do white supremacists express such hatred and fear of Jews? This ambiguity in Jewish racial identity is precisely what white supremacists find so threatening. Jewish history reveals race as a social designation, rather than a God-given or genetic endowment. Jews blur the boundaries between whites and people of color, failing to fall securely on either side of the divide. And it is ambiguity that white supremacists fear most of all.

I find it especially ironic that, today, some strict Orthodox Jewish leaders also find that ambiguity threatening. Speaking out against the high rates of intermarriage among Jews and non-Jews, they issue dire warnings. Like white supremacists, they fear assaults on the integrity of the community and fight to secure its racial boundaries, defining Jewishness as biological and restricting it only to those with Jewish mothers. For both white supremacists and such Orthodox Jews, intermarriage is tantamount to genocide.

For me, the task is no longer to resolve the ambiguity, but to embrace it. My exploration of white-supremacist ideology has revealed just how subversive doing so can be: Reading white-supremacist discourse through the lens of Jewish experience has helped me toward new interpretations. White supremacy is not a movement just about hatred, but even more about fear: fear of the vulnerability and instability of white identity and privilege. For white supremacists, the central goal is to naturalize racial identity and hierarchy, to establish boundaries.

Both my own experience and Jewish history reveal that to be an impossible task. Embracing Jewish identity and history, with all their contradictions, has given me an empowering alternative to white-supremacist conceptions of race. I have found that eliminating ambivalence does not require eliminating ambiguity.

9

Are Asian Americans Becoming "White"?

MIN ZHOU

Are Asian Americans becoming "white"? For many public officials the answer must be yes, because they classify Asian-origin Americans with European-origin Americans for equal opportunity programs. But this classification is premature and based on false premises. Although Asian Americans as a group have attained the career and financial success equated with being white, and although many have moved next to or have even married whites, they still remain culturally distinct and suspect in a white society.

At issue is how to define Asian American and white. The term "Asian American" was coined by the late historian and activist Yuji Ichioka during the ethnic consciousness movements of the late 1960s. To adopt this identity was to reject the Western-imposed label of "Oriental." Today, "Asian American" is an umbrella category that includes both U.S. citizens and immigrants whose ancestors came from Asia, east of Iran. Although widely used in public discussions, most Asian-origin Americans are ambivalent about this label, reflecting the difficulty of being American and still keeping some ethnic identity: Is one, for example, Asian American or Japanese American?

Similarly, "white" is an arbitrary label having more to do with privilege than biology. In the United States, groups initially considered nonwhite, such as the Irish and Jews, have attained "white" membership by acquiring status and wealth. It is hardly surprising, then, that nonwhites would aspire to becoming "white" as a mark of and a tool for material success. However, becoming white can mean distancing oneself from "people of color" or disowning one's ethnicity. Pan-ethnic identities—Asian American, African American, Hispanic American—are one way the politically vocal in any group try to stem defections. But these group identities may restrain individual members' aspirations for personal advancement.

VARIETIES OF ASIAN AMERICANS

Privately, few Americans of Asian ancestry would spontaneously identify themselves as Asian, and fewer still as Asian American. They instead link their identities to specific countries of origin, such as China, Japan, Korea, the Philippines,

SOURCE: Zhou, Min. 2003. "Are Asian Americans Becoming 'White'?" *Contexts*, 3(1): 29–37.

India or Vietnam. In a study of Vietnamese youth in San Diego, for example, 53 percent identified themselves as Vietnamese, 32 percent as Vietnamese American, and only 14 percent as Asian American. But they did not take these labels lightly; nearly 60 percent of these youth considered their chosen identity as very important to them.

Some Americans of Asian ancestry have family histories in the United States longer than many Americans of Eastern or Southern European origin. However, Asian-origin Americans became numerous only after 1970, rising from 1.4 million to 11.9 million (4 percent of the total U.S. population), in 2000. Before 1970, the Asian-origin population was largely made up of Japanese, Chinese and Filipinos. Now, Americans of Chinese and Filipino ancestries are the largest subgroups (at 2.8 million and 2.4 million, respectively), followed by Indians, Koreans, Vietnamese and Japanese (at more than one million). Some 20 other national-origin groups, such as Cambodians, Pakistanis, Laotians, Thai, Indonesians and Bangladeshis, were officially counted in government statistics only after 1980; together they amounted to more than two million Americans in 2000.

The sevenfold growth of the Asian-origin population in the span of 30-odd years is primarily due to accelerated immigration following the Hart-Celler Act of 1965, which ended the national origins quota system, and the historic resettlement of Southeast Asian refugees after the Vietnam War. Currently, about 60 percent of the Asian-origin population is foreign-born (the first generation), another 28 percent are U.S.-born of foreign-born parents (the second generation), and just 12 percent were born to U.S.-born parents (the third generation and beyond).

Unlike earlier immigrants from Asia or Europe, who were mostly low-skilled laborers looking for work, today's immigrants from Asia have more varied backgrounds and come for many reasons, such as to join their families, to invest their money in the U.S. economy, to fill the demand for highly skilled labor, or to escape war, political or religious persecution and economic hardship. For example, Chinese, Taiwanese, Indian, and Filipino Americans tend to be over-represented among scientists, engineers, physicians and other skilled professionals, but less educated, low-skilled workers are more common among Vietnamese, Cambodian, Laotian, and Hmong Americans, most of whom entered the United States as refugees. While middle-class immigrants are able to start their American lives with high-paying professional careers and comfortable suburban lives, low-skilled immigrants and refugees often have to endure low-paying menial jobs and live in inner-city ghettos.

Asian Americans tend to settle in large metropolitan areas and concentrate in the West. California is home to 35 percent of all Asian Americans. But recently, other states such as Texas, Minnesota and Wisconsin, which historically received few Asian immigrants, have become destinations for Asian American settlement. Traditional ethnic enclaves, such as Chinatown, Little Tokyo, Manilatown, Koreatown, Little Phnom Penh, and Thaitown, persist or have emerged in gateway cities, helping new arrivals to cope with cultural and linguistic difficulties. However, affluent and highly-skilled immigrants tend to bypass inner-city enclaves and settle in suburbs upon arrival, belying the stereotype of the "unacculturated" immigrant. Today, more than half of the Asian-origin population is spreading out in suburbs surrounding traditional gateway cities, as well as in new urban centers of Asian settlement across the country.

Differences in national origins, timing of immigration, affluence and settlement patterns profoundly inhibit the formation of a pan-ethnic identity. Recent arrivals are less likely than those born or raised in the United States to identify as Asian American. They are also so busy settling in that they have little time to think about being Asian or Asian American, or, for that matter, white. Their diverse origins include drastic differences in languages and dialects, religions, cuisines and customs. Many national groups also bring to America their histories of conflict (such as the Japanese colonization of Korea and Taiwan, Japanese attacks on China, and the Chinese invasion of Vietnam).

Immigrants who are predominantly middle-class professionals, such as the Taiwanese and Indians, or predominantly small business owners, such as the Koreans, share few of the same concerns and priorities as those who are predominantly uneducated, low-skilled refugees, such as Cambodians and Hmong. Finally, Asian-origin people living in San Francisco or Los Angeles among many other Asians and self-conscious Asian Americans develop a stronger ethnic identity than those living in predominantly Latin Miami or predominantly European Minneapolis. A politician might get away with calling Asians "Oriental" in Miami but get into big trouble in San Francisco. All of these differences create obstacles to fostering a cohesive pan-Asian solidarity. As Yen Le Espiritu shows, pan-Asianism is primarily a political ideology of U.S.-born, American-educated, middle-class Asians rather than of Asian immigrants, who are conscious of their national origins and overburdened with their daily struggles for survival.

UNDERNEATH THE MODEL MINORITY: "WHITE" OR "OTHER"

The celebrated "model minority" image of Asian Americans appeared in the mid-1960s, at the peak of the civil rights and the ethnic consciousness movements, but before the rising waves of immigration and refugee influx from Asia. Two articles in 1966—"Success Story, Japanese-American Style," by William Petersen in the *New York Times Magazine*, and "Success of One Minority Group in U.S.," by the *US News & World Report* staff—marked a significant departure from how Asian immigrants and their descendants had been traditionally depicted in the media. Both articles congratulated Japanese and Chinese Americans on their persistence in overcoming extreme hardships and discrimination to achieve success, unmatched even by U.S.-born whites, with "their own almost totally unaided effort" and "no help from anyone else." (The implicit contrast to other minorities was clear.) The press attributed their winning wealth and respect in American society to hard work, family solidarity, discipline, delayed gratification, non-confrontation and eschewing welfare.

This "model minority" image remains largely unchanged even in the face of new and diverse waves of immigration. The 2000 U.S. Census shows that Asian Americans continue to score remarkable economic and educational achievements. Their median household income in 1999 was more than $55,000—the highest of all racial groups, including whites—and their poverty rate was under 11 percent, the lowest of all racial groups. Moreover, 44 percent of all Asian Americans over 25 years of age had at least a bachelor's degree, 18 percentage points more than any other racial group. Strikingly, young Asian Americans, including both the children of foreign-born physicians, scientists, and professionals and those of uneducated and penniless refugees, repeatedly appear as high school valedictorians and academic decathlon winners. They also enroll in the freshman classes of prestigious universities in disproportionately large numbers. In 1998, Asian Americans, just 4 percent of the nation's population, made up more than 20 percent of the undergraduates at universities such as Berkeley, Stanford, MIT and Cal Tech. Although some ethnic groups, such as Cambodians, Lao, and Hmong, still trail behind other East and South Asians in most indicators of achievement, they too show significant signs of upward mobility. Many in the media have dubbed Asian Americans the "new Jews." Like the second-generation Jews of the past, today's children of Asian immigrants are climbing up the ladder by way of extraordinary educational achievement.

One consequence of the model-minority stereotype is that it reinforces the myth that the United States is devoid of racism and accords equal opportunity to all, fostering the view that those who lag behind do so because of their own poor choices and inferior culture. Celebrating "model minorities" can help impede other racial minorities' demands for social justice by pitting minority groups against each other. It can also pit Asian Americans against whites. On the surface, Asian Americans seem to be on their way to becoming white, just like the offspring of earlier European immigrants. But the model-minority image implicitly casts Asian Americans as different from whites. By placing Asian Americans above whites, this image still sets them apart from other Americans, white or nonwhite, in the public mind.

There are two other less obvious effects. The model-minority stereotype holds Asian Americans to higher standards, distinguishing them from average Americans. "What's wrong with being a model minority?" a black student once asked, in a class I taught on race, "I'd rather be in the model minority than in the downtrodden minority that nobody respects." Whether people are in a model minority or a downtrodden minority, they are still judged by standards different from average Americans. Also, the model-minority stereotype places particular expectations on members of the group so labeled, channeling them to specific avenues of success, such as science and engineering. This, in turn, makes it harder for Asian Americans to pursue careers outside these designated fields. Falling into this trap, a Chinese immigrant father gets upset when his son tells him he has changed his major from engineering to English. Disregarding his son's talent for creative writing, such a father rationalizes his concern, "You have a 90 percent chance of getting a decent job with an engineering degree, but what chance would you have of earning income as a writer?" This thinking represents more than typical parental concern; it constitutes the self-fulfilling prophecy of a stereotype.

The celebration of Asian Americans rests on the perception that their success is unexpectedly high. The truth is that unusually many of them, particularly among the Chinese, Indians and Koreans, arrive as middle-class or upper middle-class immigrants. This makes it easier for them and their children to succeed and regain their middle-class status in their new homeland. The financial resources that these immigrants bring also subsidize ethnic businesses and services, such as private after-school programs. These, in turn, enable even the less fortunate members of the groups to move ahead more quickly than they would have otherwise.

NOT SO MUCH BEING "WHITE" AS BEING AMERICAN

Most Asian Americans seem to accept that "white" is mainstream, average and normal, and they look to whites as a frame of reference for attaining higher social position. Similarly, researchers often use non-Hispanic whites as the standard against which other groups are compared, even though there is great diversity among whites, too. Like most immigrants to the United States, Asian immigrants tend to believe in the American Dream and measure their achievements materially. As a Chinese immigrant said to me in an interview, "I hope to accomplish nothing but three things: to own a home, to be my own boss, and to send my children to the Ivy League." Those with sufficient education, job skills and money manage to move into white middle-class suburban neighborhoods immediately upon arrival, while others work intensively to accumulate enough savings to move their families up and out of inner-city ethnic enclaves. Consequently, many children of Asian ancestry have lived their entire childhood in white communities, made friends with mostly white peers, and grown up speaking only English. In fact, Asian Americans are the most acculturated non-European group in the United States. By the second generation, most have lost fluency in their parents' native languages (see "English-Only Triumphs, But the Costs are High," *Contexts*, Spring 2002). David Lopez finds that in Los Angeles, more than three-quarters of second-generation Asian Americans (as opposed to one-quarter of second-generation Mexicans) speak only English at home. Asian Americans also intermarry extensively with whites and with members of other minority groups. Jennifer Lee and Frank Bean find that more than one-quarter of married Asian Americans have a partner of a different racial background, and 87 percent of those marry whites; they also find that 12 percent of all Asian Americans claim a multiracial background, compared to 2 percent of whites and 4 percent of blacks.

Even though U.S.-born or U.S.-raised Asian Americans are relatively acculturated and often intermarry with whites, they may be more ambivalent about becoming white than their immigrant parents. Many only cynically agree that "white" is synonymous with "American." A Vietnamese high school student in New Orleans told me in an interview, "An American is white. You often hear people say, hey, so-and-so is dating an 'American.' You know she's dating a white boy. If he were black, then people would say he's black." But while they recognize whites as a frame of reference, some reject the idea of becoming white themselves: "It's not so much being white as being American," commented a Korean-American student in my class on the new second generation. This aversion to becoming white is particularly common among second-generation college students who have taken ethnic studies courses, and among Asian-American community activists. However, most of the second generation continues to strive for the privileged status associated with whiteness, just like their parents. For example, most U.S.-born or U.S.-raised Chinese-American youth end up studying engineering, medicine, or law in college, believing that these areas of study guarantee a middle-class life.

Second-generation Asian Americans are also more conscious of the disadvantages associated with being nonwhite than their parents, who as immigrants tend to be optimistic about overcoming the disadvantages of this status. As a Chinese-American woman points out from her own experience, "The truth is, no matter how American you think you are or try to be, if you have almond-shaped eyes, straight black hair, and a yellow complexion, you are a foreigner by default.... You can certainly be as good as or even better than whites, but you will never become accepted as white." This remark echoes a commonly-held frustration among second-generation, U.S.-born Asians who detest being treated as immigrants or foreigners. Their experience suggests that whitening has more to do with the beliefs of white America, than with the actual situation of Asian Americans. Speaking perfect English, adopting mainstream cultural values, and even intermarrying members of the dominant group may help reduce this "otherness" for particular individuals, but it has little effect on the group as a whole. New stereotypes can emerge and un-whiten Asian Americans, no matter how "successful" and "assimilated" they have become. For example, Congressman David Wu once was invited by the Asian-American employees of the U.S. Department of Energy to give a speech in celebration of Asian-American Heritage Month. Yet, he and his Asian-American staff were not allowed into the department building, even after presenting their congressional Identification, and were repeatedly asked about their citizenship and country of origin. They were told that this was standard procedure for the Department of Energy and that a congressional ID card was not a reliable document. The next day, a congressman of Italian descent was allowed to enter the same building with his congressional ID, no questions asked.

The stereotype of the "honorary white" or model minority goes hand-in-hand with that of the "forever foreigner." Today, globalization and U.S.-Asia relations, combined with continually high rates of immigration, affect how Asian Americans are perceived in American society. Many historical stereotypes, such as the "yellow peril" and "Fu Manchu" still exist in contemporary American life, as revealed in such highly publicized incidents as the murder of Vincent Chin, a Chinese American mistaken for Japanese and beaten to death by a disgruntled white auto worker in the 1980s; the trial of Wen Ho Lee, a nuclear scientist suspected of spying for the Chinese government in the mid-1990s; the 1996 presidential campaign finance scandal, which implicated Asian Americans in funneling foreign contributions to the Clinton campaign; and most recently, in 2001, the Abercrombie & Fitch t-shirts that depicted Asian cartoon characters in stereo-typically negative ways, with slanted eyes, thick glasses and heavy Asian accents. Ironically, the ambivalent, conditional nature of their acceptance by whites prompts many Asian Americans to organize pan-ethnically to fight back—which consequently heightens their racial distinctiveness. So becoming white or not is beside the point. The bottom line is: Americans of Asian ancestry still have to constantly prove that they truly are loyal Americans.

10

Race as Class

HERBERT J. GANS

Humans of all colors and shapes can make babies with each other. Consequently most biologists, who define races as subspecies that cannot interbreed, argue that scientifically there can be no human races. Nonetheless, laypeople still see and distinguish between races. Thus, it is worth asking again why the lay notion of race continues to exist and to exert so much influence in human affairs.

Laypersons are not biologists, nor are they sociologists who argue these days that race is a social construction arbitrary enough to be eliminated if "society" chose to do so. The laity operates with a very different definition of race. They see that humans vary, notably in skin color, the shape of the head, nose, and lips, and quality of hair, and they choose to define the variations as individual races.

More important, the lay public uses this definition of race to decide whether strangers (the so-called "other") are to be treated as superior, inferior, or equal. Race is even more useful for deciding quickly whether strangers might be threatening and thus should be excluded. Whites often consider dark-skinned strangers threatening until they prove otherwise, and none more than African Americans.

Scholars believe the color differences in human skins can be traced to climatic adaptation. They argue that the high levels of melanin in dark skin originally protected people living outside in hot, sunny climates, notably in Africa and South Asia, from skin cancer. Conversely, in cold climates, the low amount of melanin in light skins enabled the early humans to soak up vitamin D from a sun often hidden behind clouds. These color differences were reinforced by millennia of inbreeding when humans lived in small groups that were geographically and socially isolated. This inbreeding also produced variations in head and nose shapes and other facial features so that Northern Europeans look different from people from the Mediterranean area, such as Italians and, long ago, Jews. Likewise, East African faces differ from West African ones, and Chinese faces from Japanese ones. (Presumably the inbreeding and isolation also produced the DNA patterns that geneticists refer to in the latest scientific revival and redefinition of race.)

Geographic and social isolation ended long ago, however, and human population movements, intermarriage, and other occasions for mixing are eroding physical differences in bodily features. Skin color stopped being adaptive too

SOURCE: Gans, Herbert J. 2005. "Race as Class." *Contexts*, 4(4): 17–21.

after people found ways to protect themselves from the sun and could get their vitamin D from the grocery or vitamin store. Even so, enough color variety persists to justify America's perception of white, yellow, red, brown, and black races.

Never mind for the moment that the skin of "whites," as well as many East Asians and Latinos is actually pink; that Native Americans are not red; that most African Americans come in various shades of brown; and that really black skin is rare. Never mind either that color differences within each of these populations are as great as the differences between them, and that, as DNA testing makes quite clear, most people are of racially mixed origins, even if they do not know it. But remember that this color palette was invented by whites. Nonwhite people would probably divide the range of skin colors quite differently.

Advocates of racial equality use these contradictions to fight against racism. However, the general public also has other priorities. As long as people can roughly agree about who looks "white," "yellow," or "black" and find that their notion of race works for their purposes, they ignore its inaccuracies, inconsistencies, and other deficiencies.

Note, however, that only some facial and bodily features are selected for the lay definition of race. Some, like the color of women's nipples or the shape of toes (and male navels), cannot serve because they are kept covered. Most other visible ones, like height, weight, hairlines, ear lobes, finger or hand sizes—and even skin texture—vary too randomly and frequently to be useful for categorizing and ranking people or judging strangers. After all, your own child is apt to have the same stubby fingers as a child of another skin color or, what is equally important, a child from a very different income level.

RACE, CLASS, AND STATUS

In fact, the skin colors and facial features commonly used to define race are selected precisely because, when arranged hierarchically, they resemble the country's class-and-status hierarchy. Thus, whites are on top of the socioeconomic pecking order as they are on top of the racial one, while variously shaded nonwhites are below them in socioeconomic position (class) and prestige (status).

The darkest people are for the most part at the bottom of the class-status hierarchy. This is no accident, and Americans have therefore always used race as a marker or indicator of both class and status. Sometimes they also use it to enforce class position, to keep some people "in their place." Indeed, these uses are a major reason for its persistence.

Of course, race functions as more than a class marker, and the correlation between race and the socioeconomic pecking order is far from statistically perfect: All races can be found at every level of that order. Still, the race-class correlation is strong enough to utilize race for the general ranking of others. It also becomes more useful for ranking dark-skinned people as white poverty declines so much that whiteness becomes equivalent to being middle or upper class.

The relation between race and class is unmistakable. For example, the 1998–2000 median household income of non-Hispanic whites was $45,500; of Hispanics (currently seen by many as a race) as well as Native Americans, $32,000; and of African Americans, $29,000. The poverty rates for these same groups were 7.8 percent among whites, 23.1 among Hispanics, 23.9 among blacks, and 25.9 among Native Americans. (Asians' median income was $52,600—which does much to explain why we see them as a model minority.)

True, race is not the only indicator used as a clue to socioeconomic status. Others exist and are useful because they can also be applied to ranking co-racials. They include language (itself a rough indicator of education), dress, and various kinds of taste, from given names to cultural preferences, among others.

American English has no widely known working-class dialect like the English Cockney, although "Brooklynese" is a rough equivalent, as is "black vernacular." Most blue-collar people dress differently at work from white-collar, professional, and managerial workers. Although contemporary American leisure-time dress no longer signifies the wearer's class, middle-income Americans do not usually wear Armani suits or French haute couture, and the people who do can spot the knockoffs bought by the less affluent.

Actually, the cultural differences in language, dress, and so forth that were socially most noticeable are declining. Consequently, race could become yet more useful as a status marker, since it is so easily noticed and so hard to hide or change. And in a society that likes to see itself as classless, race comes in very handy as a substitute.

THE HISTORICAL BACKGROUND

Race became a marker of class and status almost with the first settling of the United States. The country's initial holders of cultural and political power were mostly WASPs (with a smattering of Dutch and Spanish in some parts of what later became the United States). They thus automatically assumed that their kind of whiteness marked the top of the class hierarchy. The bottom was assigned to the most powerless, who at first were Native Americans and slaves. However, even before the former had been virtually eradicated or pushed to the country's edges, the skin color and related facial features of the majority of colonial America's slaves had become the markers for the lowest class in the colonies.

Although dislike and fear of the dark are as old as the hills and found all over the world, the distinction between black and white skin became important in America only with slavery and was actually established only some decades after the first importation of black slaves. Originally, slave owners justified their enslavement of black Africans by their being heathens, not by their skin color.

In fact, early Southern plantation owners could have relied on white indentured servants to pick tobacco and cotton or purchased the white slaves that were available then, including the Slavs from whom the term slave is derived. They also had access to enslaved Native Americans. Blacks, however, were cheaper, more plentiful, more easily controlled, and physically more able to survive the intense heat and brutal working conditions of Southern plantations.

After slavery ended, blacks became farm laborers and sharecroppers, de facto indentured servants, really, and thus they remained at the bottom of the class hierarchy. When the pace of industrialization quickened, the country needed new sources of cheap labor. Northern industrialists, unable and unwilling to recruit southern African Americans, brought in very poor European immigrants, mostly peasants. Because these people were near the bottom of the class hierarchy, they were considered nonwhite and classified into races. Irish and Italian newcomers were sometimes even described as black (Italians as "guineas"), and the eastern and southern European immigrants were deemed "swarthy."

However, because skin color is socially constructed, it can also be reconstructed. Thus, when the descendants of the European immigrants began to move up economically and socially, their skins apparently began to look lighter to the whites who had come to America before them. When enough of these descendants became visibly middle class, their skin was seen as fully white. The biological skin color of the second and third generations had not changed, but it was socially blanched or whitened. The process probably began in earnest just before the Great Depression and resumed after World War II. As the cultural and other differences of the original European immigrants disappeared, their descendants became known as white ethnics.

This pattern is now repeating itself among the peoples of the post-1965 immigration. Many of the new immigrants came with money and higher education, and descriptions of their skin color have been shaped by their class position. Unlike the poor Chinese who were imported in the 19th century to build the West and who were hated and feared by whites as a "yellow horde," today's affluent Asian newcomers do not seem to look yellow. In fact, they are already sometimes thought of as honorary whites, and later in the 21st century they may well turn into a new set of white ethnics. Poor East and Southeast Asians may not be so privileged, however, although they are too few to be called a "yellow horde."

Hispanics are today's equivalent of a "swarthy" race. However, the children and grandchildren of immigrants among them will probably undergo "whitening" as they become middle class. Poor Mexicans, particularly in the Southwest, are less likely to be whitened, however. (Recently a WASP Harvard professor came close to describing these Mexican immigrants as a brown horde.)

Meanwhile, black Hispanics from Puerto Rico, the Dominican Republic, and other Caribbean countries may continue to be perceived, treated, and mistreated as if they were African American. One result of that mistreatment is their low median household income of $35,000, which was just $1,000 more than that of non-Hispanic blacks but $4,000 below that of so-called white Hispanics.

Perhaps South Asians provide the best example of how race correlates with class and how it is affected by class position. Although the highly educated Indians and Sri Lankans who started coming to America after 1965 were often darker than African Americans, whites only noticed their economic success.

They have rarely been seen as nonwhites, and are also often praised as a model minority.

Of course, even favorable color perceptions have not ended racial discrimination against newcomers, including model minorities and other affluent ones. When they become competitors for valued resources such as highly paid jobs, top schools, housing, and the like, they also become a threat to whites. California's Japanese-Americans still suffer from discrimination and prejudice four generations after their ancestors arrived here.

AFRICAN-AMERICAN EXCEPTIONALISM

The only population whose racial features are not automatically perceived differently with upward mobility are African Americans: Those who are affluent and well educated remain as visibly black to whites as before. Although a significant number of African Americans have become middle class since the civil rights legislation of the 1960s, they still suffer from far harsher and more pervasive discrimination and segregation than nonwhite immigrants of equivalent class position. This not only keeps whites and blacks apart but prevents blacks from moving toward equality with whites. In their case, race is used both as a marker of class and, by keeping blacks "in their place," an enforcer of class position and a brake on upward mobility.

In the white South of the past, African Americans were lynched for being "uppity." Today, the enforcement of class position is less deadly but, for example, the glass ceiling for professional and managerial African Americans is set lower than for Asian Americans, and on-the-job harassment remains routine.

Why African-American upward economic mobility is either blocked or, if allowed, not followed by public blanching of skin color remains a mystery. Many explanations have been proposed for the white exceptionalism with which African Americans are treated. The most common is "racism," an almost innate prejudice against people of different skin color that takes both personal and institutional forms. But this does not tell us why such prejudice toward African Americans remains stronger than that toward other nonwhites.

A second explanation is the previously mentioned white antipathy to blackness, with an allegedly primeval fear of darkness extrapolated into a primordial fear of dark-skinned people. But according to this explanation, dark-skinned immigrants such as South Asians should be treated much like African Americans.

A better explanation might focus on "Negroid" features. African as well as Caribbean immigrants with such features—for example, West Indians and Haitians—seem to be treated somewhat better than African Americans. But this remains true only for new immigrants; their children are generally treated like African Americans.

Two additional explanations are class-related. For generations, a majority or plurality of all African Americans were poor, and about a quarter still remain so. In addition, African Americans continue to commit a proportionally greater share of the street crime, especially street drug sales—often because legitimate job opportunities are scarce. African Americans are apparently also more often arrested without cause. As one result, poor African Americans are more often considered undeserving than are other poor people, although in some parts of America, poor Hispanics, especially those who are black, are similarly stigmatized.

The second class-based explanation proposes that white exceptionalist treatment of African Americans is a continuing effect of slavery: They are still perceived as ex-slaves. Many hateful stereotypes with which today's African Americans are demonized have changed little from those used to dehumanize the slaves. (Black Hispanics seem to be equally demonized, but then they were also slaves, if not on the North American continent.) Although slavery ended officially in 1864, ever since the end of Reconstruction subtle efforts to discourage African-American upward mobility have not abated, although these efforts are today much less pervasive or effective than earlier.

Some African Americans are now millionaires, but the gap in wealth between average African Americans and whites is much greater than the gap between incomes. The African-American middle class continues to grow, but many of its members barely have a toehold in it, and some are only a few paychecks away from a return to poverty. And the African-American poor still face the most formidable obstacles to upward mobility. Close to a majority of working-age African-American men are jobless or out of the labor force. Many women, including single mothers, now work in the low-wage economy, but they must do without most of the support systems that help middle-class working mothers. Both federal and state governments have been punitive, even in recent Democratic administrations, and the Republicans have cut back nearly every antipoverty program they cannot abolish.

Daily life in a white-dominated society reminds many African Americans that they are perceived as inferiors, and these reminders are louder and more relentless for the poor, especially young men. Regularly suspected of being criminals, they must constantly prove that they are worthy of equal access to the American Dream. For generations, African Americans have watched immigrants pass them in the class hierarchy, and those who are poor must continue to compete with current immigrants for the lowest-paying jobs. If unskilled African Americans reject such jobs or fail to act as deferentially as immigrants, they justify the white belief that they are less deserving than immigrants. Blacks' resentment of such treatment gives whites additional evidence of their unworthiness, thereby justifying another cycle of efforts to keep them from moving up in class and status.

Such practices raise the suspicion that the white political economy and white Americans may, with the help of nonwhites who are not black, use African Americans to anchor the American class structure with a permanently lower-class population. In effect, America, or those making decisions in its name, could be seeking, not necessarily consciously, to establish an undercaste that cannot move out and up. Such undercastes exist in other societies: the gypsies of Eastern Europe, India's untouchables, "indigenous people," and "aborigines" in yet other places. But these are far poorer countries than the United States.

SOME IMPLICATIONS

The conventional wisdom and its accompanying morality treat racial prejudice, discrimination, and segregation as irrational social and individual evils that public policy can reduce but only changes in white behavior and values can eliminate. In fact, over the years, white prejudice as measured by attitude surveys has dramatically declined, far more dramatically than behavioral and institutional discrimination.

But what if discrimination and segregation are more than just a social evil? If they are used to keep African Americans down, then they also serve to eliminate or restrain competitors for valued or scarce resources, material and symbolic. Keeping African Americans from decent jobs and incomes as well as quality schools and housing makes more of these available to all the rest of the population. In that case, discrimination and segregation may decline significantly only if the rules of the competition change or if scarce resources, such as decent jobs, become plentiful enough to relax the competition, so that the African-American population can become as predominantly middle class as the white population. Then the stigmas, the stereotypes inherited from slavery, and the social and other arrangements that maintain segregation and discrimination could begin to lose their credibility. Perhaps "black" skin would eventually become as invisible as "yellow" skin is becoming.

THE MULTIRACIAL FUTURE

One trend that encourages upward mobility is the rapid increase in interracial marriage that began about a quarter century ago. As the children born to parents of different races also intermarry, more and more Americans will be multiracial, so that at some point far in the future the current quintet of skin colors will be irrelevant. About 40 percent of young Hispanics and two-thirds of young Asians now "marry out," but only about 10 percent of blacks now marry nonblacks—yet another instance of the exceptionalism that differentiates blacks.

Moreover, if race remains a class marker, new variations in skin color and in other visible bodily features will be taken to indicate class position. Thus, multiracials with "Negroid" characteristics could still find themselves disproportionately at the bottom of the class hierarchy. But what if at some point in the future everyone's skin color varied by only a few shades of brown? At that point, the dominant American classes might have to invent some new class markers.

If in some utopian future the class hierarchy disappears, people will probably stop judging differences in skin color and other features. Then lay Americans would probably agree with biologists that race does not exist. They might even insist that race does not need to exist.

11

Is Capitalism Gendered and Racialized?

JOAN ACKER

Capitalism is racialized and gendered in two intersecting historical processes. ... First, industrial capitalism emerged in the United States dominated by white males, with a gender- and race-segregated labor force, laced with wage inequalities, and a society-wide gender division of caring labor. The processes of reproducing segregation and wage inequality changed over time, but segregation and inequality were not eliminated. A small group of white males still dominate the capitalist economy and its politics. The society-wide gendered division of caring labor still exists. Ideologies of white masculinity and related forms of consciousness help to justify capitalist practices. In short, conceptual and material practices that construct capitalist production and markets, as well as beliefs supporting those practices, are deeply shaped through gender and race divisions of labor and power and through constructions of white masculinity.

Second, these gendered and racialized practices are embedded in and replicated through the gendered substructures of capitalism. These gendered substructures exist in ongoing incompatible organizing of paid production activities and unpaid domestic and caring activities. Domestic and caring activities are devalued and seen as outside the "main business" (Smith 1999) of capitalism. The commodification of labor, the capitalist wage form, is an integral part of this process, as family provisioning and caring become dependent upon wage labor. The abstract language of bureaucratic organizing obscures the ongoing impact on families and daily life. At the same time, paid work is organized on the assumption that reproduction is of no concern. The separations between paid production and unpaid life-sustaining activities are maintained by corporate claims that they have no responsibility for anything but returns to shareholders. Such claims are more successful in the United States, in particular, than in countries with stronger labor movements and welfare states. These often successful claims contribute to the corporate processes of establishing their interests as more important than those of ordinary people.

SOURCE: Acker, Joan. 2006. *Class Questions, Feminist Answers*, pp. 113–117.

THE GENDERED AND RACIALIZED DEVELOPMENT OF U.S. CAPITALISM

Segregations and Wage Inequalities

Industrial capitalism is historically, and in the main continues to be, a white male project, in the sense that white men were and are the innovators, owners, and holders of power. Capitalism developed in Britain and then in Europe and the United States in societies that were already dominated by white men and already contained a gender-based division of labor. The emerging waged labor force was sharply divided by gender, as well as by race and ethnicity with many variations by nation and regions within nations. At the same time, the gendered division of labor in domestic tasks was reconfigured and incorporated in a gendered division between paid market labor and unpaid domestic labor. In the United States, certain white men, unburdened by caring for children and households and already the major wielders of gendered power, buttressed at least indirectly by the profits from slavery and the exploitation of other minorities, were, in the nineteenth century, those who built the U.S. factories and railroads, and owned and managed the developing capitalist enterprise. As far as we know, they were also heterosexual and mostly of Northern European heritage. Their wives and daughters benefited from the wealth they amassed and contributed in symbolic and social ways to the perpetuation of their class, but they were not the architects of the new economy.

Recruitment of the labor force for the colonies and then the United States had always been transnational and often coercive. Slavery existed prior to the development of industrialism in the United States: Capitalism was built partly on profits from that source. Michael Omi and Howard Winant (1994, 265) contend that the United States was a racial dictatorship for 258 years, from 1607 to 1865. After the abolition of slavery in 1865, severe exploitation, exclusion, and domination of blacks by whites perpetuated racial divisions cutting across gender and some class divisions, consigning blacks to the most menial, low-paying work in agriculture, mining, and domestic service. Early industrial workers were immigrants. For example, except for the brief tenure (twenty-five years) of young, native-born white women workers in the Lowell, Massachusetts, mills, immigrant women and children were the workers in the first mass production industry in the United States, the textile mills of Massachusetts and Philadelphia, Pennsylvania (Perrow 2002). This was a gender and racial/ethnic division of labor that still exists, but now on a global basis. Waves of European immigrants continued to come to the United States to work in factories and on farms. Many of these European immigrants, such as impoverished Irish, Poles, and eastern European Jews were seen as non-white or not-quite-white by white Americans and were used in capitalist production as low-wage workers, although some of them were actually skilled workers (Brodkin 1998). The experiences of racial oppression built into industrial capitalism varied by gender within these racial/ethnic groups.

Capitalist expansion across the American continent created additional groups of Americans who were segregated by race and gender into racial and ethnic enclaves and into low-paid and highly exploited work. This expansion included the extermination and expropriation of native peoples, the subordination of Mexicans in areas taken in the war with Mexico in 1845, and the recruitment of Chinese and other Asians as low-wage workers, mostly on the west coast (Amott and Matthaei 1996; Glenn 2002).

Women from different racial and ethnic groups were incorporated differently than men and differently than each other into developing capitalism in the late nineteenth and early twentieth centuries. White Euro-American men moved from farms into factories or commercial, business, and administrative jobs. Women aspired to be housewives as the male breadwinner family became the ideal. Married white women, working class and middle class, were housewives unless unemployment, low wages, or death of their husbands made their paid work necessary (Goldin 1990, 133). Young white women with some secondary education moved into the expanding clerical jobs and into elementary school teaching when white men with sufficient education were unavailable (Cohn 1985). African Americans, both women and men, continued to be confined to menial work, although some were becoming factory workers, and even teachers and professionals as black schools and colleges were formed (Collins 2000). Young women from first- and second-generation European immigrant families worked in factories and offices. This is a very sketchy outline of a complex process (Kessler-Harris 1982), but the overall point is that the capitalist labor force in the United States emerged as deeply segregated horizontally by occupation and stratified vertically by positions of power and control on the basis of both gender and race.

Unequal pay patterns went along with sex and race segregation, stratification, and exclusion. Differences in the earnings and wealth (Keister 2000) of women and men existed before the development of the capitalist wage (Padavic and Reskin 2002). Slaves, of course, had no wages and earned little after abolition. These patterns continued as capitalist wage labor became the dominant form and wages became the primary avenue of distribution to ordinary people. Unequal wages were justified by beliefs about virtue and entitlement. A living wage or a just wage for white men was higher than a living wage or a just wage for white women or for women and men from minority racial and ethnic groups (Figart, Mutari, and Power 2002). African-American women were at the bottom of the wage hierarchy.

The earnings advantage that white men have had throughout the history of modern capitalism was created partly by their organization to increase their wages and improve their working conditions. They also sought to protect their wages against the competition of others, women and men from subordinate groups (for example, Cockburn 1983, 1991). This advantage also suggests a white male coalition across class lines (Connell 2000; Hartmann 1976), based at least partly in beliefs about gender and race differences and beliefs about the superior skills of white men. White masculine identity and self-respect were complexly involved in these divisions of labor and wages. This is another way in which capitalism is a gendered and racialized accumulation process (Connell 2000). Wage differences between white men and all other groups, as well as divisions of labor between these groups, contributed to profit and flexibility, by helping to maintain growing occupational areas, such as clerical work, as segregated and low paid. Where women worked in manufacturing or food processing, gender divisions of labor kept the often larger female work force in low-wage routine jobs, while males worked in other more highly paid, less routine, positions (Acker and Van Houten 1974). While white men might be paid more, capitalist organizations could benefit from this "gender/racial dividend." Thus, by maintaining divisions, employers could pay less for certain levels of skill, responsibility, and experience when the worker was not a white male.

This is not to say that getting a living wage was easy for white men, or that most white men achieved it. Labor-management battles, employers' violent tactics to prevent unionization, [and] massive unemployment during frequent economic depressions characterized the situation of white industrial workers as wage labor spread in the nineteenth and early twentieth centuries. During the same period, new white-collar jobs were created to manage, plan, and control the expanding industrial economy. This rapidly increasing middle class was also stratified by gender and race. The better-paid, more respected jobs went to white men; white women were secretaries and clerical workers; people of color were absent. Conditions and issues varied across industries and regions of the country. But, wherever you look, those variations contained underlying gendered and racialized divisions. Patterns of stratification and segregation were written into employment contracts in work content, positions in work hierarchies, and wage differences, as well as other forms of distribution.

These patterns persisted, although with many alterations, through extraordinary changes in production and social life. After World War II, white women, except for a brief period immediately after the war, went to work for pay in the expanding service sector, professional, and managerial fields. African Americans moved to the North in large numbers, entering industrial and service sector jobs. These processes accelerated after the 1960s, with the civil rights and women's movements, new civil rights laws, and affirmative action. Hispanics and Asian Americans, as well as other racial/ethnic groups, became larger proportions of the population, on the whole finding work in low-paid, segregated jobs. Employers continued, and still continue, to select and promote workers based on gender and racial identifications, although the processes are more subtle, and possibly less visible, than in the past (for example, Brown et al. 2003; Royster 2003). These processes continually recreate gender and racial inequities, not as cultural or ideological survivals from earlier times, but as essential elements in present capitalisms (Connell 1987, 103–106).

Segregating practices are a part of the history of white, masculine-dominated capitalism that establishes class as gendered and racialized. Images of masculinity support these practices, as they produce a taken-for-granted world in which certain men legitimately make employment and other economic decisions that affect the lives of most other people. Even though some white women and people from other-than-white groups now hold leadership positions, their actions are shaped within networks of practices sustained by images of masculinity (Wacjman 1998).

Masculinities and Capitalism

Masculinities are essential components of the ongoing male project, capitalism. While white men were and are the main publicly recognized actors in the history of capitalism, these are not just any white men. They have been, for example, aggressive entrepreneurs or strong leaders of industry and finance (Collinson and Hearn 1996). Some have been oppositional actors, such as self-respecting and tough workers earning a family wage, and militant labor leaders. They have been particular men whose locations within gendered and racialized social relations and practices can be partially captured by the concept of masculinity. "Masculinity" is a contested term. As Connell (1995, 2000), Hearn (1996), and others have pointed out, it should be pluralized as "masculinities," because in any society at any one time there are several ways of being a man. "Being a man" involves cultural images and practices. It always implies a contrast to an unidentified femininity.

Hegemonic masculinity can be defined as the taken-for-granted, generally accepted form, attributed to leaders and other influential figures at particular historical times. Hegemonic masculinity legitimates the power of those who embody it. More than one type of hegemonic masculinity may exist simultaneously, although they may share characteristics, as do the business leader and the sports star at the present time. Adjectives describing hegemonic masculinities closely follow those describing characteristics of successful business organizations, as Rosabeth Moss Kanter (1977) pointed out in the 1970s. The successful CEO and the successful organization are aggressive, decisive, competitive, focused on winning and defeating the enemy, taking territory from others. The ideology of capitalist markets is imbued with a masculine ethos. As R. W. Connell (2000, 35) observes, "The market is often seen as the antithesis of gender (marked by achieved versus ascribed status, etc.). But the market operates through forms of rationality that are historically masculine and involve a sharp split between instrumental reason on the one hand, emotion and human responsibility on the other" (Seidler 1989). Masculinities embedded in collective practices are part of the context within which certain men made and still make the decisions that drive and shape the ongoing development of capitalism. We can speculate that how these men see themselves, what actions and choices they feel compelled to make and they think are legitimate, how they and the world around them define desirable masculinity, enter into that decision making (Reed 1996). Decisions made at the very top reaches of (masculine) corporate power have consequences that are experienced as inevitable economic forces or disembodied social trends. At the same time, these decisions symbolize and enact varying hegemonic masculinities (Connell 1995). However, the embeddedness of masculinity within the ideologies of business and the market may become invisible, seen as just part of the way business is done. The relatively few women who reach the highest positions probably think and act within these strictures.

Hegemonic masculinities and violence are deeply connected within capitalist history: The violent acts of those who carried out the slave trade or organized colonial conquests are obvious examples. Of course, violence has been an essential component of power in many other socioeconomic systems, but it continues into the rational organization of capitalist economic activities. Violence is frequently a legitimate, if implicit, component of power exercised by bureaucrats as well as "robber barons." Metaphors of violence, frequently military violence, are often linked to notions of the masculinity of corporate leaders, as "defeating the enemy" suggests. In contemporary capitalism, violence and its links to masculinity are often masked by the seeming impersonality of objective conditions. For example, the masculinity of top managers, the ability to be tough, is involved in the implicit violence of many corporate decisions, such as those cutting jobs in order to raise profits and, as a result, producing unemployment. Armies and other organizations, such as the police, are specifically organized around violence. Some observers of recent history suggest that organized violence, such as the use of the military, is still mobilized at least partly to reach capitalist goals, such as controlling access to oil supplies. The masculinities of those making decisions to deploy violence in such a way are hegemonic, in the sense of powerful and exemplary. Nevertheless, the connections between masculinity, capitalism, and violence are complex and contradictory, as Jeff Hearn and Wendy Parkin (2001) make clear. Violence is always a possibility in mechanisms of control and domination, but it is not always evident, nor is it always used.

As corporate capitalism developed, Connell (1995) and others (for example, Burris 1996) argue that a hegemonic masculinity based on claims to expertise developed alongside masculinities organized around domination and control. Hegemonic masculinity relying on claims to expertise does not necessarily lead to economic organizations free of domination and violence, however (Hearn and Parkin 2001). Hearn and Parkin (2001) argue that controls relying on both explicit and implicit violence exist in a wide variety of organizations, including those devoted to developing new technology.

Different hegemonic masculinities in different countries may reflect different national histories, cultures, and change processes. For example, in Sweden in the mid-1980s, corporations were changing the ways in which they did business toward a greater participation in the international economy, fewer controls on currency and trade, and greater emphasis on competition. Existing images of dominant masculinity were changing, reflecting new business practices. This seemed to be happening in the banking sector, where I was doing research on women and their jobs (Acker 1994a). The old paternalistic leadership, in which primarily men entered as young clerks expecting to rise to managerial levels, was being replaced by young, aggressive men hired as experts and managers from outside the banks. These young, often technically trained, ambitious men pushed the idea that the staff was there to sell bank products to customers, not, in the first instance, to take care of the needs of clients. Productivity goals were put in place; nonprofitable customers, such as elderly pensioners, were to be encouraged not to come into the bank and occupy the staff's attention. The female clerks we interviewed were disturbed by these changes, seeing them as evidence that the men at the top were changing from paternal guardians of the people's interests to manipulators who only wanted riches for themselves. The confirmation of this came in a scandal in which the CEO of the largest bank had to step down because he had illegally taken money from the bank to pay for his housing. The amount of money was small; the disillusion among employees was huge. He had been seen as a benign father; now he was no better than the callous young men on the way up who were dominating the daily work in the banks. The hegemonic masculinity in Swedish banks was changing as the economy and society were changing.

Hegemonic masculinities are defined in contrast to subordinate masculinities. White working class masculinity, although clearly subordinate, mirrors in some of its more heroic forms the images of strength and responsibility of certain successful business leaders. The construction of working class masculinity around the obligations to work hard, earn a family wage, and be a good provider can be seen as providing an identity that both served as a social control and secured male advantage in the home. That is, the good provider had to have a wife and probably children for whom to provide. Glenn (2002) describes in some detail how this image of the white male worker also defined him as superior to and different from black workers.

Masculinities are not stable images and ideals, but [shift] with other societal changes. With the turn to neoliberal business thinking and globalization, there seem to be new forms. Connell (2000) identifies "global business masculinity," while Lourdes Beneria (1999) discusses the "Davos man," the global leader from business, politics, or academia who meets his peers once a year in the Swiss town of Davos to assess and plan the direction of globalization. Seeing masculinities as implicated in the ongoing production of global capitalism opens the possibility of seeing sexualities, bodies, pleasures, and identities as also implicated in economic relations.

In sum, gender and race are built into capitalism and its class processes through the long history of racial and gender segregation of paid labor and through the images and actions of white men who dominate and lead central capitalist endeavors. Underlying these processes is the subordination to production and the market of nurturing and caring for human beings, and the assignment of these responsibilities to women as unpaid work. Gender segregation that differentially affects women in all racial groups rests at least partially on the ideology and actuality of women as carers. Images of dominant masculinity enshrine particular male bodies and ways of being as different from the female and distanced from caring.... I argue that industrial capitalism, including its present neoliberal form, is organized in ways that are, at the same time, antithetical and necessary to the organization of caring or reproduction and that the resulting tensions contribute to the perpetuation of gendered and racialized class inequalities. Large corporations are particularly important in this process as they increasingly control the resources for provisioning but deny responsibility for such social goals.

REFERENCES

Acker, Joan. 1994a. The Gender Regime of Swedish Banks. *Scandinavian Journal of Management* 10, no. 2: 117–30.

Acker, Joan, and Donald Van Houten. 1974. Differential Recruitment and Control: The Sex Structuring of Organizations. *Administrative Science Quarterly* 19 (June, 1974): 152–63.

Amott, Teresa, and Julie Matthaei. 1996. *Race, Gender, and Work: A Multi-cultural Economic History of Women in the United States*. Revised edition. Boston: South End Press.

Beneria, Lourdes. 1999. Globalization, Gender and the Davos Man. *Feminist Economics* 5, no. 3: 61–83.

Brodkin, Karen. 1998. Race, Class, and Gender: The Metaorganization of American Capitalism. *Transforming Anthropology* 7, no. 2: 46–57.

Brown, Michael K., Martin Carnoy, Elliott Currie, Troy Duster, David B. Oppenheimer, Marjorie M. Shultz, and David Wellman. 2003. *White-Washing Race The Myth of a Color-Blind Society*. Berkeley: University of California Press.

Burris, Beverly H. 1996. Technocracy, Patriarchy and Management. In *Men as Managers, Managers as Men*, ed. David L. Collinson and Jeff Hearn. London: Sage.

Cockburn, Cynthia. 1983. *Brothers*. London: Pluto Press.

______. 1991. *In the Way of Women: Men's Resistance to Sex Equality in Organization*. Ithaca, N.Y: ILR Press.

Cohn, Samuel. 1985. *The Process of Occupational Sex-Typing: The Femininization of Clerical Labor in Great Britain*. Philadelphia: Temple University Press.

Collins, Patricia Hill. 2000. *Black Feminist Thought*, second edition. New York and London: Routledge.

Collinson, David L., and Jeff Hearn. 1996. Breaking the Silence: On Men, Masculinities and Managements. In *Men as Managers, Managers as Men*, ed. David L. Collinson and Jeff Hearns. London: Sage.

Connell, R. W. 1987. *Gender & Power*. Stanford, Calif.: Stanford University Press.

______. 1995. *Masculinities*. Berkeley: University of California Press.

______. 2000. *The Men and the Boys*. Berkeley: University of California Press.

Figart, Deborah M., Ellen Mutari, and Marilyn Power. 2002. *Living Wages, Equal Wages*. London and New York: Routledge.

Glenn, Evelyn Nakano. 2002. *Unequal Freedom: How Race and Gender Shaped American Citizenship and Labor*. Cambridge: Harvard University Press.

Goldin, Claudia. 1990. *Understanding the Gender Gap: An Economic History of American Women*. New York and Oxford: Oxford University Press.

Hartmann, Heidi. 1976. "Capitalism, Patriarchy, and Job Segregation by Sex," *Signs: Journal of Women in Culture and Society* 1(3), part 2, spring: 137–167.

Hearn, Jeff. 1996. Is Masculinity Dead? A Critique of the Concept of Masculinity/ Masculinities. In *Understanding Masculinities: Social Relations and Cultural Arenas*, ed. M. Mac an Ghaill. Buckingham: Oxford University Press.

______. 2004. From Hegemonic Masculinity to the Hegemony of Men. *Feminist Theory* 5, no. 1: 49–72.

Hearn, Jeff, and Wendy Parkin. 2001. *Gender, Sexuality and Violence in Organizations*. London: Sage.

Kanter, Rosabeth Moss. 1977. *Men and Women of the Corporation*. New York: Basic Books.

Keister, Lisa. 2000. *Wealth in America: Trends in Wealth Inequality*. Cambridge: Cambridge University Press.

Kessler-Harris, Alice. 1982. *Out to Work: A History of Wage-Earning Women in the United States*. New York: Oxford University Press.

Omi, Michael, and Howard Winant. 1994. *Racial Formation in the United States*. New York: Routledge.

Padavic, Irene, and Barbara Reskin. 2002. *Women and Men at Work*, second edition. Thousand Oaks, Calif.: Pine Forge Press.

Perrow, Charles. 2002. *Organizing America*. Princeton and Oxford: Princeton University Press.

Reed, Rosslyn. 1996. Entrepreneurialism and Paternalism in Australian Management: A Gender Critique of the "Self-Made" Man. In *Men as Managers, Managers as Men*, ed. David L. Collinson and Jeff Hearn. London: Sage.

Royster, Deirdre A. 2003. *Race and the Invisible Hand: How White Networks Exclude Black Men from Blue-Collar Jobs*. Berkeley: University of California Press.

Seidler, Victor J. 1989. *Rediscovering Masculinity: Reason, Language, and Sexuality*. London and New York: Routledge.

Smith, Dorothy. 1999. *Writing the Social: Critique, Theory, and Investigation*. Toronto: University of Toronto Press.

Wacjman, Judy. 1998. *Managing Like a Man*. Cambridge: Polity Press.

12

The Great Divergence

Growing Income Inequality Could Destabilize the U.S. So Why Isn't Anyone Talking about It?

TIMOTHY NOAH

In 1915, a statistician at the University of Wisconsin named Willford I. King published *The Wealth and Income of the People of the United States*, the most comprehensive study of its kind to date. The United States was displacing Great Britain as the world's wealthiest nation, but detailed information about its economy was not yet readily available; the federal government wouldn't start collecting such data in any systematic way until the 1930s. One of King's purposes was to reassure the public that all Americans were sharing in the country's newfound wealth.

King was somewhat troubled to find that the richest 1% possessed about 15% of the nation's income. (A more authoritative subsequent calculation puts the figure slightly higher, at about 18%.)

This was the era in which the accumulated wealth of America's richest families—the Rockefellers, the Vanderbilts, the Carnegies—helped prompt creation of the modern income tax, lest disparities in wealth turn the United States into a European-style aristocracy. The socialist movement was at its historic peak, a wave of anarchist bombings was terrorizing the nation's industrialists, and President Woodrow Wilson's attorney general, Alexander Palmer, would soon stage brutal raids on radicals of every stripe. In American history, there has never been a time when class warfare seemed more imminent.

That was when the richest 1% accounted for 18% of the nation's income. Today, the richest 1% of Americans account for 24% of U.S. income. What caused this to happen?

Income inequality in the United States has not worsened steadily since 1915. It dropped a bit in the late teens, then started climbing again in the 1920s, reaching its peak just before the 1929 crash. The trend then reversed itself. Incomes started to become more equal in the 1930s and then became dramatically more equal in the 1940s. Income distribution remained roughly stable through the postwar economic boom of the 1950s and 1960s. Economic historians Claudia Goldin and Robert Margo have termed this midcentury era the "Great Compression."

SOURCE: © Noah, Timothy. 2012. *The Great Divergence: America's Growing Inequality Crisis and What We Can Do about It*. New York: Bloomsbury Press, an imprint of Bloomsbury Publishing Inc.

The deep nostalgia for that period felt by the Second World War generation—the era of *Life* magazine and the bowling league—reflects something more than mere sentimentality. Assuming you were white, not of draft age and Christian, there probably was no better time to belong to America's middle class.

The Great Compression ended in the 1970s. Wages stagnated, inflation raged, and by the decade's end, income inequality had started to rise. Income inequality grew through the 1980s, slackened briefly at the end of the 1990s, and then resumed with a vengeance in the aughts. In his 2007 book *The Conscience of a Liberal*, the Nobel laureate, Princeton economist and *New York Times* columnist Paul Krugman labeled the post-1979 epoch the "Great Divergence."

It's generally understood that we live in a time of growing income inequality, but "the ordinary person is not really aware of how big it is," Krugman told me. During the late 1980s and the late 1990s, the United States experienced two unprecedentedly long periods of sustained economic growth—the "seven fat years" and the " long boom." Yet from 1980 to 2005, more than 80% of total increase in Americans' income went to the top 1%. Economic growth was more sluggish in the aughts, but the decade saw productivity increase by about 20%. Yet virtually none of the increase translated into wage growth at middle and lower incomes, an outcome that left many economists scratching their heads.

Why don't Americans pay more attention to growing income disparity? One reason may be their enduring belief in social mobility. Economic inequality is less troubling if you live in a country where any child, no matter how humble his or her origins, can grow up to be president. In a survey of 27 nations conducted from 1998 to 2001, the country where the highest proportion agreed with the statement "people are rewarded for intelligence and skill" was, of course, the United States (69%). But when it comes to real as opposed to imagined social mobility, surveys find less in the United States than in much of (what we consider) the class-bound Old World. France, Germany, Sweden, Denmark, Spain—not to mention some newer nations like Canada and Australia—are all places where your chances of rising from the bottom are better than they are in the land of Horatio Alger's *Ragged Dick*.

All my life I've heard Latin America described as a failed society (or collec tion of failed societies) because of its grotesque maldistribution of wealth. Peasants in rags beg for food outside the high walls of opulent villas, and so on. But according to the Central Intelligence Agency (whose patriotism I hesitate to question), income distribution in the United States is more unequal than in Guyana, Nicaragua and Venezuela, and roughly on par with Uruguay, Argentina and Ecuador. Income inequality is actually declining in Latin America even as it continues to increase in the United States. Economically speaking, the richest nation on Earth is starting to resemble a banana republic.

The main difference is that the United States is big enough to maintain geographic distance between the villa-dweller and the beggar. As Ralston Thorpe tells his St. Paul's classmate, the investment banker Sherman McCoy, in Tom Wolfe's 1987 novel *The Bonfire of the Vanities*: "You've got to insulate, insulate, insulate."

In 1915, King wrote, "It is easy to find a man in almost any line of employment who is twice as efficient as another employee," yet "it is very rare to find one who is 10 times as efficient. It is common, however, to see one man possessing not 10 times but a thousand times the wealth of his neighbour .... Is the middle class doomed to extinction and shall we soon find the handful of plutocrats, the modern barons of wealth, lined up squarely in opposition to the propertyless masses with no buffer between to lessen the chances of open battle? With the middle class gone and the labourer condemned to remain a lifelong wage-earner with no hope of attaining wealth or even a competence in his old age, all the conditions are ripe for a crowning class-conflict equaling in intensity and bitterness anything pictured by the most radical follower of Karl Marx. Is this condition soon coming to pass?"

In the end, King concluded it wasn't. Income distribution in the United States, he found, was more equal than in Prussia, France and the United Kingdom. King was no socialist. Redistributing income to the poor, he wrote, "would merely mean more rapid multiplication of the lowest and least desirable classes," who remained, "from the reproductive standpoint, on the low point of their four-footed ancestors." A Malthusian, he believed in population control. Income inequality in the United States could be addressed by limiting immigration (King deplored "low-standard alien invaders") and by discouraging excessive breeding among the poor ("eugenicists are just beginning to impress upon us the absurd folly of breeding great troops of paupers, defectives and criminals to be a burden upon organized society").

Today, incomes in the U.S. are more unequal than in Germany, France, and the United Kingdom, not less so. Eugenics (thankfully) has fallen out of fashion and the immigration debate has become (somewhat) more polite. As for income inequality, it's barely entered America's national political debate. Indeed, the evidence from the 2000 and 2004 presidential elections suggests that even mild economic populism was a loser for Democrats.

But income inequality is a topic of huge importance to American society and therefore a subject of large and growing interest to a host of economists, political scientists and other wonky types. Except for a few libertarian outliers, these experts agree that the country's growing income inequality is deeply worrying. Even Alan Greenspan, the former Federal Reserve Board chairman and onetime Ayn Rand acolyte, has registered concern.

"This is not the type of thing which a democratic society—a capitalist democratic society—can really accept without addressing," Greenspan said in 2005. Greenspan's Republican-appointed successor, Ben Bernanke, has also fretted about income inequality.

The Great Divergence may represent the most significant change in American society in your lifetime—and it's not a change for the better.

13

Closing the Wealth Gap

A Review of Racial and Ethnic Inequalities in Homeownership

MEGHAN KUEBLER

INTRODUCTION

Homeownership has historically been central to the economic well-being of Americans.... However, minorities have not had the same access to homeownership that Whites have.... In 1910, approximately 49 percent of Whites owned their homes compared with 37 percent of Hispanics and just 23 percent of Blacks (Ruggles et al., 2010). By 1960, these disparities continued, but overall homeownership rates went up. As of 2010, 72 percent of Whites, 58 percent of Asians, 47 percent of Hispanics, and 44 percent of Blacks owned their homes.... Such rate differentials have had harsh repercussions on the wealth of minorities. The transfer of wealth via homes contributes to intergenerational wealth disparities between Whites and minorities. Exploring the racial gap in homeownership is fundamental to understanding racial inequalities and formulating strategies and policies to help close such disparities (Wachter and Megbolugbe, 1992). The goal of this paper is to explore these gaps....

THE AMERICAN DREAM AND WEALTH

Homeownership plays an important part in the American dream.... A typical scene described in novels and in television and film depicts a home with a yard and white picket fence as the quintessential American dream. Owning one's home has been promulgated as essential to both family and financial stability in America (Jackman and Jackman, 1980, Wachter and Megbolugbe, 1992). The U.S. government, a staunch supporter of homeownership, has supported homeownership through the tax code including tax deductions for mortgage interest

SOURCE: Kuebler, Meghan. 2013. In *Sociology Compass*. Oxford: Blackwell Publishing. (7/8): pp. 670–685.

and property taxes which supply financial incentives to owning one's home. It has also created other programs and dedicated units such as the Federal Housing Administration which insures individual mortgages and the Home Affordable Modification Program which modifies mortgages to make them more affordable to qualified homeowners.... The federal government, both through Republicans and Democrats, has encouraged Americans of all backgrounds to become homeowners.... America's social and fiscal policies regarding homeownership support several economic benefits of owning over renting including (i) lower monthly costs for comparable housing because homeowners do not pay monthly costs which go towards landlords' profit, (ii) the income tax benefits which allow for mortgage interest and property tax deductions, (iii) forced savings and wealth building created by paying down one's mortgage which builds home equity (Bratt et al., 2006). On the other hand, critics argue that the mortgage interest deduction disproportionately helps middle and upper income families who can itemize deductions, whereas lower–income families may not outlay enough taxes to reap the benefits of deductions hence forgoing this annual benefit of homeownership. Homeownership also improves social status, fosters upward mobility, and helps assimilate minorities into mainstream America (Alba and Logan, 1992, Wachter and Megbolugbe, 1992)....

... Homeownership is the chief vehicle for wealth accumulation for most Americans and especially for those at the lower end of the income spectrum who have few opportunities for alternative investment.... Home equity represents a very sizable proportion of total wealth among minority homeowners.... Whites have one of the smallest gaps in homeownership between persons of low and high income suggesting that Whites of all classes typically own their homes, not just the well off (Alba and Logan, 1992). Homeownership is a key contributor to racial wealth disparities both as the proportion of each racial group who owns and also the value of the homes among each group (Squires and Kubrin, 2006, Anong et al., 2007, Oliver and Shapiro 2006, Charles and Hurst, 2002)....

... Homeownership can influence stratification because it places some groups in areas with better services, such as Whites in the suburbs and Blacks in the inner-city (Alba and Logan, 1992, Oliver and Shapiro, 2006).

Despite its reputation as a good investment and its historical value as a wealth builder, caution should be taken when interpreting homeownership as a positive investment in the U.S. in recent years. Between 2000 and 2010, increases in homeownership rates were replaced with declines (Callis and Kresin, 2012). Steadily increasing home prices stalled and home values plummeted. Predatory loans grew more prevalent than ever, exposing risk particularly in minority communities....

BLACK HOMEOWNERSHIP

Blacks have historically been disadvantaged on a variety of socioeconomic indicators compared with Whites.... On average, lower levels of education and lower salaries than Whites help to limit Blacks' entrance into homeownership. Another detriment to Black homeownership is marital status—Black adults are significantly less likely to be married and more likely to divorce than are other racial and ethnic groups.... These family formation patterns suggest Blacks are not only less likely to be homeowners but also more likely to exit from homeownership and return to renting due to higher divorce rates.

Blacks are generally more likely than Whites to report that their overall net worth is either zero or negative. Blacks who do own assets consistently report lower levels of asset value as compared with Whites (Martin, 2009, Oliver and Shapiro, 2006, Blau and Graham, 1990). Homeownership is even more important for Blacks than Whites because it accounts for a larger proportion of their wealth since Blacks are less likely to hold other investment products such as stocks (Oliver and Shapiro, 2006, Squires and Kubrin, 2006).

Owning property has been a particular challenge for many Black families.... Blacks own older, more crowded, and structurally inadequate homes relative to Whites (Bianchi et al., 1982). The value of Black-owned homes is less than that of comparable White-owned homes, and Black-owned homes experience lower levels of appreciation compared with Whites (Alba and Logan, 1992, Flippen, 2004, Krivo and Kaufman, 2004, Blau and Graham, 1990). Blacks have more difficulty initially acquiring home equity as well as recovering accumulated home equity (Bianchi et al., 1982, Krivo and Kaufman, 2004).

Black neighbors negatively affect home appreciation even when controlling for several socioeconomic factors (Flippen, 2004). Black population depresses home values beyond the relationship with poverty (Flippen, 2004). When even a small number of Black households move into a neighborhood, the homes begin to depreciate in value. When Blacks move in, the least tolerant Whites elect to move out. This is called "White flight" (Crowder, 2000, Crowder and South, 2008). Those who left are replaced with more minorities at which point a "tipping point" is reached in which no more Whites will move into the neighborhood (Schelling, 1971, Galster, 1990, Sampson, 2009). This racial turnover in which the neighborhood becomes less White and more racially diverse negatively impacts housing values and the likelihood that new homeowners can obtain mortgages (Kuebler, 2012)....

Blacks face several challenges to healthy, sustained homeownership in contrast to Whites. Homeownership in segregated neighborhoods is less beneficial than in mainstream neighborhoods because houses tend to be older, in poorer condition, multi-family versus single-family, valued lower and earn less home equity (Bianchi et al., 1982, Harris, 1999, Krivo and Kaufman, 2004). Blacks are more likely to receive subprime, predatory mortgage loans which increase cost of the loan and the chance the homeowner will default, foreclose, and/or return to renting (Wyly and Ponder, 2011, Immergluck, 2009, Rugh and Massey, 2010). Lower socioeconomic status also makes Blacks more vulnerable to financial shocks and risk of foreclosure. These factors somewhat diminish the value of homeownership for Blacks, particularly that Blacks earn home equity slower than Whites, earn less overall home equity, and across the board experience greater risk of return to renting....

HISPANIC HOMEOWNERSHIP

Hispanics, the fastest growing minority group, have surpassed Blacks as the largest minority group in America.... There is great variation among Hispanics including not only cultural differences but also considerable socioeconomic differences.... Puerto Ricans exhibit similar homeownership patterns as Blacks (Alba and Logan, 1992). Other Hispanic groups exhibit great variation in their homeownership patterns. Homeownership levels are depressed in areas with large numbers of recent Hispanic immigrants (Alba and Logan, 1992, Myers et al., 2005). There is variation in achievements relating to homeownership between Hispanic groups. For instance, comparing foreign-born and second-generation Dominicans to native U.S.-born Puerto Ricans, Kasinitz et al. (2008) report that the Dominicans are more likely to graduate high school than Puerto Ricans—a trait positively related to homeownership.

Small numbers of Hispanics in neighborhoods have been shown to increase home values compared with neighborhoods without Hispanics. This Hispanic benefit can be attributed to higher home values, newer homes, and fewer vacancies, particularly in the western U.S. states (Flippen, 2004). Hispanics, however, acquire significantly less home equity and receive fewer benefits from homeownership than Whites (Krivo and Kaufman, 2004). Segregated Hispanic neighborhoods, like segregated Black neighborhoods—but not to the same extent—experience lower average home appreciation than other neighborhoods (Flippen, 2004, Squires and Kubrin, 2006).

Recent data indicate that Hispanic home buyers on average took out larger loans compared with Blacks and Whites, placing them in a more vulnerable economic position with higher debt burdens (Kochhar et al., 2009). While new immigrant households are less likely than natives to own, their recent housing-related losses have been smaller than natives, and they have not experienced a reversal in homeownership as others did during the past decade (Kochhar et al., 2009). Regardless of income, Hispanics are more likely than Whites to take out high-cost loans, meaning for instance, greater compounded interest over the life of the loan (Kochhar et al., 2009, Calem et al., 2004, Bocian et al., 2008). Essentially, this means that Hispanics pay more for their homes than Whites.

Hispanic home buyers experience discrimination in the housing market, mortgage credit market, and the home insurance market (Denton, 2006, Squires and Kubrin, 2006, Turner et al., 2002). Similar to, but to a lesser extent than Black neighborhoods and Black mortgage applicants, mortgage loan applications made by Hispanics, or in Hispanic neighborhoods by persons of any race, are more likely to be turned down than loans in White neighborhoods even when controlling for economic and neighborhood characteristics (Kuebler, 2012, Oliver and Shapiro, 2006, Canner et al., 1991)....

Hispanic homeowners face similar challenges to healthy, sustained homeownership as do Blacks. Recently, Hispanics received a disproportionate number of subprime loans which presented large risks of default and foreclosure and proved more costly than prime, mainstream loans (Calem et al., 2004a)....

ASIAN HOMEOWNERSHIP

Prior to Immigration Act of 1965, Asian immigration to the USA was limited, and consequently, Asians were extremely underrepresented compared with European immigrants.... Typically, only male laborers were accepted and not their families due to the Chinese Exclusion Act of 1882 and the National Origins Act of 1924. ... As Asian immigrants flowed into the USA after 1965, many set up households somewhat similar to Whites, as married couple homeowners with children. In fact, Asians are the most likely racial/ethnic group to be married—a trait positively associated with homeownership (Ishii-Kuntz, 2004, Cherlin, 2009). As immigration shifted away from male laborers to middle class, educated, and urbanized families, homeownership rates grew. Despite higher rates of homeownership compared with other minorities, most Asians are less likely to own their homes than Whites.... In contrast to other minority groups, Asians on average have higher household incomes than Whites. However, as a group, they have a larger proportion of households living in poverty than Whites (Ishii-Kuntz, 2004)....

While race is a significant factor in determining the quality of housing for minorities, this issue is less relevant for Asians (Bratt et al., 2006). Fifty years ago, Asian homeownership rates were comparable with Hispanics and slightly higher than Blacks, but in recent years, Asians have made impressive gains in homeownership, quickly outpacing other minorities.... During the past decade, Asians increased their homeownership rates faster than any other group, closing the gap with Whites more rapidly than other minorities (Kochhar et al., 2009)....

Studies find that Asians, in addition to Blacks and Hispanics, have more difficulty entering physically adequate homes than Whites (Stone, ... 2006). Research using data from the year 2000 Housing Discrimination Study finds that Asians experience discrimination approximately 20 percent of the time when purchasing homes (Turner and Ross, 2003). A rental audit study conducted in Toronto, one of North America's most diverse metropolitan areas, found that in addition to Blacks, Asian men were also discriminated against in the housing market (Hogan and Berry, 2011). However, Asians do not experience segregation or housing discrimination anywhere close to that of Blacks (Massey and Denton, 1993, Denton, 2006). Asians are successful buying homes and entering White neighborhoods with little resistance compared to other ethnic groups....

CONCLUSION

... Homeownership lies at the heart of the American dream. Even after very large losses in wealth due to the housing market crash, home equity remains the largest source of wealth for American families.... Sustained homeownership plays a pivotal role in bringing families into the middle class and enabling them to pass that class status on to their children. A marker of success, the lack of long term homeownership among many minorities, positions them in a disadvantaged state with little to no wealth to pass on to their children.

Rates of American homeownership increased steadily throughout the past century peaking at 69 percent in 2004 but declined to 65.5 percent according to the most recent estimates (Callis and Kresin, 2012). Despite gains in minority ownership during the 1990s, minorities were hit particularly hard by the housing crash and remain significantly less likely to own their homes than Whites. Blacks and Hispanics are much less likely to own than Asians and face multiple barriers to homeownership such as segregation. Until minorities own their homes to the same extent that Whites do, socioeconomic disparities will continue to exist. While Whites hold wealth in various forms such as stocks and retirement accounts, minorities, especially Blacks, typically rely on home equity for their wealth and as such, homeownership is even more important for the economic livelihood of minorities than Whites.…

REFERENCES

Alba, R. and Logan, J. 1992. Assimilation and Stratification of Homeownership Patterns of Racial and Ethnic Groups. *International Migration Review*, 26, 1314–1341.

Anong, S. T., Devaney, S. A. and Yang, Y. 2007. Asset Ownership by Black and White Families. *Financial Counseling and Planning*, 18, 33–45.

Bianchi, S. M., Farley, R. and Spain, D. 1982. Racial Inequalities in Housing: An Examination of Recent Trends. *Demography*, 19, 37–51.

Blau, F. D. and Graham, J. W. 1990. Black-White Differences in Wealth and Asset Composition. *Quarterly Journal of Economics*, 105, 321–339.

Bocian, D. G., Ernst, K. S. and Li, W. 2008. Race, Ethnicity and Subprime Home Loan Pricing. *Journal of Economics and Business*, 60, 110–124.

Bratt, R. G., Stone, M. E. and Hartman, C. (eds.). 2006. *A Right to Housing*. Philadelphia, PA: Temple University Press.

Calem, P. S., Gillen, K. and Wachter, S. 2004. The neighborhood distribution of subprime mortgage lending. *The Journal of Real Estate Finance and Economics*, 29, 393–410.

Callis, R. R. and Kresin, M. 2012. *Residential Vacancies and Homeownership in the Third Quarter* 2012. U.S. Census Bureau News.

Canner, G., Gabriel, S. A. and Woolley, J. M. 1991. *Race, Default Risk and Mortgage Lending: A Study of the FHA and Conventional Loan Markets*. Southern Economic Journal, LVII1, 249–62.

Charles, K. K. and Hurst, E. 2002. The Transition to Home Ownership and the Black-White Wealth Gap. *The Review of Economics and Statistics* 84, 281–297.

Cherlin, A. J. 2009. *Public and Private Families*, New York, McGraw Hill.

Crowder, K. and South, S. J. 2008. Spatial Dynamics of White Flight: The Effects of Local and Extralocal Racial Conditions on Neighborhood Out-Migration. *American Sociological Review*, 73, 792–812.

Crowder, K. D. 2000. The Racial Context of White Mobility: An Individual-Level Assessment of the White Flight Hypothesis. *Social Science Research*, 29, 223–57.

Denton, N. A. 2006. Segregation and Discrimination in Housing. *In:* Bratt, R. G., Stone, M. E. and Hartman, C. (eds.) *A Right to Housing*. Philadelphia, PA: Temple University Press.

Flippen, C. 2004. Unequal Returns to Housing Investments? A Study of Real Housing Appreciation among Black, White and Hispanic Households. *Social Forces*, 82, 1523–1551.

Galster, G. C. 1990. White Flight from Racially Integrated Neighbourhoods in the 1970s: the Cleveland Experience, *Urban Studies*, 27, 385–399.

Harris, D. R. 1999. Property Values Drop When Blacks Move in, Because..:Racial and Socioeconomic Determinants of Neighborhood Desirability. *American Sociological Review*, 64, 461–479.

Hogan, B., and Berry, B. 2011. Racial and Ethnic Biases in Rental Housing: An Audit Study of Online Apartment Listings. *City & Community*, 10, 351–372.

Immergluck, D. 2009. *Foreclosed: high-risk lending, deregulation, and the undermining of America's mortgage market*, Itahca, NY, Cornell Univ Pr.

Ishii-Kuntz, M. 2004. Asian-American Families: Diverse History, Contemporary Trends, and the Future. *In:* Coleman, M. and Ganong, L. H. (eds.) *Handbook of Contemporary Families*. Thousand Oaks, CA: Sage.

Jackman, M. R. and Jackman, R. W. 1980. Racial Inequalities in Home Ownership. *Social Forces*, 58, 1221–1233.

Kasinitz, P., Mollenkopf, J. H. and Waters, M. C. 2008. *Inheriting the City: The Children of Immigrants Come of Age*, Russell Sage Foundation.

Kochhar, R., Gonzalez-Barrera, A. and Dockterman, D. 2009. Through Boom and Bust: Minorities, Immigrants and Homeownership. *Pew Hispanic Center, May* 12, 2009.

Krivo, L. J. and Kaufman, R. L. 2004. Housing and Wealth Inequality: Racial-Ethnic Differences in Home Equity in the United States. *Demography*, 41, 585–605.

Kuebler, M. 2012. Lending in the Modern Era: Does Racial Composition of Neighborhoods Matter When Individuals Seek Home Financing? A Pilot Study in New England. *City & Community*, 11, 31–50.

Martin, L. L. 2009. Black asset ownership: Does ethnicity matter? *Social Science Research*, 38, 312–323.

Massey, D. S. and Denton, N. A. 1993. *American Apartheld: Segregation and the Making of the Underclass*, Cambridge, MA: Harvard University Press.

Myers, D., Painter, G., Yu, Z., Ryu, S. H. and Wei, L. 2005. Regional Disparities in Homeownership Trajectories: Impacts of Affordability, New Construction, and Immigration. *Housing policy debate*, 16, 53–83.

Oliver, M. L. and Shapiro, T. M. 2006. *Black wealth, white wealth: A new perspective on racial inequality*, Routledge.

Ruggles, S. J., Alexander, T., Genadek, K., Goeken, R., Schroeder, M. B. and Sobek, M. 2010. Integrated Public Use Microdata Series: Version 5.0 [Machine-readable database]. *In*: MINNESOTA, U. O. (ed.) Minneapolis.

Rugh, J. S. and Massey, D. S. 2010. Racial Segregation and the American Foreclosure Crisis. *American Sociological Review*, 75, 629–651.

Sampson, R. J. 2009. Racial Stratification and the Durable Tangle of Neighborhood Inequality. *Annals, AAPSS*, 621, 260–279.

Schelling, T. 1971. Dynamic Models of Segregation. *Journal of Mathematical Sociology*, 1, 143–86.

Squires, G. D. and Kubrin, C. E. 2006. *Privileged Places: Race, Residence and the Structure of Opportunity*, Boulder, Colorado, Lynn Rienner Publishers.

Stone, M. E. 2006. Housing Affordability: One-Third of a Nation Shelter-Poor. *In*: Bratt, R. G., Stone, M. E. and Hartman, C. (eds.) *A Right to Housing*, Philadelphia, PA: Temple.

Turner, M. A. and Ross, S. 2003. Discrimination in Metropolitan Housing Markets: Phase II-Asian and Pacific Islanders. *In*: INSTITUTE. H. U. (ed.).

Turner, M. A., Ross, S. L., Galster, G. C., Yinger, J., Godfrey, E. B., Bednarz, B. A., Herbig, C., Lee, S. J., Hossain, A. R. and Zhao, B. 2002. *Discrimination in Metropolitan Housing Markets: National Results from Phase I HDS 2000*. The Urban Institute Metropolitan Housing and Communities Policy Center.

Wachter, S. M. and Megbolugbe, I. F. 1992. Impacts of housing and mortgage market discrimination racial and ethnic disparities in homeownership. *Housing Policy Debate*, 3, 332–370.

Wyly, E. and Ponder, C. S. 2011. Gender, age, and race in subprime America. *Housing policy debate*, 21, 529–564.

14

The Intersection of Poverty Discourses

Race, Class, Culture, and Gender

DEBRA HENDERSON AND ANN TICKAMYER

...WELFARE RACISM AND CULTURALISM

In most instances, the images associated with poverty have become synonymous with those of African Americans, although other race and ethnic minorities may be similarly vulnerable. The media plays a role in racializing welfare policy, and the subsequent perceptions of broader society, by disproportionately presenting images of African Americans in missives on poverty (Avery & Peffley, 2003). However, the pejorative race/poverty connection can be linked to a history of influential racially biased policy analysis and subsequent welfare policy.

During the 1960s' war on poverty, Moynihan (1965) systematically linked poverty in the inner cities to the deterioration of the Black family. Rather than, address structural causes of poverty, he argued that economically disadvantaged Blacks were caught in a "tangle of pathology" and blamed inner-city poverty on a Black culture that embraced broken families, illegitimacy, and intergenerational dependency on welfare. Much of Moynihan's work was rejected by social scientists; however, current welfare reform policy rests on his and subsequent analysts' "culture of poverty" arguments that blame the poor for their own poverty (Murray, 1984). Proponents of welfare reform have argued that the "problem" with welfare lies not with broader structural constraints but with the inability of poor people to take personal responsibility for themselves and their families, resulting in denial of "traditional family values," lack of a strong work ethic, inability to delay personal gratification, and, ultimately, the long term dependency on public assistance (Reese, 2005; Hays, 2003).

Blaming welfare recipients for their poverty was part of a discourse imbued with racism that represented recipients, particularly African Americans, as welfare cheats and perpetuated racial stereotypes that produced controlling images such as the "welfare queen." As a result, the ghost of Moynihan's "tangle of pathology" view of the Black family was covertly reintroduced into the debate on

SOURCE: Henderson, Debra, and Ann Tickamyer. 2009. "The Intersection of Poverty Discourses: Race, Class, Culture, and Gender." Pp. 50–72 in *Emerging Intersections: Race, Class, and Gender in Theory, Policy, and Practice*, edited by B. Thornton Dill and R. E. Zambrana. New Brunswick, NJ: Rutgers University Press.

welfare reform. This discourse "relied on and reinforced racist views of people of color in general, and African Americans in particular" (Sparks, 2003, 178).

The role of racism in the development and implementation of welfare policy indicates that the social welfare system is not color-blind, but systematically discriminates against people of color (Schram, Soss, & Fording, 2003: Roberts, 2003; Piven, 2003; Dill, Baca Zinn, & Patton, 1998). Socially and politically we "whitewash poverty" by reporting that the majority of welfare recipients are White. However, due to racialized media presentations, the persistence of explanations founded in a culture of poverty, and the strength of race-based stereotypes such as "welfare queen," the face of welfare is Black (Dill, Baca Zinn, & Patton, 1998; Schram, 2003). More importantly, racism not only informs and molds public assistance policies, but it also results in outcomes that significantly impact the life chances of poor people of color.

By disregarding race-based prejudices and discrimination in broader society, people of color are placed at a disadvantage in terms of successfully leaving public assistance programs and obtaining self-sufficiency. Welfare reform policy does not take into account that racial discrimination impacts the likelihood that recipients of color will be able to secure employment that pays a living wage or obtain safe, affordable housing and childcare (Neubeck & Cazenave, 2001). A growing body of research reports the existence of "racial disparities" in the implementation and outcomes of welfare reform policy, as well as the treatment of Black recipients in comparison to White recipients (Schram, 2005, 253). For example, it appears that as the percentage of people of color increases there is a higher likelihood of stricter welfare reform restrictions and enforcement (Soss, et al., 2001; Fellowes & Rowe, 2004). Furthermore White recipients are more likely than Black recipients to successfully leave welfare (Finegold & Staveteig, 2002), to stay off of assistance once they have left (Loprest, 1999), and to have better options and referrals for education/training (Gooden, 2003).

Current discussions of welfare racism often devolve into analyses of the Black experience. However, other racial/ethnic minorities have also experienced the impact of welfare racism via the implementation and outcomes of welfare reform legislation. This is particularly apparent among immigrant groups seeking new opportunities in the United States and those minorities that live in isolated geographic locations. Empirical evidence suggests that the welfare system has been grounded in a "history of exclusion" and that welfare reform legislation works to further the disenfranchisement of other people of color in the United States (Fujiwara, 1998, 2005; Weinberg, 1998, 2000).

Prior to the implementation of welfare reform, growing anti-immigration sentiment responded to arguments that immigrants were taking jobs away from U.S. citizens (Fujiwara, 2005) and immigrant women were taking advantage of an overly indulgent welfare system (Fujiwara, 2005; Lindsley, 2002; Roberts, 1997) by restricting the access of legal immigrants to need based public assistance such as Supplemental Security Income, food stamps, and services provided by Temporary Aid to Needy Families (Fix & Zimmerman, 1995). While there has been a partial restoration of need based assistance to noncitizen immigrants, this mandate placed them at the center of a racially charged political debate and cast them as members of the undeserving poor, further bolstering anti-immigrant bias and stereotypes (Fujiwara, 2005).

Similarly the racial politics underlying welfare reform had a significant negative impact on the life chances of other racial/ethnic minorities. The limited research on poor Hispanic recipients in the United States suggests that their welfare experiences may be similar to those of African American recipients (Dill, Baca Zinn, & Patton, 1998; McPhee & Bronstein, 2003: Lee & Abrams, 2001; Weinberg, 1998). Furthermore the image of the "welfare queen" that has haunted African American women is also a prominent controlling image for poor Hispanic women (Briggs, 2002). Hispanics also face difficulties that are distinct to their racial/ethnic background when attempting to meet the mandates of welfare reform. For example, language barriers and conflicting cultural demands play an important role in the distribution of assistance, opportunities for further education, and successful entry into the labor market (Allegro, 2005; Bok, 2004; Briggs, 2002; Burnham, 2002; Cattan & Girard, 2004).

Other minorities residing in isolated communities have also been disproportionately impacted by welfare reform policy. This is particularly the case for reservation bound Native Americans. Given the extreme poverty on reservations, the high unemployment rates, and the lack of employment opportunities, tribal governments found themselves without the social capital or economic resources necessary to successfully implement welfare reform (Pandey, et al., 1999; Stromwall, et al., 1998). Yet, when recipients were unsuccessful in entering the labor force the stereotypical image of the "lazy, alcoholic Indian" was evoked as explanation for the inability to achieve personal responsibility.

In summary, welfare reform has consistently been tied to racial politics and has significantly impacted the life chances of racial/ethnic minorities across the country. Missing from this discussion is an analysis of how White poverty fits into the system of welfare racism. There is a need to examine White women's experiences with the welfare system, especially in racially homogeneous rural areas such as Appalachia where clear spatial inequalities exist (Tickamyer, et al., 2006) and cultural stereotypes that create stigma and controlling images (Billings, Norman, & Ledford, 1999). Appalachian scholarship suggests that the construction of an Appalachian identity and a rigid two-class system often parallels, and is as enduring, as race disadvantage (Duncan, 2000). Thus the construction of a rural Appalachian identity creates a dimension of difference that is denigrated by policy makers and represented as culturally inferior (Billings, Norman, & Ledford, 1999), which in turn has a material reality for poor women in the region (Duncan, 2000; Billings & Blee, 2000; Lichter & Jensen, 2002; Tickamyer, et al., 2004).

We synthesize theory and research on welfare racism produced by "intersectional analysis" (Dill, 2002; Dill, Jones-Deweever, & Schram, 2004) with our empirical research on Appalachian poverty (Tickamyer, et al., 2006, 2000) to show that the facade of a universal social welfare system and policies disguises discourses that reinforce the real disadvantages that different race and ethnic groups experience, including White Appalachian women. Appalachian women, as a cultural minority and a segment of the lower socioeconomic class, are twice exploited by the stereotypical images of welfare recipients. First, in isolated rural Appalachia the stereotypical images of White poverty—"hillbilly" and "White trash"—and belief in a culture of poverty serve to define poverty in the area and foster hostility from working- and middle-class Whites. These images influence economic development within the region with impacts on employment opportunities. At an individual level they stigmatize recipients, making it difficult for women to leave welfare and enter the workforce. Second, at the same time that poor Appalachian women are denigrated as yet another example of the deficiencies attributed to welfare dependency, their Whiteness becomes the means to argue that the welfare system is bias-free regarding race. Thus these stereotypical images serve to control the opportunities of both White Appalachian women and women of color as they attempt to negotiate the welfare system.

The existence of rural White poverty reinforces the claim that public policy is a universal that does not discriminate by gender, race, or any other salient category. The realities are that these dimensions of difference are reinforced by the pretense that poverty policy is unbiased and gender/race neutral. Poverty policy is neither culturally nor gender/race neutral, but the existence of pockets of poor White women in rural Appalachia provides a convenient way for policy makers to argue that neutrality exists with detrimental outcomes for all poor women....

APPALACHIAN CULTURE AND CONTROLLING IMAGES

The Appalachian region is one of the poorest geographic locations in the nation (Billings & Blee, 2000; The State of Poverty in Ohio, 2004), and lags other regions in economic development, employment rates, and per capita income (Appalachian Regional Commission, 2005). Rural Appalachia has a long history of high unemployment and persistent, severe poverty that is equal to, or worse than, that in urban areas (Rural Sociological Task Force on Persistent Rural Poverty, 1993; Lichter & Jensen, 2002). While those receiving public assistance in both rural and urban areas may confront similar problems in terms of making ends meet and complying with the mandates of welfare reform, empirical evidence indicates that the constraints experienced by welfare recipients in rural areas are distinct from those in urban areas (U.S. Department of Health and Human Services, 2002).

Rural communities are often perceived by broader society as idyllic settings that offer a stable, wholesome environment in which to live and raise a family (Brown & Swanson, 2003; Logan, 1996; National Rural Electric Cooperative Association, 1992; Seebach, 1992). However, these views rarely acknowledge the true nature or conditions of the region (Logan, 1996; Brown & Swanson, 2003). The extreme level of spatial inequality in rural Appalachia, particularly in isolated communities, has a significant impact on the ability of poor people to leave welfare and successfully enter the labor market (Tickamyer, et al., 2006; Zimmerman & Hirschi, 2003). Leaving welfare and achieving self-sufficiency through employment is hindered by the lack of job opportunities and the social/human capital required to facilitate the transition (Parisi, et al., 2003; Tickamyer, et al., 2006; Weber, Duncan, & Whitener, 2002). Unlike their urban counterparts, those in these isolated rural communities often confront an absolute lack of resources such as economic assets, childcare, transportation, healthcare, and housing (Henderson, et al., 2002; Rural Policy Research Institute, 1999; Zimmerman & Hirschi, 2003). Consequently, in comparison to those in urban areas, poor women living in rural communities often encounter more hardship (Brown & Lichter, 2004; Snyder & McLaughlin, 2004)….

… What are the implications for poverty and welfare policy? First, is the obvious point that political action and sound policies and programs to benefit the poor are never an easy sell, but they are made virtually impossible when successful divide and conquer strategies are applied to both the poor themselves and to how other citizens view them. Convincing the majority of voters that poverty is someone else's problem, that it is particularly a problem of a racial "other," and that it is primarily a problem created by individual and collective group failure, rather than the outcome of structural impediments and system failure has been a successful strategy for advancing the agenda of those whose goal has been to radically alter public service provision.

… The question remains of how practically this can happen? How can we overcome the scripts that gender, race, class, and culture create? Here we come full circle. Guinier and Torres (2002) argue that race is the "canary in the coal mine," the first warning that the atmosphere is too poisonous to support life. What is fatal to the canary ultimately will harm all who breathe the poisoned air. They further argue that forms of resistance forged in racial struggle can play a leading role in seeking progressive social change, but this will require recognizing racial difference and harnessing the knowledge this produces. They see the possibility of transcending racial barriers not by pretending they do not exist, but by recognizing the costs, realizing that what is bad for the canary ultimately touches everyone, and using this knowledge as the means to organize for change….

REFERENCES

Allegro, L. (2005). Welfare Use and Political Response: Urban Narratives From First and Second-Generation Puerto Ricans and Dominicans in New York City. *Centro Journal, 17*(1), 221–241.

Appalachian Regional Commission. (2005). Retrieved April 14, 2005, from www.arc.gov/index.jsp.

Avery, J. M., & Peffley, M. (2003). Race Matters: The Impact of News Coverage of Welfare Reform on Public Policy. In S. Schram, J. Soss, & R. C. Fording (Eds.). *Race and the Politics of Welfare Reform* (pp. 131–150). Ann Arbor: University of Michigan Press.

Billings, D. B., & Blee, K. M. (2000). *The Road to Poverty: The Making of Wealth and Hardship in Appalachia.* Cambridge: Cambridge University Press.

Billings, D. B., Norman, G., & Ledford, K. (1999). *Back Talk from Appalachia: Confronting Stereotypes.* Lexington: University of Kentucky Press.

Bok, M. (2004). Education and Training for Low-Income Women: An Elusive Goal. *Affilia, 19*(1), 39–52.

Briggs, L. (2002). La Vida, Moynihan, and Other Libels: Migration, Social Science, and the Making of the Puerto Rican Welfare Queen. *Centro Journal, 14*(1), 75–101.

Brown, D. L., & Swanson, L. E. (2003). Introduction: Rural America Enters the New Millennium. In D. L. Brown & L. E. Swanson (Eds.), *Challenges for Rural America in the Twenty-First Century* (pp. 1–15). University Park: Pennsylvania State University Press.

Brown, J. B., & Lichter, D. (2004). Poverty, Welfare, and the Livelihood Strategies of Non-Metropolitan Single Mothers. *Rural Sociology, 63*(2), 282–301.

Burnham, L. (2002). Welfare Reform, Family Hardship, and Women of Color. In R. Albelda & K. Byron (Eds.), *Losing Ground* (pp. 43–56). Boston: South End Press.

Cattan, P., & Girard, C. (2004). Recent Cuban Immigrants and Native-Born African Americans Leaving Welfare. *Hispanic Journal of Behavioral Sciences, 26*(3), 312–332.

Dill, B. T. (2002). Intersections, Identities & Inequalities in Higher Education. *Robin Williams Jr. Lecture.* Eastern Sociological Society.

Dill, B. T., Baca Zinn, M., & Patton, S. (1998). Valuing Families Differently: Race, Poverty and Welfare Reform. *Sage Race Relations Abstracts, 23*(3), 4–30.

Dill, B. T., Jones-Deweever, A., & Schram, S. (2004). Racial, Ethnic, and Gender Disparities in Access to Jobs, Education and Training under Welfare Reform. *Research and Action Briefs 1.* College Park, MD: The Consortium on Race, Gender and Ethnicity.

Duncan, C. M. (2000). *Worlds Apart Why Poverty Persists in Rural America.* New Haven, CT: Yale University Press.

Fellowes, M. C., & Rowe, G. (2004). Politics and the New American Welfare State. *American Journal of Political Science, 48*(2), 362–373.

Finegold, K., & Staveteig, S. (2002). Race, Ethnicity, and Welfare Reform. In A. Well & K. Finegold (Eds.), *Welfare Reform: The Next Act* (pp. 203–244). Washington, DC: Urban Institute Press.

Fix, M., & Zimmerman, W. (1995). When Should Immigrants Receive Public Benefits? In I. V. Sawhill (Ed.), *Welfare Reform: An Analysis of the Issues* (ch. 13). Retrieved April 30, 2005, from http://www.urban.org/url.cml?lD=306620.

Fujiwara, I. H. (1998). The Impact of Welfare Reform on Asian Immigrant Communities, *Social Justice, 25*(1), 82–104.

Fujiwara, L. H. (2005). Immigrant Rights are Human Rights: The Reframing of Immigrant Entitlement and Welfare. *Social Problems, 52*(1), 79–101.

Gooden, S. (2003). Contemporary Approaches to Enduring Challenges: Using Performance Measures to Promote Racial Equality under TANF. In S. Schram, J. Soss, & R. C. Fording (Eds.), *Race and the Politics of Welfare Reform* (pp. 254–275). Ann Arbor University of Michigan Press.

Guinier, L., & Torres, G. (2002). *The Miner's Canary: Enlisting Race Resisting Power, Transforming Democracy.* Cambridge, MA: Harvard University Press.

Hays, S. (2003). *Flat Broke with Children: Women in the Age of Welfare Reform.* New York: Oxford University Press.

Lee, S. Z., & Abrums, L. S. (2001). Challenging Depictions of Dependency: TANF Recipients React to Welfare Reform. *Journal of Poverty, 5*(1), 91–111.

Lichter, D., & Jensen, L. (2002). Rural America in Transition: Poverty and Welfare at the Turn of the Twenty-First Century. In B. Weber, G. Duncan, & L Whitener (Eds.), *Rural Dimensions of Welfare Reform* (pp. 77–110). Kalamazoo, MI: Upjohn.

Lindsley, S. (2002). The Gendered Assault on Immigrants. In J. Silliman & A. Bhattacharjee (Eds.), *Policing the National Body: Race, Gender and Criminalization* (pp. 175–196). Boston: South End Press.

Logan, J. R. (1996). Rural America as a Symbol of American Values. *Rural Development Perspectives, 12*(1), 19–21.

Loprest, P. (1999). How Families That Left Welfare are Doing: A National Picture. *Assessing the New Federalism, B-01.* Washington, DC: Urban Institute.

McPhee, D. M., & Bronstein, L. R. (2003). The Journey From Welfare to Work: Learning From Women Living in Poverty. *Affilia., 18*(1), 34–48.

Moynihan, D. P. (1965). The Negro Family: The Case for National Action. *Office of Family Planning and Research, Department of Labor.* Washington, DC.

Murray, C. (1984). *Losing Ground.* New York: Basic Books.

National Rural Electric Cooperative Association. (1992). Public Attitudes toward Rural America. CD3–46. Washington, DC: NRECA Public Relations Division.

Neubeck, K. J., & Cazenave, N. A., (2001). *Welfare Racism: Playing the Race Card against America's Poor.* New York: Routledge Press.

Pandey, S., Brown, E. F., Scheuler-Whitaker, L., Gundersen, B., & Eyrich. K. (1999). Promise of Welfare Reform, Development through Devolution on Indian Reservations. *Journal of Poverty, 3*(4), 37–61.

Parisi, D., McLaughlin, D., Grice, S., Taquino, M., & Gill, D. (2003). TANF Participation Rates: Do Community Conditions Matter? *Rural Sociology, 68*(4), 491–512.

Piven, F. F. (2003). Why Welfare is Racist. In S. Schram, J. Soss, & R. C. Fording (Eds.), *Race and the Politics of Welfare Reform* (pp. 323–335). Ann Arbor: University of Michigan Press.

Reese, E. (2005). *Backlash against Welfare Mothers Past and Present.* Berkeley: University of California Press.

Roberts, D. (1997). Who May Give Birth to Citizens? Reproduction, Eugenics, and immigration. In J. F. Pera (Ed.), *Immigrants Out: The New Nativism and the Anti-Immigrant Impulse in the United States* (pp. 205–219). New York: New York University Press.

Roberts, D. (2003). *Shattered Bonds: The Color of Child Welfare.* New York: Basic Books.

Rural Policy Research Institute (RUPRI). (1999). *Rural America and Welfare Reform: An Overview Assessment.* Policy paper pp. 99–103, February 10, Rural Welfare Reform Initiative. Retrieved April 24, 2005, from www.rupri.org/publications/.

Rural Sociological Task Force on Persistent Rural Poverty. (1993). *Poverty in Rural America.* Boulder, CO: Westview Press.

Schram, S. (2003). Putting a Black Face on Welfare: The Good and the Bad. In S. Schram, J. Soss, and R. C. Fording (Eds.), *Race and the Politics of Welfare Reform* (pp. 195–221). Ann Arbor: University of Michigan Press.

Schram, S. (2005, June). Contextualizing Racial Disparities in American Welfare Reform: Toward a New Poverty Research. *Perspectives on Politics, 3*(2), 253–268.

Schram, S., Soss, J., & Fording, R. C. (2003). *Race and the Politics of Welfare Reform*. Ann Arbor: University of Michigan Press.

Seebach, M. (1992). Small Towns Have a Rosy Image. *American Demographics, 14*(10), 19.

Snyder, A., & Mclaughlin, D. (2004). Female-Headed Families and Poverty in Rural America. *Rural Sociology, 69*(1), 127–149.

Soss, J., Schram, S. F., Vartanian, T. P., & O'Brien, E. (2001). The Hard Line and the Color Line: Race, Welfare, and the Roots of Get-Tough Reform. In S. Schram, J. Soss, & R. C., Fording (Eds.), *Race and the Politics of Welfare Reform* (pp. 225–249). Ann Arbor: University of Michigan Press.

Sparks, H. (2003). Queens, Teens, and Model Mothers: Race, Gender, and the Discourse of Welfare Reform. In S. Schram, J. Soss, & R. C. Fording (Eds.), *Race and the Politics of Welfare Reform* (pp. 171–195). Ann Arbor: University of Michigan Press.

Stromwall, L. K., Brzuzy, S., Sharp, P., & Anderson, C. (1998). The Implications of "Welfare Reform" for American Indian Families and Communities. *Journal of Poverty, 2*(4), 1–15.

The State of Poverty in Ohio. (2004). Ohio Association of Community Action Agencies. *Community Action Partnership*. Retrieved August 21, 2005, from www.ceoge.org/research/contentonly.pdf.

Tickamyer, A., Henderson, D., Tadlock, B., & White, J. (2004). *The Impact of Welfare Reform on Rural Livelihood Practices*. Presented at the Rural Sociological Society Annual Meetings. Sacramento. CA.

Tickamyer, A. R., Henderson, D., White, J. A., & Tadlock, B. (2000). Voices of Welfare Reform: Bureaucratic Rationality Versus the Perceptions of Welfare Participants. *Affilia: Journal of Women and Social Work, 15*(2), 173–192.

Tickamyer, A. R., White, J., Tadlock, B., & Henderson, D. A. (2006). The Spatial Politics of Public Policy: Devolution, Development and Welfare Reform. In L. Lobeo, G. Hooks, & A. Tickamyer (Eds.), *Spatial Inequality* (pp. 113–139). New York: SUNY Press.

U.S. Department of Health and Human Services [USDHHS]. (2002). One Department Serving Rural America" HHS Rural Task Force Report to the Secretary. Retrieved from ftp://ftp.hrsa.gov/ruralhealth/PublicReportJune2002.pdf.

Weber, B. A., Duncan, G. J., & Whitener, L A. (2002). *Rural Dimensions of Welfare Reform*. Kalamazo, MI: UpJohn Institute.

Weber, L. (2001). *Understanding Race, Class, Gender, and Sexuality*. Boston: McGraw Hill.

Weinberg, S. B. (1998). Mexican American Mothers and the Welfare Debate: A History of Exclusion. *Journal of Poverty, 2*(3), 53–75.

Weinberg, S. B. (2000). Welfare Reform and Mutual Family Support Effects on Mother-Led Mexican American Families. *Affilia: Journal of Women and Social Work, 15*(2), 204–223.

Zimmerman, J. N., & Hirshl, T. A. (2003). Welfare Reform in Rural Areas: A Voyage Through Uncharted Waters. In D. L. Brown & L. E. Swanson (Eds.), *Challenges for Rural America in the Twenty-First Century* (pp. 363–374). University Park: Pennsylvania State University Press.

15

Health and Wealth

Our Appalling Health Inequality Reflects and Reinforces Society's Other Gaps

LAWRENCE R. JACOBS AND JAMES A. MORONE

A look at Americans' health reveals the astonishing inequalities in our society. American girls are born with a life expectancy that ranks 19th in the world (in another survey they fall to 28th). Male babies rank 31st—in a dead tie with Brunei. Among the 13 wealthiest countries, the United States ranks last or nearly so in almost every way we measure health: infant mortality, low birth weight, life expectancy at birth, life expectancy for infants. The average American boy lives three and a half fewer years than the average Japanese baby, despite higher rates of cigarette smoking in Japan. The American adolescent death rate is twice as high as, say, England's.

These dismal American averages mask vast differences across our population. A male born in some sections of Washington, D.C., for example, has a life expectancy 40 years lower than a woman born in many wealthy neighborhoods. In short, great differences in wealth match up to—indeed, they create—terrible differences in health.

Why do Americans come out so badly in the cross-national health statistics? Why are our infants more likely to die than those in, say, Croatia? Our health troubles have three interrelated causes: inequality, poverty, and the way we organize our health-care system.

Let's start with inequality. A famous study of the British civil service found that with each rung up the ladder of success, people suffered fewer fatal heart attacks—the clerks and messengers at the bottom were four times more likely to die than the executives at the top. Researchers following up this study reached a surprising conclusion that seems to hold up in one nation after another: The wider the inequality, the worse the nation's overall health.

Why should this be so? For one thing, falling behind in the race to make ends meet generates stress and physiological harm—the results are depression, hypertension, other illnesses, and high mortality rates. In addition, the middle-class scramble to get ahead erodes neighborly feelings, frays our communities, and lowers trust in institutions like churches and governments. All of these are

SOURCE: Jacobs, Lawrence R., and James A. Morone. 2004. "Health and Wealth: Our Appalling Health Inequality Reflects and Reinforces Society's Other Gaps." *The American Prospect* (May 17) Volume 15, Issue 6. Reprinted by permission.

factors in other countries. But most industrial nations buffer their citizens against economic uncertainty and lost jobs. In the United States, only the market winners get security.

Of course, American health problems go beyond inequality and are closely correlated with the poverty in which more than one in 10 Americans now live. Of our 34.6 million "poor" citizens, according to the U.S. Census Bureau, more than 14 million are "severely poor," meaning they don't even make it halfway to the federal poverty line. The numbers are worse for minorities, with nearly a quarter of blacks and more than a fifth of Hispanics living in poverty.

And poverty brings troubles like hunger (33 million Americans live with "food insecurity," as defined by the Department of Agriculture) and homelessness (perhaps as many as 3.5 million a year), which disproportionately fall on kids. Poor neighborhoods face high crime, inferior schools, few good jobs, and inadequate health-care facilities. Instead, poverty attracts danger—too much alcohol and tobacco, illegal drugs, and fast foods. One observer after another has gone off to study poor communities and come back with the same report. The lives of the poor are full of stress and the struggle to get by.

People die younger in Harlem than in Bangladesh. Why? It is not what most people think—homicide, drug abuse, and AIDS are far down the list. Rather, as *The New England Journal of Medicine* reports, the leading causes of death in poor black neighborhoods are "unrelenting stress," "cardiovascular disease," "cancer," and "untreated medical conditions."

Finally, beyond the fundamentals—inequality and poverty—there is that stubborn American policy dilemma: No other industrial nation tolerates such yawning gaps in health insurance. According to the Congressional Budget Office, 43.6 million people were uninsured in 2002, with 19.9 million coming from the ranks of full-time workers; 74.7 million Americans under 65 were without health insurance for all or part of 2001 and 2002. Part of the problem is that workplace coverage is unraveling as more employers shift costs like premiums, co-payments, and coverage limitations onto their workers. Meanwhile, medical costs are rising faster than personal-income growth.

Simple medical care—annual check-ups, screenings, vaccinations, eyeglasses, dentistry—saves lives, improves well-being, and is shockingly uneven. Well-insured people get assigned hospital beds; the uninsured get patched up and sent back to the streets. From diagnostic procedures—prostate screenings, mammograms, and Pap smears—to treatment for asthma, the uninsured get less care, they get it later in their illnesses, and they are roughly three times more likely to have an adverse health outcome. The Institute of Medicine recently blamed gaps in insurance coverage for 17,000 preventable deaths a year.

Even middle-class parents worry about the next medical emergency or, in many cases, the routine trip to the doctor's office. Life without health insurance means constantly measuring aches and fevers against the next payday. Changing jobs brings a new set of anxieties about shifts in medical coverage. Health bills are the largest cause of personal bankruptcy in the United States.

Of course, no health-care system treats everyone the same way. But in America, our disparities are unusually wide and deep.

How can we reverse these trends and begin to build the good society? Recent experience counsels incremental reform that builds on past successes while pushing bold new proposals for the future.

As recent history shows, even half steps—like adding amendments to bipartisan legislation—can add up to something important. Back when the Reagan administration was attacking poverty programs while cutting taxes and running up enormous deficits, California Congressman Henry Waxman oversaw bipartisan support for a series of minor expansions in Medicaid eligibility. The result: In the late 1980s, the program grew to cover an additional 5 million children and 500,000 pregnant women.

While Bill Clinton's failure to pass national health insurance got most of the press, his administration quietly enacted the Children's Health Insurance Program for states in 1997. Using federal matching funds as a prod, the program pushed states to widen coverage to uninsured children, helping Medicaid reach 20 million kids by 2000 and funding non-Medicaid programs to cover an additional 2 million.

Even further below the national radar screen, the Robert Wood Johnson Foundation induced state governments to place health-care clinics directly in schools. Families in underserved neighborhoods suddenly—and usually for the first time—found it easy for their kids to get into a physician's office. Despite strong initial opposition from the cultural right over birth control, teachers, public-health advocates, parents, and community organizers have managed to open 1,498 school centers from Maine to California.

Reforms beyond medical care can also improve general living conditions and boost American health. The Earned Income Tax Credit, for example, has lifted millions of low-income workers and their children out of poverty. To be sure, making Americans healthy means addressing the economic insecurity that threatens these struggling families, forcing middle-class Americans to work double shifts and the poor to confront hunger and homelessness.

Making Americans healthy also means casting off the political torpor of this new Gilded Age and reclaiming a long-standing commitment to our neighbors and communities. Only great aspirations will galvanize a new populist politics and leverage our reluctant state.

There is not much mystery about what works. Other industrial countries rely on three familiar paths to good health. First, government plays an important role through such policies as family and housing allowances, universal health care, pensions, and tax credits. The generous welfare states of northern Europe and nations with more modest programs like France, Germany, and Canada all have poor, middle-class, and wealthy populations. However, all these nations achieve much narrower income gaps among groups than now exist in the United States.

A second type of policy fosters opportunity. Governments invest in education to expand the supply of skilled labor and help workers help themselves. Lowering the barriers to college education and worker retraining reduces the

high premium for skilled labor. In addition, European governments collaborate with businesses by regularly adjusting the minimum wage and overseeing the negotiations between business and labor.

Finally, most wealthy nations maintain taxes. The new global economy was expected to spark dramatic tax cuts as governments competed with one another to create an attractive business climate and lure investment and skilled labor. In Europe and Canada, international pressures did not eviscerate the government's capacity to raise revenues. Instead, domestic support to maintain programs (and international pressure to limit deficits) barred governments from plunging into tax-cut wars.

In short, America's allies have tried to defend all their citizens from the worst effects of a global economy. The results across the industrial world are powerful: Policies that moderate income disparities turn out to be good for your health.

American public policy, has, on balance, gone the other way: Tax cuts, deregulation, and unmediated markets sabotage our incremental stabs at fostering real opportunity. Some individuals have grown fantastically wealthy; most struggle to make ends meet. The dirty policy secret lies in the health consequences: Our population suffers more illness and dies younger.

Our call to reform is simple: A civilized society should not accept gaping disparities in life and death, health and disability. Americans are too generous and fair-minded to tolerate so much preventable suffering. This moral vision undergirds a hardheaded analysis of the rapidly changing global economy that has reshuffled the distribution of money in American society and unsettled the life circumstances that nurture and protect the health of the country. The solutions are no mystery. Other nations successfully protect their people. So can we.

16

"Is This a White Country, or What?"

LILLIAN B. RUBIN

"They're letting all these coloreds come in and soon there won't be any place left for white people," broods Tim Walsh, a thirty-three-year-old white construction worker. "It makes you wonder: Is this a white country, or what?"

It's a question that nags at white America, one perhaps that's articulated most often and most clearly by the men and women of the working class. For it's they who feel most vulnerable, who have suffered the economic contractions of recent decades most keenly, who see the new immigrants most clearly as direct competitors for their jobs.

It's not whites alone who stew about immigrants. Native-born blacks, too, fear the newcomers nearly as much as whites—and for the same economic reasons. But for whites the issue is compounded by race, by the fact that the newcomers are primarily people of color. For them, therefore, their economic anxieties have combined with the changing face of America to create a profound uneasiness about immigration—a theme that was sounded by nearly 90 percent of the whites I met, even by those who are themselves first-generation, albeit well-assimilated, immigrants.

Sometimes they spoke about this in response to my questions; equally often the subject of immigration arose spontaneously as people gave voice to their concerns. But because the new immigrants are dominantly people of color, the discourse was almost always cast in terms of race as well as immigration, with the talk slipping from immigration to race and back again as if these are not two separate phenomena. "If we keep letting all them foreigners in, pretty soon there'll be more of them than us and then what will this country be like?" Tim's wife, Mary Anne, frets. I mean, this is *our* country, but the way things are going, white people will be the minority in our own country. Now does that make any sense?"

Such fears are not new. Americans have always worried about the strangers who came to our shores, fearing that they would corrupt our society, dilute our culture, debase our values. So I remind Mary Anne, "When your ancestors came here, people also thought we were allowing too many foreigners into the country. Yet those earlier immigrants were successfully integrated into the American society. What's different now?"

SOURCE: Rubin, Lillian B. 1994. *Families on the Fault Line.* pp. 172–196. New York: HarperCollins Publishers.

"Oh, it's different, all right," she replies without hesitation. "When my people came, the immigrants were all white. That makes a big difference."...

Listening to Mary Anne's words I was reminded again how little we Americans look to history for its lessons, how impoverished is our historical memory.

For, in fact, being white didn't make "a big difference" for many of those earlier immigrants. The dark-skinned Italians and the eastern European Jews who came in the late nineteenth and early twentieth centuries didn't look very white to the fair-skinned Americans who were here then. Indeed, the same people we now call white—Italians, Jews, Irish—were seen as another race at that time. Not black or Asian, it's true, but an alien other, a race apart, although one that didn't have a clearly defined name. Moreover, the racist fears and fantasies of native-born Americans were far less contained then than they are now, largely because there were few social constraints on their expression.

When, during the nineteenth century, for example, some Italians were taken for blacks and lynched in the South, the incidents passed virtually unnoticed. And if Mary Anne and Tim Walsh, both of Irish ancestry, had come to this country during the great Irish immigration of that period, they would have found themselves defined as an inferior race and described with the same language that was used to characterize blacks: "low-browed and savage, grovelling and bestial, lazy and wild, simian and sensual."[1] Not only during that period but for a long time afterward as well, the U.S. Census Bureau counted the Irish as a distinct and separate group, much as it does today with the category it labels "Hispanic."

But there are two important differences between then and now, differences that can be summed up in a few words: the economy and race. Then, a growing industrial economy meant that there were plenty of jobs for both immigrant and native workers, something that can't be said for the contracting economy in which we live today. True, the arrival of the immigrants, who were more readily exploitable than native workers, put Americans at a disadvantage and created discord between the two groups. Nevertheless, work was available for both.

Then, too, the immigrants—no matter how they were labeled, no matter how reviled they may have been—were ultimately assimilable, if for no other reason than that they were white. As they began to lose their alien ways, it became possible for Native Americans to see in the white ethnics of yesteryear a reflection of themselves. Once this shift in perception occurred, it was possible for the nation to incorporate them, to take them in, chew them up, digest them, and spit them out as Americans—with subcultural variations not always to the liking of those who hoped to control the manners and mores of the day, to be sure, but still recognizably white Americans.

Today's immigrants, however, are the racial other in a deep and profound way.... And integrating masses of people of color into a society where race consciousness lies at the very heart of our central nervous system raises a whole new set of anxieties and tensions....

The increased visibility of other racial groups has focused whites more self-consciously than ever on their own racial identification. Until the new immigration shifted the complexion of the land so perceptibly, whites didn't think of themselves as white in the same way that Chinese know they're Chinese and African-Americans know they're black. Being white was simply a fact of life, one that didn't require any public statement, since it was the definitive social value against which all others were measured. "It's like everything's changed and I don't know what happened," complains Marianne Bardolino. "All of a sudden you have to be thinking all the time about these race things. I don't remember growing up thinking about being white like I think about it now. I'm not saying I didn't know there was coloreds and whites; it's just that I didn't go along thinking, *Gee, I'm a white person.* I never thought about it at all. But now with all the different colored people around, you have to think about it because they're thinking about it all the time."

"You say you feel pushed now to think about being white, but I'm not sure I understand why. What's changed?" I ask.

"I told you," she replies quickly, a small smile covering her impatience with my question. "It's because they think about what they are, and they want things their way, so now I have to think about what I am and what's good for me and my kids." She pauses briefly to let her thoughts catch up with her tongue, then continues. "I mean, if somebody's always yelling at you about being black or Asian or something, then it makes you think about being white. Like, they want the kids in school to learn about their culture, so then I think about being white and being Italian and say: What about my culture? If they're going to teach about theirs, what about mine?"

To which America's racial minorities respond with bewilderment. "I don't understand what white people want," says Gwen Tomalson. "They say if black kids are going to learn about black culture in school, then white people want their kids to learn about white culture. I don't get it. What do they think kids have been learning about all these years? It's all about white people and how they live and what they accomplished. When I was in school you wouldn't have thought black people existed for all our books ever said about us."

As for the charge that they're "thinking about race all the time," as Marianne Bardolino complains, people of color insist that they're forced into it by a white world that never lets them forget. "If you're Chinese, you can't forget it, even if you want to, because there's always something that reminds you," Carol Kwan's husband, Andrew, remarks tartly. "I mean, if Chinese kids get good grades and get into the university, everybody's worried and you read about it in the papers."

While there's little doubt that racial anxieties are at the center of white concerns, our historic nativism also plays a part in escalating white alarm. The new immigrants bring with them a language and an ethnic culture that's vividly expressed wherever they congregate. And it's this also, the constant reminder of an alien presence from which whites are excluded, that's so troublesome to them.

The nativist impulse isn't, of course, given to the white working class alone. But for those in the upper reaches of the class and status hierarchy—those whose children go to private schools, whose closest contact with public transportation is the taxi cab—the immigrant population supplies a source of cheap labor, whether as nannies for their children, maids in their households, or workers in their businesses. They may grouse and complain that "nobody speaks English anymore," just as working-class people do. But for the people who use immigrant labor, legal or illegal, there's a payoff for the inconvenience—a payoff that doesn't exist for the families in this study but that sometimes costs them dearly. For while it may be true that American workers aren't eager for many of the jobs immigrants are willing to take, it's also true that the presence of a large immigrant population—especially those who come from developing countries where living standards are far below our own—helps to make these jobs undesirable by keeping wages depressed well below what most American workers are willing to accept....

It's not surprising, therefore, that working-class women and men speak so angrily about the recent influx of immigrants. They not only see their jobs and their way of life threatened, they feel bruised and assaulted by an environment that seems suddenly to have turned color and in which they feel like strangers in their own land. So they chafe and complain: "They come here to take advantage of us, but they don't really want to learn our ways," Beverly Sowell, a thirty-three-year old white electronics assembler, grumbles irritably. "They live different than us; it's like another world how they live. And they're so clannish. They keep to themselves, and they don't even *try* to learn English. You go on the bus these days and you might as well be in a foreign country; everybody's talking some other language, you know, Chinese or Spanish or something. Lots of them have been here a long time, too, but they don't care; they just want to take what they can get."

But their complaints reveal an interesting paradox, an illuminating glimpse into the contradictions that beset native-born Americans in their relations with those who seek refuge here. On the one hand, they scorn the immigrants; on the other, they protest because they "keep to themselves." It's the same contradiction that dominates black-white relations. Whites refuse to integrate blacks but are outraged when they stop knocking at the door, when they move to sustain the separation on their own terms' [*sic*]—in black theme houses on campuses, for example, or in the newly developing black middle-class suburbs.

I wondered, as I listened to Beverly Sowell and others like her, why the same people who find the life ways and languages of our foreign-born population offensive also care whether they "keep to themselves."

"Because like I said, they just shouldn't, that's all," Beverly says stubbornly. "If they're going to come here, they should be willing to learn our ways—you know what I mean, be real Americans. That's what my grandparents did, and that's what they should do."

"But your grandparents probably lived in an immigrant neighborhood when they first came here, too," I remind her.

"It was different," she insists. "I don't know why; it was. They wanted to be Americans; these here people now, I don't think they do. They just want to take advantage of this country["]....

"Everything's changed, and it doesn't make sense. Maybe you get it, but I don't. We can't take care of our own people and we keep bringing more and more foreigners in. Look at all the homeless. Why do we need more people here when our own people haven't got a place to sleep?"

"Why do we need more people here?"—a question Americans have asked for two centuries now. Historically, efforts to curb immigration have come during economic downturns, which suggests that when times are good, when American workers feel confident about their future, they're likely to be more generous in sharing their good fortune with foreigners. But when the economy falters, as it did in the 1990s, and workers worry about having to compete for jobs with people whose standard of living is well below their own, resistance to immigration rises. "Don't get me wrong; I've got nothing against these people," Tim Walsh demurs. "But they don't talk English, and they're used to a lot less, so they can work for less money than guys like me can. I see it all the time; they get hired and some white guy gets left out."

It's this confluence of forces—the racial and cultural diversity of our new immigrant population; the claims on the resources of the nation now being made by those minorities who, for generations, have called America their home; the failure of some of our basic institutions to serve the needs of our people; the contracting economy, which threatens the mobility aspirations of working-class families—all these have come together to leave white workers feeling as if everyone else is getting a piece of the action while they get nothing. "I feel like white people are left out in the cold," protests Diane Johnson, a twenty-eight-year-old white single mother who believes she lost a job as a bus driver to a black woman. "First it's the blacks; now it's all those other colored people, and it's like everything always goes their way. It seems like a white person doesn't have a chance anymore. It's like the squeaky wheel gets the grease, and they've been squeaking and we haven't," she concludes angrily.

Until recently, whites didn't need to think about having to "squeak"—at least not specifically as whites. They have, of course, organized and squeaked at various times in the past—sometimes as ethnic groups, sometimes as workers. But not as whites. As whites they have been the dominant group, the favored ones, the ones who could count on getting the job when people of color could not. Now suddenly there are others—not just individual others but identifiable groups, people who share a history, a language, a culture, even a color—who lay claim to some of the rights and privileges that formerly had been labeled for whites only." And whites react as if they've been betrayed, as if a sacred promise has been broken. They're white, aren't they? They're *real* Americans, aren't they? This is their country, isn't it?

The answers to these questions used to be relatively unambiguous. But not anymore. Being white no longer automatically assures dominance in the politics of a multiracial society. Ethnic group politics, however, has a long and fruitful history. As whites sought a social and political base on which to stand, therefore, it was natural and logical to reach back to their ethnic past. Then they, too, could be "something"; they also would belong to a group; they would have a name, a history, a culture, and a voice. "Why is it only the blacks or Mexicans or Jews that are 'something'?" asks Tim Walsh. "I'm Irish, isn't that something, too? Why doesn't that count?"

In reclaiming their ethnic roots, whites can recount with pride the tribulations and transcendence of their ancestors and insist that others take their place in the line from which they have only recently come. "My people had a rough time, too. But nobody gave us anything, so why do we owe them something? Let them pull their share like the rest of us had to do," says Al Riccardi, a twenty-nine-year-old white taxi driver.

From there it's only a short step to the conviction that those who don't progress up that line are hampered by nothing more than their own inadequacies or, worse yet, by their unwillingness to take advantage of the opportunities offered them. "Those people, they're hollering all the time about discrimination," Al continues, without defining who "those people" are. "Maybe once a long time ago that was true, but not now. The problem is that a lot of those people are lazy. There's plenty of opportunities, but you've got to be willing to work hard."

He stops a moment, as if listening to his own words, then continues, "Yeah, yeah, I know there's a recession on and lots of people don't have jobs. But it's different with some of those people. They don't really want to work, because if they did, there wouldn't be so many of them selling drugs and getting in all kinds of trouble."

"You keep talking about 'those people' without saying who you mean," I remark.

"Aw c'mon, you know who I'm talking about," he says, his body shifting uneasily in his chair. "It's mostly the black people, but the Spanish ones, too."

In reality, however, it's a no-win situation for America's people of color, whether immigrant or native born. For the industriousness of the Asians comes in for nearly as much criticism as the alleged laziness of other groups. When blacks don't make it, it's because, whites like Al Riccardi insist, their culture doesn't teach respect for family; because they're hedonistic, lazy, stupid, and/or criminally inclined. But when Asians demonstrate their ability to overcome the obstacles of an alien language and culture, when the Asian family seems to be the repository of our most highly regarded traditional values, white hostility doesn't disappear. It just changes its form. Then the accomplishments of Asians, the speed with which they move up the economic ladder, aren't credited to their superior culture, diligence, or intelligence—even when these are granted—but to the fact that they're "single minded," "untrustworthy," "clannish drones," "narrow people" who raise children who are insufficiently "well rounded."[2]

Not surprisingly, as competition increases, the various minority groups often are at war among themselves as they press their own particular claims, fight over turf, and compete for an ever-shrinking piece of the pie. In several African-American communities, where Korean shopkeepers have taken the place once held by Jews, the confrontations have been both wrenching and tragic. A Korean grocer in Los Angeles shoots and kills a fifteen-year-old black girl for allegedly trying to steal some trivial item from the store.[3] From New York City to Berkeley, California, African-Americans boycott Korean shop owners who, they charge, invade their neighborhoods, take their money, and treat them disrespectfully.[4] But painful as these incidents are for those involved, they are only symptoms of a deeper malaise in both communities—the contempt and distrust in which the Koreans hold their African-American neighbors, and the rage of blacks as they watch these new immigrants surpass them.

Latino-black conflict also makes headlines when, in the aftermath of the riots in South Central Los Angeles, the two groups fight over who will get the lion's share of the jobs to rebuild the neighborhood. Blacks, insisting that they're being discriminated against, shut down building projects that don't include them in satisfactory numbers. And indeed, many of the jobs that formerly went to African-Americans are now being taken by Latino workers. In an article entitled "Black vs. Brown," Jack Miles, an editorial writer for the *Los Angeles Times*, reports that janitorial firms serving downtown Los Angeles have almost entirely replaced their unionized black work force with non-unionized immigrants.[5]...

But the disagreements among America's racial minorities are of little interest or concern to most white working-class families. Instead of conflicting groups, they see one large mass of people of color, all of them making claims that endanger their own precarious place in the world. It's this perception that has led some white ethnics to believe that reclaiming their ethnicity alone is not enough, that so long as they remain in their separate and distinct groups, their power will be limited. United, however, they can become a formidable countervailing force, one that can stand fast against the threat posed by minority demands. But to come together solely as whites would diminish their impact and leave them open to the charge that their real purpose is simply to retain the privileges of whiteness. A dilemma that has been resolved, at least for some, by the birth of a new entity in the history of American ethnic groups—the "European-Americans."[6]...

At the University of California at Berkeley, for example, white students and their faculty supporters insisted that the recently adopted multicultural curriculum include a unit of study of European-Americans. At Queens College in New York City, where white ethnic groups retain a more distinct presence, Italian-American students launched a successful suit to win recognition as a disadvantaged minority and gain the entitlements accompanying that status, including special units of Italian-American studies.

White high school students, too, talk of feeling isolated and, being less sophisticated and wary than their older sisters and brothers, complain quite openly that there's no acceptable and legitimate way for them to acknowledge a white identity. "There's all these things for all the different ethnicities, you know, like clubs for black kids and Hispanic kids, but there's nothing for me and my friends to join," Lisa Marshall, a sixteen-year-old white high school student, explains with exasperation. "They won't let us have a white club because that's supposed to be racist. So we figured we'd just have to call it something else, you know, some ethnic thing, like Euro-Americans. Why not? They have African-American clubs."

Ethnicity, then, often becomes a cover for "white," not necessarily because these students are racist but because racial identity is now such a prominent feature of the discourse in our social world. In a society where racial consciousness is so high, how else can whites define themselves in ways that connect them to a community and, at the same time, allow them to deny their racial antagonisms?

Ethnicity and race—separate phenomena that are now inextricably entwined. Incorporating newcomers has never been easy, as our history of controversy and violence over immigration tells us.[7] But for the first time, the new immigrants are also people of color, which means that they tap both the nativist and racist impulses that are so deeply a part of American life. As in the past, however, the fear of foreigners, the revulsion against their strange customs and seemingly unruly ways, is only part of the reason for the anti-immigrant attitudes that are increasingly being expressed today. For whatever xenophobic suspicions may arise in modern America, economic issues play a critical role in stirring them up.

REFERENCES

Alba, Richard D. *Ethnic Identity*. New Haven: Yale University Press, 1990.

Roediger, David R. *The Wages of Whiteness*. New York: Verso, 1991.

NOTES

1. David R. Roediger, *The Wages of Whiteness* (New York: Verso, 1991), p. 133.
2. These were, and often still are, the commonly held stereotypes about Jews. Indeed, the Asian immigrants are often referred to as "the new Jews."
3. Soon Ja Du, the Korean grocer who killed fifteen-year-old Latasha Harlins, was found guilty of voluntary manslaughter and sentenced to four hundred hours of community service, a $500 fine, reimbursement of funeral costs to the Harlins family, and five years' probation.
4. The incident in Berkeley didn't happen in the black ghetto, as most of the others did. There, the Korean grocery store is near the University of California campus, and the woman involved in the incident is an African-American university student who was maced by the grocer after an argument over a penny.
5. Jack Miles, "Blacks vs. Browns," *Atlantic Monthly* (October 1992), pp. 41–68.
6. For an interesting analysis of what he calls "the transformation of ethnicity," see Richard D. Alba, *Ethnic Identity* (New Haven, CT: Yale University Press, 1990).
7. In the past, many of those who agitated for a halt to immigration were immigrants or native-born children of immigrants. The same often is true today. As anti-immigrant sentiment grows, at least some of those joining the fray are relatively recent arrivals. One man in this study, for example—a fifty-two-year-old immigrant from Hungary—is one of the leaders of an anti-immigration group in the city where he lives.

17

Must-See TV

South Asian Characterizations in American Popular Media

BHOOMI K. THAKORE

INTRODUCTION

Why are there so many Indians on TV all of a sudden?
–Nina Shen Rastogi (2010)

In June 2010, Rastogi pondered the above question in her article, "Beyond Apu: Why are there suddenly so many Indians on Television?" for the online magazine, *Slate.* The increasing number of Indians and South Asians... in American popular television has been hard not to notice.... For example, among *TV Guide's* top 15 television shows of 2010 (the year Rastogi wrote this article); four had a South Asian character or actor—*Community, Glee, The Good Wife*, and *Parks and Recreation.* Since then, more characters have entered the fold—including those in such shows as *The Big Bang Theory*, *Royal Pains, Outsourced*, and *The Mindy Project.*

As Rastogi (2010) noted, the popularity of the 2008 film *Slumdog Millionaire* helped propel Indians to become a noteworthy ethnic group in popular media. These days, South Asian characters tend to be presented as the minority alongside majority-white characters. South Asians are also a good stand-in for Arab and Muslim characters in this post-9/11 reality of fear....

Contemporary South Asian media characters tend to reflect the characteristics of one of the two South Asian demographic groups. In the 1960s and 1970s, highly educated South Asians were allowed to immigrate to the United States after the passage of the 1965 Immigration and Nationality Act.... Soon after that, less educated family members of these immigrants arrived to the United States and worked in service positions, including behind the counters of convenience stores, franchises, and motels.... To date, there has been relatively little scholarship addressing the reasons behind this increasing trend of South Asian characters in the media and, more specifically, the ways in which these characters

SOURCE: Thakore, Bhoomi K. 2014. "Must-See TV: South Asian Characterizations in American Popular Media." *Sociology Compass*, 8(2): 1751–9020.

have been created, written, and produced. In this review, I discuss these representations by bridging the gaps between discussions in the fields of immigration studies, race/ethnicity studies, and critical media studies. In my discussion, I identify the concept of "(ethnic) characterization" and illustrate how studying the media representations of this ethnic group are in line with the concerns of sociology....

SOUTH ASIAN IMMIGRATION AND ASSIMILATION

The immigration and assimilation experiences of South Asians are relevant when understanding their influence on the representation of South Asians in American media. Those ethnic characteristics that media producers (and most Americans) know as "South Asian characteristics" will be used in the media characterization of these characters. This is evident in those media examples of South Asian characters that rely on overt stereotypes of this group, such as convenience store clerks or cab drivers. Thus, it is important to understand how South Asians, as a new and growing demographic, have assimilated in their new society and negotiated their own hyphenated-American identity.

Like all other immigrant and ethnic groups before them, Indians and South Asians have experienced an uphill battle in conceptualizing their identity in the United States and claiming their place in the American racial hierarchy. After President Johnson signed the Immigration and Nationality Act in 1965 that allowed technical professionals from Asia to immigrate to the United States, most South Asians immigrated in order to achieve financial and professional success. Many were able to experience upward social mobility as a result of the educational capital they brought with them. Additionally, their success influenced the success of their children and proceeding generations.

Experiences of assimilation are particularly salient for the second-generation American born children of first-generation South Asian immigrants, many of whom are portrayed in American media, and also happen to be the actors of these media characters. As Alba and Nee (2003) argued, while immigrants of the late 20th century overall have equal chances for social success as compared to their non-immigrant counterparts, the experiences of these immigrants collectively are not always the same. Their experiences are influenced by the various forms of capital a particular first or second-generation immigrant possesses, and the extent to which that capital can be useful within economic and labor markets. These experiences of segmented assimilation explain not only differences in the social mobility of immigrant groups, but also the differences by individuals within an immigrant group.

Segmented assimilation, as it relates to the social mobility between first-generation parents and second-generation children is noteworthy for all post-1965 South Asian immigrant families in the United States today (e.g. Alba and Nee 2003; Haller et al. 2011; Zhou and Xiong 2007). Additionally, it is important to note that the characterizations of South Asians in American media do not occur in a vacuum, but are informed in large part by the extent to which they assimilate into American society. However, traditional theories of assimilation fail to take into account the everyday experiences of racism that occur in the labor market and throughout society. These experiences inform the level at which ethnic groups can integrate into their new society, which in turn inform their acceptance by mainstream (white) Americans. Both dynamics are influenced by ideologies inherent in the pre-existing U.S. racial hierarchy.

THE RACIALIZATION OF SOUTH ASIANS

In the history of the United States, race relations have been fluid in order to serve particular political or social interests. The extent to which immigrants are able to assimilate into mainstream American culture will influence their place within the American racial hierarchy. Their place within the American racial hierarchy will also determine their social success in American society. As I argue, racial perceptions play a significant role in the characterization of South Asians in the media.

Some contemporary race scholars have argued that the U.S. racial hierarchy is developing into a three-tiered system consisting of Whites at the top, Blacks at the bottom, and (South) Asians as honorary Whites in the middle (Bonilla-Silva 2004; Feagin 2001; Kim 1999). As Kim (1999) argued, Asian Americans are triangulated between Blacks and Whites in terms of perceived superiority but outside of both groups for their perceived foreignness. Tuan (1999) identified these physical differences as the "forever foreigner" syndrome that Asians in the United States are subjected to regardless of their immigrant or citizenship status. However, as Bonilla-Silva (2004) argued, light skin tone and high class can help "whiten" the position and experiences of South Asians in the United States. These dynamics are particularly influential when considering the "types" of South Asian characters found in American popular media.

Historically, immigrants and racial minorities (including South Asians) have been subjected to a fluid racial hierarchy in the United States. As Omi and Winant (1994) suggested, these racial formations are the result of social, political, and economic forces that determine the social status of racial and ethnic minorities. These statuses have formed over time and are dependent on various social and historical circumstances. Examples of such racial formations include everything from Jim Crow slavery to changes in immigration policy.

Racial formations are further influenced by the level of racialization that immigrants and minorities experience. Racialization is a process by which individuals are categorized into racial groups based on their physical appearance. Additionally, these racial categorizations are then used as units of analysis to explain social relations.... Bonilla-Silva (1997) develops upon the idea of racialization in his racialized social systems theory, which supposes that political, economic, and social structures are dependent on a racialized society and the racialization of individuals into it.

... The racialization of Indian and South Asian Americans tends to be a negative process, specifically through the perpetuation of those negative stereo-types and assumptions associated with this group.... While slavery and Jim Crow shaped the racialization of African-Americans in the United States, immigration and legislative policy have shaped the racialization of South Asians and determined the extent to which they compare to Whites in America.... Contemporary ideologies, including the post-9/11 rhetoric in America and the global West, have contributed to a racialization of South Asians that portray them in popular media as foreigners and "others." These dynamics are further intersected by overt racialized perceptions that use obvious differences in skin color, religion, and ethnicity as markers of difference....

... Whites "commit" racialization of South Asians through the negative, stereotypical, and secondary ways in which they perceive them. Not only are they seen as "others" in American society, but they are also, seen as "less than" in the American racial hierarchy.... Racialization exists separate from the segmented assimilation based on skin color and class to which South Asians are subjected. Racialization is a process imposed upon all Indians and South Asians as another way to maintain the perceptions of this group as outside of American norms. As Kibria (1998) suggested, the development of such hyphenated umbrella identities as "South Asian American" is a result of racialization and these racialized experiences.

South Asian media characterizations are informed by the degree at which South Asians are racialized in society. This is evident in the examples of South Asian characters that are characterized solely around overt stereotypes. As I argue in the next section, these stereotypes not only serve the purpose of maintaining the perception of South Asians as foreigners in society, but are used consciously by media producers in their characterizations and ultimately reflect how South Asians are already perceived in the United States.

SOUTH ASIAN CHARACTERIZATIONS IN AMERICAN POPULAR MEDIA AND SOCIETY

Racial and ethnic minorities have historically been stereotyped in American popular media.... While these representations have generally improved in recent years, insofar as there are significantly fewer examples of overt stereotypes, reflections of covert and subtle stereotypes remain. As I argue, the characterization of South Asians and other minorities in the media is informed by an intentional characterization that is dependent in large part on the racial ideologies that are reproduced in the representations. This is evident in the historical trajectory of South Asian characterizations in American popular media.

South Asian media characters began appearing sporadically in films throughout the 20th century.... These early examples generally consisted of savage Indians in India who were defeated by the White star and savior (e.g. Vera and Gordon 2003). What was unique about these early representations was the location of the story itself—most were represented as Indians in India. During the 1980s, Indian and South Asian characters began appearing sporadically as tertiary or non-speaking characters cast in an American, usually urban, environment. Examples of representations included the generic and stereotypical cab driver, convenient store owner, and high-achieving student.

On the one hand, this stereotype of South Asians as a low-level service employee runs counter to the historical realities of South Asian demographics in the United States. Immigrants who arrived from South Asia in the 1960s and 1970s were highly educated.... [Eighty-three percent] of Indian immigrants between 1966 and 1977 had backgrounds in the STEM fields, including approximately 20,000 science PhDs, 40,000 engineers, and 25,000 physicians. However, after these early migrants and their immediate families were settled, many chose to invest in franchises and small businesses, including fast food stores, convenience store, and motels.... Once these businesses became established, Indian American owners took advantage of family reunification immigration policies of the 1970s and 1980s to bring over relatives to work for them. As more and more extended family members were able to settle in the United States, these individuals with few skills took other blue collar jobs working in factories or driving taxi cabs.

Media producers in positions to create media characterizations do so based on their personal experiences. While they may have been less likely to run into South Asians who were scientists, professors, or even doctors, they were more likely to run into South Asians who were behind the counter of a local convenience store or driving their cab in an urban city. Additionally, such representations proved useful to them in the context of the stories they were producing—well-to-do White American characters encountering bumbling, foreign, South Asian immigrants working in jobs most identifiable to viewers, often with ensuing hilarity. These stereotypes proved useful for the story and for the characterization of South Asian characters. As critical media studies scholars argue, White media executives have total control over the major media outlets and consciously reproduce upper class ideologies, which in turn subjugate racial minorities....

In the early 21st century, there were more noteworthy examples of Indian and South Asian media characters that were cast in such roles as highly skilled scientists or medical professionals. It is difficult to identify what led to this change, but it is likely due to the increased awareness of the high economic capital possessed by South Asian Americans, thus identifying them as a group to be coveted by advertisers (the financiers of network television). These new representations were more in line with the "model minority" stereotype, which was originally used in the 1960s to characterize East Asians in the United States.... While it is assumed that the model minority stereotype is a positive one, many scholars have identified its problematic nature. The assumption that Asian Americans are the model minority presupposes that they experience no discrimination in the United States. In fact, South Asians are subjected to the same discriminatory experiences of not being White as are other ethnic minorities in the United States.

One key example of such discrimination is through skin tone, particularly for women. All women of color deal with hegemonic skin tone ideologies in their racial/ethnic communities, with lighter skin tone and Caucasian facial features considered more appealing and attractive.... These same beauty ideals are also reproduced in the media.... This is evident in the examples of South Asian women in the media, particularly those created and cast by White, American producers.... As media producers favor casting women who are attractive, so too do the same media producers favor casting women of color who are attractive in terms of their proximity to White physical characteristics. Not only is this another reproduction of hegemonic ideology that favors one particular type of physical appearance over others, but it creates a social assumption around what an attractive South Asian can "look like."

CONCLUSION

... It is important to acknowledge a few points. First, ethnic media characterizations are intentional decisions made by media producers. These characterizations reflect hegemonic ideologies and also reproduce commonly understood stereotypes. These stereotypes are the by-products of the U.S. racial hierarchy. The racial hierarchy in turn is developed alongside immigration and assimilation trends, which further determine the qualities that an individual needs to become "American." All of these dynamics inform and influence 21st century representations in American popular media....

REFERENCES

Alba, Richard and Victor Nee. 2003. *Remaking the American Mainstream.* Cambridge, MA: Harvard University Press.

Bonilla-Silva, Eduardo, 1997. 'Rethinking Racism: Toward a Structural Interpretation.' *American Sociological Review* 62(3): 465–80.

Bonilla-Silva, Eduardo. 2004. 'From Bi-Racial to Tri-Racial: Towards a New System of Racial Stratification in the USA.' *Ethnic and Racial Studies* 27(6): 931–50.

Feagin, Joe, 2001. *Racist America: Roots, Current Realities, and Future Reparation.* New York: Routledge.

Haller, William, Alejandro Portes, and Scott M. Lynch. 2011. 'Dreams Fulfilled, Dreams Shattered: Determinants of Segmented Assimilation in the Second Generation.' *Social Forces* 89(3): 733–62.

Kibria, Nazli. 1998. "The Contested Meaning of 'Asian American': Racial Dilemmas in the Contemporary U.S.' *Ethnic and Racial Studies* 21(5): 939–58.

Kim, Claire Jean. 1999. 'The Racial Triangulation of Asian Americans.' *Politics and Society* 27(1): 105–38.

Omi, Michael and Howard Winant. 1994. *Racial Formation in the United States: From the 1960s to the 1990s,* 2nd ed. New York: Routledge.

Rastogi, Nina, 2010. Beyond Apu: Why are There Suddenly So Many Indians on Television? *Slate* June 9. Accessed July 7, 2013 (http://www.slate.com/id/2255937/).

Tuan, Mia. 1999. *Forever Foreigners or Honorary Whites? The Asian Ethnic Experience Today*. New Brunswick: Rutgers University Press.

Vera, Hernán and Andrew Gordon. 2003. *Screen Saviors: Hollywood Fictions of Whiteness*. Lanham, MD: Rowman and Littlefield.

Zhou, Min and Yang Sao Xiong. 2007. 'The Multifaceted American Experiences of the Children of Asian Immigrants: Lessons for Segmented Assimilation.' *Ethnic and Racial Studies* 28(6): 1119–52.

18

Optional Ethnicities

For Whites Only?

MARY C. WATERS

What does it mean to talk about ethnicity as an option for an individual? To argue that an individual has some degree of choice in their ethnic identity flies in the face of the commonsense notion of ethnicity many of us believe in—that one's ethnic identity is a fixed characteristic, reflective of blood ties and given at birth. However, social scientists who study ethnicity have long concluded that while ethnicity is based on a *belief* in a common ancestry, ethnicity is primarily a *social* phenomenon, not a biological one (Alba 1985, 1990; Barth 1969; Weber [1921] 1968, p. 389). The belief that members of an ethnic group have that they share a common ancestry may not be a fact. There is a great deal of change in ethnic identities across generations through intermarriage, changing allegiances, and changing social categories. There is also a much larger amount of change in the identities of individuals over their lives than is commonly believed. While most people are aware of the phenomenon known as "passing"—people raised as one race who change at some point and claim a different race as their identity—there are similar life course changes in ethnicity that happen all the time and are not given the same degree of attention as racial "passing."

White Americans of European ancestry can be described as having a great deal of choice in terms of their ethnic identities. The two major types of options White Americans can exercise are (1) the option of whether to claim any specific ancestry, or to just be "White" or American, [Lieberson (1985) called these people "unhyphenated Whites"] and (2) the choice of which of their European ancestries to choose to include in their description of their own identities. In both cases, the option of choosing how to present yourself on surveys and in everyday social interactions exists for Whites because of social changes and societal conditions that have created a great deal of social mobility, immigrant assimilation, and political and economic power for Whites in the United States. Specifically, the option of being able to not claim any ethnic identity exists for Whites of European background in the United States because they are the majority group—in terms of holding political and social power, as well as being a numerical majority. The option of choosing among different ethnicities in their

SOURCE: Waters, Mary C. 1996. "Optional Ethnicities: For Whites Only." Pp. 444–454 in *Origins and Destinies: Immigration, Race and Ethnicity in America*, edited by S. Pedraza and R.G. Rumbaut. Belmont, CA: Cengage Learning.

family backgrounds exists because the degree of discrimination and social distance attached to specific European backgrounds has diminished over time....

SYMBOLIC ETHNICITIES FOR WHITE AMERICANS

What do these ethnic identities mean to people and why do they cling to them rather than just abandoning the tie and calling themselves American? My own field research with suburban Whites in California and Pennsylvania found that later-generation descendants of European origin maintain what are called "symbolic ethnicities." Symbolic ethnicity is a term coined by Herbert Gans (1979) to refer to ethnicity that is individualistic in nature and without real social cost for the individual. These symbolic identifications are essentially leisure-time activities, rooted in nuclear family traditions and reinforced by the voluntary enjoyable aspects of being ethnic (Waters 1990). Richard Alba (1990) also found later-generation Whites in Albany, New York, who chose to keep a tie with an ethnic identity because of the enjoyable and voluntary aspects to those identities, along with the feelings of specialness they entailed. An example of symbolic ethnicity is individuals who identify as Irish, for example, on occasions such as Saint Patrick's Day, on family holidays, or for vacations. They do not usually belong to Irish American organizations, live in Irish neighborhoods, work in Irish jobs, or marry other Irish people. The symbolic meaning of being Irish American can be constructed by individuals from mass media images, family traditions, or other intermittent social activities. In other words, for later-generation White ethnics, ethnicity is not something that influences their lives unless they want it to. In the world of work and school and neighborhood, individuals do not have to admit to being ethnic unless they choose to. And for an increasing number of European-origin individuals whose parents and grandparents have intermarried, the ethnicity they claim is largely a matter of personal choice as they sort through all of the possible combinations of groups in their genealogies....

RACE RELATIONS AND SYMBOLIC ETHNICITY

However much symbolic ethnicity is without cost for the individual, there is a cost associated with symbolic ethnicity for the society. That is because symbolic ethnicities of the type described here are confined to White Americans of European origin. Black Americans, Hispanic Americans, Asian Americans, and American Indians do not have the option of a symbolic ethnicity at present in the United States. For all of the ways in which ethnicity does not matter for White Americans, it does matter for non-Whites. Who your ancestors are does affect your choice of spouse, where you live, what job you have, who your friends are, and what your chances are for success in American society, if those ancestors happen not to be from Europe. The reality is that White ethnics have a lot more choice and room for maneuver than they themselves think they do.

The situation is very different for members of racial minorities, whose lives are strongly influenced by their race or national origin regardless of how much they may choose not to identify themselves in terms of their ancestries.

When White Americans learn the stories of how their grandparents and great-grandparents triumphed in the United States over adversity, they are usually told in terms of their individual efforts and triumphs. The important role of labor unions and other organized political and economic actors in their social and economic successes are left out of the story in favor of a generational story of individual Americans rising up against communitarian, Old World intolerance, and New World resistance. As a result, the "individualized" voluntary, cultural view of ethnicity for Whites is what is remembered.

One important implication of these identities is that they tend to be very individualistic. There is a tendency to view valuing diversity in a pluralist environment as equating all groups. The symbolic ethnic tends to think that all groups are equal; everyone has a background that is their right to celebrate and pass on to their children. This leads to the conclusion that all identities are equal and all identities in some sense are interchangeable—"I'm Italian American, you're Polish American. I'm Irish American, you're African American." The important thing is to treat people as individuals and all equally. However, this assumption ignores the very big difference between an individualistic symbolic ethnic identity and a socially enforced and imposed racial identity.

My favorite example of how this type of thinking can lead to some severe misunderstandings between people of different backgrounds is from the *Dear Abby* advice column. A few years back a person wrote in who had asked an acquaintance of Asian background where his family was from. His acquaintance answered that this was a rude question and he would not reply. The bewildered White asked Abby why it was rude, since he thought it was a sign of respect to wonder where people were from, and he certainly would not mind anyone asking HIM about where his family was from. Abby asked her readers to write in to say whether it was rude to ask about a person's ethnic background. She reported that she got a large response, that most non-Whites thought it was a sign of disrespect, and Whites thought it was flattering:

> Dear Abby,
> I am 100 percent American and because I am of Asian ancestry I am often asked "What are you?" It's not the personal nature of this question that bothers me, it's the question itself. This query seems to question my very humanity. "What am I? Why I am a person like everyone else!"
> *Signed, A REAL AMERICAN*

> Dear Abby,
> Why do people resent being asked what they are? The Irish are so proud of being Irish, they tell you before you even ask. Tip O'Neill has never tried to hide his Irish ancestry.
> *Signed, JIMMY.*

(SOURCE: Published by Universal Press Syndicate)

In this exchange Jimmy cannot understand why Asians are not as happy to be asked about their ethnicity as he is, because he understands his ethnicity and theirs to be separate but equal. Everyone has to come from somewhere—his family from Ireland, another's family from Asia—each has a history and each should be proud of it. But the reason he cannot understand the perspective of the Asian American is that all ethnicities are not equal; all are not symbolic, costless, and voluntary. When White Americans equate their own symbolic ethnicities with the socially enforced identities of non-White Americans, they obscure the fact that the experiences of Whites and non-Whites have been qualitatively different in the United States and that the current identities of individuals partly reflect that unequal history.

In the next section I describe how relations between Black and White students on college campuses reflect some of these asymmetries in the understanding of what a racial or ethnic identity means. While I focus on Black and White students in the following discussion, you should be aware that the myriad other groups in the United States—Mexican Americans, American Indians, Japanese Americans—all have some degree of social and individual influences on their identities, which reflect the group's social and economic history and present circumstance.

RELATIONS ON COLLEGE CAMPUSES

Both Black and White students face the task of developing their race and ethnic identities. Sociologists and psychologists note that at the time people leave home and begin to live independently from their parents, often ages eighteen to twenty-two, they report a heightened sense of racial and ethnic identity as they sort through how much of their beliefs and behaviors are idiosyncratic to their families and how much are shared with other people. It is not until one comes in close contact with many people who are different from oneself that individuals realize the ways in which their backgrounds may influence their individual personality. This involves coming into contact with people who are different in terms of their ethnicity, class, religion, region, and race. For White students, the ethnicity they claim is more often than not a symbolic one—with all of the voluntary, enjoyable, and intermittent characteristics I have described above.

Black students at the university are also developing identities through interactions with others who are different from them. Their identity development is more complicated than that of Whites because of the added element of racial discrimination and racism, along with the "ethnic" developments of finding others who share their background. Thus Black students have the positive attraction of being around other Black students who share some cultural elements, as well as the need to band together with other students in a reactive and oppositional way in the face of racist incidents on campus.

Colleges and universities across the country have been increasing diversity among their student bodies in the last few decades. This has led in many cases to strained relations among students from different racial and ethnic backgrounds. The 1980s and 1990s produced a great number of racial incidents and high racial tensions on campuses. While there were a number of racial incidents that were due to bigotry, unlawful behavior, and violent or vicious attacks, much of what happens among students on campuses involves a low level of tension and awkwardness in social interactions.

Many Black students experience racism personally for the first time on campus. The upper-middle-class students from White suburbs were often isolated enough that their presence was not threatening to racists in their high schools. Also, their class background was known by their residence and this may have prevented attacks being directed at them. Often Black students at the university who begin talking with other students and recognizing racial slights will remember incidents that happened to them earlier that they might not have thought were related to race.

Black college students across the country experience a sizeable number of incidents that are clearly the result of racism. Many of the most blatant ones that occur between students are the result of drinking. Sometimes late at night, drunken groups of White students coming home from parties will yell slurs at single Black students on the street. The other types of incidents that happen include being singled out for special treatment by employees, such as being followed when shopping at the campus bookstore, or going to the art museum with your class and the guard stops you and asks for your I.D. Others involve impersonal encounters on the street—being called a nigger by a truck driver while crossing the street, or seeing old ladies clutch their pocketbooks and shake in terror as you pass them on the street. For the most part these incidents are not specific to the university environment, they are the types of incidents middle-class Blacks face every day throughout American society, and they have been documented by sociologists (Feagin 1991).

In such a climate, however, with students experiencing these types of incidents and talking with each other about them, Black students do experience a tension and a feeling of being singled out. It is unfair that this is part of their college experience and not that of White students. Dealing with incidents like this, or the ever-present threat of such incidents, is an ongoing developmental task for Black students that takes energy, attention, and strength of character. It should be clearly understood that this is an asymmetry in the "college experience" for Black and White students. It is one of the unfair aspects of life that results from living in a society with ongoing racial prejudice and discrimination. It is also very understandable that it makes some students angry at the unfairness of it all, even if there is no one to blame specifically. It is also very troubling because, while most Whites do not create these incidents, some do, and it is never clear until you know someone well whether they are the type of person who could do something like this. So one of the reactions of Black students to these incidents is to band together.

In some sense then, as Blauner (1992) has argued, you can see Black students coming together on campus as both an "ethnic" pull of wanting to be together to share common experiences and community, and a "racial" push of banding together defensively because of perceived rejection and tension from Whites. In this way the ethnic identities of Black students are in some sense similar to, say, Korean students wanting to be together to share experiences. And it is an ethnicity that is generally much stronger than, say, Italian Americans. But for Koreans who come together there is generally a definition of themselves as "different from" Whites. For Blacks reacting to exclusion, there is a tendency for the coming together to involve both being "different from" but also "opposed to" Whites.

The anthropologist John Ogbu (1990) has documented the tendency of minorities in a variety of societies around the world, who have experienced severe blocked mobility for long periods of time, to develop such oppositional identities. An important component of having such an identity is to describe others of your group who do not join in the group solidarity as devaluing and denying their very core identity. This is why it is not common for successful Asians to be accused by others of acting "White" in the United States, but it is quite common for such a term to be used by Blacks and Latinos. The oppositional component of a Black identity also explains how Black people can question whether others are acting "Black enough." On campus, it explains some of the intense pressures felt by Black students who do not make their racial identity central and who choose to hang out primarily with non-Blacks. This pressure from the group, which is partly defining itself by not being White, is exacerbated by the fact that race is a physical marker in American society. No one immediately notices the Jewish students sitting together in the dining hall, or the one Jewish student sitting surrounded by non-Jews, or the Texan sitting with the Californians, but everyone notices the Black student who is or is not at the "Black table" in the cafeteria.

An example of the kinds of misunderstandings that can arise because of different understandings of the meanings and implications of symbolic versus oppositional identities concerns questions students ask one another in the dorms about personal appearances and customs. A very common type of interaction in the dorm concerns questions Whites ask Blacks about their hair. Because Whites tend to know little about Blacks, and Blacks know a lot about Whites, there is a general asymmetry in the level of curiosity people have about one another. Whites, as the numerical majority, have had little contact with Black culture; Blacks, especially those who are in college, have had to develop bicultural skills—knowledge about the social worlds of both Whites and Blacks. Miscommunication and hurt feelings about White students' questions about Black students' hair illustrate this point. One of the things that happens freshman year is that White students are around Black students as they fix their hair. White students are generally quite curious about Black students' hair—they have basic questions such as how often Blacks wash their hair, how they get it straightened or curled, what products they use on their hair, how they comb it, etc. Whites often wonder to themselves whether they should ask these questions. One thought experiment Whites perform is to ask themselves whether a particular question would upset them. Adopting the "do unto others" rule, they ask themselves, "If a Black person was curious about my hair would I get upset?" The answer usually is "No, I would be happy to tell them." Another example is an Italian American student wondering to herself, "Would I be upset if someone asked me about calamari?" The answer is no, so she asks her Black roommate about collard greens, and the roommate explodes with an angry response such as, "Do you think all Black people eat watermelon too?" Note that if this Italian American knew her friend was Trinidadian American and asked about peas and rice the situation would be more similar and would not necessarily ignite underlying tensions.

Like the debate in *Dear Abby,* these innocent questions are likely to lead to resentment. The issue of stereotypes about Black Americans and the assumption that all Blacks are alike and have the same stereotypical cultural traits has more power to hurt or offend a Black person than vice versa. The innocent questions about Black hair also bring up a number of asymmetries between the Black and White experience. Because Blacks tend to have more knowledge about Whites than vice versa, there is not an even exchange going on, the Black freshman is likely to have fewer basic questions about his White roommate than his White roommate has about him. Because of the differences historically in the group experiences of Blacks and Whites there are some connotations to Black hair that don't exist about White hair. (For instance, is straightening your hair a form of assimilation, do some people distinguish between women having "good hair" and "bad hair" in terms of beauty and how is that related to looking "White"?) Finally, even a Black freshman who cheerfully disregards or is unaware that there are these asymmetries will soon slam into another asymmetry if she willingly answers every innocent question asked of her. In a situation where Blacks make up only 10 percent of the student body, if every non-Black needs to be educated about hair, she will have to explain it to nine other students. As one Black student explained to me, after you've been asked a couple of times about something so personal you begin to feel like you are an attraction in a zoo, that you are at the university for the education of the White students.

INSTITUTIONAL RESPONSES

Our society asks a lot of young people. We ask young people to do something that no one else does as successfully on such a wide scale—that is to live together with people from very different backgrounds, to respect one another, to appreciate one another, and to enjoy and learn from one another. The successes that occur every day in this endeavor are many, and they are too often overlooked. However, the problems and tensions are also real, and they will not vanish on their own. We tend to see pluralism working in the United States in much the same way some people expect capitalism to work. If you put together people with various interests and abilities and resources, the "invisible hand" of capitalism is supposed to make all the parts work together in an economy for the common good.

There is much to be said for such a model—the invisible hand of the market can solve complicated problems of production and distribution better than any "visible hand" of a state plan. However, we have learned that unequal power relations among the actors in the capitalist marketplace, as well as "externalities" that the market cannot account for, such as long-term pollution, or collusion between corporations, or the exploitation of child labor, means that state regulation is often needed. Pluralism and the relations between groups are very similar. There is a lot to be said for the idea that bringing people who belong to different ethnic or racial groups together in institutions with no interference will have good consequences. Students from different backgrounds will make friends if they share a dorm room or corridor, and there is no need for the institution to do any more than provide the locale. But like capitalism, the invisible hand of pluralism does not do well when power relations and externalities are ignored. When you bring together individuals from groups that are differentially valued in the wider society and provide no guidance, there will be problems. In these cases the "invisible hand" of pluralist relations does not work, and tensions and disagreements can arise without any particular individual or group of individuals being "to blame." On college campuses in the 1990s some of the tensions between students are of this sort. They arise from honest misunderstandings, lack of a common background, and very different experiences of what race and ethnicity mean to the individual.

The implications of symbolic ethnicities for thinking about race relations are subtle but consequential. If your understanding of your own ethnicity and its relationship to society and politics is one of individual choice, it becomes harder to understand the need for programs like affirmative action, which recognize the ongoing need for group struggle and group recognition, in order to bring about social change. It also is hard for a White college student to understand the need that minority students feel to band together against discrimination. It also is easy, on the individual level, to expect everyone else to be able to turn their ethnicity on and off at will, the way you are able to, without understanding that ongoing discrimination and societal attention to minority status makes that impossible for individuals from minority groups to do. The paradox of symbolic ethnicity is that it depends upon the ultimate goal of a pluralist society, and at the same time makes it more difficult to achieve that ultimate goal. It is dependent upon the concept that all ethnicities mean the same thing, that enjoying the traditions of one's heritage is an option available to a group or an individual, but that such a heritage should not have any social costs associated with it.

As the Asian Americans who wrote to *Dear Abby* make clear, there are many societal issues and involuntary ascriptions associated with non-White identities. The developments necessary for this to change are not individual but societal in nature. Social mobility and declining racial and ethnic sensitivity are closely associated. The legacy and the present reality of discrimination on the basis of race or ethnicity must be overcome before the ideal of a pluralist society, where all heritages are treated equally and are equally available for individuals to choose or discard at will, is realized.

REFERENCES

Alba, Richard D. 1985. *Italian Americans: Into the of Twilight Ethnicity*. Englewood Cliffs, NJ: Prentice-Hall.

Alba, Richard D. 1990. *Ethnic Identity: The Transformation of White America*. New Haven: Yale University Press.

Barth, Frederick. 1969. *Ethnic Groups and Boundaries*. Boston: Little, Brown.

Blauner, Robert. 1992. "Talking Past Each Other: Black and White Languages of Race." *American Prospect* (Summer): 55–64.

Feagin, Joe R. 1991. "The Continuing Significance of Race: Anti-Black Discrimination in Public Places." *American Sociological Review* 56: 101–17.

Gans, Herbert. 1979. "Symbolic Ethnicity: The Future of Ethnic Groups and Cultures in America." *Ethnic and Racial Studies* 2: 1–20.

Lieberson, Stanley. 1985. *Making It Count: The Improvement of Social Research and Theory*. Berkeley: University of California Press.

Ogbu, John. 1990. "Minority Status and Literacy in Comparative Perspective." *Daedalus* 119: 141–69.

Waters, Mary C. 1990. *Ethnic Options: Choosing Identities in America*. Berkeley: University of California Press.

Weber, Max. [1921]/1968. *Economy and Society: An Outline of Interpretive Sociology*. Eds. Guenther Roth and Claus Wittich, trans. Ephraim Fischoff. New York: Bedminister Press.

19

Living "Illegal"

The Human Face of Unauthorized Immigration

MARIE FRIEDMANN MARQUARDT, TIMOTHY J. STEIGENGA, PHILIP J. WILLIAMS, AND MANUEL A. VÁSQUEZ

"Waves of illegal aliens swarming across our border, joining violent gangs, forcing families to live in fear." As the narrator speaks, a crowd of dark figures passes through a hole in the soaring fence. Crouching low and holding flashlights, they grin as they move furtively beyond it. The fence gives way to a shadowy alley, where a group of young men dressed in the fashion of gang members advances menacingly. The television advertisement continues, and these images, intended to portray "illegal aliens," are juxtaposed with images that aim to portray American citizens. In one version of the television commercial, two working-class American men in hard hats express bafflement and annoyance. In another version, the two workers are replaced by a white family fretting together, as scowling young Latino men wearing bandanas and sporting tattooed chests fill the screen. While these dark images flash across the television, viewers are told that federal legislators who support immigration reform aim to "give tax breaks and social security benefits to illegal aliens" and support "a plan that gives illegals a pathway to amnesty and even special college tuition rates." A final image of young white schoolchildren gathering eagerly around their teacher emerges, as the narrator decries the Senate majority leader's decision to vote "against making English the national language."[1]

At the height of the 2010 midterm election campaigns, the American public was inundated with powerfully charged depictions of unauthorized immigrants and with strong claims about the devastating impact of unauthorized immigration on the life of the nation. But as this and other widely publicized campaign ads circulated throughout the United States, entering into the spotlight of national media attention and the consciousness of many Americans, other stories more quietly unfolded.

Until the morning of March 29, 2010, Jessica Colotl lived the rather uneventful life of a typical hardworking college student. Her parents brought her from Mexico in 1996 at the age of seven, and she studied hard, eventually earning a 3.8 grade point average and graduating with academic honors from

SOURCE: Marquardt, Marie Friedmann, Timothy J. Steigenga, Philip J. Williams, and Manuel A. Vásquez. 2011. Excerpt from *Living "Illegal": The Human Face of Unauthorized Immigration*.

Lakeside High School in DeKalb County, Georgia. According to Lila Parra, a close friend and sorority sister, when Colotl discovered that her parents had brought her into the United States without the proper documentation, she filed papers to regularize her and her younger sibling's status. Parra explained, "She took it upon herself to do that for herself and her younger sibling. She was like, 'I've been here forever, I consider America my country.'"[2]

Undeterred by the lack of resolution on her legal status, Colotl applied to and was accepted into Kennesaw State University (KSU), where she became a political science major, with the hope of eventually going to law school. To support her studies, Colotl worked with her mother, cleaning office buildings in Atlanta until late at night. Despite the long working hours, Jessica found the time to establish a Latina sorority that has been active in the local community.

Colotl's plans were derailed on that morning in March, when she was stopped by a campus police officer for a minor traffic infraction. Unable to produce a valid driver's license, she was turned over to the Cobb County sheriff's department, which has a 287(g) agreement with Immigration and Customs Enforcement (ICE). This arrangement allowed the sheriff to check her immigration status. After authorities confirmed that Colotl was in the country without authorization, they sent her to an ICE detention center in Etowah, a small rural community in the northeastern corner of Alabama.

Upon learning about Colotl's situation, a myriad of local organizations, including her own sorority and the president of KSU, quickly mobilized to secure her release. After spending more than a month in the Etowah detention center, she was released, and ICE deferred action on her case for a year to enable her to finish her degree at KSU before being sent back to Mexico, a country that she has not visited since her parents brought her to the United States. But this was not the end of Colotl's ordeal. Cobb County sheriff Neil Warren obtained a new warrant for her arrest, adducing that she had lied about her address during her first arrest. Under Georgia law, making false statements to a law enforcement official is a felony. Colotl, who turned herself in voluntarily to the Cobb County jail, is awaiting a decision from the judge on the felony charge. Reflecting on Jessica's case, Lila Parra expressed admiration for her sorority sister: "I've never seen anybody fight so hard for their education.... [Jessica] pays for it all on her own and pays out-of-state tuition. She doesn't want to just get by—she wants to get that 4.0 GPA.... We want other students to not get discouraged by situations like this, and for them to move forward.... So many students, they just want to be educated, because they realize their family is not."[3]

Jessica's situation is not unique. Recently, the *Los Angeles Times* reported the case of Cal State Fresno student body president Pedro Ramirez, who was brought from Mexico by his undocumented parents at the age of three. Under a law that allows anyone who attended high school in California for three years to pay in-state tuition, Ramirez was able to attend college. However, he could not receive any financial aid from the federal government nor could he work legally, often having to resort to working with his father mowing lawns and with his mother cleaning houses. He also refused to accept a $9,000 stipend that comes with the office of student body president, volunteering to serve without pay. When an anonymous tip to the college newspaper revealed his undocumented status, Ramirez declared: "In a way, I'm relieved.... I don't want to be a liability or cost the school donations. I never really thought this was going to happen. But now that it's out there, I finally feel ready to say 'Yes, it's me. I'm one of the thousands.'"[4]

Although Jessica Colotl and Pedro Ramirez do not know each other and live on opposite sides of the country, when they were asked about their aspirations, their answers were remarkably similar. Summoning a version of the narrative that has inspired countless immigrants to come to America, from the heyday of Ellis Island to the present, each of them stated: "I'm just trying to live the American dream and finish my education."[5]

These stories lay bare the emotional intensity surrounding the issue of unauthorized immigration, as well as the enormous gulf between the potent images circulating in our media and the complex reality of life as an unauthorized immigrant in the United States today. On one side of the gulf, such potent imagery has helped doom attempts to pass comprehensive federal immigration reform in 2006 and 2007 and has since shut down all alternatives beyond the enforcement of a system that most politicians and scholars agree is broken. The arguments underlying these images is best summarized by the expression popular among groups opposing immigration reform: "What part of illegal don't you understand?" After all, a lawbreaker is a lawbreaker. There is nothing to discuss. To advocate anything but punishment—in this case deportation—simply amounts to aiding and abetting criminals, opening the way for others to commit the same offense with total impunity.

The simple contrast between "legal" and "illegal" is bolstered through the repeated use of four broadly articulated claims about unauthorized immigrants:

1. Unauthorized immigrants flood across the U.S.—Mexico border to take advantage of public benefits and social services, while contributing very little to U.S. society. Thus, the solution to the problem of unauthorized immigration is to seal U.S. borders. If they want to come to America, they should get in line and do it legally.
2. Unauthorized immigrants are a burden on the U.S. economy. They take jobs from U.S. citizens, exacerbating unemployment and depressing wages for working-class Americans. In addition, taxpayers pay heavily for the government services they use. For this reason, the solution to the problem of unauthorized immigration is to deny them any access to social services and public benefits.
3. Unauthorized immigrants are closely associated with criminality, violence, drugs, and gangs. They threaten the safety and stability of local communities. Therefore, the solution to the problem of unauthorized immigration lies in vigorous local and state enforcement. Any other approach represents amnesty for lawbreakers.
4. Unauthorized immigrants cannot be integrated into U.S. society because they bring values that are contrary to the values of this nation. Furthermore, these immigrants do not want to integrate. They choose instead to retain

their language and to have dual national allegiances facilitated by connections with their countries of origin. Therefore, unauthorized immigrants threaten the sovereignty and the future of the nation.

These broad claims obscure the complex human stories that lie behind the phenomenon of unauthorized immigration. Furthermore, … they generate a patchwork of policies that simply fail to solve the complex problems associated with unauthorized immigration.

From the other side of the enormous gulf, we recognize that there is a lot about illegality that we do not fully understand. The reality on the ground is much more complicated than the simple contrast between "legal" and "illegal" that characterizes mainstream media and political discourse. Jessica Colotl's and Pedro Ramirez's cases suggest that if we take a closer look into the lives of unauthorized immigrants, we might find that these immigrants share many of the core values that have shaped the history of the United States: hard work, individual initiative, willingness to take risks in the quest for self-improvement, dedication to families and communities. Colotl's and Ramirez's cases also point to the strong desire of unauthorized immigrants to integrate and contribute to their adopted society. As Colotl attempted to do without success, these immigrants would readily regularize their status if given the opportunity, paying fines, enlisting in the army, or doing community service to make up for having entered the country without proper authorization. Drawing from a helpful distinction that scholars of citizenship and immigration make, we may say that although Colotl and Ramirez are not citizens in the liberal sense, since they do not have legal membership in the U.S. polity, they certainly behave like citizens in the civic republican mode, for which engagement in the life and well-being of the community is key. The trouble lies in the fact that lack of formal citizenship limits the kinds of civic engagement these immigrants can have. This tension may have negative consequences not only for Colotl and Ramirez, who carry the stigma of illegality, but also for this nations' democracy, which fails to tap into the civic values and energy they bring.…

… [W]e recognize that unauthorized immigration is an issue that raises valid concerns and that has real costs for our society, as well as for immigrants and their communities of origin. In particular, communities in the United States that have witnessed the growing presence of unauthorized immigrants as part of their economic boom have legitimate concerns about the strain put on school systems, emergency units, hospitals, and other local services at a time when resources are scare. Furthermore, there are understandable reasons to worry about the presence of a large group of unskilled and vulnerable workers in the labor market. As we will see, many Americans are ambivalent about unauthorized immigration. For instance, they recognize the needs that lead many Latin Americans to come to the United States without legal status and admire the work and family-centered ethics of Latino immigrants. Yet they also worry that, because these immigrants are willing to work under any conditions and to be paid less than the minimum wage, the native born will be "underbid" and displaced by unauthorized Latino workers.

The people and communities raising these valid concerns should not be labeled and dismissed as racists or xenophobes. These concerns certainly call for rational, open-minded, and careful reflection on the causes, costs, and benefits of unauthorized immigration, as well as thoughtful debate and discussion of potential solutions. We might disagree in the end on these solutions, but such a debate should be informed by the realities on the ground, by moral considerations, and by a robust knowledge of the historical and contemporary forces that have led to unauthorized immigration. Americans must account for the ways in which actions by the U.S. government and even American lifestyles and patterns of consumption have contributed to this phenomenon. Instead, the valid concerns and struggles of local communities have been overshadowed and profoundly distorted by an increasingly shrill discourse surrounding immigration....

... To avoid the negative and polarizing emotional baggage that has become attached to the word *illegal*, we will refrain from using it. Instead, we will use the term *unauthorized immigrant*, since it is factually correct. This term indicates that the immigrants who are the subjects of this [article] are out of status, having entered the country or remained in the country without following proper procedures. In other words, they might have crossed the border at a place other than a designated port of entry or they might have overstayed their tourist or work-based visas. Occasionally we will also use the term *undocumented immigrant*, which is far more common in everyday discourse but is not always accurate. Many unauthorized immigrants do have documents, such as passports from their home countries, driver's licenses from their state of residence in the United States (as in New Mexico, which allows the issuance of driver's licenses without proof of legal residence), or *matrículas consulares*, identification cards distributed by consulates. What they do not have is valid social security numbers, green cards, or U.S. passports, documents that mark their formal status as members of our polity.

Debates about unauthorized immigration, however, are not merely about semantics. They are shaped by a widespread lack of knowledge of the root causes of this immigration and, most important, of the situation unauthorised immigrants face in their daily lives....

... Currently, many politicians use unauthorized immigration as a hot-button issue and link it with racial stereotypes and violent criminality. This indicates that they and their media consultants feel this strategy will resonate with the frustrations and fears of voters. Yet, the word *illegal* has become so emotionally charged that it dehumanizes not only unauthorized immigrants, who are objectified as nothing more than faceless criminals, but even those who use the term uncritically. The widespread use of the term *illegal* leaves no room to consider the moral and policy contradictions that are behind the need for people to leave their homes and risk their lives crossing the border without authorization or to overstay visas and live in a precarious status. When a congressman can say that "illegal" immigrant women "multiply like rats" and not have to apologize for such a derogatory remark, it is clear that the public discourse on the topic has reached its nadir.[6]

The media have tended to magnify the raw passions elicited by the term *illegal*, contributing to an overall climate of mistrust, hostility, and incivility, which ... stymies constructive public debate, impeding the search for rational, pragmatic, and long-term repairs to the broken immigration system. Furthermore, the federal government has not been able to articulate a compelling narrative of why the United States needs comprehensive immigration reform that is in line with the values of the nation and its evolving place in the world. In the absence of immigration reform at the federal level, states and communities that are directly experiencing the contradictions of immigration are left to cope with the issue on their own, often with dwindling resources. In response, these states and localities have passed an incongruent patchwork of ordinances and laws. Many of these do not address the root causes of unauthorized migration but instead seek short-term solutions that at best merely deflect the problem and at worst threaten civil liberties and increase local tensions. These laws put an especially heavy burden on families, often separating unauthorized parents from their U.S.-born children. They also may lead to racial profiling, eroding the overall sense of community, interethnic trust, openness, and tolerance to pluralism that is at the heart of American civil society. Moreover, these laws have in effect criminalized unauthorized immigrants. What is often lost in our immigration debates is the simple fact that immigration law is administrative law, and illegal entry and presence are in violation of administrative procedures. The growing patchwork of state and local ordinances, in practice, turn unauthorized presence from civil infraction into felony.

Equally troubling has been the rapid growth of a culture of enforcement that threatens to spin out of control.... Immigration enforcement increasingly is being placed in the hands of local law enforcement agencies, and immigrants increasingly are being detained in for-profit, privately run prisons. In Cobb County, Georgia, where we conducted some of our field work, the ACLU reports a situation of "policing run amok [as] ... law enforcement and jail personnel routinely abuse their power under 287(g)." Of the 3,180 inmates the county jail processed for ICE detention in 2008, almost 69 percent were arrested for traffic violations, belying the avowed focus on removing criminals from local communities.[7] This trend will likely increase, since the Secure Communities Program, which also relies on local and state law enforcement agencies for implementation, is scheduled to be in all U.S. jurisdictions by 2013. The increasing criminalization and incarceration of unauthorized immigrants distracts local communities and the nation, drawing attention and efforts away from addressing the forces that propel unauthorized immigration....

NOTES

1. This television commercial originally aired in 2010 to support the candidacy of Sharron Angle, a Republican candidate opposing Harry Reid, the Democratic senator and majority leader from Nevada. See "Sharron Angle TV ad: 'At Your Expense,'" YouTube, http://www.youtube.com/watch?v=uJC_RmcO7Ts&feature=channel (accessed December 8, 2010). See also "Another Fear Mongering and Anti-Latino Ad from Sharron Angle," YouTube, http://www.youtube.com/watch?v=_wdcxvP4tyE&feature=related (accessed December 8, 2010).

2. Kathyrn Dobies, "Traffic Stop Puts KSU Student in Jail as an Illegal Immigrant," *Marietta Daily Journal*, May 1, 2010, http://www.mdjonline.com/view/full_story/7265546/article-Traffic-stop-puts-KSU-student-in-jail-as-an-illegal-immigrant.
3. Ibid.
4. See Diana Marcum, "He's the Cal State Fresno Student Body President—and an Illegal Immigrant," *Los Angeles Times*, http://www.latimes.com/news/local/la-me-1118-illegal-immigrant-presiden20101118,0,5635027.story (accessed December 8, 2010).
5. Mark David and Helena Oliviero, "New Face on an Old Debate: Colotl Case Spotlights Illegal Immigration Sage in Cobb County," *Atlanta Journal Constitution*, May 16, 2010.
6. These comments were made by Rep. Curry Todd of Collierville, Tennessee, when told during a hearing that Cover Kid, a state-funded health program, does not require proof of citizenship for a mother to receive prenatal care, since any child born in the United States is a citizen. See Erik Schelzig, "Bredesen Slams Rep. Curry Todd's Immigrant 'Rats' Remark," Associated Press, November 13, 2010, http://www.tennessean.com/article/20101113/NEWS02/11130324/Bredesen-slams-Rep-Curry-Todd-s-immigrant-rats-remark (accessed December 16, 2010).
7. *Terror and Isolation in Cobb: How Unchecked Police Power Under 287(g) Has Torn Families Apart and Threatened Public Safety*, American Civil Liberties Union Foundation of Georgia, October 2009, http://www.acluga.org/287gReport.pdf. A recent article in the *New York Times* focusing on neighboring Gwinnett County reported that 45 percent of those arrested under the 287(g) program were detained for traffic offenses other than driving under the influence. See Julia Preston and Robert Gebeloff, "Some Unlicensed Drivers Risk More than a Fine," *New York Times*, December 9, 2010.

20

A Dream Deferred

Undocumented Students at CUNY

CAROLINA BANK MUÑOZ

I first became aware of the difficulties for undocumented students at the City University of New York (CUNY) when I started teaching a course at Brooklyn College, a CUNY campus, on the sociology of immigration. On the first day of class, five students requested appointments to speak with me in private. This was extremely unusual to say the least. All five students were undocumented and had family members who were undocumented. They were hoping I could help. As one student put it, "I'm hoping you can teach me how to get my papers." I had to explain that I was not a lawyer, nor was the class about how to immigrate "legally," but about the social process of immigration. Needless to say, the students were deeply disappointed, but nevertheless stayed enrolled in the course. One student in particular made a tremendous impression on me.

Luisa came to the United States when she was five years old.[1] Her father was diagnosed with a rare and serious illness and they initially migrated so that he could be treated. Like many other immigrants, they obtained a visa to visit the United States. Once Luisa's father was treated and recovering, they decided to remain in the United States. They overstayed their visa, and from one night to the next became undocumented. Luisa attended public school while both of her parents worked in the garment industry. After Proposition 187 passed in California, Luisa's parents decided that it was time to leave California and move to New York, where the anti-immigrant climate was less intense.

most overstay their visas

During high school, Luisa worked after school as a seamstress in the factory where her mother worked. Her father was now a union janitor and their financial situation had stabilized substantially. In her last year of high school, Luisa started researching colleges. At that point she realized that there were very few opportunities for undocumented immigrants. She had been in this country for twelve years. She had learned English, worked hard, and made good grades. Yet, she was not going to be able to simply apply to college like many of her classmates. Despite her 3.8 GPA, Luisa would have to attend a community college because she simply could not afford to pay full tuition at a 4-year college and she was not eligible for any federal loans. After working full-time and attending school part-time for three years, Luisa had finally saved enough

SOURCE: Muñoz, Carolina Bank. 2008. "A Dream Deferred: Undocumented Students at CUNY." *Radical Teacher* 84 (Spring): 8–17.

money to enroll at Brooklyn College. During her first year at BC, Luisa's brother was deported. She used her entire savings to bring her brother back to the United States across the U.S.-Mexico border and was forced to drop out. She was devastated to have to delay her education, but her family was the priority. After a two year hiatus from school, Luisa was able to re-enroll at Brooklyn College. That very semester she enrolled in my immigration course. While Luisa's story is incredible, it is not exceptional.

Hundreds of undocumented students across the country have similar stories.[2]

In fact, over 60,000 undocumented students, the vast majority of whom are people of color, graduate from high school every year (UCLA Labor Center 2007). Most of these students migrated to the United States at a young age along with parents or other family members. Yet they are subject to the same harsh immigration policies as their parents who predominantly work in the low wage sector. The United States is the only home that most of these students know, but they are forced to live in the shadows of American society, living in fear of Immigration and Customs Enforcement (ICE) with marginal access to good jobs or a college education.

For these students, a college education is usually only a dream. In fact, only five to ten percent of these students make it to college (UCLA Labor Center 2007, NILC 2006). Undocumented college students have no access to federal and state student aid, work study programs, or many scholarships. Furthermore, since the 1990s but especially since September 11th, 2001, access to higher education for undocumented students has been severely curtailed. Many states passed laws that required colleges and universities to charge non-resident tuition to undocumented students (Gonzales 2007). Nonresident tuition is often 2 to 3 times more expensive than in-state tuition, making it nearly impossible for undocumented students to attend college.

The half-dozen undocumented students in my class and the more than two thousand undocumented students at CUNY (according to the CUNY Immigration and Citizenship Project) have had to overcome tremendous adversity to be at the university. Over the course of the semester, several of my students saw their family members deported, one student successfully evaded a workplace raid by Immigration and Customs Enforcement, and two students had to drop out because they simply could not afford to stay in school. Since my first experience with these five students, I have run into dozens of undocumented college students across CUNY who have had to drop out of college, find work, save money, and return to college a few years later. Many never return to school because they simply do not earn enough in the low wage sector or underground economy to afford a college education.

Undocumented students are systematically denied access to a college education by a flawed immigration system that has roots in institutionalized racism. At its root, contemporary immigration policy is inherently flawed because it seeks an individual solution to a structural problem. In the latest round of immigration reform, legislators have focused on either blocking the flow of migration through "solutions" such as a border fence or severely limiting it through guest worker programs and other means. These policies treat immigration as a faucet that can be turned on and off. In fact, immigration is far more complex. There are structural conditions and policies that *force* people to migrate. The North American Free Trade Agreement (NAFTA), structural adjustment policies, and war all impact migration. NAFTA, in particular, has been instrumental in increased forced migration from Mexico. NAFTA resulted in removing tariffs that were protecting Mexican farmers without removing U.S. subsidies to U.S. producers. As a result, the Mexican market was flooded with underpriced agricultural goods from the United States, especially corn. Unable to compete with these underpriced goods, Mexican farmers had no choice but to leave their land and seek employment in other parts of Mexico. Many displaced farmers migrate to large cities in Mexico to work in factories. As those factory jobs disappear or are exported to other countries, they have nowhere to go but the United States (Bank Muñoz 2008). Ironically, then, U.S. economic and trade policies are significantly responsible for the increase in immigration from Mexico and other Latin American countries.

The ongoing backlash against immigration disproportionately affects immigrants of color (militias and vigilante groups such as the Minutemen for the most part are not violently attacking "illegal" Germans). As Ngai (2005) aptly puts it "restrictive immigration laws produced new categories of racial difference.... The legal racialization of these ethnic groups' national origin cast them as permanently foreign and unassimilable to the nation" (7–8). A perfect example is that in the contemporary immigration debate, Latina/o immigrants are racialized as "illegal" even if they were born in the United States or otherwise hold "legal" status (Bank Muñoz 2008). This blanket racialization falls on undocumented students in very particular ways.

The crisis for access to higher education for undocumented students affects not only the students who are in college or trying to get into college now, but also younger undocumented students who drop out of high school because they see that they have no opportunities for upward mobility. We are facing the possibility of a lost generation of extraordinarily bright and talented students....

TEACHING AND WORKING WITH IMMIGRANT STUDENTS

Working with immigrant students, and particularly undocumented students of color, offers various opportunities and challenges. On the one hand, their life experience provides them with an intuitive sense of the global economy and racial and class disparities. Many of them come from the Global South and have experienced poverty, racism, and exploitation. These students also tend to have a greater understanding of world politics. Needless to say, their knowledge and experiences contribute tremendously to a vibrant classroom environment. I recall a particularly intense classroom discussion over the idea of reparations. Native born Blacks were arguing that only Black people who can prove a link to slavery in the United States should benefit from reparations while Caribbean

Blacks argued that they were also entitled to reparations because they were forced to migrate to the United States due to the devastations of globalization and colonization. All students in the class, immigrants and native born alike, benefited tremendously from this exchange.

In my experience, undocumented students are among the most self-motivated and focused students I have had, perhaps because they have the most to gain or lose. There are very few paths to obtain permanent residence and a green card. One can either acquire it through a family member (spouse, parent, etc.) who is a U.S. citizen or through employment. Employment is often the best option for undocumented students. In this case, employers have to make a case for why a foreign national (instead of a U.S. citizen) is better suited for the position (USCIS 2008). A college degree gives undocumented students, especially in high demand fields, some hope of finding a job in which an employer will be able to help secure their immigration status. On the other hand, having to drop out of school minimizes the chances that these students will find good jobs and a road to citizenship. Therefore, college recruitment and retention of undocumented students is imperative to enhancing their life chances.

Unfortunately, undocumented students face extreme barriers to succeeding in school and completing their college education even when they overcome the barriers to getting into college. Immigrant students in general and undocumented students in particular often live in poor neighborhoods with under-funded public schools. This is also true for other students of color. However, immigrant students have the additional barriers of having had to transition from schools in their native countries to a new method of U.S. education and learn U.S. English.[3] As a result they often have weaker writing and public speaking skills than other students. Additionally, as I have already mentioned, many of these students have significant barriers outside the classroom, which limit their prospects for campus based *[sic]* activism.

WHAT CAN WE DO? *SUPPORT THE DREAM ACT*

It is not good enough to rely on states to change their policies regarding undocumented students. We need a federal policy that would affect all states so that all undocumented students have access to higher education in the United States. To this end, lawmakers have been trying to pass the Dream Act, which would give undocumented students a road to citizenship. Several variations of the Dream Act have been introduced in Congress since 2001. While no form of the legislation has passed, it has gained significant momentum. The Dream Act would make two major changes to current law. It would "permit certain immigrant students who have grown up in the U.S. to apply for temporary legal status and eventually obtain permanent status and become eligible for citizenship if they go to college or serve in the U.S. military" (NILC 2007). It would also "eliminate a federal provision that penalizes states that provide in-state tuition without regard to immigration status" (NILC 2007).

Under the Dream Act, students of "good moral character" who came to the United States at age fifteen or younger would obtain conditional permanent resident status (six years) upon acceptance to college, graduation from a U.S. high school, or receiving a GED. Students would also be able to qualify for the federal work study program and for student loans. At the end of the six-year conditional period, students would be granted unrestricted lawful permanent resident status if "during the conditional period the immigrant has maintained good moral character, avoided lengthy trips abroad, and either 1) graduated from a 2 year college or studied for at least 2 years towards a B.A. or higher degree, or 2) served in the U.S. Military for 2 years" (NILC 2007).

The Dream Act is far from perfect. The condition of "good moral character," for example, is troubling. How is moral character defined? How would gay students, activist students, students who have been arrested for acts of civil disobedience, and students in left organizations fare under this conditon? What kind of invasive investigations into their moral character would they be subjected to? These are all important questions, and as a result of the vagueness of the concept, a significant layer of students would not be eligible to reap the benefits of the Dream Act. Furthermore, the option of participating in military service to obtain permanent resident status is deeply problematic. It gives military recruiters who already prey on communities of color further ammunition to convince these students to participate in military service instead of going to college. States would have an incentive to encourage young undocumented immigrants to go into the military since it would save them the costs of granting instate tuition and save federal government Pell grant money. Moreover, given that recruiters use free college tuition as a carrot for potential recruits, joining the military might look especially attractive to young undocumented immigrants.

Moreover, the Dream Act would not require (or prohibit) states to provide instate tuition, nor would students qualify for federal Pell grants. They would, however, be eligible for federal work study and student loans. In short, the Dream Act has many problems. However, it would offer undocumented students a path to citizenship. Most importantly, states would not be restricted from providing their own financial aid to students. Given that we are unlikely to see progressive immigration reform in the near future, it is important to support the Dream Act as a first step towards change.

CONCLUSION

Why is higher education for undocumented immigrant students so important? It's a fundamental issue of rights. [Today], we would be hard pressed to find individuals who do not believe that women and native born minorities should be given access to higher education. What is so different about someone who at the age of five was brought to this country by their parents? The United States is the only country that a majority of these students know. They are going to stay, work, pay taxes, and possibly raise families in this country. As I have mentioned, earning a B.A. degree significantly improves the life chances of all citizens.

An education means access to better jobs, which means access to savings, which means access to the accumulation of wealth that is passed on from one generation to the next. A majority of these students will be contributing to our economy, culture, and collective conscience. As a nation, we should want undocumented students to be empowered through education.

Currently, our immigration policy is sending the message that undocumented students who have been raised in the United States are disposable. Undocumented students' dreams and aspirations are shattered every year, as they realize that they have few possibilities for obtaining a college degree. Many lose hope for the future and begin the slow decline into accepting their fate. Others turn to crime and violence as a method of releasing their frustrations with inequality. Undocumented students who migrated to the country with their parents should not be expected to pay the price of a flawed immigration system....

NOTES

1. All names have been changed to protect students' identities.
2. For an excellent resource on this issue see *Underground Undergrads: UCLA Undocumented Immigrant Students Speak Out*, reviewed in this issue.
3. I use the term U.S. English, because many immigrants from the Caribbean already speak English, but the writing norms and vocabulary for U.S. English are different. So while these students speak and write English, they often have to relearn it to reflect U.S. norms.

REFERENCES

Bank Muñoz, Carolina. 2008. *Transnational Tortillas: Race, Gender and Shop Floor Politics in Mexico and the United States*. Ithaca: Cornell University Press.

Gonzales, Roberto. 2007. "Wasted Talent and Broken Dreams: The Lost Potential of Undocumented Students." *Immigration Policy in Focus*, v5 (13). Immigration Policy Center. www.immigrationpolicy.org

National Immigration Law Center. 2006. "Basic Facts About In-State Tuition for Undocumented Immigrant Students." www.nilc.org

National Immigration Law Center. 2007. "The Dream Act: Basic Facts." www.nilc.org

Ngai, Mae M. 2005. *Impossible Subjects: Illegal Aliens and the Making of Modern America*. Princeton: Princeton University Press.

UCLA Labor Center. 2007. "Undocumented Students, Unfulfilled Dreams ..." Report. www.labor.ucla.edu

United States Citizenship and Immigration Services. (2008). www.uscis.gov

birdcage

21

Sex and Gender through the Prism of Difference

MAXINE BACA ZINN, PIERRETTE HONDAGNEU-SOTELO, AND MICHAEL MESSNER

"Men can't cry." "Women are victims of patriarchal oppression." "After divorces, single mothers are downwardly mobile, often moving into poverty." "Men don't do their share of housework and child care." "Professional women face barriers such as sexual harassment and a 'glass ceiling' that prevent them from competing equally with men for high-status positions and high salaries." "Heterosexual intercourse is an expression of men's power over women." Sometimes, the students in our sociology and gender studies courses balk at these kinds of generalizations. And they are right to do so. After all, some men are more emotionally expressive than some women, some women have more power and success than some men, some men do their share—or more—of housework and child care, and some women experience sex with men as both pleasurable and empowering. Indeed, contemporary gender relations are complex and changing in various directions, and as such, we need to be wary of simplistic, if handy, slogans that seem to sum up the essence of relations between women and men.

On the other hand, we think it is a tremendous mistake to conclude that "all individuals are totally unique and different," and that therefore all generalizations about social groups are impossible or inherently oppressive. In fact, we are convinced that it is this very complexity, this multifaceted nature of contemporary gender relations, that fairly begs for a sociological analysis of gender. We use the image of "the prism of difference" to illustrate our approach to developing this sociological perspective on contemporary gender relations. The *American Heritage Dictionary* defines "prism," in part, as "a homogeneous transparent solid, usually with triangular bases and rectangular sides, used to produce or analyze a continuous spectrum." Imagine a ray of light—which to the naked eye appears to be only one color—refracted through a prism onto a white wall. To the eye, the result is not an infinite, disorganized scatter of individual colors. Rather, the refracted light displays an order, a structure of relationships among the different colors—a rainbow. Similarly, we propose to use the "prism of difference" ... to analyze a continuous spectrum of people, in order to show how gender is organized and experienced differently when refracted through the prism of

SOURCE: Baca Zinn, Maxine, Pierrette Hondagneu-Sotelo, and Michael Messner, eds. 2005. *Gender through the Prism of Difference*, 2nd ed., pp. 1–6. Copyright © 2005. By permission of Oxford University Press, USA.

sexual, racial/ethnic, social class, physical abilities, age, and national citizenship differences.

EARLY WOMEN'S STUDIES: CATEGORICAL VIEWS OF "WOMEN" AND "MEN"

... It is possible to make good generalizations about women and men. But these generalizations should be drawn carefully, by always asking the questions "*which* women?" and "*which* men?" Scholars of sex and gender have not always done this. In the 1960s and 1970s, women's studies focused on the differences *between* women and men rather than *among* women and men. The very concept of gender, women's studies scholars demonstrated, is based on socially defined difference between women and men. From the macro level of social institutions such as the economy, politics, and religion, to the micro level of interpersonal relations, distinctions between women and men structure social relations. Making men and women *different* from one another is the essence of gender. It is also the basis of men's power and domination. Understanding this was profoundly illuminating. Knowing that difference produced domination enabled women to name, analyze, and set about changing their victimization.

In the 1970s, riding the wave of a resurgent feminist movement, colleges and universities began to develop women's studies courses that aimed first and foremost to make women's lives visible. The texts that were developed for these courses tended to stress the things that women shared under patriarchy—having the responsibility for housework and child care, the experience or fear of men's sexual violence, a lack of formal or informal access to education, and exclusion from high-status professional and managerial jobs, political office, and religious leadership positions (Brownmiller, 1975; Kanter, 1977).

The study of women in society offered new ways of seeing the world. But the 1970s approach was limited in several ways. Thinking of gender primarily in terms of differences *between* women and men led scholars to overgeneralize about both. The concept of patriarchy led to a dualistic perspective of male privilege and female subordination. Women and men were cast as opposites. Each was treated as a homogeneous category with common characteristics and experiences. This approach *essentialized* women and men. Essentialism, simply put, is the notion that women's and men's attributes and indeed women and men themselves are categorically different. From this perspective, male control and coercion of women produced conflict between the sexes. The feminist insight originally introduced by Simone De Beauvoir in 1953—that women, as a group, had been socially defined as the "other" and that men had constructed themselves as the subjects of history, while constructing women as their objects—fueled an energizing sense of togetherness among many women. As college students read books such as *Sisterhood Is Powerful* (Morgan, 1970), many of them joined organizations that fought—with some success—for equality and justice for women.

THE VOICES OF "OTHER" WOMEN

Although this view of women as an oppressed "other" was empowering for certain groups of women, some women began to claim that the feminist view of universal sisterhood ignored and marginalized their major concerns. It soon became apparent that treating women as a group united in its victimization by patriarchy was biased by too narrow a focus on the experiences and perspectives of women from more privileged social groups. "Gender" was treated as a generic category, uncritically applied to women. Ironically, this analysis, which was meant to unify women, instead produced divisions between and among them. The concerns projected as "universal" were removed from the realities of many women's lives. For example, it became a matter of faith in second-wave feminism that women's liberation would be accomplished by breaking down the "gendered public-domestic split." Indeed, the feminist call for women to move out of the kitchen and into the workplace resonated in the experiences of many of the college-educated white women who were inspired by Betty Friedan's 1963 book, *The Feminine Mystique*. But the idea that women's movement into workplaces was itself empowering or liberating seemed absurd or irrelevant to many working-class women and women of color. They were already working for wages, as had many of their mothers and grandmothers, and did not consider access to jobs and public life "liberating." For many of these women, liberation had more to do with organizing in communities and workplaces—often alongside men—for better schools, better pay, decent benefits, and other policies to benefit their neighborhoods, jobs, and families. The feminism of the 1970s did not seem to address these issues.

As more and more women analyzed their own experiences, they began to address the power relations that created differences among women and the part that privileged women played in the oppression of others. For many women of color, working-class women, lesbians, and women in contexts outside the United States (especially women in non-Western societies), the focus on male domination was a distraction from other oppressions. Their lived experiences could support neither a unitary theory of gender nor an ideology of universal sisterhood. As a result, finding common ground in a universal female victimization was never a priority for many groups of women.

Challenges to gender stereotypes soon emerged. Women of varied races, classes, national origins, and sexualities insisted that the concept of gender be broadened to take their differences into account (Baca Zinn et al., 1986; Hartmann, 1976; Rich, 1980; Smith, 1977). Many women began to argue that their lives were affected by their location in a number of different hierarchies: as African Americans, Latinas, Native Americans, or Asian Americans in the race hierarchy; as young or old in the age hierarchy; as heterosexual, lesbian, or bisexual in the sexual orientation hierarchy; and as women outside the Western industrialized nations, in subordinated geopolitical contexts. These arguments made it clear that women were not victimized by gender alone but by the historical and systematic denial of rights and privileges based on other differences as well.

MEN AS GENDERED BEINGS

As the voices of "other" women in the mid- to late 1970s began to challenge and expand the parameters of women's studies, a new area of scholarly inquiry was beginning to stir—a critical examination of men and masculinity. To be sure, in those early years of gender studies, the major task was to conduct studies and develop courses about the lives of women in order to begin to correct centuries of scholarship that rendered invisible women's lives, problems, and accomplishments. But the core idea of feminism—that "femininity" and women's subordination is a social construction—logically led to an examination of the social construction of "masculinity" and men's power. Many of the first scholars to take on this task were psychologists who were concerned with looking at the social construction of "the male sex role" (e.g., Pleck, 1976). By the late 1980s, there was a growing interdisciplinary collection of studies of men and masculinity, much of it by social scientists (Brod, 1987; Kaufman, 1987; Kimmel, 1987; Kimmel & Messner, 1989).

Reflecting developments in women's studies, the scholarship on men's lives tended to develop three themes: First, what we think of as "masculinity" is not a fixed, biological essence of men, but rather is a social construction that shifts and changes over time as well as between and among various national and cultural contexts. Second, power is central to understanding gender as a relational construct, and the dominant definition of masculinity is largely about expressing difference from—and superiority over—anything considered "feminine." And third, there is no singular "male sex role." Rather, at any given time there are various masculinities. R. W. Connell (1987; 1995; 2002) has been among the most articulate advocates of this perspective. Connell argues that hegemonic masculinity (the dominant form of masculinity at any given moment) is constructed in relation to femininities *as well as* in relation to various subordinated or marginalized masculinities. For example, in the United States, various racialized masculinities (e.g., as represented by African American men, Latino immigrant men, etc.) have been central to the construction of hegemonic (white middle-class) masculinity. This "othering" of racialized masculinities helps to shore up the privileges that have been historically connected to hegemonic masculinity. When viewed this way, we can better understand hegemonic masculinity as part of a system that includes gender as well as racial, class, sexual, and other relations of power.

The new literature on men and masculinities also begins to move us beyond the simplistic, falsely categorical, and pessimistic view of men simply as a privileged sex class. When race, social class, sexual orientation, physical abilities, immigrant, or national status are taken into account, we can see that in some circumstances, "male privilege" is partly—sometimes substantially—muted (Kimmel & Messner, 2004). Although it is unlikely that we will soon see a "men's movement" that aims to undermine the power and privileges that are connected

with hegemonic masculinity, when we begin to look at "masculinities" through the prism of difference, we can begin to see similarities and possible points of coalition between and among certain groups of women and men (Messner, 1998).

Certain kinds of changes in gender relations—for instance, a national family leave policy for working parents—might serve as a means of uniting particular groups of women and men.

GENDER IN INTERNATIONAL CONTEXTS

It is an increasingly accepted truism that late twentieth-century increases in transnational trade, international migration, and global systems of production and communication have diminished both the power of nation-states and the significance of national borders. A much more ignored issue is the extent to which gender relations—in the United States and elsewhere in the world—are increasingly linked to patterns of global economic restructuring. Decisions made in corporate headquarters located in Los Angeles, Tokyo, or London may have immediate repercussions on how women and men thousands of miles away organize their work, community, and family lives (Sassen, 1991). It is no longer possible to study gender relations without giving attention to global processes and inequalities.…

Around the world, women's paid and unpaid labor is key to global development strategies. Yet it would be a mistake to conclude that gender is molded from the "top down." What happens on a daily basis in families and workplaces simultaneously constitutes and is constrained by structural transnational institutions. For instance, in the second half of the twentieth century young, single women, many of them from poor rural areas, were (and continue to be) recruited for work in export assembly plants along the U.S.-Mexico border, in East and Southeast Asia, in Silicon Valley, in the Caribbean, and in Central America. While the profitability of these multinational factories depends, in part, on management's ability to manipulate the young women's ideologies of gender, the women … do not respond passively or uniformly, but actively resist, challenge, and accommodate. At the same time, the global dispersion of the assembly line has concentrated corporate facilities in many U.S. cities, making available myriad managerial, administrative, and clerical jobs for college educated women. Women's paid labor is used at various points along this international system of production. Not only employment but also consumption embodies global interdependencies. There is a high probability that the clothing you are wearing and the computer you use originated in multinational corporate headquarters and in assembly plants scattered around third world nations. And if these items were actually manufactured in the United States, they were probably assembled by Latin American and Asian-born women.

Worldwide, international labor migration and refugee movements are creating new types of multiracial societies. While these developments are often discussed and analyzed with respect to racial differences, gender typically remains absent. As several commentators have noted, the white feminist movement in the United States has not addressed issues of immigration and nationality. Gender, however, has been fundamental in shaping immigration policies (Chang,

1994; Hondagneu-Sotelo, 1994). Direct labor recruitment programs generally solicit either male or female labor (e.g., Filipina nurses and Mexican male farm workers), national disenfranchisement has particular repercussions for women and men, and current immigrant laws are based on very gendered notions of what constitutes "family unification." As Chandra Mohanty suggests, "analytically these issues are the contemporary metropolitan counterpart of women's struggles against colonial occupation in the geographical third world" (1991:23). Moreover, immigrant and refugee women's daily lives often challenge familiar feminist paradigms. The occupations in which immigrant and refugee women concentrate—paid domestic work, informal sector street vending, assembly or industrial piece work performed in the home—often blur the ideological distinction between work and family and between public and private spheres (Hondagneu-Sotelo, 2001; Parrenas, 2001).

FROM PATCHWORK QUILT TO PRISM

All of these developments—the voices of "other" women, the study of men and masculinities, and the examination of gender in transnational contexts—have helped redefine the study of gender. By working to develop knowledge that is inclusive of the experiences of all groups, new insights about gender have begun to emerge. Examining gender in the context of other differences makes it clear that nobody experiences themselves as solely gendered. Instead, gender is configured through cross-cutting forms of difference that carry deep social and economic consequences.

By the mid-1980s, thinking about gender had entered a new stage, which was more carefully grounded in the experiences of diverse groups of women and men. This perspective is a general way of looking at women and men and understanding their relationships to the structure of society. Gender is no longer viewed simply as a matter of two opposite categories of people, males and females, but a range of social relations among differently situated people. Because centering on difference is a radical challenge to the conventional gender framework, it raises several concerns. If we think of all the systems that converge to simultaneously influence the lives of women and men, we can imagine an infinite number of effects these interconnected systems have on different women and men. Does the recognition that gender can be understood only contextually (meaning that there is no singular "gender" per se) make women's studies and men's studies newly vulnerable to critics in the academy? Does the immersion in difference throw us into a whirlwind of "spiraling diversity" (Hewitt, 1992:316) whereby multiple identities and locations shatter the categories "women" and "men"? ...

We take a position directly opposed to an empty pluralism. Although the categories "woman" and "man" have multiple meanings, this does not reduce gender to a "postmodern kaleidoscope of lifestyles. Rather, it points to the *relational* character of gender" (Connell, 1992:736). Not only are masculinity and femininity relational, but different *masculinities* and *femininities* are interconnected through other social structures such as race, class, and nation. The concept of relationality suggests that the lives of different groups are interconnected even without face-to-face relations (Glenn, 2002:14). The meaning of "woman" is defined by the existence of women of different races and classes. Being a white woman in the United States is meaningful only insofar as it is set apart from and in contradistinction to women of color.

Just as masculinity and femininity each depend on the definition of the other to produce domination, differences *among* women and *among* men are also created in the context of structured relationships. Some women derive benefits from their race and class position and from their location in the global economy, while they are simultaneously restricted by gender. In other words, such women are subordinated by patriarchy, yet their relatively privileged positions within hierarchies of race, class, and the global political economy intersect to create for them an expanded range of opportunities, choices, and ways of living. They may even use their race and class advantage to minimize some of the consequences of patriarchy and/or to oppose other women. Similarly, one can become a man in opposition to other men. For example, "the relation between heterosexual and homosexual men is central, carrying heavy symbolic freight. To many people, homosexuality is the *negation* of masculinity.... Given that assumption, antagonism toward homosexual men may be used to define masculinity" (Connell, 1992:736).

In the past decade, viewing gender through the prism of difference has profoundly reoriented the field (Acker, 1999; Glenn, 1999, 2002; Messner, 1996; West & Fenstermaker, 1995). Yet analyzing the multiple constructions of gender does not just mean studying groups of women and groups of men as different. It is clearly time to go beyond what we call the "patchwork quilt" phase in the study of women and men—that is, the phase in which we have acknowledged the importance of examining differences within constructions of gender, but do so largely by collecting together a study here on African American women, a study there on gay men, a study on working-class Chicanas, and so on. This patchwork quilt approach too often amounts to no more than "adding difference and stirring." The result may be a lovely mosaic, but like a patchwork quilt, it still tends to overemphasize boundaries rather than to highlight bridges of inter-dependency. In addition, this approach too often does not explore the ways that social constructions of femininities and masculinities are based on and reproduce relations of power. In short, we think that the substantial quantity of research that has now been done on various groups and subgroups needs to be analyzed within a framework that emphasizes differences and inequalities not as discrete areas of separation, but as interrelated bands of color that together make up a spectrum....

REFERENCES

Acker, Joan. 1999. "Rewriting Class, Race and Gender: Problems in Feminist Rethinking." Pp. 44–69 in Myra Marx Ferree, Judith Lorber, and Beth B. Hess (eds.), *Revisioning Gender*. Thousand Oaks, CA: Sage Publications.

Baca Zinn, M., L., Weber Cannon, E., Higgenbotham, & B., Thornton Dill. 1986. "The Costs of Exclusionary Practices in Women's Studies," *Signs: Journal of Women in Culture and Society* 11: 290–303.

Brod, Harry (ed.). 1987. *The Making of Masculinities: The New Men's Studies*. Boston: Allen & Unwin.

Brownmiller, Susan. 1975. *Against Our Will: Men, Women, and Rape*. New York: Simon & Schuster.

Chang, Grace. 1994. "Undocumented Latinas: The New 'Employable Mothers.'" Pp. 259–285 in Evelyn Nakano Glenn, Grace Chang, and Linda Rennie Forcey (eds.), *Mothering, Ideology, Experience, and Agency*. New York and London: Routledge.

Connell, R. W. 1987. *Gender and Power*. Stanford, CA: Stanford University Press.

Connell, R. W. 1992. "A Very Straight Gay: Masculinity, Homosexual Experience, and the Dynamics of Gender," *American Sociological Review* 57: 735–751.

Connell, R. W. 1995. *Masculinities*. Berkeley: University of California Press.

Connell, R. W. 2002. *Gender*. Cambridge: Polity.

De Beauvoir, Simone. 1953. *The Second Sex*. New York: Knopf.

Glenn, Evelyn Nakano. 1999. "The Social Construction and Institutionalization of Gender and Race: An Integrative Framework." Pp. 3–43 in Myra Marx Ferree, Judith Lorber, and Beth B. Hess (eds.), *Revisioning Gender*. Thousand Oaks. CA: Sage Publications.

Glenn, Evelyn Nakano. 2002. *Unequal Sisterhood: How Race and Gender Shaped American Citizenship and Labor*. Cambridge, MA: Harvard University Press.

Hartmann, Heidi. 1976. "Capitalism, Patriarchy, and Job Segregation by Sex," *Signs: Journal of Women in Culture and Society* 1(3), part 2, spring: 137–167.

Hewitt, Nancy A. 1992. "Compounding Differences," *Feminist Studies* 18: 313–326.

Hondagneu-Sotelo, Pierrette. 1994. *Gendered Transitions: Mexican Experiences of Immigration*. Berkeley: University of California Press.

Hondagneu-Sotelo, Pierrette. 2001. *Doméstica: Immigrant Workers Cleaning and Caring in the Shadows of Affluence*. Berkeley: University of California Press.

Kanter, Rosabeth Moss. 1977. *Men and Women of the Corporation*. New York: Basic Books.

Kaufman, Michael. 1987. *Beyond Patriarchy: Essays by Men on Pleasure, Power, and Change*. Toronto and New York: Oxford University Press.

Kimmel, Michael S. (ed.). 1987. *Changing Men: New Directions in Research on Men and Masculinity*. Newbury Park, CA: Sage.

Kimmel, Michael S. 1996. *Manhood in America: A Cultural History*. New York: Free Press.

Kimmel, Michael S. & Michael A. Messner (eds.). 1989. *Men's Lives*. New York: Macmillan.

Kimmel, Michael S. & Michael A. Messner (eds.). 2004. *Men's Lives*, 6th ed. Boston: Pearson.

Messner, Michael A. 1996. "Studying Up on Sex," *Sociology of Sport Journal* 13: 221–237.

Messner, Michael A. 1998. *Politics of Masculinities: Men in Movements*. Thousand Oaks, CA: Sage Publications.

Mohanty, Chandra Talpade. 1991. "Cartographies of Struggle: Third World Women and the Politics of Feminism." Pp. 51–80 in Chandra Talpade Mohanty, Ann Russo, and Lourdes Torres, (eds.), *Third World Women and the Politics of Feminism.* Bloomington: Indiana University Press.

Morgan, Robin. 1970. *Sisterhood Is Powerful: An Anthology of Writing from the Women's Liberation Movement.* New York: Vintage Books.

Parrenas, Rhacel Salazar. 2001. *Servants of Globalization: Women, Migration and Domestic Work.* Stanford: Stanford University Press.

Pleck, J. H. 1976. "The Male Sex Role: Definitions, Problems, and Sources of Change," *Journal of Social Issues* 32: 155–164.

Rich, Adrienne. 1980. "Compulsory Heterosexuality and the Lesbian Experience," *Signs: Journal of Women in Culture and Society* 5: 631–660.

Sassen, Saskia. 1991. *The Global City: New York, London, Tokyo.* Princeton: Princeton University Press.

Smith, Barbara. 1977. *Toward a Black Feminist Criticism.* Freedom, CA: Crossing Press.

West, Candace & Sarah Fenstermaker. 1995. "Doing Difference," *Gender & Society* 9: 8–37.

22

Seeing Privilege Where It Isn't

Marginalized Masculinities and the Intersectionality of Privilege

BETHANY M. COSTON AND MICHAEL KIMMEL

Systems of privilege exist worldwide, in varying forms and contexts, and while this examination of privilege focuses on only one specific instance (the United States), the theorizing of said privilege is intended to be universal. This is because no matter the context, the idea that "privilege is invisible to those who have it" has become a touchstone epigram for work on the "super-ordinate"—in this case, White people, men, heterosexuals, and the middle class.... When one is privileged by class, or race or gender or sexuality, one rarely sees exactly how the dynamics of privilege work. Thus, efforts to make privilege visible, such as McIntosh's (1988) "invisible knapsack" and the "Male Privilege Checklist" or the "heterosexual questionnaire" have become staples in college classes.

Yet unlike McIntosh's autobiographical work, some overly-simple pedagogical tools like the "heterosexual questionnaire" or "Male Privilege checklist" posit a universal and dichotomous understanding of privilege: one either has it or one does not. It's as if all heterosexuals are white; all "males" are straight. The notion of intersectionality complicates this binary understanding. Occasionally, a document breaks through those tight containers, such as Woods' (2010) "Black Male Privilege Checklist," but such examples are rare.

We propose to investigate sites of inequality within an overall structure of privilege. Specifically, we look at three groups of men—disabled men, gay men, and working class men—to explore the dynamics of having privilege in one sphere but being unprivileged in another arena. What does it mean to be privileged by gender and simultaneously marginalized by class, sexuality, or bodily status?

This is especially important, we argue, because, for men, the dynamics of removing privilege involve assumptions of emasculation—exclusion from that category that would confer privilege. Gender is the mechanism by which the marginalized are marginalized. That is, gay, working class, or disabled men are seen as "not-men" in the popular discourse of their marginalization. It is their masculinity—the site of privilege—that is specifically targeted as the grounds for

SOURCE: Coston, Bethany M., and Michael Kimmel. 2012. *Journal of Social Issues* by Society for the Psychological Study of Social Issues 68: 97–111.

exclusion from privilege. Thus, though men, they often see themselves as reaping few, if any, of the benefits of their privileged status as men (Pratto & Stewart, 2012).

Of course, they do reap those benefits. But often, such benefits are less visible, since marginalized men are less likely to see a reduced masculinity dividend as much compensation for their marginalization. This essay will explore these complex dynamics by focusing on three groups of marginalized men: working class, disabled, and gay men.

DOING GENDER AND THE MATRIX OF OPPRESSION

In the United States, there is a set of idealized standards for men. These standards include being brave, dependable, and strong, emotionally stable, as well as critical, logical, and rational. The ideal male is supposed to be not only wealthy, but also in a position of power over others. Two words sum up the expectations for men: hegemonic masculinity (cf. Connell, 1995). That is, the predominant, overpowering concept of what it is to be a "real man."

The idealized notion of masculinity operates as both an ideology and a set of normative constraints. It offers a set of traits, attitudes and behaviors (the "male role") as well as organizing institutional relationships among groups of women and men. Gender operates at the level of interaction (one can be said to "do" gender through interaction) as well as an identity (one can be said to "have" gender, as in the sum total of socialized attitudes and traits). Gender can also be observed within the institutionally organized sets of practices at the macrolevel—states, markets, and the like all express and reproduce gender arrangements. One of the more popular ways to see gender is as an accomplishment; an everyday, interactional activity that reinforces itself via our activities and relationships. "Doing gender involves a complex of socially guided perceptual, interactional, and micropolitical activities that cast particular pursuits as expressions of masculine and feminine 'natures'" (West & Zimmerman, 1987).

These "natures," or social *norms* for a particular gender, are largely internalized by the men and women who live in a society, consciously and otherwise. In other words, these social norms become personal identities. Moreover, it is through the intimate and intricate process of daily interaction with others that we fully achieve our gender, and are seen as valid and appropriate gendered beings. For men, masculinity often includes preoccupation with proving gender to others. Indeed, "In presenting ourselves as a gendered person, we are making ourselves accountable—we are purposefully acting in such a way as to be able to be recognized as gendered" (West & Fenstermaker, 1995).

Society is full of men who have embraced traditional gender ideologies—even those who might otherwise be marginalized. While the men we discuss below may operate within oppression in one aspect of their lives, they have access to alternate sites of privilege via the rest of their demographics (e.g., race, physical ability, sexual orientation, gender, sex, age, social class, religion). A working class man, for example, may also be White and have access to white privilege and male privilege. What is interesting is how these men choose to navigate and access their privilege within the confines of a particular social role that limits, devalues, and often stigmatizes them as not-men.

Marginalization requires the problematization of the category (in this case masculinity) so that privilege is rendered invisible. At the same time, marginalization also frames power and privilege from an interesting vantage point; it offers a seemingly existential choice: to overconform to the dominant view of masculinity as a way to stake a claim to it or to resist the hegemonic and develop a masculinity of resistance.

The commonalities within the somewhat arbitrary categories (race, class, sexuality, etc.) are often exaggerated and the behavior of the most dominant group within the category (e.g., rich, straight, White men) becomes idealized as the only appropriate way to fulfill one social role. "This conceptualization is then employed as a means of excluding and stigmatizing those who do not or cannot live up to these standards. This process of "doing difference" is realized in constant interpersonal interactions that reaffirm and reproduce social structure" (West & Fenstermaker, 1995).

It is important to realize that masculinity is extremely diverse, not homogenous, unchanging, fixed, or undifferentiated. Different versions of masculinities coexist at any given historical period and can coexist within different groups. However, it is this diversity and coexistence that creates a space for marginalization. "The dominant group needs a way to justify its dominance—that difference is inferior" (Cheng, 2008).

DYNAMICS OF MARGINALIZATION AND STIGMA

Marginalization is both gendered and dynamic. How do marginalized men respond to the problematization of their masculinity as they are marginalized by class, sexuality or disability status?...

Disabled Men

Discrimination against men with disabilities is pervasive in American society, and issues of power, dominance, and hegemonic masculinity are the basis. Over time, hegemonic masculinity has grown to encompass all aspects of social and cultural power, and the discrimination that arises from this can have an alarmingly negative affect on a man and his identity. Disabled men do not meet the unquestioned and idealized standards of appearance, behavior, and emotion for men. The values of capitalist societies based on male dominance are dedicated to warrior values, and a frantic able-bodiedness represented through aggressive sports and risk-taking activities, which do not make room for those with disabilities.

For example, one man interviewed by Robertson (2011) tells the story of his confrontations with those who discriminate against him. Frank says,

> If somebody doesn't want to speak to me 'cause I'm in a chair, or they shout at me 'cause I'm in a chair, I wanna know why, why they feel they have to shout. I'm not deaf you know. If they did it once and I told them and they didn't do it again, that'd be fair enough. But if they keep doing it then that would annoy me and if they didn't know that I could stand up then I'd put me brakes on and I'd stand up and I'd tell them face-to-face. If they won't listen, then I'll intimidate them, so they will listen, because it's important. (p. 12)

Scholars seem to agree that terms such as "disability" and "impairment" refer to limitations in function resulting from physiological, psychological and anatomical dysfunction of bodies (including minds), causing restrictions in a person's ability to perform culturally defined normal human activities (World Health Organization). Normal life activities are defined as walking, talking, using any of the senses, working, and/or caring for oneself.

Men with physical disabilities have to find ways to express themselves within the role of "disabled." Emotional expression is not compatible with the aforementioned traits because it signifies vulnerability; in this way, men, especially disabled men, must avoid emotional expression. If they fail in stoicism, discrimination in the form of pejorative words ("cripple," "wimp," "retard") are sometimes used to suppress or condemn the outward expressions vulnerability.

But, men with disabilities don't need verbal reminders of their "not-men" status. Even without words, their social position, their lack of power over themselves (let alone others), leads them to understand more fully their lacking masculinity. One man, Vernon, detailed these feelings specifically,

> Yeah, 'cause though you know you're still a man, I've ended up in a chair, and I don't feel like a red-blooded man. I don't feel I can handle 10 pints and get a woman and just do the business with them and forget it, like most young people do. You feel compromised and still sort of feeling like 'will I be able to satisfy my partner.' Not just sexually, other ways, like DIY, jobs round the house and all sorts. (Robertson, 2011, pp. 8–9)

It seems that in the presence of their disability, these men are often left with three coping strategies: they can reformulate their ideas of masculinity; ... rely on and promote certain hegemonic ideals of masculinity; ... or reject the mass societal norms and deny the norms' importance, creating another set of standards for themselves... (Gerschick & Miller, 1995).

When reformulating ideas of masculinity, these men usually focus on personal strengths and abilities, regardless of the ideal standards. This can include maneuvering an electric wheelchair or driving a specially equipped vehicle, tasks that would be very difficult for other people. Men who rely on hegemonic ideals are typically very aware of others' opinions of masculinity. These men internalized ideals such as physical and sexual prowess, and athleticism, though it can be nearly impossible for them to meet these standards. Then there are men who reject hegemonic masculinity. These men believe that masculine norms are wrong; they sometimes form their own standards for masculinity, which often go against what society thinks is right for men. Some men [tried] devaluing masculinity's importance altogether. The operative word is *try* because despite men's best efforts to reformulate or reject hegemonic masculinity, the expectations and ideals for men are far more pervasive than can be controlled. Many men trying to reformulate and reject masculine standards often end up "doing" gender appropriately in one aspect of life or another.

Indeed, some men find that hypermasculinity is the best strategy....

In today's world, men with disabilities fight an uphill battle against hegemonic masculinity—their position in the social order—and its many enforcers. Men with disabilities seem to scream, "I AM A STILL A MAN!" They try to make up for their shortcoming by overexaggerating the masculine qualities they still have, and society accommodates this via their support of disabled men's sexual rights and the sexist nature of medical rehabilitation programs and standards.

Gay Men

Male homosexuality has long been associated with effeminacy (i.e. not being a real man) throughout the history of Western societies; the English language is fraught with examples equating men's sexual desire for other men with femininity: molly and nancy-boy in 18th-century England, buttercup, pansy, and she-man of early 20th-century America, and the present-day sissy, fairy, queen, and faggot (Chauncey, 1994; Edwards, 1994; Pronger, 1990). Moreover, the pathologization of male homosexuality in the early 20th century led to a rhetoric of de-masculinization. By the 1970s, a number of psychiatric theorists referred to male homosexuality as "impaired masculine self-image" (Bieber, 1965), "a flight from masculinity" (Kardiner, 1963), "a search for masculinity" (Socarides, 1968), and "masculine failure" (Ovesey & Person, 1973).

Today in the United States, gay men continue to be marginalized by gender—that is, their masculinity is seen as problematic. In a survey of over 3,000 American adults (Levitt & Klassen, 1976), 69% believed homosexuals acted like the opposite sex, and that homosexual men were suitable only to the "unmasculine" careers of artist, beautician, and florist, but not the "masculine" careers of judges, doctors, and ministers. Recent studies have found similar results, despite the changing nature of gay rights in America (Blashill & Powlishta, 2009; Wright & Canetto, 2009; Wylie, Corliss, Boulanger, Prokop, & Austin, 2010).

The popular belief that gay men are not real men is established by the links among sexism (the systematic devaluation of women and "the feminine"), homophobia (the deep-seated cultural discomfort and hatred felt towards same-sex sexuality); and compulsory heterosexuality. Since heterosexuality is integral to the way a society is organized, it becomes a naturalized, "learned" behavior. When a man decides he is gay (if this "deciding" even occurs), he is rejecting the *compulsion* toward a heterosexual lifestyle and orientation (Rich, 1980).

More than this, though, compulsory heterosexuality is a mandate; society demands heterosexuality; our informal and formal policies and laws all reflect this (Fingerhut, Riggle, & Rostosky, 2011). And, in response, men find that one of the key ways to prove masculinity is to demonstrate sexual prowess....

The gay men who conform to hegemonic norms, secure their position in the power hierarchy by adopting the heterosexual masculine role and subordinating both women and effeminate gay men. Having noted that hypermasculine gay men have been accused of being "collaborators with patriarchy," Messner (1997) pointed out the prominence of hegemonic masculinity in gay culture: "it appears that the dominant tendency in gay culture eventually became an attempt to claim, eroticize, and display the dominant symbols of hegemonic masculinity" (p. 83).

Historically, camp and drag were associated with minstrelizers, those who exaggeratedly expressed stereotypic constructions of homosexual masculinity. The 1950s hairdresser, interior decorator and florist of classic cultural stereotype were embraced as lifestyle choices, if not yet a political position. Minstrelizers embraced the stereotypes; their effeminacy asked the question "who wants to be butch all the time anyway? It's too much work."

On the other hand, there was a group of effeminists who were explicitly political. As a political movement, effeminism emerged in the first years of the modern post-Stonewall Gay Liberation movement, but unlike it's normifying brethren, effeminists explicitly and politically rejected mainstream heterosexual masculinity. Largely associated with the work of Steve Dansky, effeminists published a magazine, Double F, and three men issued "The Effeminist Manifesto" (Dansky, Knoebel, & Pitchford, 1977).

The effeminists pointed to the possibilities for a liberated masculinity offered by feminism. Effeminism, they argued, is a positive political position, aligning anti-sexist gay men with women, instead of claiming male privilege by asserting their difference from women. Since, as Dansky et al. (1977) argued, male supremacy is the root of all other oppressions, the only politically defensible position was to renounce manhood itself, to refuse privilege. Dansky and his effeminist colleagues were as critical of mainstream gay male culture (and the denigration of effeminacy by the normifiers) as by the hegemomnic dominant culture....

Working Class Men

Working class men are, perhaps, an interesting reference group when compared to disabled men and gay men. The way(s) in which they are discriminated against or stigmatized seem very different. These men, in fact, are often seen as incredibly masculine; strong, stoic, hard-workers, there is something particularly masculine about what they have to do day-in and day-out. Indeed, the masculine virtues of the working class are celebrated as the physical embodiment of what all men should embrace (Gagnon & Simon, 1973; Sanders & Mahalingam, 2012).

Working-class White males may work in a system of male privilege, but they are not the main beneficiaries; they are in fact expendable. The working class is set apart from the middle and upper classes in that the working class is defined by jobs that require less formal education, sometimes (not always) less skill, and often low pay. For men, these jobs often include manual labor such as construction, automotive work, or factory work. The jobs these men hold are typically men-dominant.

If the stereotypic construction of masculinity among the working class celebrates their physical virtues, it also problematizes their masculinity by imagining them as dumb brutes. Working class men are the male equivalent of the "dumb blonde"—endowed with physical virtues, but problematized by intellectual shortcomings....

...[T]hose in the working, or blue-collar, class form a network of relationships with other blue-collar workers that serve to support them and give them a sense of status and worth, regardless of actual status or worth in the outside world (Cohen & Hodges, 1963). In fact, because those in the working class cannot normally exercise a great amount of power in their jobs or in many other formal relationships, they tend to do so in their relationships with other working class members....

However, for those who want to minimize the apparent differences between them and the more dominant masculine ideal, a site of normification could be the focus on all men's general relationship to women and the family. Those involved in the union movement, for example, stake claims to manhood and masculinity by organizing around the principal of men as breadwinners. The basic job that all "real men" should share is to provide for their wives and children. This would explain the initial opposition to women's entry into the workplace, and also now the opposition to gay men's and lesbian women's entrance. There is a type of White, male, working-class solidarity vis-à-vis privilege that these men have constructed and maintained, that promotes and perpetuates racism, sexism, and homophobia—the nexus of beliefs that all men are supposed to value (Embrick, Walther, & Wickens, 2007).

This power in the workplace translates directly to the home, as well. In the absence of legitimated hierarchical benefits and status, working class husbands and partners are more likely to "produce hypermasculinity by relying on blatant, brutal, and relentless power strategies in their marriages, including spousal abuse" (Pyke, 1996). However, violence can also extend outside the home....

CONCLUSION

Privilege is not monolithic; it is unevenly distributed and it exists worldwide in varying forms and contexts. Among members of one privileged class, other mechanisms of marginalization may mute or reduce privilege based on another status. Thus, a White gay man might receive race and gender privilege, but will be marginalized by sexuality.... [We] described these processes for three groups of men in the United States—men with disabilities, gay men, and working class men—who see their gender privilege reduced and their masculinity questioned, not confirmed, through their other marginalized status. We described strategies these men might use to restore, retrieve, or resist that loss. Using Goffman's discussion of stigma, we described three patterns of response. It is through these strategies—minsterlization, normification, and militant chauvinism—that a person's attempt to access privilege can be viewed, and, we argue, that we can better see the standards, ideals, and norms by which any society measures a man and his masculinity, and the benefits or consequences of his adherence or deviance.

REFERENCES

Bieber, I. (1965). Clinical aspects of male homosexuality. In J. Marmor (Ed.), *Sexual inversion: The multiple roots of homosexuality* (pp. 248–267). New York, NY: Basic Books.

Blashill, A. J., & Powlishta, K. K. (2009). The impact of sexual orientation and gender role on evaluations of men. *Psychology of Men & Masculinity, 10*(2), 160. doi: 10.1037/a0014583

Chauncey, George. (1994). *Gay New York: Gender, urban culture and the making of the gay male world*. New York, NY: Basic Books.

Cheng, C. (2008). Marginalized masculinities and hegemonic masculinity: An introduction. *The Journal of Men's Studies*, 7(3), 295–315.

Cohen, A. K., & Hodges Jr, H. M. (1963). Characteristics of the lower-blue-collar-class. *Social Problems, 10*(4), 303–334.

Connell, R. W. (1995). *Masculinities*. Berkeley: University of California Press.

Dansky, S., Knoebel, J., & Pitchford, K. (1977). The effeminist manifesto. In J. Snodgrass (Ed.), *A book of readings: For men against sexism* (pp. 116–120). Albion, CA: Times Change Press.

Edwards, T. (1994). *Erotics & politics: Gay male sexuality, masculinity, and feminism*. New York, NY: Routledge.

Embrick, D. G., Walther, C. S., & Wickens, C. M. (2007). Working class masculinity: Keeping gay men and lesbians out of the workplace. *Sex Roles, 56*(11), 757–766.

Fingerhut, A. W., Riggle, E. D. B., & Rostosky, S. S. (2011). Same-sex marriage: The soial and psychological implications of policy and debates. *Journal of Social Issues, 67*(2), 225–241.

Gagnon, J. H., & Simon, W. (1973). *Sexual conduct: The social origins of human sexuality*. Chicago, IL: Aldine.

Gerschick, T. J., & Miller, A. S. (1995). Coming to terms: Masculinity and physical disability. In D. Sabo & D. F. Gordon (Eds.), *Men's health and illness: Gender, power, and the body, Research on men and masculinities series* (Vol. 8, pp. 183–204). Thousand Oaks, CA: Sage Publications.

Kardiner, A. (1963). The flight from masculinity. In H. M. Ruisenbeck (Ed.), *The problem of homosexuality in modern society* (pp. 17–39). New York, NY: Dutton.

Levitt, E. E., & Klassen, A. D. (1976). Public attitudes toward homosexuality. *Journal of Homosexuality, 1*(1), 29–43. doi: 10.1300/J082v01n01_03.

McIntosh, P. (1988). *White privilege and male privilege: A personal account of coming to see correspondences through work in women's studies.* Working Paper no. 189. Wellesley, MA: Wellesley College Center for Research on Women.

Messner, M. A. (1997). *Politics of masculinities: Men in movements.* New York, NY: Sage.

Ovesey, L., & Person, E. (1973). Gender identity and sexual psychopathology in men: A psychodynamic analysis of homosexuality, transsexualism, and transvestism. *Journal of the American Academy of Psychoanalysis and Dynamic Psychiatry, 1*(1), 53–72.

Pratto, F., & Stewart, A. L. (2012). Group dominance and the half-blindness of privilege. *Journal of Social Issues, 68*(1), 28–45. doi: 10.1111/j.1540-4560.2011.01734.x.

Pronger, B. (1990). Gay jocks: A phenomenology of gay men in athletics. *Sport, men, and the gender order: Critical feminist perspectives*, 141–152.

Pyke, K. D. (1996). Class-based masculinities: The interdependence of gender, class, and interpersonal power. *Gender & Society, 10*(5), 527–549.

Rich, A. (1980). Compulsory heterosexuality and lesbian existence. *Signs, 5*(4), 631–660.

Robertson, Steve. (2011). *Disabled men conceptualising health*, Unpublished paper.

Sanders, M. R., & Mahalingam, R. (2012). Under the radar: The role of invisible discourse in understanding class-based privilege. *Journal of Social Issues, 68*(1), 112–127. doi: 10.1111/j.1540-4560.2011.01739.x.

Socarides, C. W. (1968). A provisional theory of aetiology in male homosexuality—a case of preoedipal origin. *International Journal of Psycho-Analysis, 49*, 27–37.

West, C., & Fenstermaker, S. (1995). Doing difference. *Gender & Society, 9*(1), 8–37.

West, C., & Zimmerman, D. (1987). Doing gender. *Gender & Society, 1*(2), 125–151.

Woods, J. (2010). The black male privileges checklist. In M. Kimmel & A. L. Ferber (Eds.), *Privilege: A Reader* (2nd ed., pp. 27–37). Boulder, CO: Westview Press.

Wright, S. L., & Canetto, S. (2009). Stereotypes of older lesbians and gay men. *Educational Gerontology, 35*(5), 424–452. doi: 10.1080/03601270802505640.

Wylie, S. A., Corliss, H. L., Boulanger, V., Prokop, L. A., & Austin, S. B. (2010). Socially assigned gender nonconformity: A brief measure for use in surveillance and investigation of health disparities. *Sex Roles*, 1–13.

23

The Myth of the Latin Woman

I Just Met a Girl Named María

JUDITH ORTIZ COFER

On a bus trip to London from Oxford University, where I was earning some graduate credits one summer, a young man, obviously fresh from a pub, spotted me and as if struck by inspiration went down on his knees in the aisle. With both hands over his heart he broke into an Irish tenor's rendition of "María" from *West Side Story*. My politely amused fellow passengers gave his lovely voice the round of gentle applause it deserved. Though I was not quite as amused, I managed my version of an English smile: no show of teeth, no extreme contortions of the facial muscles—I was at this time of my life practicing reserve and cool. Oh, that British control, how I coveted it. But María had followed me to London, reminding me of a prime fact of my life: you can leave the Island, master the English language, and travel as far as you can, but if you are a Latina, especially one like me who so obviously belongs to Rita Moreno's gene pool, the Island travels with you.

This is sometimes a very good thing—it may win you that extra minute of someone's attention. But with some people, the same things can make *you* an island—not so much a tropical paradise as an Alcatraz, a place nobody wants to visit. As a Puerto Rican girl growing up in the United States and wanting like most children to "belong," I resented the stereotype that my Hispanic appearance called forth from many people I met.

Our family lived in a large urban center in New Jersey during the sixties, where life was designed as a microcosm of my parents' casas on the island. We spoke in Spanish, we ate Puerto Rican food bought at the bodega, and we practiced strict Catholicism complete with Saturday confession and Sunday mass at a church where our parents were accommodated into a one-hour Spanish mass slot, performed by a Chinese priest trained as a missionary for Latin America.

As a girl I was kept under strict surveillance, since virtue and modesty were, by cultural equation, the same as family honor. As a teenager I was instructed on how to behave as a proper señorita. But it was a conflicting message girls got, since the Puerto Rican mothers also encouraged their daughters to look and act like women and to dress in clothes our Anglo friends and their mothers found too "mature" for our age. It was, and is, cultural, yet I often felt humiliated when

SOURCE: Cofer, Judith Ortiz. 1993. *The Latin Deli: Prose & Poetry*. Pp. 148–154. Athens, GA: The University of Georgia Press.

I appeared at an American friend's party wearing a dress more suitable to a semi-formal than to a playroom birthday celebration. At Puerto Rican festivities, neither the music nor the colors we wore could be too loud. I still experience a vague sense of letdown when I'm invited to a "party" and it turns out to be a marathon conversation in hushed tones rather than a fiesta with salsa, laughter, and dancing—the kind of celebration I remember from my childhood.

I remember Career Day in our high school, when teachers told us to come dressed as if for a job interview. It quickly became obvious that to the barrio girls, "dressing up" sometimes meant wearing ornate jewelry and clothing that would be more appropriate (by mainstream standards) for the company Christmas party than as daily office attire. That morning I had agonized in front of my closet, trying to figure out what a "career girl" would wear because, essentially, except for Marlo Thomas on TV, I had no models on which to base my decision. I knew how to dress for school: at the Catholic school I attended we all wore uniforms; I knew how to dress for Sunday mass, and I knew what dresses to wear for parties at my relatives' homes. Though I do not recall the precise details of my Career Day outfit, it must have been a composite of the above choices. But I remember a comment my friend (an Italian-American) made in later years that coalesced my impressions of that day. She said that at the business school she was attending the Puerto Rican girls always stood out for wearing "everything at once." She meant, of course, too much jewelry, too many accessories. On that day at school, we were simply made the negative models by the nuns who were themselves not credible fashion experts to any of us. But it was painfully obvious to me that to the others, in their tailored skirts and silk blouses, we must have seemed "hopeless" and "vulgar." Though I now know that most adolescents feel out of step much of the time, I also know that for the Puerto Rican girls of my generation that sense was intensified. The way our teachers and classmates looked at us that day in school was just a taste of the culture clash that awaited us in the real world, where prospective employers and men on the street would often misinterpret our tight skirts and jingling bracelets as a come-on.

Mixed cultural signals have perpetuated certain stereotypes—for example, that of the Hispanic woman as the "Hot Tamale" or sexual firebrand. It is a one-dimensional view that the media have found easy to promote. In their special vocabulary, advertisers have designated "sizzling" and "smoldering" as the adjectives of choice for describing not only the foods but also the women of Latin America. From conversations in my house I recall hearing about the harassment that Puerto Rican women endured in factories where the "boss men" talked to them as if sexual innuendo was all they understood and, worse, often gave them the choice of submitting to advances or being fired.

It is custom, however, not chromosomes, that leads us to choose scarlet over pale pink. As young girls, we were influenced in our decisions about clothes and colors by the women—older sisters and mothers who had grown up on a tropical island where the natural environment was a riot of primary colors, where showing your skin was one way to keep cool as well as to look sexy. Most important of all, on the island, women perhaps felt freer to dress and move more provocatively, since, in most cases, they were protected by the traditions, mores, and laws of a Spanish/Catholic system of morality and machismo whose main rule was: *You may look at my sister, but if you touch her I will kill you.* The extended family and church structure could provide a young woman with a circle of safety in her small pueblo on the Island; if a man "wronged" a girl, everyone would close in to save her family honor.

This is what I have gleaned from my discussions as an adult with older Puerto Rican women. They have told me about dressing in their best party clothes on Saturday nights and going to the town's plaza to promenade with their girlfriends in front of the boys they liked. The males were thus given an opportunity to admire the women and to express their admiration in the form of *piropos:* erotically charged street poems they composed on the spot. I have been subjected to a few piropos while visiting the Island, and they can be outrageous, although custom dictates that they must never cross into obscenity. This ritual, as I understand it, also entails a show of studied indifference on the woman's part; if she is "decent," she must not acknowledge the man's impassioned words. So I do understand how things can be lost in translation. When a Puerto Rican girl dressed in her idea of what is attractive meets a man from the mainstream culture who has been trained to react to certain types of clothing as a sexual signal, a clash is likely to take place. The line I first heard based on this aspect of the myth happened when the boy who took me to my first formal dance leaned over to plant a sloppy overeager kiss painfully on my mouth, and when I didn't respond with sufficient passion said in a resentful tone: "I thought you Latin girls were supposed to mature early"—my first instance of being thought of as a fruit or vegetable—I was supposed to *ripen,* not just grow into womanhood like other girls.

It is surprising to some of my professional friends that some people, including those who should know better, still put others "in their place." Though rarer, these incidents are still commonplace in my life. It happened to me most recently during a stay at a very classy metropolitan hotel favored by young professional couples for their weddings. Late one evening after the theater, as I walked toward my room with my new colleague (a woman with whom I was coordinating an arts program), a middle-aged man in a tuxedo, a young girl in satin and lace on his arm, stepped directly into our path. With his champagne glass extended toward me, he exclaimed, "Evita!"

Our way blocked, my companion and I listened as the man half-recited, half-bellowed "Don't Cry for Me, Argentina." When he finished, the young girl said: "How about a round of applause for my daddy?" We complied, hoping this would bring the silly spectacle to a close. I was becoming aware that our little group was attracting the attention of the other guests. "Daddy" must have perceived this too, and he once more barred the way as we tried to walk past him. He began to shout-sing a ditty to the tune of "La Bamba"—except the lyrics were about a girl named María whose exploits all rhymed with her name and gonorrhea. The girl kept saying "Oh, Daddy" and looking at me with pleading eyes. She wanted me to laugh along with the others. My companion and I stood silently waiting for the man to end his offensive song. When he finished, I looked not at him but at his daughter. I advised her calmly never to ask her father what he had done in the army. Then I walked between them and to my room. My friend complimented me on my cool handling of the situation. I confessed to her that I really had wanted to push the jerk into the swimming pool. I knew that this same man—probably a corporate executive, well educated, even worldly by most standards—would not have been likely to regale a white woman with a dirty song in public. He would perhaps have checked his impulse by assuming that she could be somebody's wife or mother, or at least *somebody* who might take offense. But to him, I was just an Evita or a María: merely a character in his cartoon-populated universe.

Because of my education and my proficiency with the English language, I have acquired many mechanisms for dealing with the anger I experience. This was not true for my parents, nor is it true for the many Latin women working at menial jobs who must put up with stereotypes about our ethnic group such as: "They make good domestics." This is another facet of the myth of the Latin women in the United States. Its origin is simple to deduce. Work as domestics, waitressing, and factory jobs are all that's available to women with little English and few skills. The myth of the Hispanic menial has been sustained by the same media phenomenon that made "Mammy" from *Gone with the Wind* America's idea of the black woman for generations; María, the housemaid or counter girl, is now indelibly etched into the national psyche. The big and the little screens have presented us with the picture of the funny Hispanic maid, mispronouncing words and cooking up a spicy storm in a shiny California kitchen.

This media-engendered image of the Latina in the United States has been documented by feminist Hispanic scholars, who claim that such portrayals are partially responsible for the denial of opportunities for upward mobility among Latinas in the professions. I have a Chicana friend working on a Ph.D. in philosophy at a major university. She says her doctor still shakes his head in puzzled amazement at all the "big words" she uses. Since I do not wear my diplomas around my neck for all to see, I too have on occasion been sent to that "kitchen," where some think I obviously belong.

One such incident that has stayed with me, though I recognize it as a minor offense, happened on the day of my first public poetry reading. It took place in Miami in a boat-restaurant where we were having lunch before the event. I was nervous and excited as I walked in with my notebook in hand. An older woman motioned me to her table. Thinking (foolish me) that she wanted me to autograph a copy of my brand new slender volume of verse, I went over. She ordered a cup of coffee from me, assuming that I was the waitress. Easy enough to mistake my poems for menus, I suppose. I know that it wasn't an intentional act of cruelty, yet with all the good things that happened that day, I remember that scene most clearly, because it reminded me of what I had to overcome before anyone would take me seriously. In retrospect I understand that my anger gave my reading fire, that I have almost always taken doubts in my abilities as a challenge—and that the result is, most times, a feeling of satisfaction at having won a convert when I see the cold, appraising eyes warm to my words, the body language change, the smile that indicates that I have opened some avenue for communication. That day I read to that woman and her lowered eyes told me that she was embarrassed at her little faux pas, and when I willed her to look up to me, it was my victory, and she graciously allowed me to punish her with my full attention. We shook hands at the end of the reading, and I never saw her again. She has probably forgotten the whole thing but maybe not.

Yet I am one of the lucky ones. My parents made it possible for me to acquire a stronger footing in the mainstream culture by giving me the chance at an education. And books and art have saved me from the harsher forms of ethnic and racial prejudice that many of my Hispanic *compañeras* have had to endure. I travel a lot around the United States, reading from my books of poetry and my novel, and the reception I most often receive is one of positive interest by people who want to know more about my culture. There are, however, thousands of Latinas without the privilege of an education or the entrée into society that I have. For them life is a struggle against the misconceptions perpetuated by the myth of the Latina as whore, domestic, or criminal. We cannot change this by legislating the way people look at us. The transformation, as I see it, has to occur at a much more individual level. My personal goal in my public life is to try to replace the old pervasive stereotypes and myths about Latinas with a much more interesting set of realities. Every time I give a reading, I hope the stories I tell, the dreams and fears I examine in my work, can achieve some universal truth which will get my audience past the particulars of my skin color, my accent, or my clothes.

I once wrote a poem in which I called us Latinas "God's brown daughters." This poem is really a prayer of sorts, offered upward, but also, through the human-to-human channel of art, outward. It is a prayer for communication, and for respect. In it, Latin women pray "in Spanish to an Anglo God/with a Jewish heritage," and they are "fervently hoping/that if not omnipotent,/at least He be bilingual."

24

Keep Your "N" in Check

African American Women and the Interactive Effects of Etiquette and Emotional Labor

MARLESE DURR AND ADIA M. HARVEY WINGFIELD

INTRODUCTION

Perhaps more than any other First Lady since Hillary Clinton, Michelle Obama faces considerable pressure to transform, change, and adapt her persona to become more palatable to a broad spectrum of voters. She has been alternately cast as unpatriotic and as an angry, dangerous black woman (Blitt, 2008). Even while she is admired for her fashion sense, she is simultaneously carefully scrutinized and often perceived as too aggressive and pushy.

While many of the same criticisms were applied to Hillary Clinton in 1992, intersections of gender and race create a particular and somewhat unique situation for Michelle Obama. Hillary Clinton was cast as an overbearing, emasculating feminist, but Michelle Obama faces gendered racial stereotypes of the "angry black woman," an image grounded in the Sapphire stereotype of black women as domineering, vociferous, and curt. Like the image of the angry black woman, Sapphire serves to reinforce ideas of black women's inherent lack of femininity and worth (Collins, 2004). Thus, Michelle Obama has struggled to distance herself from these stereotypical images and behaviors in hopes of altering the way she is perceived.

Her challenge is familiar to many professional black women, who like her must transform or alter themselves to be welcomed and accepted in their workplaces. The nature of social relationships in the office is dictated by historical customs, which have been a traditionally white male citadel. Not until after Executive Order 11246 mandating affirmative action was implemented... did employers actively seek ways to include women and racial minorities within organizations' and agencies' hierarchies. Men, the carriers of organizational culture and authority (Acker, 1990; Kanter, 1977) created this new bureaucratic arm

SOURCE: Durr, Marlese, and Adia M. Harvey Wingfield. 2011. "Keep Your 'N' in Check." *Critical Sociology* 37: 557–571.

of professional and managerial expansion, making decisions regarding acceptable behavior, communication, skin color, style and dress. (e.g. dark suits, conservative dresses, white shirts, low heels, and no flashy jewelry, hair, or make-up) concentrating on who their clients are and their cultural tastes. This has had implications for women of all races who become employed within this new-found bureaucratic configuration of organizational norms, but has had particular consequences for women of color, who do not fit either the gendered or racialized norms of these environments.

By establishing an implicitly gendered and racialized culture, obstacles remain for women. Acker (1990) and Kanter (1977) argue that organizations are not the gender-neutral bureaucracies they purport to be. Rather, they are gendered in ways that often locate women in dead-end jobs, with exposure to the organizational hierarchy as tokens. Yet this hierarchy is two-tiered. For women of color, especially those in professional posts, these disadvantages are further complicated by race. These women often explain that upon entering the labor market as professionals, they alter their behavior by changing their look, conversation content, and style to fit in, but also to be promoted (Jones and Shorter-Gooden, 2003). They often speak of performance weariness when describing their spoken and unspoken communicative interaction exchanges with white colleagues. Many simply state that they feel they are in a "parade" where they are being judged for appearance, personal decorum, communication skills, and emotion management in addition to work productivity.

To address this, senior professional women like Kenya counsel: "That's not professional. Remember they got the s[hit] that'll get you bit! Keep your Negro in check! Don't let it jump up and show anger, disapproval, or difference of opinion. They have to like you and think that you are as close to them as possible in thought, ideas, dress, and behavior." Her advice discloses the appropriate etiquette, behavior, and emotion management, but also instructs other black women to blend manners, behavior, and reaction to fashion satisfactory workplace deportment. The counseling given to these women is directly linked to handling stress and alienation while balancing a need for survival and safety in the workplace or remaining employed without a row.

Many of these women have stated, "The work is too much. I get tired of being 'on' for [white] colleagues who scrutinize every behavior. So every now and then, I lose it." Others have said, "They [white supervisors] make deals about the next position or talk about my future being bright. They say I have time, so the next promotion available is mine. But it never happens. For some reason an organizational change erases the promised promotion. It just never happens, unless there is pressure to promote black." Even then, according to many of these women, African American "institutional gatekeepers" are consulted, who are often black women and men that possess institutional acceptance, but are not advisors that navigate new employees through their frosty work environs. Consequently, the challenges these women face involve doing the necessary work to fit in, and managing their feelings in an often inhospitable workplace in hopes of capturing some degree of professional success in the form of promotion.

Working in predominantly white agencies, organizations, and institutions, while living and working as "black," may cause part of these women's apprehension and estrangement. The ever present reminder of their master status—skin color—makes the work week a bit more difficult and requires a bit more strength. Many believe they continue to carry stereotypical and media-based depictions of them as domineering, unaccomplished breeders, whores, welfare queens, as well as confrontational. Bebe reports an administrator at her school saying "I am not afraid of aggressive women." So, "performance" becomes their safety mechanism.

These workplace contexts exact an additional toll as black women also must manage the demands of emotional labor in these work environments. Coined by Hochschild (1983), emotional labor describes women's work experiences in service economies as producing emotions in themselves or others. Emotions, which are typically gendered, become commodified and sold for a wage. Doing emotional labor, women are often expected to recreate gender appropriate feelings, e.g. paralegals are expected to make men attorneys feel cared for (Pierce, 1999), while male bill collectors induce fear and intimidation (Hochschild, 1983). However, for African American women, performance related to whites' gendered and racialized expectations is rarely accounted for or described.

Research that describes the down-to-earth contexts and contests for African American women in organizations, institutions, and agencies continues to be limited and comparative when describing mobility outcomes for black and white women. We seek to situate black women's voices and experiences at the center of discussions and research on occupational mobility by examining the interactive effects of manner, behavior, and reaction or etiquette and emotional labor. We argue that this aperture in the literature lacks a definition of informal and/or formal race-based boundary maintenance in the workplace for black women. Etiquette and emotional labor for African American women is defined as performance to describe two levels of personal deportment:

1. a generalized bureaucratic passive aggressive level; and
2. a race-based set of expectations grounded in survival strategies to cope with challenges they face in environments that are unwelcoming and possess concrete ceilings across organizations and occupations....

DATA AND METHODS

Data for this article was collected through two sources. One involved using direct participant observation from 2005 to 2007 while in conversation with 20 African American women over occupational mobility issues. Because conversations addressed issues in the workplace and their feelings about executives, managers, supervisors, and co-workers, one author listened attentively to concerns presented, speaking only when directly pulled into the discussions, as a sociologist who examined race/gender and would write about these tete-a-tetes. These women were employed as lower-level executives, middle managers, and administrators in public and private organizations, state and city government, and universities.

The use of a snowball or probabilistic sampling procedure was not possible, because many women did not wish to complete surveys and stated their desire to remain anonymous, and in many cases used coded responses or vernacular terms which shielded bureaucratic personalities they spoke of (e.g., "boyfriend," "girlfriend"), while ever so briefly and scantily describing issues in their workplaces. Participant observation allowed one author to capture etiquette and emotional labor responses in their work-social relationships. Conversations took place at sorority and fraternity dinner dances, church and community luncheons, occasional meetings at work or while shopping at local grocery or department stores. Pseudonyms will be used for confidentiality.

All respondents were college graduates, with master's and doctoral degrees, ranging in age from 30 to 55 years of age, five were married, five were divorced, and 10 had never married. Respondents were from southern New England, mid-Atlantic, and Midwestern states and employed in professional or managerial positions that contained less than 10 black employees throughout the levels of administration or management, and required working effectively in workgroups as an integral factor for employment success and /or future promotability.

Additional data came from a larger sample of 25 in-depth, semi-structured interviews with African American professionals. These respondents were located through a snowball sample. Interviews focused on the use of emotional labor in predominantly white workplaces. These respondents were employed in various professional posts in work environments where they estimated African Americans constituted 10 percent or fewer of professional employees. Our questions centered on whether and when respondents had to control emotions at work, in response to what occurrences, the frequency of this practice, and the nature of the emotions being controlled. Respondents also discussed producing emotions in others and the context in which this occurred....

FINDINGS

For professional black women, the performances that they feel compelled to give are shaped by the ways intersections of race and gender isolate them and place them under greater scrutiny. As they take stock of their work environments and perceive colleagues' stereotypes, beliefs, and preconceptions, these women learn that, like Michelle Obama, they must repackage themselves in ways that are more palatable to their white co-workers. As these colleagues' goodwill and collegiality is necessary for advancement and occupational stability, black women professionals find themselves doing both surface acting and emotional labor in order to successfully integrate their work spaces (Hochschild, 1983)....

Many of these women work in administrative roles. To an extent, this shapes the ways in which they must engage in careful self-presentation. Their positions may not originate within the organization, or link to its mission and objectives, but the series of administrative posts they have held or hold, regardless of their nature, provides an avenue and motivation for advancement. Tandy states, "I got my job as a manager because they had no black managers, but guess what I manage? I manage compensatory education staff and programs." Elizabeth says, "I do the same, but it's the only way to become a manager at ... This place does not care about their students, just the federal aid and visibility of their darkies. That's when they get them and if they stay. You are on for them all the time. But you take what you can get."

When, like Michelle Obama, these women advance, their self-presentation and communication style are scrutinized. Barbara reports, "Being direct and speaking your mind is never encouraged. In fact if you do, you encounter a world of silence and avoidance, which is one of the most severe penalties. You are placed outside of the loop, and you may stay there for a long time. Quite possibly—permanently. So despite the fact you may have a contribution, it is not welcomed." So, as they learn the verbal and body language of bureaucracy, they must negate values and styles of communication developed as a survival skill in their community. Most say they feel defenseless.

Nevertheless, for many, negotiation within this environment in most cases occurs at a cost. Sharon reports, "You learn to remain quiet and speak when spoken to and never verbalize your thoughts on an issue or policy." Others, such as Rhonda, suggest that "You go with the flow, since you realize this may not be a battle you can win, despite the fact you may be correct in assumptions and remedies you have in mind." They apply a survival-safety analysis and render their verbal participation to a lower-level of priority. They smile on cue, remaining expressionless, unmoved by the content of conversation even if the subject is distressing or controversial, and carefully couching responses in the language of the workplace. The emotion that is concealed is evident in their voices, body language, and style of conversation. Keena states, "But you have to endure this if you want to get ahead, regardless of where you might be promoted to." Mary says, "Promotions aren't everything, surviving them is. I got kids in college, a mortgage, and my hair to keep up." Laughter follows Mary's comments, but all understand what she means and silently accept her pronouncement.

Black women also engage in etiquette and emotional labor to cope with feelings of alienation and loneliness that stem from being the only, or one of few. Harlow (2003) has documented the existence of emotion management among black college and university professors. Feelings of anger, frustration, and aggravation are often stifled to conform to colleague expectations. Janice, an assistant Dean at a small liberal arts college, states, "I've dealt with people who were so dismissive, and you knew race was at the core of it. But I would have to grin and bear it, because I needed to work ... very rarely do you see me expressing my true feelings, and when I do, my reaction tells me it scares the hell out of [my white colleagues]." Irene offers a very astute assessment of the importance of concealing emotions: "If you don't play by their rules in terms of your behavior ... modulating your emotions [and] if you don't do certain things the way they want you to do them, it has a direct impact on your career and your economic stability." Similarly, Gina states, "I have to be very congenial. Sometimes I don't want to deal with [racism]. But I just have to hide my real emotions."

For many of these women, "performance" becomes their safety mechanism. In some cases, these performances are conscious and intentional—what Hochschild (1983) describes as "surface acting," where the individual is well aware that they are putting on a show. In an example of this, Barbara states, "I always prepare because I want to make sure my temper is in check. You know, I'm mellow. I have unhooked from personal feelings, previous conversations to deal with what's ahead. I do this because so few [black] women are in managerial and executive level posts." Barbara's performance is purposely crafted, specifically designed for the racial parameters of her work setting.

More often than not, workplace dynamics place these women in positions where routine interaction with co-workers heightens these performances. For instance, Deidre, a faculty member at a southern New England university, says "There is a "double standard" in departmental decisions regarding promotion and tenure." She reports, despite her service on the departmental committee to determine promotion procedures, the procedures have been used arbitrarily by those in authority. She notes, "African Americans are denied promotion while whites with fewer publications and lower scholarly profiles were promoted without question." Sheila reports slights and loud verbal assaults in the hallway by department colleagues and secretarial staff. She felt isolated and alienated by colleagues who made a concerted effort to have her removed from the teaching staff. Like Sheila, Janice, a bi-racial woman found herself in a verbal confrontation with her white female department chair in the doorway of her office. This confrontation was the culmination of several months of what she perceived as constant harassment regarding petty issues. During the verbal confrontation, she was poked in the chest by the chair as she loudly chastised her publicly.

When incidents like these occur, emotional labor becomes a key part of the resulting performance. Charlotte, the only black attorney at a mid-sized firm, states, "I think the one that stands out the most is that I am a black professional woman in [names town]. And as much as I'd like to be upset about it, I can't ever do that in my job because it would come off the wrong way. So I have to be happy-go-lucky, everything's great, and the fact that I don't have a life is wonderful when it really sucks!" Similarly, Giselle states, "At one point I was the only person of color who was not cooking, cleaning, or maintaining the grounds. It got so bad that the [black] men, they had their social support group and they invited me to join because they knew how that felt." This respondent's comments led us to revisit Rosenbaum's (1979) tournament mobility thesis through conversation. Giselle stated, "So you mean we need to be seen and supported as one of the stars." We say, "Yes." She says, "Right, that will take until I retire." Shana remarked that she feels she will never be promoted. For black women, internal promotions are a primary means of career mobility since most organizational vacancies are filled from within organizations, often as a matter of official and unofficial policy, making promotions a high reward for white collar workers (Markham et al., 1985)....

CONCLUSION

The number of African American women who now try to penetrate the "concrete ceiling" recognize promotions as their primary avenue of career mobility, given that a great many organizational vacancies are filled within and/or across organizations, often as a matter of official organizational policy. A great majority of the time these considerations take place without regard for their race and gender. They work even harder to climb the ladder to receive each position's resources and rewards.

Generally, emotional labor enables black women to present the appropriate emotional veneer that allows them to fit in and enhance their compatibility with organizational norms. This is particularly useful when confronted with racial issues. Their desire to move up the ladder in many ways represents a voluntary job shift with the same employer and acts as a proxy for vertical advance in the firm or with different employers. They are aware that occupational mobility also depends upon personal, social, and cultural characteristics. More important, they understand that the process of occupational mobility differs depending upon whether it occurs within a particular organization or involves a change of employers (Femlee, 1982). It may be argued that this is true for all women, but nuanced when it comes to African American women.

Or is it that many African Americans know what Bell and Hartmann (2007) suggest: that race is ever present in workplace diversity discussion, but missing in action? The discussions about race do not necessarily translate into decisive action intended to minimize racial inequality. African American women's need to present the appropriate etiquette and emotional labor is likely shaped by tenuous commitment to racial diversity in professional workplaces.

The levels of behavioral expectation and exceptions are both boundary maintenance mechanisms for whites and social-psychological issues of safety for African American women. While calls for inclusiveness and diversity weave seamlessly at work and in the larger society, rejection of difference remains as well. Our analysis reveals that social expectations shape individuals' fit at work and in society, while prescribing the conditions and consequences for integration within employment and social communities and while strengthening ethnic group solidarity.

They believe their social, cultural, and occupational location remains beleaguered by stereotypes beached in psychological needs. Nowhere is this seen more clearly than in a professional community where interaction is close and constant, varied but integrated, but laced with a strong sense of propagandized social acceptance. In such communities, individuals learn more about themselves based on exchange relationships when a sense of "community" and "belonging" is initiated and achieved for most of its members through working to build the community. However, in some instances, becoming part of such a community, especially for persons of color, is a journey into remembering, as well as understanding, that who we are, what we are, where we fit, and how we are received marks our continued journey.

Moreover, these results suggest that the challenges of the professional workplace are shaped in important ways by race and gender. For black women workers, attempting to perform the appropriate emotional labor while simultaneously conforming to etiquette norms creates specific issues that may not be present for other race/gender groups. Future research should consider whether these social and professional expectations pose the same challenges for Latinas, black men, Asian American women, and others. The expected norms, sanctions, and rules of the professional workplace are not neutral, but raced and gendered in ways that may have a different impact on various groups. Future research should consider how this plays out for other populations....

REFERENCES

Acker J (1990) Hierarchies, jobs, bodies: a theory of gendered organizations. *Gender and Society* 4(2): 139–158.

Bell JM and Hartmann D (2007) Diversity in everyday discourse: the cultural ambiguities and consequences of 'happy talk'. *American Sociological Review* 72(6): 895–912.

Blitt B (2008) The polities of fear. *New Yorker Magazine* 84(21): cover.

Collins PH (2004) *Black Sexual Politics*. New York, NY: Routledge.

Femlee DH (1984) The dynamics of women's job mobility. *Work and Occupations* 11(3): 259–281.

Harlow R (2003) Race doesn't matter but...: the effect of race on professors' experiences and emotion management in the undergraduate college classroom. *Social Psychology Quarterly* 66(4): 348–363.

Hochschild AR (1983) Emotional work, feeling rules, and social structure. *American Journal of Sociology* 85(3): 551–575.

Jones C and Shorter-Gooden K (2003) *Shifting: The Double Lives of African American Women in America*. Darby, PA: Diane Publishing Company.

Kanter RM (1977) *Men and Women of the Corporation*. New York, NY: Basic Books.

Markham WT, South SJ, Bonjean CM, et al. (1985) Gender and opportunity in the Federal bureaucracy. *American Journal of Sociology* 9(1): 129–150.

Pierce JL (1999) Emotional labor among paralegals. *Annals of the American Academy of Political and Social Sciences* 561(1): 127–142.

Rosenbaum JE (1979) Tournament mobility: career patterns in a corporation. *Administrative Science Quarterly* 24(2): 220–241.

25

The Gendered Rice Bowl

The Sexual Politics of Service Work in Urban China

AMY HANSER

In the 1960s, China's "iron girl brigades," crack teams of women agricultural and industrial workers, were models of a socialist future. Pictured with vigorous bodies, strong arms, and robust physiques, these women's youth and gender enabled them to represent "the forefront of history" and "the springtime of socialism" (Chen 2003a, 276). They were portrayed as key actors in forging China's "iron rice bowl," the guaranteed employment and material sustenance that state socialism promised China's urban workers. By the 1980s, however, and in the context of China's economic reforms, this image of women workers was ridiculed as an "unnatural" product of the country's failed experiment with socialism (Honig and Hershatter 1988, 24–25).

By contrast, in China today, an admired woman worker is likely to be a model—a fashion model. Increasingly, young urban women draw on the "rice bowl of youth," converting their youth and beauty into employment opportunities. Young women whose youth might once have symbolized the potential of socialist revolution are instead called on to represent a new, capitalist modernity. Meanwhile, socialism has been remapped onto the bodies of middle-aged women, who now represent the inefficiencies and backwardness of state socialism. This is especially true in the service sector, where the most exclusive stores in particular expect their salesclerks to exemplify a new, modern version of femininity associated with affluence, luxury, and deferential service....

This article considers the implications of these new conceptions of gender and sexuality for China's service workers. It asks how and why conceptions of gender and sexuality have become so integral to the organization of new service work regimes in China. Drawing insights from the literatures on gender and class, sexuality and service work, this article shows how essentialized conceptions of gender and sexuality powerfully communicate class distinctions in service settings through associations with the imagery of China's shift from socialism to a marketized society. In particular, I show how elite service organizations in China

---

AUTHOR'S NOTE: *This research was conducted with generous financial support from the Fulbright IIE, the Social Science Research Council's International Dissertarion Research Fellowship program, and a predissertation fellowship from the National Science Foundation.*

rely on new gender norms to choreograph the behaviors and dispositions of their female workforce, in the process distinguishing themselves from other service settings located further down the social hierarchy....

GENDER, CLASS, AND EMBODIED LABOR IN THE POSTSOCIALIST SERVICE SECTOR

...[G]ender and sexuality, are integral to the organization of work and the reproduction of inequality....

Sexuality often operates in tandem with workplace gender norms and expectations, and in service work settings in particular, sexualized femininity can be a key part of the mechanisms of workplace control (Adkins 2002; Loe 1996; Pringle 1988)....

In service settings, not only do sexuality and gender operate as mechanisms for control; they also can be key aspects of what workers are expected to produce. The woman service worker may be expected to reproduce feminine norms of caretaking and emotion work (Hochschild 1983), to produce an aura of sexualized femininity for the consumption of male patrons (Paules 1991; Loe 1996) or bosses (Pringle 1988, 1992), or to convey an image of normative feminine beauty for female customers (Lan 2003). Employers draw not only workers' physical labor activities but also their personalities, sexual identities, and bodily dispositions into a circuit of both material and symbolic production (Lan 2003)....

...[A] consideration of class relations has largely dropped out of examinations of gendered service work, even though gender formulations are fundamentally linked to class distinctions. Feminist scholars have reminded us that performances of femininity are always class coded (Bettie 2003; Steedman 1987)....

The production of class distinctions is also an integral element of many forms of service work. Rachel Sherman's (2003) research shows how luxury hotels act as sites for the enactment and legitimization of inequality. I argue here that in many service work settings, managers deploy seemingly "natural" gender and sexuality categories to communicate class distinctions, and that this "naturalness" simultaneously obscures the class distinctions conveyed. This is why, in part, configurations of gender and sexuality are so central to the organization of service work (see, e.g., Benson 1986).

Second, I suggest that an inattention to the interconnectedness of gender and class in service work settings makes it impossible to see how gender norms and sexual identities directed at women in one work setting are constructed in relation to other groups of women. In other words, gender norms and normative sexuality operate comparatively and serve to create distinctions among women as much as between women and men. The feminized, sexualized service worker communicates class distinctions by performing class-coded forms of femininity governed by relational gender and sexual norms.

Given China's context of rapid and dramatic social change, my case studies shed light on the relationship among class, gender, and sexuality in the service work setting. In particular, I argue that gender and the powerful imagery of China's transition from a socialist planned economy to a market economy have become key avenues through which class distinctions in urban China are conveyed. In contemporary China, the production of class distinctions in service settings, and by extension competitive relations among service organizations, are communicated through the imagery of gender and, in particular, the shift from socialism, the iron rice bowl, and proletarian female bodies to a marketized society, the rice bowl of youth, and feminized, sexualized female bodies.

There are a number of reasons why this is so. First, while discourses about class have not historically been muted or absent in China, during the reform era beginning in 1979, a new ideology of individual enterprise and achievement (Hanser 2002; Hoffman 2001; Won 2004; see also Croll 1991) has gradually displaced class-based analyses of Chinese society, to the point where class understandings of Chinese society have come to seem as anachronistic and dysfunctional as the socialist planned economy. As a result, class may increasingly be spoken through other categories of social difference.

Second, the rise of a naturalized, biologized understanding of gender in the reform era facilitates the expression of new class differences through gendered meanings. Numerous scholars have identified a trend toward the sexualization and commodification of women's bodies in China (Brownell 2001; Schein 2000; Yang 1999), a trend viewed as a departure from both the rhetoric and practices of the Maoist era (Chen 2003a; Croll 1995; Rofel 1999) and more traditional conceptions of gender difference (Barlow 1994; Furth 2002). Lisa Rofel (1999, 217) characterized this rise of essentialized notions of gender as "an allegory of postsocialism" that portrays newly sexualized gender relations as a return to the natural and inevitable. With Mao-era gender neutrality now viewed as unnatural and ludicrous, this naturalized understanding of gender and sexuality is powerfully associated with everything socialism was not—especially an affluent market society and a new, modern future for China. In many Chinese work settings, essentialized gender categories become means of justifying and masking inequalities between women and men (Lee 1998; Ong 1997; Rofel 1999; Woo 1994).

...[T]he rice bowl of youth imagery and its associations with sexualized femininity and capitalist modernity have become powerful means for conveying class distinctions in China's service sector. In what follows, I will demonstrate how a Chinese department store serving the urban elite draws on the rice bowl of youth imagery as it seeks to secure a position at the top of the retailing hierarchy. Managers in an elite service setting rely on gendered meanings to distinguish their workers from the women who work in other, less distinguished spaces. The good worker is a young, well-disciplined, and attractive woman who is distinct from both the blue-collar, middle-aged service workers who labor in state-owned stores and the highly sexualized young women employed in street markets and clothing bazaars. She is, in other words, a woman whose embodied and classed gender performances aid the store and its management in their own bid for distinction in China's retail sector and, in particular, their efforts to attract China's most elite and wealthy customers to the department store's luxury-filled sales floors....

METHOD

This article is based on data gathered in the northeastern city of Harbin, China, between 2001 and 2002. Starting in the fall of 2001, I conducted participant observation in three market settings representing three locations in the urban Chinese retailing hierarchy. In each site, I spent about two and a half months working seven-hour days, six days a week. I also spent lengths of time observing in a number of other stores, markets, and service work settings elsewhere in the city. I supplemented ethnographic work with more than 40 interviews with workers, store managers, merchandise suppliers, and other industry experts and conducted archival research on institutional changes to China's retail sector since the introduction of economic reforms in 1979....

FINDINGS

Service Work and the Production of Distinction

When I started work as a salesclerk at a cashmere sweater boutique in the Sunshine Department Store in the winter of 2002, one of the managers at the luxury department store assured me that I would find his store much better than Harbin No. X, the state-owned department store I had also studied. He suggested that the undisciplined salespeople there delivered poor service, whereas at Sunshine there were specific expectations about how clerks greet a customer. The manager added, "You will have to spend a long time at Sunshine to understand the culture here."

A key part of that culture, it turned out, was a class-coded femininity through which Sunshine workers were expected to distinguish themselves from service workers in less prestigious settings. To understand the construction of proper femininity in this high-class department store, we must first examine the two groups of women against whom elite service institutions such as Sunshine distinguish their own female workforce. A brief look at the middle-aged women working in a state-owned department store and the young women working in a clothing bazaar reveals why gender has become such a powerful way of "speaking," or marking, class distinctions.

The Middle-Aged Women Workers of Harbin No. X

In 2001, when I worked there, the Harbin No. X Department Store was a setting easily identified both with state socialism and with China's urban working class. Like a traditional socialist work unit, the state-owned store continued to provide guaranteed employment and benefits to its workers. The store opened its doors every morning to a broadcast of the store song, steeped in revolutionary fervor and sung in military chorus style, titled "Soar, Harbin No. X!" The store was also distinguished by its workforce who, quite unlike the highly selective workforces of newer department stores, ranged from their early 20s to their late

40s, with men sprinkled throughout the predominantly middle-aged, female workforce. It was here that I took up a position selling winter coats to a markedly proletarian clientele.

Space constraints preclude a full explanation of the organization of service work in this state-owned department store. Suffice it to say that because managers in this socialist department store continued to view themselves as primarily situated within the bureaucratic hierarchy of the state, and not within a market in which they must compete for customers, they did not structure sales work around the production and marking of social distinctions, and they did not cultivate either the physical space of the store or the workforce itself to convey a class-specific market position....

An important result was that workers' bodies were subject to little regulation or standardization at Harbin No. X. Management there made no organized attempts to create a gendered display of sexualized women's bodies, and minimal managerial attempts to regulate worker appearance applied equally to men and women. Uniforms were bulky and unisex, although workers might freely embellish on these basics in what often amounted to, for women especially, forms of gender display. Permed and dyed hair sat piled atop heads, and some women salesclerks skittered about in impossibly tall high-heeled shoes and stylish, tight pants. Nevertheless, such displays were incidental to our work and the business of the department store. Workers did not rely on, and were not measured by, the rice bowl of youth....

Women workers of all ages were conscious of the beauty standards against which their bodies were judged and, as they progressed through their lives, were increasingly likely to violate. Yet the department store provided a space in which these women, as workers, might be positively measured by their increasing years in the form of work experience, allowing many women clerks to reject Sunshine's construction of feminized labor. The senior salesclerk on my counter would even incorporate this idea into her sales pitches, reminding customers, "Think how much experience I have with these things, having worked the counter for so many years!" Both managers and workers believed salesclerk expertise and competence was grounded in sales floor experience and not embodied in the worker's physical form. The productive worker could easily take on the form of a middle-aged body....

...[W]hereas the middle-aged women workers at Harbin No. X labored in a workplace characterized by nonsexualized understandings of gender associated with Chinese socialism, the women who sold merchandise in The Underground, the city's largest clothing bazaar, performed a hypersexualized femininity that linked them with the moral deviance of the unregulated marketplace and the lack of culture associated with rural people. As a result, the appearances of young saleswomen served as a powerful way of identifying these women's, and their marketplace's, low class location.

In many ways, The Underground was like another world. Stairwells, immediately adjacent to or even inside stores such as Sunshine, led shoppers from the realm of closely regulated, uniformed salesclerks to a dramatically different space of highly sexualized gender displays. Most of the women working in The

Underground spent the days in languorous poses modeling the sexy clothing they offered for sale, their faces thickly painted and eyebrows carefully plucked. Young saleswomen colored their eyelids bright purple or lime green, and whenever there was a lull in business, they began touching up mascara, reapplying lipstick, or redoing hair. In the mornings, used Q-tips littered the market floor, testament to daily cosmetic rituals.

Here in The Underground, the focus on women's bodies reached new extremes, in part because young saleswomen modeled their wares, clothing renowned for its skimpiness and sex appeal. My first day at the market, I found my host, Xiao Li, dressed to sell: She wore a black sleeveless top she was selling, paired with tight jeans and a pair of red suede pumps with gold spike heels. Xiao Li changed outfits twice that day, switching her black top for a white one and then exchanging that one for a red lacy top, which she sold right off her back.

Many Harbin residents I spoke with viewed the gaudy appearances of the young women who sold clothing in The Underground as markers of morally questionable character and low social position. This point was clearly illustrated for me during a conversation with a teenaged girl and her mother. The girl, Wang Juan, expressed her disdain for the young women working in the underground market. She asked me, "Have you noticed how much makeup they wear? Really, really thick!" She added that many of those young women were "people who've given up school.... There isn't much else they can do." Her mother agreed, adding, "Those girls lead a bad lifestyle, all they want to do is fool around (*war*). After they get off work, they go out to party, dance, whatever, out until who knows what hour."

A sales manager from a cashmere sweater company even suggested to me that working in The Underground could taint a young woman permanently. He told me that if a young woman had previously worked in the clothing bazaar, he would not allow her to staff his department store sales areas. He explained that such women would be unwilling to "stay put" and would import bad habits from the market into the department store, especially the use of "uncivilized language" (*bu wenming hua*). Indeed, many Harbin residents associated Underground women with the uncultured or morally suspect behaviors found in the market, such as shouting, swearing, and the angry verbal exchanges that were a daily occurrence in the market.

The portrayal of the young women working in The Underground as bad women with no other options was in fact talk about a segment of the urban working class and rural migrant women with relatively little chance for upward mobility. These young women usually had only middle school education, a level of achievement one young saleswoman in the market described as "useless." They frequently came from families with few economic and social resources, their parents often members of Harbin's swelling numbers of laid-off state-sector workers or rural laborers....

These young women regularly asserted their claims to dignity and respect while working in The Underground, actively challenging the distinctions made between their market and highly standardized settings such as Sunshine. They were nevertheless regarded by others through the disparaging lens of popular opinion. In a marketplace both serving and employing rural people and the lower echelons of urban society, the undisciplined, hypersexualized bodies of Underground sales-women came to represent both unregulated capitalism and a low-class position. Just below the surface of gender distinctions were class ones.

Sunshine Salesclerks and the Rice Bowl of Youth

Let me now return to the Sunshine Department store, a privately owned and operated luxury retailer that was essentially born of China's economic reforms. Opening its doors in the early 1990s, the department store explicitly catered to Harbin's wealthiest consumers, its target marking being the top 3 percent income bracket in the city (Wang n.d., 11). This luxury store's managers were, as a result, acutely attuned to Harbin's evolving social hierarchy and the field of retailers who increasingly mirrored it. It was also in this setting where hiring and labor control practices were constructed in relation to the two retail settings and sets of workers I have just described, pitched to produce distinctions between them.

These distinctions were in part necessary because the Sunshine Department Store was recognized by managers, workers, and customers alike through what it clearly was not. For if Harbin No. X was regarded as a relic from the socialist past, replete with an aging, undisciplined workforce, Sunshine by contrast endeavored to represent a new China populated not by the masses but by a newly rich and upwardly mobile class served by an obedient army of attractive young women. Sunshine's management was equally anxious to distinguish both its workforce and its merchandise from the chaotic, morally suspect Underground. In this luxury department store, salesclerks and their labor were expected to represent capitalist modernity, and to do this, workers there were required to distinguish themselves symbolically from other workers—older, socialist workers and young, free market ones. These symbolic distinctions were simultaneously class distinctions targeted not only at the store's workers but also at their customers. Only a particular type of woman—young, attractive, and obedient—was suited to staff a department store servicing the upper echelons of Chinese society.

Workers' bodies served as a primary means for making these distinctions. Managers at Sunshine deployed two forms of control designed to produce and maintain clear symbolic distances between the store's workforce and other women service workers. The first was an intense level of bodily discipline, which was aimed at standardizing bodies, and the second was exposure to what might be called managerial practices of "subjectification," which were aimed at making workers manage themselves....

Various forms of disciplinary power pervaded the organization of work at Sunshine, ranging from photo ID badges to individualized sales figures. Store disciplinary routines, however, aimed to control workers' bodies by means of a norm of youthful femininity. This notion of proper femininity was expressed through rules and regulations that focused on appearance, posture, physical deportment, and demeanor. These rules applied only to salesclerks and not to the considerable numbers of women managers in the store.

I first encountered the rice bowl of youth norm when, to my surprise, Sunshine's human resources staff applied their standards to my own, clearly nonworker, body. They carefully measured my height (1.6 meters required) and scanned my personnel form (high school education required, no previous work experience necessary). The upper age limit for entering salesclerks of 25 was graciously waived, yet I wondered if the store would have done so had I deviated from the rice bowl of youth norm in more visible ways—if, for example, I had been deemed ugly or fat, or had looked older than age 30, the standard age at which salesclerks were retired from the sales floor.

These minimum requirements were expanded on during the store's initial, prework training session. At the class that a number of other new clerks and I attended, the instructor devoted more than a quarter of her lecture to issues of bodily deportment and conduct, reinforcing the connection between a disciplined body and a worthy body, valued as sufficiently high-class, Sunshine salesclerk material. The instructor did this by labeling incorrect bodily practices "disgusting" or immature. The instructor produced elaborate descriptions of outrageous salesclerks who had blown snot from their noses directly onto the floor or who had arrived at work in wrinkled or soiled uniforms....

Proper salesclerk behavior was also explicitly linked to a notion of proper feminine deportment and disciplined sexuality, highlighted through more examples of "bad" clerk behavior. For example, at one point, the instructor suddenly and quite dramatically performed a woman plucking her eyebrows. "Oooo! Ow! Ei!!" she cried, twisting her face with each imaginary pluck. She then informed the class that one time a male customer had spotted a salesclerk plucking her eyebrows while at her workstation. The man immediately went to the service desk to complain. Another example, which illustrated the importance of doing up all the buttons on your uniform, involved a young woman who exposed her chest when bending over, to the delight of some male customers (but to her own extreme embarrassment). On another occasion, a salesclerk who bent over instead of demurely crouching down in her short-skirted summer uniform reportedly caused a male customer to exclaim, "That young woman is quite interesting—she's wearing strawberry underwear!" In the latter two cases, the reported male gaze disciplined sexualized female salesclerks through humiliation; Sunshine salesclerks were not "that" kind of girl....

CONCLUSION

I have argued here that the production of class meanings and distinctions can be a key aspect of what service work is organized to produce. Managers of service organizations recognize themselves in a competition for customers who themselves seek distinction in the service setting. In urban China today, the wealthiest customers, and by extension the most elite retailers, engage in the most vigorous practices of distinction making as they stake out their positions at the top of a reconstituted social hierarchy.

But the marking of social distinctions is a fundamentally relational practice. As I have shown, managers at the Sunshine Department Store could identify their store as modern, high-class, and exclusive only by engaging in a symbolic dialogue with other, less prestigious retail settings in the city. Managers hired, trained, and monitored their salesclerks' work activities to ensure that these clerks successfully distinguished themselves from other service workers. To do so, managers drew on discourses of gender and postsocialist transition, as class distinctions within the working class and across the broader class hierarchy in urban China are spoken through the naturalness both of gender and of the progression from failed socialism to a prosperous and modern capitalist society.

However, because in China's newest, most expensive marketplaces, a "good" worker is identified with a particular kind of femininity couched in an essentializing discourse of gender, the class distinctions that are simultaneously produced are easily obscured from view. Hiring and labor control decisions focus so intensely on bodily dispositions and bodily practices—which Bourdieu (1977, 1984) has shown are deeply shaped by social location and class habitus—that performances of class distinction are masked by essentializing conceptions of gender and sexuality. Much as Julie Bettie (2003) has demonstrated for working-class sexuality in the United States, class-inflected gender performances in China are easily misrecognized as natural expressions of sex and sexuality. These conceptions of gender mark some Chinese women as appropriately, and productively, feminine by drawing on a discourse of transition in which the socialist past is devalued in relation to a particular capitalist modernity.

One result is that new class inequalities in urban China are legitimated through discourses of gender and of economic transition. When middle-aged, working-class women are associated with the inefficiencies and absurdities of the Mao era, their downward mobility and shrinking employment opportunities in the reform era seem self-evident. When young women working in free markets and clothing bazaars are associated with moral deviance and unrefined culture, their lack of upward mobility seems justified and self-imposed. At the top of the retail hierarchy, the organization of work produces class meanings that serve as a powerful public buttress to these broader social inequalities, helping to create, as Barbara Ehrenreich has said of service work in the U.S. context, "not just an economy but a culture of ... inequality" (2001, 212).

REFERENCES

Adkins, Lisa. 2002. *Revisions: Gender and sexuality in late modernity*. Buckingham, UK: Open University Press.

Benson, Susan Porter. 1986. *Counter cultures: Saleswomen, managers, and customers in American department stores, 1890–1940*. Urabana: University of Illinois Press.

Bettie, Julie. 2003. *Women without class: Girls, race, and identity*. Berkeley: University of California Press.

Bourdieu, Pierre. 1977. *Outline of a theory of practice*. Cambridge, UK: Cambridge University Press.

———. 1984. *Distinction: A social critique of the judgement of taste*. Cambridge, MA: Harvard University Press.

Chen, Tina Mai. 2003b. Proletarian white and working bodies in Mao's China. *Positions: East Asia Cultures Critique* 11:361–93.

Ehrenreich, Barbara. 2001. *Nickel and dimed: On (not) getting by in America*. New York: Owl Books.

Honig, Emily, and Gail Hershatter. 1988. *Personal voices: Chinese women in the 1980s*. Stanford, CA: Stanford University Press.

Hochschild, Arlie Russell. 1983. *The managed heart: Commercialization of human feeling*. Berkeley: University of California Press.

Lan, Pie-Chia. 2001. The body as contested terrain for labor control: Cosmetics retailers in department stores and direct selling In *The critical study of work: Labor, technology and global production*, edited by R. Baldoz, C. Koeber, and P. Kraft. Philadelphia: Temple University Press.

———. 2003. Working in a neon cage: Bodily labor of cosmetics saleswomen in Taiwan. *Feminist Studies* 29:21–45.

Lee, Ching Kwan. 1998. *Gender and the south China miracle: Two worlds of factory women*. Berkeley: University of California Press.

Loe, Mieka. 1996. Working for men—At the intersection of power, gender, and sexuality. *Sociological Inquiry* 66:339–421.

Paules, Greta Foff. 1991. *Dishing it out: Power and resistance among waitresses in a New Jersey restaurant*. Philadelphia: Temple University Press.

Pringle, Rosemary. 1988. *Secretaries talk: Sexuality, power and work*. London: Verso.

———. 1992. What is a secretary? In *Defining women: Social institutions and gender divisions*, edited by L. McDowell and R. Pringle. Cambridge, UK: Polity Press.

Sherman, Rachel E. 2003. Class acts: Producing and consuming luxury service in hotels. Ph.D. diss., University of California, Berkeley.

Steedman, Carolyn Kay. 1987. *Landscape for a good woman*. New Brunswick, NJ: Rutgers University Press.

Wang, Yanhong. n.d. Sectoral position analysis. Unpublished manuscript.

26

Prisons for Our Bodies, Closets for Our Minds

Racism, Heterosexism, and Black Sexuality

PATRICIA HILL COLLINS

> *White fear of black sexuality is a basic ingredient of white racism.*
>
> *Cornel West*

For African Americans, exploring how sexuality has been manipulated in defense of racism is not new. Scholars have long examined the ways in which "white fear of black sexuality" has been a basic ingredient of racism. For example, colonial regimes routinely manipulated ideas about sexuality in order to maintain unjust power relations. Tracing the history of contact between English explorers and colonists and West African societies, historian Winthrop Jordan contends that English perceptions of sexual practices among African people reflected preexisting English beliefs about Blackness, religion, and animals. American historians point to the significance of sexuality to chattel slavery. In the United States, for example, slaveowners relied upon an ideology of Black sexual deviance to regulate and exploit enslaved Africans. Because Black feminist analyses pay more attention to women's sexuality, they too identify how the sexual exploitation of women has been a basic ingredient of racism. For example, studies of African American slave women routinely point to sexual victimization as a defining feature of American slavery. Despite the important contributions of this extensive literature on race and sexuality, because much of the literature assumes that sexuality means heterosexuality, it ignores how racism and heterosexism influence one another.

In the United States, the assumption that racism and heterosexism constitute two separate systems of oppression masks how each relies upon the other for meaning. Because neither system of oppression makes sense without the other, racism and heterosexism might be better viewed as sharing one history with similar yet disparate effects on all Americans differentiated by race, gender, sexuality, class, and nationality. People who are positioned at the margins of both systems

SOURCE: Hill Collins, Patricia. 2004. *Black Sexual Politics: African Americans and the New Racism*. Pp. 87–88, 95–105, 114–116. New York: Routledge.

and who are harmed by both typically raise questions about the intersections of racism and heterosexism much earlier and/or more forcefully than those people who are in positions of privilege. In the case of intersections of racism and heterosexism, Black lesbian, gay, bisexual, and transgendered (LGBT) people were among the first to question how racism and heterosexism are interconnected. As African American LGBT people point out, assuming that all Black people are heterosexual and that all LGBT people are White distorts the experiences of LGBT Black people. Moreover, such comparisons misread the significance of ideas about sexuality to racism and race to heterosexism.

Until recently, questions of sexuality in general, and homosexuality in particular, have been treated as crosscutting, divisive issues within antiracist African American politics. The consensus issue of ensuring racial unity subordinated the allegedly crosscutting issue of analyzing sexuality, both straight and gay alike. This suppression has been challenged from two directions. Black women, both heterosexual and lesbian, have criticized the sexual politics of African American communities that leave women vulnerable to single motherhood and sexual assault. Black feminist and womanist projects have challenged Black community norms of a sexual double standard that punishes women for behaviors in which men are equally culpable. Black gays and lesbians have also criticized these same sexual politics that deny their right to be fully accepted within churches, families, and other Black community organizations. Both groups of critics argue that ignoring the heterosexism that underpins Black patriarchy hinders the development of a progressive Black sexual politics....

Developing a progressive Black sexual politics requires examining how racism and heterosexism mutually construct one another.

MAPPING RACISM AND HETEROSEXISM: THE PRISON AND THE CLOSET

... Racism and heterosexism, the prison and the closet, appear to be separate systems, but LGBT African Americans point out that *both* systems affect their everyday lives. If racism and heterosexism affect Black LGBT people, then these systems affect *all* people, including heterosexual African Americans. Racism and heterosexism certainly converge on certain key points. For one, both use similar state-sanctioned institutional mechanisms to maintain racial and sexual hierarchies. For example, in the United States, racism and heterosexism both rely on segregating people as a mechanism of social control. For racism, segregation operates by using race as a visible marker of group membership that enables the state to relegate Black people to inferior schools, housing, and jobs. Racial segregation relies on enforced membership in a visible community in which racial discrimination is tolerated. For heterosexism, segregation is enforced by pressuring LGBT individuals to remain closeted and thus segregated from one another. Before social movements for gay and lesbian liberation, sexual segregation meant that refusing to claim homosexual identities virtually eliminated any group-based political action to resist heterosexism. For another, the state has played a very important role in sanctioning both forms of oppression. In support of racism, the state sanctioned laws that regulated where Black people could live, work, and attend school. In support of heterosexism, the state maintained laws that refused to punish hate crimes against LGBT people, that failed to offer protection when LGBT people were stripped of jobs and children, and that generally sent a message that LGBT people who came out of the closet did so at their own risk.

Racism and heterosexism also share a common set of practices that are designed to discipline the population into accepting the status quo. These disciplinary practices can best be seen in the enormous amount of attention paid both by the state and organized religion to the institution of marriage. If marriage were in fact a natural and normal occurrence between heterosexual couples and if it occurred naturally within racial categories, there would be no need to regulate it. People would naturally choose partners of the opposite sex and the same race. Instead, a series of laws have been passed, all designed to regulate marriage. For example, for many years, the tax system has rewarded married couples with tax breaks that have been denied to single taxpayers or unmarried couples. The message is clear—it makes good financial sense to get married. Similarly, to encourage people to marry within their assigned race, numerous states passed laws banning interracial marriage. These restrictions lasted until the landmark Supreme Court decision in 1967 that overturned state laws. The state has also passed laws designed to keep LGBT people from marrying. In 1996, the U.S. Congress passed the Federal Defense of Marriage Act that defined marriage as a "legal union between one man and one woman." In all of these cases, the state perceives that it has a compelling interest in disciplining the population to marry and to marry the correct partners.

Racism and heterosexism also manufacture ideologies that defend the status quo. When ideologies that defend racism and heterosexism become taken-for-granted and appear to be natural and inevitable, they become hegemonic. Few question them and the social hierarchies they defend. Racism and heterosexism both share a common cognitive framework that uses binary thinking to produce hegemonic ideologies. Such thinking relies on oppositional categories. It views race through two oppositional categories of Whites and Blacks, gender through two categories of men and women, and sexuality through two oppositional categories of heterosexuals and homosexuals. A master binary of normal and deviant overlays and bundles together these and other lesser binaries. In this context, ideas about "normal" race (Whiteness, which ironically, masquerades as racelessness), "normal" gender (using male experiences as the norm), and "normal" sexuality (heterosexuality, which operates in a similar hegemonic fashion) are tightly bundled together. In essence, to be completely "normal," one must be White, masculine, and heterosexual, the core hegemonic White masculinity. This mythical norm is hard to see because it is so taken-for-granted. Its antithesis, its Other, would be Black, female, and lesbian, a fact that Black lesbian feminist Audre Lorde pointed out some time ago.

Within this oppositional logic, the core binary of normal/deviant becomes ground zero for justifying racism and heterosexism. The deviancy assigned to race and that assigned to sexuality becomes an important point of contact between the two systems. Racism and heterosexism both require a concept of sexual deviancy for meaning, yet the form that deviance takes within each system differs. For racism, the point of deviance is created by a *normalized White heterosexuality* that depends on a *deviant Black heterosexuality* to give it meaning. For heterosexism, the point of deviance is created by this very same *normalized White heterosexuality* that now depends on a *deviant White homosexuality*. Just as racial normality requires the stigmatization of the sexual practices of Black people, heterosexual normality relies upon the stigmatization of the sexual practices of homosexuals. In both cases, installing White heterosexuality as normal, natural, and ideal requires stigmatizing alternate sexualities as abnormal, unnatural, and sinful.

The purpose of stigmatizing the sexual practices of Black people and those of LGBT people may be similar, but the content of the sexual deviance assigned to each differs. Black people carry the stigma of *promiscuity* or excessive or unrestrained heterosexual desire. This is the sexual deviancy that has both been assigned to Black people and been used to construct racism. In contrast, LGBT people carry the stigma of *rejecting* heterosexuality by engaging in unrestrained homosexual desire. Whereas the deviancy associated with promiscuity (and, by implication, with Black people as a race) is thought to lie in an *excess* of heterosexual desire, the pathology of homosexuality (the invisible, closeted sexuality that becomes impossible within heterosexual space) seemingly resides in the *absence* of it.

While analytically distinct, in practice, these two sites of constructed deviancy work together and both help create the "sexually repressive culture" in America.... Despite their significance for American society overall, here I confine my argument to the challenges that confront Black people. Both sets of ideas frame a hegemonic discourse of *Black sexuality* that has at its core ideas about an assumed promiscuity among heterosexual African American men and women and the impossibility of homosexuality among Black gays and lesbians. How have African Americans been affected by and reacted to this racialized system of heterosexism (or this sexualized system of racism)?

AFRICAN AMERICANS AND THE RACIALIZATION OF PROMISCUITY

Ideas about Black promiscuity that produce contemporary sexualized spectacles such as Jennifer Lopez, Destiny's Child, Ja Rule, and the many young Black men on the U.S. talk show circuit have a long history. Historically, Western science, medicine, law, and popular culture reduced an African-derived aesthetic concerning the use of the body, sensuality, expressiveness, and spirituality to an ideology about *Black sexuality*. The distinguishing feature of this ideology was its reliance on the idea of Black promiscuity. The possibility of distinctive and worthwhile African-influenced worldviews on anything, including sexuality, as well as the heterogeneity of African societies expressing such views, was collapsed into an imagined, pathologized Western discourse of what was thought to be essentially African. To varying degrees, observers from England, France, Germany, Belgium, and other colonial powers perceived African sensuality, eroticism, spirituality, and/or sexuality as deviant, out of control, sinful, and as an essential feature of racial difference....

With all living creatures classified in this way, Western scientists perceived African people as being more natural and less civilized, primarily because African people were deemed to be closer to animals and nature, especially the apes and monkeys whose appearance most closely resembled humans. Like African people, animals also served as objects of study for Western science because understanding the animal kingdom might reveal important insights about civilization, culture, and what distinguished the human "race" from its animal counterparts as well as the human "races" from one another....

Those most proximate to animals, those most lacking civilization, also were those humans who came closest to having the sexual lives of animals. Lacking the benefits of Western civilization, people of African descent were perceived as having a biological nature that was inherently more sexual than that of Europeans. The primitivist discourse thus created the category of "beast" and the sexuality of such beasts as "wild." The legal classification of enslaved African people as chattel (animal-like) under American slavery that produced controlling images of bucks, jezebels, and breeder women drew meaning from this broader interpretive framework.

Historically, this ideology of Black sexuality that pivoted on a Black heterosexual promiscuity not only upheld racism but it did so in gender-specific ways. In the context of U.S. society, beliefs in Black male promiscuity took diverse forms during distinctive historical periods. For example, defenders of chattel slavery believed that slavery safely domesticated allegedly dangerous Black men because it regulated their promiscuity by placing it in the service of slave owners. Strategies of control were harsh and enslaved African men who were born in Africa or who had access to their African past were deemed to be the most dangerous. In contrast, the controlling image of the rapist appeared after emancipation because Southern Whites feared that the unfettered promiscuity of Black freedmen constituted a threat to the Southern way of life....

The events themselves may be over, but their effects persist under the new racism. This belief in an inherent Black promiscuity reappears today. For example, depicting poor and working-class African American inner-city neighborhoods as dangerous urban jungles where SUV-driving White suburbanites come to score drugs or locate prostitutes also invokes a history of racial and sexual conquest. Here sexuality is linked with danger, and understandings of both draw upon historical imagery of Africa as a continent replete with danger and peril to the White explorers and hunters who penetrated it. Just as contemporary safari tours in Africa create an imagined Africa as the "White man's playground" and mask its economic exploitation, jungle language masks social relations of hyper-segregation that leave working-class Black communities isolated, impoverished, and dependent on a punitive welfare state and an illegal international drug trade. Under this logic, just as wild animals (and the proximate African natives) belong in nature preserves (for their own protection), unassimilated, undomesticated poor and working-class African Americans belong in racially segregated neighborhoods....

African American women also live with ideas about Black women's promiscuity and lack of sexual restraint. Reminiscent of concerns with Black women's fertility under slavery and in the rural South, contemporary social welfare policies also remain preoccupied with Black women's fertility. In prior eras, Black women were encouraged to have many children. Under slavery, having many children enhanced slave owners' wealth and a good "breeder woman" was less likely to be sold. In rural agriculture after emancipation, having many children ensured a sufficient supply of workers. But in the global economy of today, large families are expensive because children must be educated. Now Black women are seen as producing too many children who contribute less to society than they take. Because Black women on welfare have long been seen as undeserving, long-standing ideas about Black women's promiscuity become recycled and redefined as a problem for the state....

RACISM AND HETEROSEXISM REVISITED

On May 11, 2003, a stranger killed fifteen-year-old Sakia Gunn who, with four friends, was on her way home from New York's Greenwich Village. Sakia and her friends were waiting for the bus in Newark, New Jersey, when two men got out of a car, made sexual advances, and physically attacked them. The women fought back, and when Gunn told the men that she was a lesbian, one of them stabbed her in the chest.

Sakia Gunn's murder illustrates the connections among class, race, gender, sexuality, and age. Sakia lacked the protection of social class privilege. She and her friends were waiting for the bus in the first place because none had access to private automobiles that offer protection for those who are more affluent. In Gunn's case, because her family initially did not have the money for her funeral, she was scheduled to be buried in a potter's grave. Community activists took up a collection to pay for her funeral. She lacked the gendered protection provided by masculinity. Women who are perceived to be in the wrong place at the wrong time are routinely approached by men who feel entitled to harass and proposition them. Thus, Sakia and her friends share with all women the vulnerabilities that accrue to women who negotiate public space. She lacked the protection of age—had Sakia and her friends been middle-aged, they may not have been seen as sexually available. Like African American girls and women, regardless of sexual orientation, they were seen as approachable. Race was a factor, but not in a framework of interracial race relations. Sakia and her friends were African American, as were their attackers. In a context where Black men are encouraged to express a hyper-heterosexuality as the badge of Black masculinity, women like Sakia and her friends can become important players in supporting patriarchy. They challenged Black male authority, and they paid for the transgression of refusing to participate in scripts of Black promiscuity. But the immediate precipitating catalyst for the violence that took Sakia's life was her openness about her lesbianism. Here, homophobic violence was the prime factor. Her death illustrates how deeply entrenched homophobia can be among many African American men and women, in this case, beliefs that resulted in an attack on a teenaged girl.

How do we separate out and weigh the various influences of class, gender, age, race, and sexuality in this particular incident? Sadly, violence against Black girls is an everyday event. What made this one so special? Which, if any, of the dimensions of her identity got Sakia Gunn killed? There is no easy answer to this question, because *all* of them did. More important, how can any Black political agenda that does not take *all* of these systems into account, including sexuality, ever hope adequately to address the needs of Black people as a collectivity? One expects racism in the press to shape the reports of this incident. In contrast to the 1998 murder of Matthew Shepard, a young, White, gay man in Wyoming, no massive protests, nationwide vigils, and renewed calls for federal hate crimes legislation followed Sakia's death. But what about the response of elected and appointed officials? The African American mayor of Newark decried the crime, but he could not find the time to meet with community activists who wanted programmatic changes to retard crimes like Sakia's murder. The principal of her high school became part of the problem. As one activist described it, "students at Sakia's high school weren't allowed to hold a vigil. And the kids wearing the rainbow flag were being punished like they had on gang colors."

Other Black leaders and national organizations spoke volumes through their silence. The same leaders and organizations that spoke out against the police beating of Rodney King by Los Angeles area police, the rape of immigrant Abner Louima by New York City police, and the murder of Timothy Thomas by Cincinnati police said nothing about Sakia Gunn's death. Apparently, she was just another unimportant little Black girl to them. But to others, her death revealed the need for a new politics that takes the intersections of racism and heterosexism as well as class exploitation, age discrimination, and sexism into account. Sakia was buried on May 16 and a crowd of approximately 2,500 people attended her funeral. The turnout was unprecedented: predominantly Black, largely high school students, and mostly lesbians. Their presence says that as long as African American lesbians like high school student Sakia Gunn are vulnerable, then every African American woman is in danger; and if all Black women are at risk, then there is no way that any Black person will ever be truly safe or free.

27

"Dude, You're a Fag"

Adolescent Masculinity and the Fag Discourse

C. J. PASCOE

… *Homophobia* is too facile a term with which to describe the deployment of *fag* as an epithet. By calling the use of the word *fag* homophobia—and letting the argument stop there—previous research has obscured the gendered nature of sexualized insults (Plummer 2001). Invoking homophobia to describe the ways boys aggressively tease each other overlooks the powerful relationship between masculinity and this sort of insult. Instead, it seems incidental, in this conventional line of argument, that girls do not harass each other and are not harassed in this same manner. This framing naturalizes the relationship between masculinity and homophobia, thus obscuring that such harassment is central to the formation of a gendered identity for boys in a way that it is not for girls.

Fag is not necessarily a static identity attached to a particular (homosexual) boy. Fag talk and fag imitations serve as a discourse with which boys discipline themselves and each other through joking relationships. Any boy can temporarily become a fag in a given social space or interaction. This does not mean that boys who identify as or are perceived to be homosexual aren't subject to intense harassment. Many are. But becoming a fag has as much to do with failing at the masculine tasks of competence, heterosexual prowess, and strength or in any way revealing weakness or femininity as it does with a sexual identity. This fluidity of the fag identity is what makes the specter of the fag such a powerful disciplinary mechanism. It is fluid enough that boys police their behaviors out of fear of having the fag identity permanently adhere and definitive enough so that boys recognize a fag behavior and strive to avoid it.

An analysis of the fag discourse also indicates ways in which gendered power works through racialized selves. The fag discourse is invoked differently by and in relation to white boys' bodies than it is by and in relation to African American boys' bodies. While certain behaviors put all boys at risk for becoming temporarily a fag, some behaviors can be enacted by African American boys without putting them at risk of receiving the label. The racialized meanings of the fag discourse suggest that something more than simple homophobia is involved in these sorts of interactions. It is not that gendered homophobia does not exist in African American communities. Indeed, making fun of "negro faggotry seems to be a rite of passage among contemporary black male rappers and filmmakers"

SOURCE: Pascoe, C.J. 2011. *"Dude, You're a Fag:" Masculinity and Sexuality in High School.* Berkeley, CA: University of California Press.

(Riggs 1991, 253). However, the fact that "white women and men, gay and straight, have more or less colonized cultural debates about sexual representation" (Julien and Mercer 1991, 167) obscures varied systems of sexualized meanings among different racialized ethnic groups (Almaguer 1991). Thus far, male homophobia has primarily been written about as a racially neutral phenomenon. However, as D. L. King's (2004) recent work on African American men and same-sex desire pointed out, homophobia is characterized by racial identities as well as sexual and gendered ones.

WHAT IS A FAG? GENDERED MEANINGS

"Since you were little boys you've been told, 'Hey, don't be a little faggot,'" explained Darnell, a football player of mixed African American and white heritage, as we sat on a bench next to the athletic field. Indeed, both the boys and girls I interviewed told me that *fag* was the worst epithet one guy could direct at another. Jeff, a slight white sophomore, explained to me that boys call each other fag because "gay people aren't really liked over here and stuff." Jeremy, a Latino junior, told me that this insult literally reduced a boy to nothing, "To call someone *gay* or *fag* is like the lowest thing you can call someone. Because that's like saying that you're nothing."

Most guys explained their or others' dislike of fags by claiming that homophobia was synonymous with being a guy. For instance, Keith, a white soccer-playing senior, explained, "I think guys are just homophobic." However, boys were not equal-opportunity homophobes. Several students told me that these homophobic insults applied only to boys and not to girls. For example, while Jake, a handsome white senior, told me that he didn't like gay people, he quickly added, "Lesbians, okay, that's *good*." Similarly Cathy, a popular white cheerleader, told me, "Being a lesbian is accepted because guys think, 'Oh that's cool.'" Darnell, after telling me that boys were warned about becoming faggots, said, "They [guys] are fine with girls. I think it's the guy part that they're like ewwww." In this sense, it was not strictly homophobia but a gendered homophobia that constituted adolescent masculinity in the culture of River High. It is clear, according to these comments, that lesbians were "good" because of their place in heterosexual male fantasy, not necessarily because of some enlightened approach to same-sex relationships. A popular trope in heterosexual pornography depicts two women engaging in sexual acts for the purpose of male titillation. The boys at River High are not unique in making this distinction; adolescent boys in general dislike gay men more than they dislike lesbians (Baker and Fishbein 1998). The fetishizing of sex acts between women indicates that using only the term *homophobia* to describe boys' repeated use of the word *fag* might be a bit simplistic and misleading.

Girls at River High rarely deployed the word *fag* and were never called fags. I recorded girls uttering *fag* only three times during my research. In one instance, Angela, a Latina cheerleader, teased Jeremy, a well-liked white senior involved in student government, for not ditching school with her: "You wouldn't 'cause you're a faggot." However, girls did not use this word as part of their regular lexicon. The sort of gendered homophobia that constituted adolescent masculinity did not constitute adolescent femininity. Girls were not called dykes or lesbians in any sort of regular or systematic way. Students did tell me that *slut* was the worst thing a girl could be called. However, my field notes indicate that the word *slut* (or its synonym *ho*) appeared one time for every eight times the word *fag* appeared.

Highlighting the difference between the deployment of *gay* and *fag* as insults brings the gendered nature of this homophobia into focus. For boys and girls at River High *gay* was a fairly common synonym for "stupid." While this word shared the sexual origins of *fag,* it didn't *consistently* have the skew of gender-loaded meaning. Girls and boys often used *gay* as an adjective referring to inanimate objects and male or female people, whereas they used *fag* as a noun that denoted only unmasculine males. Students used *gay* to describe anything from someone's clothes to a new school rule that they didn't like. For instance, one day in auto shop, Arnie pulled out a large older version of a black laptop computer and placed it on his desk. Behind him Nick cried, "That's a gay laptop! It's five inches thick!" The rest of the boys in the class laughed at Arnie's outdated laptop. A laptop can be gay, a movie can be gay, or a group of people can be gay. Boys used *gay* and *fag* interchangeably when they referred to other boys, but *fag* didn't have the gender-neutral attributes that *gay* frequently invoked.

Surprisingly, some boys took pains to say that the term *fag* did not imply sexuality. Darnell told me, "It doesn't even have anything to do with being gay." Similarly, J. L., a white sophomore at Hillside High (River High's crosstown rival), asserted, "*Fag*, seriously, it has nothing to do with sexual preference at all. You could just be calling somebody an idiot, you know?" I asked Ben, a quiet, white sophomore who wore heavy-metal T-shirts to auto shop each day, "What kind of things do guys get called a fag for?" Ben answered, "Anything … literally, anything. Like you were trying to turn a wrench the wrong way, 'Dude, you're a fag.' Even if a piece of meat drops out of your sandwich, 'You fag!'" Each time Ben said, "You fag," his voice deepened as if he were imitating a more masculine boy. While Ben might rightly *feel* that a guy could be called a fag for "anything … literally, anything," there were actually specific behaviors that, when enacted by most boys, could render them more vulnerable to a *fag* epithet. In this instance Ben's comment highlights the use of *fag* as a generic insult for incompetence, which in the world of River High, was central to a masculine identity. A boy could get called a fag for exhibiting any sort of behavior defined as unmasculine (although not necessarily behaviors aligned with femininity): being stupid or incompetent, dancing, caring too much about clothing, being too emotional, or expressing interest (sexual or platonic) in other guys. However, given the extent of its deployment and the laundry list of behaviors that could get a boy in trouble, it is no wonder that Ben felt a boy could be called fag for "anything." These nonsexual meanings didn't replace sexual meanings but rather existed alongside them….

RACIALIZING THE FAG

While all groups of boys, with the exception of the Mormon boys, used the word *fag* or fag imagery in their interactions, the fag discourse was not deployed consistently or identically across social groups at River High. Differences between white boys' and African American boys' meaning making, particularly around appearance and dancing, reveal ways the specter of the fag was racialized. The specter of the fag, these invocations reveal, was consistently white. Additionally, African American boys simply did not deploy it with the same frequency as white boys. For both groups of boys, the *fag* insult entailed meanings of emasculation, as evidenced by Darnell's earlier comment. However, African American boys were much more likely to tease one another for being white than for being a fag. Precisely because African American men are so hypersexualized in the United States, white men are, by default, feminized, so *white* was a stand-in for *fag* among many of the African American boys at River High. Two of the behaviors that put a white boy at risk for being labeled a fag didn't function in the same way for African American boys.

Perhaps because they are, by necessity, more invested in symbolic forms of power related to appearance (much like adolescent girls), a given African American boy's status is not lowered but enhanced by paying attention to clothing or dancing. Clean, oversized, carefully put together clothing is central to a hip-hop identity for African American boys who identify with hip-hop culture. Richard Majors (2001) calls this presentation of self a "cool pose" consisting of "unique, expressive and conspicuous styles of demeanor, speech, gesture, clothing, hairstyle, walk, stance and handshake," developed by African American men as a symbolic response to institutionalized racism (211). Pants are usually several sizes too big, hanging low on the hips, often revealing a pair of boxers beneath. Shirts and sweaters are similarly oversized, sometimes hanging down to a boy's knees. Tags are frequently left on baseball hats worn slightly askew and perched high on the head. Meticulously clean, unlaced athletic shoes with rolled-up socks under the tongue complete a typical hip hop outfit. In fact, African American men can, without risking a fag identity, sport styles of self and interaction frequently associated with femininity for whites, such as wearing curlers (Kelley 2004). These symbols, at River High, constituted a "cool pose."...

As in many places in the United States, racial divisions in Riverton line up relatively easily with class divisions. Darnell grabbed me at lunch one day to point this out to me, using school geography as an example. He sauntered up and whispered in my ear, "Notice the separation? There's the people who hang out in there (pointing toward the cafeteria), the people who hang out in the quad. And then the people who leave." He smashed one hand against the other in frustration: "I talk to these people in class. Outside we all separate into our groups. We don't talk to each other. Rich people are not here. They got cars and they go out." He told me that the "ball players" sat in the cafeteria. And he was right: there were two tables at the rear of the cafeteria populated by African American boys on the basketball and football teams, the guys whom Darnell described to me as his "friends." He said there were "people who leave, people who stay and the people over there [in the quad]. The people who stay are ghetto." He added, "*Ghetto* come to mean 'niggerish.' That reflects people who are poor or urban."...

None of this is to say that the sexuality of boys of color wasn't policed. In fact, because African American boys were regarded as so hypersexual, in the few instances I documented in which boys were punished for engaging in the fag discourse, African American boys were policed more stringently than white boys. It was as if when they engaged in the fag discourse the gendered insult took on actual combative overtones, unlike the harmless sparring associated with white boys' deployments. The intentionality attributed to African American boys in their sexual interactions with girls seemed to occur as well in their deployment of the fag discourse. One morning as I waited with the boys on the asphalt outside the weight room for Coach Ramirez to arrive, I chatted with Kevin and Darrell. The all-male, all-white wrestling team walked by, wearing gold and black singlets. Kevin, an African American sophomore, yelled out, "Why are you wearing those faggot outfits? Do you wear those tights with your balls hanging out?" The weight-lifting students stopped their fidgeting and turned to watch the scene unfold. The eight or so members of the wrestling team stopped at their SUV and turned to Kevin. A small redhead whipped around and yelled aggressively, "Who said that?" Fingers from wrestling team members quickly pointed toward Kevin. Kevin, angrily jumping around, yelled back as he thrust his chest out, "Talk about jumping me, nigger?" He strutted over, advancing toward the small redhead. A large wrestler sporting a cowboy hat tried to block Kevin's approach. The redhead meanwhile began to jump up and down, as if warming up for a fight. Finally the boy in the cowboy hat pushed Kevin away from the team and they climbed in the truck, while Kevin strutted back to his classmates, muttering, "All they know how to do is pick somebody up. Talk about jumping me ... weak-ass wrestling team. My little bro could wrestle better than any of those motherfuckers."

It would seem, based on the fag discourse scenarios I've described thus far, that this was, in a sense, a fairly routine deployment of the sexualized and gendered epithet. However, at no other time did I see this insult almost cause a fight. Members of the white wrestling team presumably took it so seriously that they reported the incident to school authorities. This in itself is stunning. Boys called each other fag so frequently in everyday discussion that if it were always reported most boys in the school would be suspended or at least in detention on a regular basis....

REFRAMING HOMOPHOBIA

Homophobia is central to contemporary definitions of adolescent masculinity. Unpacking multilayered meanings that boys deploy through their uses of homophobic language and joking rituals makes clear that it is not just homophobia but a gendered and racialized homophobia. By attending to these meanings, I reframe the discussion as a fag discourse rather than simply labeling it as homophobia. The fag is an "abject" (Butler 1993) position, a position outside masculinity that actually constitutes masculinity. Thus masculinity, in part, becomes the daily interactional work of repudiating the threatening specter of the fag.

The fag extends beyond a static sexual identity attached to a gay boy. Few boys are permanently identified as fags; most move in and out of fag positions. Looking at fag as a discourse in addition to a static identity reveals that the term can be invested with different meanings in different social spaces. *Fag* may be used as a weapon with which to temporarily assert one's masculinity by denying it to others. Thus the fag becomes a symbol around which contests of masculinity take place.

Researchers who look at the intersection of sexuality and masculinity need to attend to how racialized identities may affect how *fag* is deployed and what it means in various social situations. While researchers have addressed the ways in which masculine identities are racialized (Bucholtz 1999; Connell 1995; J. Davis 1999; Ferguson 2000; Majors 2001; Price 1999; Ross 1998), they have not paid equal attention to the ways *fag* might be a racialized epithet. Looking at when, where, and with what meaning *fag* is deployed provides insight into the processes through which masculinity is defined, contested, and invested in among adolescent boys....

REFERENCES

Almaguer, Tomas, 1991. "Chicano Men: A Cartography of Homosexual Identity and Behavior." *Differences* 3 (2): 75–100.

Baker, Janet, G., and Harold D. Fishbein. 1998. "The Development of Prejudice towards Gays and Lesbians by Adolescents." *Journal of Homosexuality* 36 (1): 89–100.

Bucholtz, Mary. 1999. "You Da Man: Narrating the Racial Other in the Production of White Masculinity." *Journal of Sociolinguistics* 3 (4): 443-460.

Butler, Judith. 1993. *Bodies that Matter: On the Discursive Limits of "Sex."* New York: Routledge.

Butler, Judith. 1995 "Melancholy Gender/Refused Identification." Pp. 21–36 in *Constructing Masculinity,* edited by Maurice Berger, Brian Wallis, and Simon Watson. New York: Routledge.

Butler, Judith. 1999. *Gender Trouble: Feminism and the Subversion of Identity*. New York: Routledge.

Connell, R. W. 1995. *Masculinities*. Berkeley, CA: University of California Press.

Davis, James Earl. 1999. "Forbidden Fruit: Black Males' Constructions of Transgressive Sexualities in Middle School." Pp. 49-59 in *Queering Elementary Education: Advancing the Dialogue about Sexualities and Schooling*, edited by William J. Letts IV and James T. Sears. Lanham, MD: Rowman and Littlefield.

Ferguson, Ann. 2000. *Bad Boys: Public Schools in the Making of Black Masculinity*. Ann Arbor, MI: University of Michigan Press.

Julien, Isaac, and Kobena Mercer. 1991. "True Confessions: A Discourse on Images of Black Male Sexuality." Pp. 167–173 in *Brother to Brother: New Writings by Black Gay Men*, edited by Essex Hemphill. Boston, MA: Alyson Publications.

King, J. L. 2004. *On the Down Low: A Journey into the Lives of Straight Black Men Who Sleep with Men*. New York: Broadway Books.

Majors, Richard. 2001. "Cool Pose: Black Masculinity and Sports." Pp. 208-217 In *The Masculinities Reader*, edited by Stephen Whitehead and Frank Barrett. Cambridge: Polity Press.

Plummer, David C. 2001. "The Quest for Modern Manhood: Masculine Stereotypes, Peer Culture and the Social Significance of Homophobia." *Journal of Adolescence* 24 (1): 15–23.

Price, Jeremy. 1999. "Schooling and Racialized Masculinities: The Diploma, Teachers and Peers in the Lives of Young, African American Men." *Youth and Society* 31 (2): 224–63.

Riggs, Marlon. 1991. "Black Macho Revisited: Reflections of a Snap! Queen." Pp. 253-260 in *Brother to Brother: New Writings by Black Gay Men*, edited by Essex Hemphill. Boston, MA: Alyson Publications.

Ross, Marlon B. 1998. "In Search of Black Men's Masculinities." *Feminist Studies* 24 (3): 599–626.

28

The Invention of Heterosexuality

JONATHAN NED KATZ

Heterosexuality is old as procreation, ancient as the lust of Eve and Adam. That first lady and gentleman, we assume, perceived themselves, behaved, and felt just like today's heterosexuals. We suppose that heterosexuality is unchanging, universal, essential: ahistorical.

Contrary to that common sense conjecture, the concept of heterosexuality is only one particular historical way of perceiving, categorizing, and imagining the social relations of the sexes. Not ancient at all, the idea of heterosexuality is a modern invention, dating to the late nineteenth century. The heterosexual belief, with its metaphysical claim to eternity, has a particular, pivotal place in the social universe of the late nineteenth and twentieth centuries that it did not inhabit earlier. This essay traces the historical process by which the heterosexual idea was created as a historical and taken-for-granted....

By not studying the heterosexual idea in history, analysts of sex, gay and straight, have continued to privilege the "normal" and "natural" at the expense of the "abnormal" and "unnatural." Such privileging of the norm accedes to its domination, protecting it from questions. By making the normal the object of a thoroughgoing historical study, we simultaneously pursue a pure truth and a sex-radical and subversive goal: we upset basic preconceptions. We discover that the heterosexual, the normal, and the natural have a history of changing definitions. Studying the history of the term challenges its power.

Contrary to our usual assumption, past Americans and other peoples named, perceived, and socially organized the bodies, lusts, and intercourse of the sexes in ways radically different from the way we do. If we care to understand this vast past sexual diversity, we need to stop promiscuously projecting our own hetero and homo arrangement. Though lip service is often paid to the distorting, ethnocentric effect of such conceptual imperialism, the category heterosexuality continues to be applied uncritically as a universal analytical tool. Recognizing the time-bound and culturally specific character of the heterosexual category can help us begin to work toward a thoroughly historical view of sex....

SOURCE: Katz, Jonathan Ned. 1990. "The Invention of Heterosexuality." *Socialist Review* 20 (January–March): 7–34. Copyright © 2009. Reprinted by permission of the author.

BEFORE HETEROSEXUALITY: EARLY VICTORIAN TRUE LOVE, 1820–1860

In the early nineteenth-century United States, from about 1820 to 1860, the heterosexual did not exist. Middle-class white Americans idealized a True Womanhood, True Manhood, and True Love, all characterized by "purity"—the freedom from sensuality.[1] Presented mainly in literary and religious texts, this True Love was a fine romance with no lascivious kisses. This ideal contrasts strikingly with late nineteenth- and twentieth-century American incitements to a hetero sex.[2]

Early Victorian True Love was only realized within the mode of proper procreation, marriage, the legal organization for producing a new set of correctly gendered women and men. Proper womanhood, manhood, and progeny—not a normal male-female eros—was the main product of this mode of engendering and of human reproduction.

The actors in this sexual economy were identified as manly men and womanly women and as procreators, not specifically as erotic beings or heterosexuals. Eros did not constitute the core of a heterosexual identity that inhered, democratically, in both men and women. True Women were defined by their distance from lust. True Men, though thought to live closer to carnality, and in less control of it, aspired to the same freedom from concupiscence.

Legitimate natural desire was for procreation and a proper manhood or womanhood; no heteroerotic desire was thought to be directed exclusively and naturally toward the other sex; lust in men was roving. The human body was thought of as a means towards procreation and production; penis and vagina were instruments of reproduction, not of pleasure. Human energy, thought of as a closed and severely limited system, was to be used in producing children and in work, not wasted in libidinous pleasures.

The location of all this engendering and procreative labor was the sacred sanctum of early Victorian True Love, the home of the True Woman and True Man—a temple of purity threatened from within by the monster masturbator, an archetypal early Victorian cult figure of illicit lust. The home of True Love was a castle far removed from the erotic exotic ghetto inhabited most notoriously then by the prostitute, another archetypal Victorian erotic monster....

LATE VICTORIAN SEX-LOVE: 1860–1892

"Heterosexuality" and "homosexuality" did not appear out of the blue in the 1890s. These two eroticisms were in the making from the 1860s on. In late Victorian America and in Germany, from about 1860 to 1892, our modern idea of an eroticized universe began to develop, and the experience of a hetero-lust began to be widely documented and named....

In the late nineteenth-century United States, several social factors converged to cause the eroticizing of consciousness, behavior, emotion, and identity that became typical of the twentieth-century Western middle class. The transformation of the family from producer to consumer unit resulted in a change in family members' relation to their own bodies; from being an instrument primarily of work, the human body was integrated into a new economy, and began more commonly to be perceived as a means of consumption and pleasure. Historical work has recently begun on how the biological human body is differently integrated into changing modes of production, procreation, engendering, and pleasure so as to alter radically the identity, activity, and experience of that body.[3]

The growth of a consumer economy also fostered a new pleasure ethic. This imperative challenged the early Victorian work ethic, finally helping to usher in a major transformation of values. While the early Victorian work ethic had touted the value of economic production, that era's procreation ethic had extolled the virtues of human reproduction. In contrast, the late Victorian economic ethic hawked the pleasures of consuming, while its sex ethic praised an erotic pleasure principle for men and even for women.

In the late nineteenth century, the erotic became the raw material for a new consumer culture. Newspapers, books, plays, and films touching on sex, "normal" and "abnormal," became available for a price. Restaurants, bars, and baths opened, catering to sexual consumers with cash. Late Victorian entrepreneurs of desire incited the proliferation of a new eroticism, a commoditized culture of pleasure.

In these same years, the rise in power and prestige of medical doctors allowed these upwardly mobile professionals to prescribe a healthy new sexuality. Medical men, in the name of science, defined a new ideal of male-female relationships that included, in women as well as men, an essential, necessary, normal eroticism. Doctors, who had earlier named and judged the sex-enjoying woman a "nymphomaniac," now began to label women's *lack* of sexual pleasure a mental disturbance, speaking critically, for example, of female "frigidity" and "anesthesia."[4]

By the 1880s, the rise of doctors as a professional group fostered the rise of a new medical model of Normal Love, replete with sexuality. The new Normal Woman and Man were endowed with a healthy libido. The new theory of Normal Love was the modern medical alternative to the old Cult of True Love. The doctors prescribed a new sexual ethic as if it were a morally neutral, medical description of health. The creation of the new Normal Sexual had its counterpart in the invention of the late Victorian Sexual Pervert. The attention paid the sexual abnormal created a need to name the sexual normal, the better to distinguish the average him and her from the deviant it.

HETEROSEXUALITY: THE FIRST YEARS, 1892–1900

In the periodization of heterosexual American history suggested here, the years 1892 to 1900 represent "The First Years" of the heterosexual epoch, eight key years in which the idea of the heterosexual and homosexual were initially and tentatively formulated by U.S. doctors. The earliest-known American use of the word "heterosexual" occurs in a medical journal article by Dr. James G. Kiernan of Chicago, read before the city's medical society on March 7, 1892, and published that May—portentous dates in sexual history.[5] But Dr. Kiernan's heterosexuals were definitely not exemplars of normality. Heterosexuals, said Kiernan, were defined by a mental condition, "psychical hermaphroditism." Its symptoms were "inclinations to both sexes." These heterodox sexuals also betrayed inclinations "to abnormal methods of gratification," that is, techniques to insure pleasure without procreation. Dr. Kiernan's heterogeneous sexuals did demonstrate "traces of the normal sexual appetite" (a touch of procreative desire). Kiernan's normal sexuals were implicitly defined by a monolithic other-sex inclination and procreative aim. Significantly, they still lacked a name.

Dr. Kiernan's article of 1892 also included one of the earliest-known uses of the word "homosexual" in American English. Kiernan defined "Pure homosexuals" as persons whose "general mental state is that of the opposite sex." Kiernan thus defined homosexuals by their deviance from a gender norm. His heterosexuals displayed a double deviance from both gender and procreative norms.

Though Kiernan used the new words heterosexual and homosexual, an old procreative standard and a new gender norm coexisted uneasily in his thought. His word heterosexual defined a mixed person and compound urge, abnormal because they wantonly included procreative and non-procreative objectives, as well as same-sex and different-sex attractions.

That same year, 1892, Dr. Krafft-Ebing's influential *Psychopathia Sexualis* was first translated and published in the United States.[6] But Kiernan and Krafft-Ebing by no means agreed on the definition of the heterosexual. In Krafft-Ebing's book, "hetero-sexual" was used unambiguously in the modern sense to refer to an erotic feeling for a different sex. "Homosexual" referred unambiguously to an erotic feeling for a "same sex." In Krafft-Ebing's volume, unlike Kiernan's article, heterosexual and homosexual were clearly distinguished from a third category, a "psycho-sexual hermaphroditism," defined by impulses toward both sexes.

Krafft-Ebing hypothesized an inborn "sexual instinct" for relations with the "opposite sex," the inherent "purpose" of which was to foster procreation. Krafft-Ebing's erotic drive was still a reproductive instinct. But the doctor's clear focus on a different-sex versus same-sex sexuality constituted a historic, epochal move from an absolute procreative standard of normality toward a new norm. His definition of heterosexuality as other-sex attraction provided the basis for a revolutionary, modern break with a centuries-old procreative standard.

It is difficult to overstress the importance of that new way of categorizing. The German's mode of labeling was radical in referring to the biological sex, masculinity or femininity, and the pleasure of actors (along with the procreant purpose of acts). Krafft-Ebing's heterosexual offered the modern world a new norm that came to dominate our idea of the sexual universe, helping to change it from a mode of human reproduction and engendering to a mode of pleasure. The heterosexual category provided the basis for a move from a production-oriented, procreative imperative to a consumerist pleasure principle—an institutionalized pursuit of happiness....

Only gradually did doctors agree that heterosexual referred to a normal, "other-sex" eros. This new standard-model heterosex provided the pivotal term for the modern regularization of eros that paralleled similar attempts to standardize masculinity and femininity, intelligence, and manufacturing.[7] The idea of heterosexuality as the master sex from which all others deviated was (like the idea of the master race) deeply authoritarian. The doctors' normalization of a sex that was hetero proclaimed a new heterosexual separatism—an erotic apartheid that forcefully segregated the sex normals from the sex perverts. The new, strict boundaries made the emerging erotic world less polymorphous—safer for sex normals. However, the idea of such creatures as heterosexuals and homosexuals emerged from the narrow world of medicine to become a commonly accepted notion only in the early twentieth century. In 1901, in the comprehensive *Oxford English Dictionary*, "heterosexual" and "homosexual" had not yet made it.

THE DISTRIBUTION OF THE HETEROSEXUAL MYSTIQUE: 1900–1930

In the early years of this heterosexual century the tentative hetero hypothesis was stabilized, fixed, and widely distributed as the ruling sexual orthodoxy: The Heterosexual Mystique. Starting among pleasure-affirming urban working-class youths, southern blacks, and Greenwich Village bohemians as defensive subculture, heterosex soon triumphed as dominant culture.[8]

In its earliest version, the twentieth-century heterosexual imperative usually continued to associate heterosexuality with a supposed human "need," "drive," or "instinct" for propagation, a procreant urge linked inexorably with carnal lust as it had not been earlier. In the early twentieth century, the falling birth rate, rising divorce rate, and "war of the sexes" of the middle class were matters of increasing public concern. Giving vent to heteroerotic emotions was thus praised as enhancing baby-making capacity, marital intimacy, and family stability. (Only many years later, in the mid-1960s, would heteroeroticism be distinguished completely, in practice and theory, from procreativity and male-female pleasure sex justified in its own name.)

The first part of the new sex norm—hetero—referred to a basic gender divergence. The "oppositeness" of the sexes was alleged to be the basis for a universal, normal, erotic attraction between males and females. The stress on the sexes' "oppositeness," which harked back to the early nineteenth century, by no means simply registered biological differences of females and males. The early twentieth-century focus on physiological and gender dimorphism reflected the deep anxieties of men about the shifting work, social roles, and power of men over women, and about the ideals of womanhood and manhood. That gender anxiety is documented, for example, in 1897, in *The New York Times'* publication of the Reverend Charles Parkhurst's diatribe against female "andro-maniacs," the preacher's derogatory, scientific-sounding name for women who tried to "minimize distinctions by which manhood and womanhood are differentiated."[9] The stress on gender difference was a conservative response to the changing social-sexual division of activity and feeling which gave rise to the independent "New Woman" of the 1880s and eroticized "Flapper" of the 1920s.

The second part of the new hetero norm referred positively to sexuality. That novel upbeat focus on the hedonistic possibilities of male-female conjunctions also reflected a social transformation—a revaluing of pleasure and procreation, consumption and work in commercial, capitalist society. The democratic attribution of a normal lust to human females (as well as males) served to authorize women's enjoyment of their own bodies and began to undermine the early Victorian idea of the pure True Woman—a sex-affirmative action still part of women's struggle. The twentieth-century Erotic Woman also undercut the nineteenth-century feminist assertion of women's moral superiority, cast suspicions of lust on women's passionate romantic friendships with women, and asserted the presence of a menacing female monster, "the lesbian."[10]...

In the perspective of heterosexual history, this early twentieth-century struggle for the more explicit depiction of an "opposite-sex" eros appears in a curious new light. Ironically, we find sex-conservatives, the social purity advocates of censorship and repression, fighting against the depiction not just of sexual perversity but also of the new normal hetero-sexuality. That a more open depiction of normal sex had to be defended against forces of propriety confirms the claim that heterosexuality's predecessor, Victorian True Love, had included no legitimate eros....

THE HETEROSEXUAL STEPS OUT: 1930–1945

In 1930, in *The New York Times*, heterosexuality first became a love that dared to speak its name. On April 20th of that year, the word "heterosexual" is first known to have appeared in *The New York Times Book Review*. There, a critic described the subject of André Gide's *The Immoralist* proceeding "from a heterosexual liaison to a homosexual one." The ability to slip between sexual categories was referred to casually as a rather unremarkable aspect of human possibility. This is also the first known reference by *The Times* to the new hetero/homo duo.[11]

In September the second reference to the hetero/homo dyad appeared in *The New York Times Book Review*, in a comment on Floyd Dell's *Love in the Machine Age*. This work revealed a prominent antipuritan of the 1930s using the dire threat of homosexuality as his rationale for greater heterosexual freedom. *The Times* quoted Dell's warning that current abnormal social conditions kept the young dependent on their parents, causing "infantilism, prostitution and homosexuality." Also quoted was Dell's attack on the inculcation of purity" that "breeds distrust of the opposite sex." Young people, Dell said, should be "permitted to develop normally to heterosexual adulthood." "But," *The Times* reviewer emphasized, "such a state already exists, here and now." And so it did. Heterosexuality, a new gender-sex category, had been distributed from the narrow, rarified realm of a few doctors to become a nationally, even internationally, cited aspect of middle-class life.[12]...

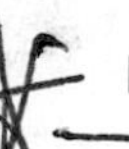

HETEROSEXUAL HEGEMONY: 1945–1965

The "cult of domesticity" following World War II—the reassociation of women with the home, motherhood, and child-care; men with fatherhood and wage work outside the home—was a period in which the predominance of the hetero norm went almost unchallenged, an era of heterosexual hegemony. This was an age in which conservative mental-health professionals reasserted the old link between heterosexuality and procreation. In contrast, sex-liberals of the day strove, ultimately with success, to expand the heterosexual ideal to include within the boundaries of normality a wider-than-ever range of nonprocreative, premarital, and extramarital behaviors. But sex-liberal reform actually helped to extend and secure the dominance of the heterosexual idea, as we shall see when we get to Kinsey.

The postwar sex-conservative tendency was illustrated in 1947, in Ferdinand Lundberg and Dr. Marynia Farnham's books, *Modern Woman: The Lost Sex.* Improper masculinity and femininity was exemplified, the authors decreed, by "engagement in heterosexual relations ... with the complete intent to see to it that they do not eventuate in reproduction."[13] Their procreatively defined heterosex was one expression of a postwar ideology of fecundity that, internalized and enacted dutifully by a large part of the population, gave rise to the postwar baby boom.

The idea of the feminine female and masculine male as prolific breeders was also reflected in the stress, specific to the late 1940s, on the homosexual as sad symbol of "sterility"—that particular loaded term appears incessantly in comments on homosex dating to the fecund forties.

In 1948, in *The New York Times Book Review*, sex liberalism was in ascendancy. Dr. Howard A. Rusk declared that Alfred Kinsey's just published report on *Sexual Behavior in the Human Male* had found "wide variations in sex concepts and behavior." This raised the question: "What is 'normal' and 'abnormal'?" In particular, the report had found that "homosexual experience is much more common than previously thought," and "there is often a mixture of both homo and hetero experience."[14]

Kinsey's counting of orgasms indeed stressed the wide range of behaviors and feelings that fell within the boundaries of a quantitative, statistically accounted heterosexuality. Kinsey's liberal reform of the hetero/homo dualism widened the narrow, old hetero category to accord better with the varieties of social experience. He thereby contradicted the older idea of a monolithic, qualitatively defined, natural procreative act, experience, and person.[15]

Though Kinsey explicitly questioned "whether the terms 'normal' and 'abnormal' belong in a scientific vocabulary," his counting of climaxes was generally understood to define normal sex as majority sex. This quantified norm constituted a final, society-wide break with the old qualitatively defined reproductive standard. Though conceived of as purely scientific, the statistical definition of the normal as the-sex-most-people-are-having substituted a new, quantitative moral standard for the old, qualitative sex ethic—another triumph for the spirit of capitalism.

Kinsey also explicitly contested the idea of an absolute, either/or antithesis between hetero and homo persons. He denied that human beings "represent two discrete populations, heterosexual and homosexual." The world, he ordered, "is not to be divided into sheep and goats." The hetero/homo division was not nature's doing: "Only the human mind invents categories and tries to force facts into separated pigeon-holes. The living world is a continuum."[16]

With a wave of the taxonomist's hand, Kinsey dismissed the social and historical division of people into heteros and homos. His denial of heterosexual and homosexual personhood rejected the social reality and profound subjective force of a historically constructed tradition which, since 1892 in the United States, had cut the sexual populaton in two and helped to establish the social reality of a heterosexual and homosexual identity.

On the one hand, the social construction of homosexual persons has led to the development of a powerful gay liberation identity politics based on an ethnic group model. This has freed generations of women and men from a deep, painful, socially induced sense of shame, and helped to bring about a society-wide liberalization of attitudes and responses to homosexuals.[17] On the other hand, contesting the notion of homosexual and heterosexual persons was one early, partial resistance to the limits of the hetero/homo construction. Gore Vidal, rebel son of Kinsey, has for years been joyfully proclaiming:

> ...there is no such thing as a homosexual or a heterosexual person. There are only homo- or heterosexual acts. Most people are a mixture of impulses if not practices, and what anyone does with a willing partner is of no social or cosmic significance.
>
> So why all the fuss? In order for a ruling class to rule, there must be arbitrary prohibitions. Of all prohibitions, sexual taboo is the most useful because sex involves everyone.... We have allowed our governors to divide the population into two teams. One team is good, godly, straight; the other is evil, sick, vicious.[18]

HETEROSEXUALITY QUESTIONED: 1965–1982

By the late 1960s, anti-establishment counter culturalists, fledgling feminists, and homosexual-rights activists had begun to produce an unprecedented critique of sexual repression in general, of women's sexual repression in particular, of marriage and the family—and of some forms of heterosexuality....

Heterosexual History: Out of the Shadows

Our brief survey of the heterosexual idea suggests a new hypothesis. Rather than naming a conjunction old as Eve and Adam, heterosexual designates a word and concept, a norm and role, an individual and group identity, a behavior and feeling, and a peculiar sexual-political institution particular to the late nineteenth and twentieth centuries.

Because much stress has been placed here on heterosexuality as word and concept, it seems important to affirm that heterosexuality (and homosexuality) came into existence before it was named and thought about. The formulation of the heterosexual idea did not create a heterosexual experience or behavior; to suggest otherwise would be to ascribe determining power to labels and concepts. But the titling and envisioning of heterosexuality did play an important role in consolidating the construction of the heterosexual's social existence. Before the wide use of the word "heterosexual," I suggest, women and men did not mutually lust with the same profound, sure sense of normalcy that followed the distribution of "heterosexual" as universal sanctifier.

According to this proposal, women and men make their own sexual histories. But they do not produce their sex lives just as they please. They make their sexualities within a particular mode of organization given by the past and altered by their changing desire, their present power and activity, and their vision of a better world. That hypothesis suggests a number of good reasons for the immediate inauguration of research on a historically specific heterosexuality.

The study of the history of the heterosexual experience will forward a great intellectual struggle still in its early stages. This is the fight to pull heterosexuality, homosexuality, and all the sexualities out of the realm of nature and biology [and] into the realm of the social and historical. Feminists have explained to us that anatomy does not determine our gender destinies (our masculinities and femininities). But we've only recently begun to consider that *biology does not settle our erotic fates.* The common notion that biology determines the object of sexual desire, or that physiology and society together cause sexual orientation, are determinisms that deny the break existing between our bodies and situations and our desiring. Just as the biology of our hearing organs will never tell us why we take pleasure in Bach or delight in Dixieland, our female or male anatomies, hormones, and genes will never tell us why we yearn for women, men, both, other, or none. That is because desiring is a self-generated project of individuals within particular historical cultures. Heterosexual history can help us see the place of values and judgments in the construction of our own and others' pleasures, and to see how our erotic tastes—our aesthetics of the flesh—are socially institutionalized through the struggle of individuals and classes.

The study of heterosexuality in time will also help us to recognize the *vast historical diversity of sexual emotions and behaviors*—a variety that challenges the monolithic heterosexual hypothesis. John D'Emilio and Estelle Freedman's *Intimate Matters: A History of Sexuality in America* refers in passing to numerous substantial changes in sexual activity and feeling: for example, the widespread use of contraceptives in the nineteenth century, the twentieth-century incitement of the female orgasm, and the recent sexual conduct changes by gay men in response to the AIDS epidemic. It's now a commonplace of family history that people in particular classes feel and behave in substantially different ways under different, historical conditions. Only when we stop assuming an invariable essence of heterosexuality will we begin the research to reveal the full variety of sexual emotions and behaviors.

The historical study of the heterosexual experience can help us *understand the erotic relationships of women and men in terms of their changing modes of social organization.* Such model analysis actually characterizes a sex history well underway. This suggests that the eros-gender-procreation system (the social ordering of lust, femininity and masculinity, and baby-making) has been linked closely to a society's particular organization of power and production. To understand the subtle history of heterosexuality we need to look carefully at correlations between (1) society's organization of eros and pleasure; (2) its mode of engendering persons as feminine or masculine (its making of women and men); (3) its ordering of human reproduction; and (4) its dominant political economy. This General Theory of Sexual Relativity proposes that substantial historical changes in the social organization of eros, gender, and procreation have basically altered the activity and experience of human beings within those modes.

A historical view locates heterosexuality and homosexuality in time, helping us distance ourselves from them. This distancing can help us formulate new questions that clarify our long-range sexual-political goals: What has been and is the social function of sexual categorizing? Whose interests have been served by the division of the world into heterosexual and homosexual? Do we dare not draw a line between those two erotic species? Is some sexual naming socially necessary? Would human freedom be enhanced if the sex-biology of our partners in lust was of no particular concern, and had no name? In what kind of society could we all more freely explore our desire and our flesh?

As we move [into the present], a new sense of the historical making of the heterosexual and homosexual suggests that these are ways of feeling, acting, and being with each other that we can together unmake and radically remake according to our present desire, power, and our vision of a future political-economy of pleasure.

NOTES

1. Barbara Welter, "The Cult of True Womanhood: 1820–1860," *American Quarterly*, vol. 18 (Summer 1966); Welter's analysis is extended here to include True Men and True Love.
2. Some historians have recently told us to revise our idea of sexless Victorians: their experience and even their ideology, it is said, were more erotic than we previously thought. Despite the revisionists, I argue that "purity" was indeed the dominant, early Victorian, white middle-class standard. For the debate on Victorian sexuality see John D'Emilio and Estelle Freedman, *Intimate Matters: A History of Sexuality in America* (New York: Harper & Row, 1988), p. xii.

3. See, for example, Catherine Gallagher and Thomas Laqueur, eds., "The Making of the Modern Body: Sexuality and Society in the Nineteenth Century," *Representations*, no. 14 (Spring 1986) (republished, Berkeley: University of California Press, 1987).
4. This reference to females reminds us that the invention of heterosexuality had vastly different impacts on the histories of women and men. It also differed in its impact on lesbians and heterosexual women, homosexual and heterosexual men, the middle class and working class, and on different religious, racial, national, and geographic groups.
5. Dr. James G. Kieman, "Responsibility in Sexual Perversion," *Chicago Medical Recorder*, vol. 3 (May 1892), pp. 185–210.
6. R. von Krafft-Ebing, *Psychopathia Sexualis, with Especial Reference to Contrary Sexual Instinct: A Medico-Legal Study*, trans. Charles Gilbert Chaddock (Philadelphia: F. A. Davis, 1892), from the 7th and revised German ed. Preface, November 1892.
7. For the standardization of gender see Lewis Terman and C. C. Miles, *Sex and Personality, Studies in Femininity and Masculinity* (New York: McGraw Hill, 1936). For the standardization of intelligence see Lewis Terman, *Stanford-Binet Intelligence Scale* (Boston: Houghton Mifflin, 1916). For the standardization of work, see "scientific management" and "Taylorism" in Harry Braverman, *Labor and Monopoly Capital: The Degradation of Work in the Twentieth Century* (New York: Monthly Review Press, 1974).
8. See D'Emilio and Freedman, *Intimate Matters*, pp. 194–201, 231, 241, 295–96; Ellen Kay Trimberger, "Feminism, Men, and Modern Love: Greenwich Village, 1900–1925," in *Powers of Desire: The Politics of Sexuality*, ed. Ann Snitow, Christine Stansell, and Sharon Thompson (New York: Monthly Review Press, 1983), pp. 131–52; Kathy Peiss, " 'Charity Girls' and City Pleasures: Historical Notes on Working Class Sexuality, 1880–1920," in *Powers of Desire*, pp. 74–87; and Mary P. Ryan, "The Sexy Saleslady: Psychology, Heterosexuality, and Consumption in the Twentieth Century," in her *Womanhood in America*, 2nd ed. (New York: Franklin Watts, 1979), pp. 151–82.
9. [Rev. Charles Parkhurst], "Woman. Calls Them Andromaniacs. Dr. Parkhurst So Characterizes Certain Women Who Passionately Ape Everything That Is Mannish. Woman Divinely Preferred. Her Supremacy Lies in Her Womanliness, and She Should Make the Most of It—Her Sphere of Best Usefulness the Home," *The New York Times*, May 23, 1897, p. 16:1.
10. See Lisa Duggan, "The Social Enforcement of Heterosexuality and Lesbian Resistance in the 1920s," in *Class, Race, and Sex: The Dynamics of Control*, ed. Amy Swerdlow and Hanah Lessinger (Boston: G. K. Hall, 1983), pp. 75–92; Rayna Rapp and Ellen Ross, "The Twenties Backlash: Compulsory Heterosexuality, the Consumer Family, and the Waning of Feminism," in *Class, Race, and Sex*; Christina Simmons, "Companionate Marriage and the Lesbian Threat," *Frontiers*, vol. 4, no. 3 (Fall 1979), pp. 54–59; and Lillian Faderman, *Surpassing the Love of Men* (New York: William Morrow, 1981).
11. Louis Kronenberger, review of André Gide, *The Immoralist, New York Times Book Review*, April 20, 1930, p. 9.
12. Henry James Forman, review of Floyd Dell, *Love in the Machine Age* (New York: Farrar & Rinehart), *New York Times Book Review*, September 14, 1930, p. 9.
13. Ferdinand Lundberg and Dr. Marynia F. Farnham, *Modern Woman: The Lost Sex* (New York: Harper, 1947).

14. Dr. Howard A. Rusk, *New York Times Book Review*, January 4, 1948, p. 3.
15. Alfred Kinsey, Wardell B. Pomeroy, and Clyde E. Martin, *Sexual Behavior in the Human Male* (Philadelphia: W. B. Saunders, 1948), pp. 199–200.
16. Kinsey, *Sexual Behavior*, pp. 637, 639.
17. See Steven Epstein, "Gay Politics, Ethnic Identity: The Limits of Social Constructionism," *Socialist Review* 93/94 (1987), pp. 9–54.
18. Gore Vidal, "Someone to Laugh at the Squares With" [Tennessee Williams], *New York Review of Books*, June 13, 1985; reprinted in his *At Home: Essays, 1982–1988* (New York: Random House, 1988), p. 48.

29

Straight

The Surprisingly Short History of Heterosexuality

HANNE BLANK

Every time I go to the doctor, I end up questioning my sexual orientation. On some of its forms, the clinic I visit includes five little boxes, a small matter of demographic bookkeeping. Next to the boxes are the options "gay," "lesbian," "bisexual," "transgender," or "heterosexual." You're supposed to check one.

You might not think this would pose a difficulty. I am a fairly garden-variety female human being, after all, and I am in a long-term monogamous relationship, well into our second decade together, with someone who has male genitalia. But does this make us, or our relationship, straight? This turns out to be a good question, because there is more to my relationship—and much, much more to hetero-sexuality—than easily meets the eye.

There's biology, for one thing. My partner was diagnosed male at birth because he was born with, and indeed still has, a fully functioning penis. But, as the ancient Romans used to say, *barba non facit philosophum*—a beard does not make one a philosopher. Neither does having a genital outie necessarily make one male. Indeed, of the two sex chromosomes—XY—which would be found in the genes of a typical male, and XX, which is the hallmark of the genetically typical female—my partner's DNA has all three: XXY, a pattern that is simultaneously male, female, and neither.

This particular genetic pattern, XXY, is the signature of Kleinfelter Syndrome, one of the most common sex-chromosome anomalies. XXY often goes undiagnosed because the people who have it often look perfectly normal from the outside. In many cases, XXY individuals do not find out about their chromosomal anomaly unless they try to have children and end up seeing a fertility doctor, who ultimately orders an image called a karyotype, essentially a photo of the person's chromosomes made with a very powerful microscope. In a karyotype, the trisomy, or three-chromosome grouping, is instantly revealed. As genetic anomalies go, this particular trisomy is not a cause for major alarm (aside from infertility, it causes few significant problems), which is a good thing, since it is fairly common. The estimates vary, in part because diagnosis is so haphazard, but it is believed that as many as one in every two thousand people who are declared male at birth may in fact be XXY. At minimum, there are about half a million Americans whose genetics are this way, most of whom will never know it.

SOURCE: Blank, Hanne. 2012. *Straight: The Surprisingly Short History of Heterosexuality*. Boston: Beacon Press. Reprinted by permission of Beacon Press, Boston.

What does an unusual sexual biology mean for sexual orientation? Is it even *possible* for XXY people to have a sexual orientation in the way we usually think about sexual orientation? What about their lovers, partners, and spouses? "Heterosexual," "homosexual," and "bisexual" are all dependent on the idea that there are two, and only two, biological sexes. What happens when biology refuses to fit neatly into this scheme? If I'm attracted to, and in love with, someone who is technically speaking neither male nor female, does that make me heterosexual, homosexual, bisexual, or something else altogether? Who gets to decide? And, more to the point, on what grounds?

Some would argue that genetics aren't as important as anatomy and bodily functions. After all, you can't see chromosomes with the naked eye. But here, too, I run into problems. Part of what makes a man, as we are all taught from childhood, is that he has a penis and testicles that produce sperm, which in turn are necessary to fertilize a female's egg cells and conceive a fetus. The ability to sire a child has been considered proof of masculinity for thousands of years. This is something my partner cannot do. His external plumbing looks and acts pretty much like any genetically typical male's, but, in the words of one of my partner's vasectomied coworkers, "he shoots blanks." In my partner's case, no vasectomy was required. His testicles do not produce viable sperm. They never have and never will. This is part of the territory for most people who have XXY sex chromosomes.

So if heterosexuality is by definition, as some of our right-wing brothers and sisters like to claim, about the making of babies, then there is no possible way for my partner and me to be construed as heterosexual. But even the Bible recognizes that infertility exists. The notoriously procreation-fixated Catholic Church permits marriage, and marital sex, between people known to be infertile. Curiously, whether or not reproduction is a cornerstone of heterosexuality seems to depend on whom you ask, and in what circumstances.

Not that it really matters in practice. At this point in time contraception is more the rule than the exception for sexual activity between different-sex partners throughout the first world. Many people, including members of committed male/female couples, don't have children or plan to have them, yet somehow this doesn't stop them from feeling quite certain that they know what their sexual orientations are. They consider sexual orientation as being rooted in a calculus of subjective attraction and biological sameness. The Greek "hetero" means "other" or "different," after all, and biological men and women do differ from one another. We make use of these biological differences every day without thinking every time we look at people and identify them as either male or female, ask whether a baby or a dog is a boy or a girl, or determine the sexes of the members of a couple we spot on the street and assign them sexual orientations in our minds.

Surely such informal, man-in-the-street diagnostics ought to apply just as well to my partner and me. Or perhaps not. As an XXY individual who has chosen not to take hormone supplements, my partner's naturally occurring sex hormones take a middle path. His estrogen levels hang out a little lower than mine, his testosterone levels a little higher. As a result, my partner, like other

XXY people who don't take exogenous hormones, has an androgynous appearance, with little to no facial and body hair, a fine smooth complexion, and a tendency to develop small breasts and slightly rounded hips if he puts on a little weight.... When we lived in a LGBT-heavy neighborhood in Boston, my partner and I were often identified by others as lesbians. We were regularly referred to as "ladies" by shopkeepers, door-to-door Mormons, and parents trying to prevent their kids from crowding us at the zoo. Lesbian couples we encountered in passing often shot us little conspiratorial smiles of recognition. (We always smiled back. Still do.) But it wasn't all pleasantries. Once while walking together we had bottles thrown at us from a car, its occupants screaming "Fuckin' dykes!" out the windows as they sped away. Assumptions of sexual orientation are never merely innocent perceptions, because these perceptions shape behavior.

Assumptions about biology and gender are complicated, fraught, and by no means clear or unambiguous. The ways people have identified my partner's biological sex, and therefore not only the nature of our relationship to one another but also our respective sexual orientations, have run an extraordinary gamut that might be distressing if we hadn't long ago learned to laugh at it all. My partner's physical androgyny—the minimal facial hair, refined complexion, and elegant, long-limbed build that are common side effects of his genetic anomaly—has led some people to assume that he is a female-to-male transsexual who is early in the transition process, still more hormonally female than male. I have heard him identified as a "passing butch." Once, at a party, I overheard a woman stating with assurance that my partner was a very feminine gay man who had "made an exception" for me. At other times I have been assumed to be the one making the exception, a "hasbian" who turned from dating women to seeing a gentle, feminine straight man. By the same token, these reactions have changed as we've aged and our styles of dress and grooming have changed. For the past several years, with my partner usually dressing in corporate-office menswear and sporting a dashing haircut modeled on the young Cole Porter's, we have typically, though not always, been read as heterosexual. If the range of responses we've had can tell me anything about what my sexual orientation is supposed to be, it's that other people don't necessarily know what box I should be checking off on those clinic forms either.

My own sense of sexual identity is, incidentally, no help. I have no deep personal attachment to labeling myself in terms of sexual orientation, nor do I have the sensation of "being" heterosexual or homosexual or anything but a human being who loves and desires other human beings. I have been romantically and sexually involved with people of a variety of biological sexes and social genders over the course of my adult life. When pressed, I am most likely to declare my "sexual identity" as "taken." This option, however much it might be the best fit, is not available to me on most forms that ask this sort of question.

I could, I suppose, resort to legal documents to sort out the question of what my orientation is, and what the orientation of my relationship with my partner might be. Here at last it is uncomplicated. Based on our birth certificates, my partner and I and our relationship could be defined as uncomplicatedly heterosexual. But there's a caveat, and it's a big one: our sexes were diagnosed at birth on the basis of a visual check of our genitals, on the assumption that external genitals are an infallible indicator of biological sex. This is the assumption behind every "it's a boy" or "it's a girl," not just historically but every day around the world. Thanks to the publicity given to cases like that of intersex South African athlete Caster Semenya in 2009, and indeed to the ink I am spilling here, however, mainstream culture is gradually becoming aware that this assumption is not necessarily warranted. Many biologists, including Brown University biologist Anne Fausto-Sterling, have eloquently testified that humans have at least five major sexes—of which typical male and typical female are merely the most numerous—and that furthermore, human chromosomes, gonads, internal sexual organs, external genitals, sex hormones, and secondary sexual characteristics can appear in many different guises.

The law, however, still acknowledges only two sexes. It does not always or necessarily acknowledge sexual orientation at all. On the occasions when it does recognize sexual orientation, it typically acknowledges only two of them as well, heterosexual and homosexual. (Once in a while bisexuality is included, but often not.) All of these sexual orientations are wholly dependent upon—and could not be conceptualized without—the general consensus that there are two and only two human biological sexes. But as we now know, and as is demonstrated so charmingly in the person of my very own beloved, this is not necessarily so. Rather, the convenient sorting of human beings into two biological sexes and a correspondingly limited number of sexual orientations is an artifact of a historical system that was formed at a time when medicine, biology, and social theory were capable of far less than they are now. We are still using a very limited nineteenth-century set of ideas and terminology to talk about a decidedly more expansive twenty-first-century landscape of biology, medicine, law, social theory, and human behavior.

It has, in point of actual fact, only been possible to be a heterosexual since 1869.... Prior to that time, men and women got married, had sex, had children, formed families, and sometimes even fell in love, but they were categorically not heterosexuals. They didn't identify themselves as "being" something called "heterosexual." They didn't think of themselves as having a "straight" sexual identity, or indeed have any awareness that something called a "sexual identity" even existed. They couldn't have. Neither the terms nor the ideas that they express existed yet.

"Heterosexual" and "heterosexuality" are creations of a particular, distinct, well-documented time and place. They are words, and ideas, developed by people whose names are known to us and whose handwritten letters we can still read. Their adoption and integration into Western culture was a remarkable process that historian Jonathan Ned Katz, the first to chronicle it, has aptly called "the invention of heterosexuality."

It was an invention whose time had clearly come, for it took less than a century for "heterosexual" and "heterosexuality" to leap out of the honestly rather obscure medical and legal backwaters where they were born and become part of a vast and opaque umbrella sheltering an enormous amount of social, economic, scientific, legal, political, and cultural activity. Exactly how this happened is a complicated, diffuse story that takes place on many different stages at roughly the same time, over a span of time measured in decades.

We need not, however, labor under the delusion that "heterosexual" became such a culture-transforming success because it represented the long-awaited discovery of a vital and inescapable scientific truth. It wasn't. As we shall see, the original creation of "heterosexual" and "homosexual" had nothing to do with scientists or science at all. Nor did it have anything to do with biology or medicine. "Heterosexual" (and "homosexual") originated in a quasi-legal context, a term of art designed to argue a philosophical point of legislature.

Perhaps this should not surprise us. Indeed, it can be argued that the biomedical business of sexuality has nothing to do with sexual orientation or sexual identity anyway. The materials and physiology of sexual activities are, on a strictly mechanical level, a separate problem from the subjective mechanics of attraction or desire, as rape—something that can and does happen to people without regard to biological sex, age, condition, or consent—attests with such brutal efficiency. Separate from human sexual orientation or identity in a different way are the chemistry and alchemy of human conception, which can, after all, take place in a petri dish. There is, biomedically speaking, nothing about what human beings do sexually that requires that something like what we now think of as "sexual orientation" exists. If there were, and the attribute we now call "heterosexuality" were a prerequisite for people to engage in sex acts or procreate, chances are excellent that we would not have waited until the late nineteenth century to figure out that it was there.

"Heterosexual" became a success, in other words, not because it represented a new scientific verity or capital-T Truth. It succeeded because it was *useful*. At a time when moral authority was shifting from religion to the secular society at a precipitous pace, "heterosexual" offered a way to dress old religious priorities in immaculate white coats that looked just like the ones worn among the new power hierarchy of scientists. At a historical moment when the waters of anxiety about family, nation, class, gender, and empire were at a rather hysterical high, "heterosexual" seemed to offer a dry, firm place for authority to stand. This new concept, gussied up in a mangled mix of impressive-sounding dead languages,... gave old orthodoxies a new and vibrant lease on life by suggesting, in authoritative tones, that science had effectively pronounced them natural, inevitable, and innate.

What does all this have to do with me, my partner, and the unanswered question of which multiple-choice box I should tick? Plenty. The history of "the heterosexual" lurks unexamined not just in our beliefs about our inmost private selves, but also in our beliefs about our bodies, our social interactions, our romances, our family lives, the way we raise our children, and, of course, in our sex lives. Virtually everyone alive today, especially in the developed world, has lived their entire lives in a culture of sexuality that assumes that "heterosexual" and "homosexual" are objectively real elements of nature.

As a result of this pervasiveness, heterosexuality is like air, all around us and yet invisible. But as we all know, the fact that we can see through air doesn't mean it can't exert force, push things around, and create friction. In the process of asking questions about my own life, I have had to learn to think about heterosexuality like an aircraft pilot thinks about the air: as something with a real, tangible presence, something that is not only capable of but is constantly in the process of influencing if not dictating thoughts, actions, and reactions. If I, or any of us, are to be able to decide whether or not we or our relationships qualify as "heterosexual," it behooves us to understand what that means. This history represents the attempt to begin to comprehend what exactly this invisible wind is, where it comes from, what it's made of, and where it might be pushing you and me and all of us....

30

Selling Sex for Visas

Sex Tourism as a Stepping-Stone to International Migration

DENISE BRENNAN

On the eve of her departure for Germany to marry her German client-turned-boyfriend, Andrea, a Dominican sex worker, spent the night with her Dominican boyfriend. When I dropped by the next morning to wish her well, her Dominican boyfriend was still asleep. She stepped outside, onto her porch. She could not lie about her feelings for her soon-to-be husband. "No," she said, "it's not love." But images of an easier life for herself and her two daughters compelled her to migrate off the island and out of poverty. She put love aside—at least temporarily.

Andrea, like many Dominican sex workers in Sosúa, a small town on the north coast of the Dominican Republic, makes a distinction between marriage *por amor* (for love) and marriage *por residencia* (for visas). After all, why waste a marriage certificate on romantic love when it can be transformed into a visa to a new land and economic security?

Since the early 1990s, Sosúa has been a popular vacation spot for male European sex tourists, especially Germans. Poor women migrate from throughout the Dominican Republic to work in Sosúa's sex trade; there, they hope to meet and marry foreign men who will sponsor their migration to Europe. By migrating to Sosúa, these women are engaged in an economic strategy that is both familiar and altogether new: they are attempting to capitalize on the very global linkages that exploit them. These poor single mothers are not simply using sex work in a tourist town with European clients as a survival strategy; they are using it as an *advancement* strategy.

The key aims of this strategy are marriage and migration off the island. But even short of these goals, Sosúa holds out special promise to its sex workers, who can establish ongoing transnational relationships with the aid of technologies such as fax machines at the phone company in town (the foreign clients and the women communicate about the men's return visits in this manner) and international money wires from clients overseas. Sosúa's sex trade also stands apart from that of many other sex-tourist destinations in the developing world in

SOURCE: Brennan, Denise. 2004. "Selling Sex for Visas: Sex Tourism as a Stepping-Stone to International Migration." Pp. 158–168 in *Global Woman: Nannies, Maids, and Sex Workers in the New Economy*, edited by B. Ehrenreich and A. Russell Hochschild. New York: Holt.

that it does not operate through pimps, nor is it tied to the drug trade; young women are not trafficked to Sosúa, and as a result they maintain a good deal of control over their working conditions.

Certainly, these women still risk rape, beatings, and arrest; the sex trade is dangerous, and Sosúa's is no exception. Nonetheless, Dominican women are not coerced into Sosúa's trade but rather end up there through networks of female family members and friends who have worked there. Without pimps, sex workers keep all their earnings; they are essentially working freelance. They can choose the bars and nightclubs in which to hang out, the number of hours they work, the clients with whom they will work, and the amount of money to charge.

There has been considerable debate over whether sex work can be anything but exploitative. The stories of Dominican women in Sosúa help demonstrate that there is a wide range of experiences within the sex trade, some of them beneficial, others tragic.... I have been particularly alarmed at the media's monolithic portrayal of sex workers in sex-tourist destinations, such as Cuba, as passive victims easily lured by the glitter of consumer goods. These overly simplistic and implicitly moralizing stories deny that poor women are capable of making their own labor choices. The women I encountered in Sosúa had something else to say.

SEX WORKERS AND SEX TOURISTS

Sex workers in Sosúa are at once independent and dependent, resourceful and exploited. They are local agents caught in a web of global economic relations. To the extent that they can, they try to take advantage of the men who are in Sosúa to take advantage of them. The European men who frequent Sosúa's bars might see Dominican sex workers as exotic and erotic because of their dark skin color; they might pick one woman over another in the crowd, viewing them all as commodities for their pleasure and control. But Dominican sex workers often see the men, too, as readily exploitable—potential dupes, walking visas, means by which the women might leave the island, and poverty, behind.

Even though only a handful of women have actually married European men and migrated off the island, the possibility of doing so inspires women to move to Sosúa from throughout the island and to take up sex work. Once there, however, Dominican sex workers are beholden to their European clients to deliver visa sponsorships, marriage proposals, and airplane tickets. Because of the differential between sex workers and their clients in terms of mobility, citizenship, and socioeconomic status, these Dominican sex workers might seem to occupy situations parallel to those that prevail among sex workers throughout the developing world. Indeed, I will recount stories here of disappointment, lies, and unfulfilled dreams. Yet some women make modest financial gains through Sosúa's sex trade—gains that exceed what they could achieve working in export-processing zones or domestic service, two common occupations among poor Dominican women. These jobs, on average, yield fewer than 1,000 pesos ($100) a month, whereas sex workers in Sosúa charge approximately 500 pesos for each encounter with a foreign client.

Sex tourism, it is commonly noted, is fueled by the fantasies of white, First World men who exoticize dark-skinned "native" bodies in the developing world, where they can buy sex for cut-rate prices. These two components—racial stereotypes and the economic disparity between the developed and the developing worlds—characterize sex-tourist destinations everywhere. But male sex tourists are not the only ones who travel to places like Sosúa to fulfill their fantasies. Many Dominican sex workers look to their clients as sources not only of money, marriage, and visas, but also of greater gender equity than they can hope for in the households they keep with Dominican men. Some might hope for romance and love, but most tend to fantasize about greater resources and easier lives.

Yet even for the women with the most pragmatic expectations, there are few happy endings. During the time I spent with sex workers in Sosúa, I, too, became invested in the fantasies that sustained them through their struggles. Although I learned to anticipate their return from Europe, disillusioned and divorced, I continued to hope that they would find financial security and loving relationships. Similarly, Sosúa's sex workers built their fantasies around the stories of their few peers who managed to migrate as the girlfriends or wives of European tourists—even though nearly all of these women returned, facing downward mobility when they did so. Though only a handful of women regularly receive money wires from clients in Europe, the stories of those who do circulate among sex workers like Dominicanized versions of Hollywood's *Pretty Woman*.

The women who pursue these fantasies in Sosúa tend to be pushed by poverty and single motherhood. Of the fifty women I interviewed and the scores of others I met, only two were not mothers. The practice of consensual unions (of not marrying but living together), common among the poor in the Dominican Republic, often leads to single motherhood, which then puts women under significant financial pressure. Typically, these women receive no financial assistance from their children's fathers. I met very few sex workers who had sold sex before migrating to Sosúa, and I believe that the most decisive factor propelling these women into the sex trade is their status as single mothers. Many women migrated to Sosúa within days of their partners' departure from the household and their abandonment of their financial obligations to their children.

Most women migrated from rural settings with meager job opportunities, among them sporadic agricultural work, low-wage hairstyling out of one's home, and waitressing. The women from Santo Domingo, the nation's capital, had also held low-paying jobs, working in domestic service or in *zonas francas* (export-processing zones). Women who sell sex in Sosúa earn more money, more quickly than they can in any other legal job available to poor women with limited educations (most have not finished school past their early teens) and skill bases. These women come from *los pobres,* the poorest class in the

Dominican Republic, and they simply do not have the social networks that would enable them to land work, such as office jobs, that offer security or mobility. Rather, their female-based social networks can help them find factory jobs, domestic work, restaurant jobs, or sex work.

Sex work offers women the possibility of making enough money to start a savings account while covering their own expenses in Sosúa and their children's expenses back home. These women tend to leave their children in the care of female family members, but they try to visit and to bring money at least once a month. If their home communities are far away and expensive to get to, they return less frequently. Those who manage to save money use it to buy or build homes back in their home communities. Alternatively, they might try to start small businesses, such as *colmados* (small grocery stores), out of their homes.

While saving money is not possible in factory or domestic work, sex workers, in theory at least, make enough money to build up modest savings. In practice, however, it is costly to live in Sosúa. Rooms in boardinghouses rent for 30 to 50 pesos a day, while apartments range from 1,500 to 3,000 pesos a month, and also incur start-up costs that most women cannot afford (such as money for a bed and cooking facilities). Since none of the boardinghouses have kitchens, women must spend more for take-out or restaurant meals. On top of these costs, they must budget for bribes to police officers (for release from jail), since sex workers usually are arrested two to five times a month. To make matters worse, the competition for clients is so fierce, particularly during the low-volume tourist seasons, that days can go by before a woman finds a client. Many sex workers earn just enough to cover their daily expenses in Sosúa while sending home modest remittances for their children. Realizing this, and missing their children, most women return to their home communities in less than a year, just as poor as when they first arrived....

The sex workers I interviewed, who generally have no immediate family members abroad, have never had reliable transnational resources available to them. Not only do they not receive remittances but they cannot migrate legally through family sponsorship. Sex workers' transnational romantic ties act as surrogate family-migration networks. Consequently, migration to Sosúa from other parts of the Dominican Republic can be seen as both internal and international, since Sosúa is a stepping-stone to migration to other countries. For some poor young women, hanging out in the tourist bars of Sosúa is a better use of their time than waiting in line at the United States embassy in Santo Domingo. Carla, a first-time sex worker, explained why Sosúa draws women from throughout the country: "We come here because we dream of a ticket," she said, referring to an airline ticket. But without a visa—which they can obtain through marriage—that airline ticket is of little use.

If sex workers build their fantasies around their communities' experiences of migration, the fantasies sex tourists hope to enact in Sosúa are often first suggested through informal networks of other sex tourists. Sosúa first became known among European tourists by word of mouth. Most of the sex tourists I met in Sosúa had been to other sex-tourist destinations as well. These seasoned sex tourists, many of whom told me that they were "bored" with other destinations, decided to try Sosúa and Dominican women based on the recommendations of friends. This was the case for a group of German sex tourists who were drinking at a bar on the beach. They nodded when the German bar owner explained, "Dominican girls like to fuck." One customer chimed in, "With German women it's over quickly. But Dominican women have fiery blood.... When the sun is shining it gives you more hormones."

The Internet is likely to increase the traffic of both veteran and first-time sex tourists to previously little-known destinations like Sosúa. Online travel services provide names of "tour guides" and local bars in sex-tourism hot spots. On the World Sex Guide, a Web site on which sex tourists share information about their trips, one sex tourist wrote that he was impressed by the availability of "dirt cheap colored girls" in Sosúa, while another gloated, "When you enter the discos, you feel like you're in heaven! A tremendous number of cute girls and something for everyone's taste (if you like colored girls like me)!"

As discussions and pictures of Dominican women proliferate on the Internet sites—for "travel services" for sex tourists, pen-pal services, and even cyber classified advertisements in which foreign men "advertise" for Dominican girlfriends or brides—Dominican women are increasingly often associated with sexual availability. A number of articles in European magazines and newspapers portray Dominican women as sexually voracious. The German newspaper *Express* even published a seven-day series on the sex trade in Sosúa, called "Sex, Boozing, and Sunburn," which included this passage: "Just going from the street to the disco—there isn't any way men can take one step alone. Prostitutes bend over, stroke your back and stomach, and blow you kisses in your ear. If you are not quick enough, you get a hand right into the fly of your pants. Every customer is fought for, by using every trick in the book." A photo accompanying one of the articles in this series shows Dieter, a sex tourist who has returned to Sosúa nine times, sitting at a German-owned bar wearing a T-shirt he bought in Thailand; the shirt is emblazoned with the words SEX TOURIST.

With all the attention in the European press and on the Internet associating Dominican women with the sex industry, fear of a stigma has prompted many Dominican women who never have been sex workers to worry that the families and friends of their European boyfriends or spouses might wonder if they once were. And since Dominican women's participation in the overseas sex trade has received so much press coverage in the Dominican Republic, women who have lived or worked in Europe have become suspect at home. "I know when I tell people I was really with a folk-dance group in Europe, they don't believe me," a former dancer admitted. When Sosúans who were not sex workers spoke casually among themselves of a woman working overseas as a domestic, waitress, or dancer, they inevitably would raise the possibility of sex work, if only to rule it out explicitly. One Dominican café owner cynically explained why everyone assumes that Dominican women working overseas must be sex workers: "Dominican women have become known throughout the world as prostitutes. They are one of our biggest exports."...

Marginalized women in a marginalized economy can and do fashion creative strategies to control their economic lives. Globalization and the accompanying transnational phenomena, including sex tourism, do not simply shape everything in their paths. Individuals react and resist. Dominican sex workers use sex, romance, and marriage as means of turning Sosúa's sex trade into a site of opportunity and possibility, not just exploitation and domination. But exits from poverty are rarely as permanent as the sex workers hope; relationships sour, and subsequently, an extended family's only lifeline from poverty disintegrates. For every promise of marriage a tourist keeps, there are many more stories of disappointment. Dominican women's attempts to take advantage of these "walking visas" call attention, however, to the savviness and resourcefulness of the so-called powerless.

PART III

The Structure of Social Institutions

MARGARET L. ANDERSEN AND PATRICIA HILL COLLINS

Social institutions exert a powerful influence on everyday life. Social institutions are also important channels for societal privileges and penalties. The type of work you do, the structure of your family, whether your religion will be recognized or suppressed, the kind of education you receive, the images you see of yourself in the media, how you are treated by the state, and even how your body is perceived are all shaped by social institutions. People rely on institutions to meet their needs, although social institutions treat some people better than others. When a specific institution (such as the economy) fails them, people often appeal to another institution (such as the state) for redress. In this sense, institutions are both sources of support and sources of repression.

The concept of an institution is abstract because there is no thing or object that one can point to as an institution. *Social institutions* are the established societal patterns of behavior organized around particular purposes. The economy is an institution, as are the family, the media, education, and the state. Each is organized around a specific purpose. The economy, for example, is for the production, distribution, and consumption of goods and services. Within a given institution, there may be various other social structures, such as different work organizations. Institutions change over time because societal conditions evolve and groups may challenge specific institutional structures. Yet, institutions are also enduring and persistent, even in the face of active efforts to change them. Institutions confront people from birth and live on after people die. Although institutions may change internally, it often takes a movement from outside the institution to make substantial change.

Across all societies, institutions are patterned by intersections of race, class, gender, sexuality, age, ethnicity, and disability (among other social characteristics). As a result, the effects of these systems of power on social institutions differ from one society to the next. Each society has a distinctive history and different institutional configurations of social inequalities. In the United States, race, class, and gender are fundamental elements that shape social institutions, making them conduits for people's experiences of social inequality. Social institutions are also interconnected. Understanding the interconnections among institutions helps us see that we are all part of one historically created system—the structural form of which is the interrelationship among social institutions.

The dominant American ideology portrays institutions as neutral in their treatment of different groups. Indeed, the liberal framework of the law allegedly makes access to public institutions (such as education and work) gender- and race-blind. Still, institutions differentiate on the basis of race, class, and gender. As the readings included here show, institutions are actually structured based on race, class, and gender relations.

As an example, think of the economy. Economic institutions in American society are founded on capitalism—an economic system based on the pursuit of profit and the principle of private ownership. Such a system creates class inequality because, in simple terms, the profits of some stem from the exploitation of the labor of others. Race and gender further divide the U.S. capitalist economy, resulting in labor and consumer markets that routinely advantage some and disadvantage others. In particular, corporate and government structures create jobs for some while leaving others underemployed or without work.

Those with jobs encounter a dual labor market that includes: (1) a *primary labor market* characterized by relatively high wages, opportunities for advancement, employee benefits, and rules of due process that protect workers' rights, and (2) a *secondary labor market* (where most women and minorities are located) characterized by low wages, little opportunity for advancement, few benefits, and little protection for workers. One result of the dual labor market is the persistent wage gap between men and women and across racial-ethnic groups (see Figure 1).

Women also tend to be clustered in jobs that employ mostly women workers; such jobs have been economically devalued as a result. Gender segregation and race segregation intersect in the dual labor market, so women of color are most likely to be working in occupations where most of the other workers are also women of color. At the same time, men of color are also segregated into

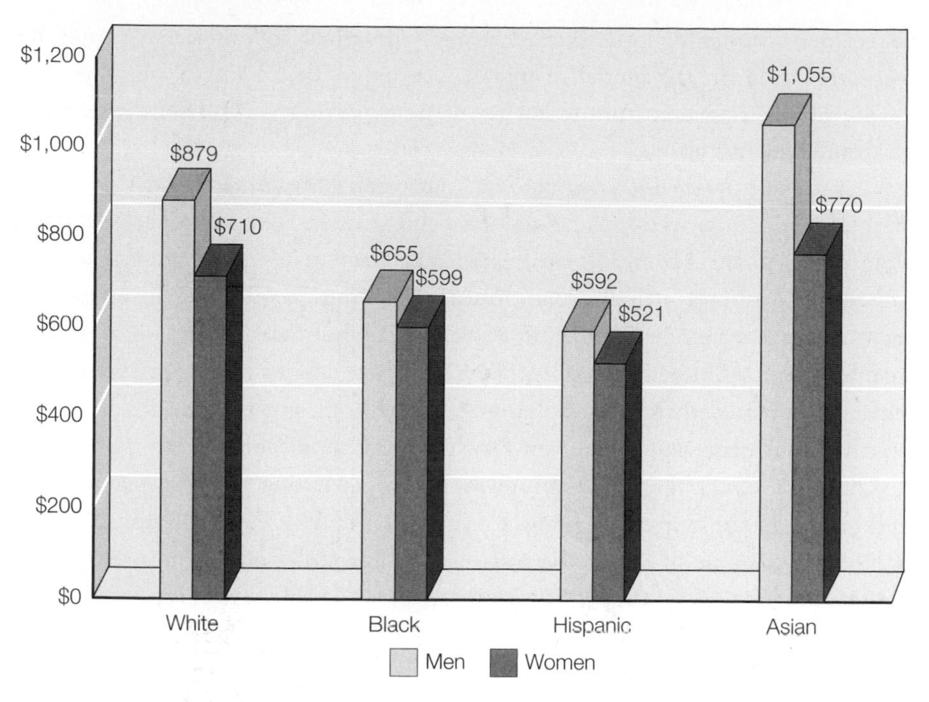

FIGURE 1 Median Weekly Earnings of Full-time Earners, 2012

NOTE: These data are based on full-time workers. Note that the data are highly aggregated—that is, they cannot reflect wages of different subgroups within a given racial-ethnic category (such as Chinese American versus Vietnamese Americans workers). These data also cannot show differences that exist within particular occupations.

SOURCE: Bureau of Labor Statistics. 2014. *Employment and Earnings Online*. Washington, DC: U.S. Department of Labor. www.bls.gov

particular segments of the market. Indeed, there is a direct connection between gender and race segregation and wages because wages are lowest in occupations where women of color predominate (U.S. Department of Labor 2014). This is what it means to say that institutions are structured by intersections of race, class, and gender: Institutions are built from and then reflect the historical and contemporary patterns of race, class, and gender relations in society.

As a second example, think of the state. The state refers to the organized system of power and authority in society. This includes the government, the police, the military, and the law (as reflected in both social policy and the civil and criminal justice system). The state is supposed to protect all citizens, regardless of their race, class, and gender (as well as other characteristics, such as disability and age), yet state policies routinely privilege White men. This means not only that the majority of powerful people in the state are men (such as elected officials, judges, police, and the military) but, just as important, that the state works to protect men's interests. Policies about reproductive rights provide a good example. They are largely enacted by men but have an important effect on women. Similarly, social welfare policies designed to encourage people to work are based on the model of men's experiences because they presume that staying home to care for one's children is not working. In this sense, the state is a gendered institution.

Most people do not examine the institutional structure of society when thinking about intersections of race, class, and gender. The individualist framework of dominant American culture sees race, class, gender, ethnicity, sexuality, and age as attributes of individuals, instead of seeing them as systems of power and inequality that are embedded in institutional structures. People do, of course, have identities of race, class, and gender, and these identities have an enormous impact on individual experience. However, seeing race, class, and gender solely from an individual viewpoint overlooks how profoundly embedded these identities are in the structure of American institutions. Moving historically marginalized groups to the center of analysis clarifies the importance of social institutions as links between individual experience and larger structures of race, class, and gender.

In this section of the book, we examine how social institutions structure systems of power and inequality of race, class, gender, ethnicity, and sexuality that, in turn, structure those very same social institutions. We look at six important social institutions: Work and the economic system; families; cultural institutions such as the media that reproduce ideas; education; the state; and—new to this edition—the institutional structures that shape bodies, beauty, and sports. Each institution can be shown to have a unique impact on different groups (such as the discriminatory treatment of African American men by the criminal justice system). At the same time, each institution and its interrelationship with other institutions can be seen as specifically structured through intersecting dynamics of race, class, and gender.

WORK AND ECONOMIC TRANSFORMATION

Structural transformations in the global economy have dramatically changed the conditions under which all people work, as Margaret L. Andersen points out in "Seeing in 3D: A Race, Class, and Gender Lens on the Economic Downturn," Understanding the intersections of race, class, and gender is critical to understanding the serious economic problems facing our nation, which are rooted in how the economy is structured.

Several major transformations are affecting the current character of work. First, the economy is now global; thus, the work experiences in any one nation (the United States included) are deeply linked to the work experiences of those who may be in remote regions of the world. Among other things, this has meant that employers have often exported jobs as a way of cutting costs of labor and to increase profitability. The jobs remaining are, in many cases, either low-wage jobs or jobs requiring a very high degree of education—thus bifurcating the labor market into two classes: Those with a high degree of skill and education and those with little opportunity for job mobility.

Second, the consequences of the global economy are exacerbated by the development of new technologies that have made some jobs obsolete while other jobs require a high degree of technological and/or scientific skill—clearly demonstrated here in Matt Vidal's research, "Inequality and the Growth of Bad Jobs." Race, class, and gender frame how individual workers encounter these dual processes of "deskilling" and job export.

Third, the economy in the United States has shifted from being one primarily based on manufacturing jobs to one based on the broadly defined "service" industry—a term used to include jobs that not only provide direct service (such as cashiers, food preparation workers, janitors, and so forth), but also jobs that are based on information services, not the actual production of goods. Occupations like teaching, financial services, even professional jobs like physicians, lawyers, or scientists, are based more on the production and transmission of information than they are on the actual manufacture of goods.

Altogether, these changes have resulted in a very different workforce than one would have seen forty or fifty years ago. The transformation has been underway for some time, but it is having dramatic effects on the character of work in the United States—and other parts of the world. Someone entering this economy as a young person has to have the skills to adapt to these changes or may face a lifetime of dead-end work or no work at all—a fact that has plagued many racial-ethnic minorities, especially young, poor, African American men who simply are not positioned in such an economy to get a foot in the door (Wilson 1997).

The growth of the service sector fuels the dual economy described earlier. The dual economy operates by creating high-paying service work for skilled, college-educated workers (accountants, marketing representatives, and so on), and low-paying service work for everyone else (food service workers, nursing home aides, child care workers, and domestic workers, for example). A young man who years ago might have found a decent-paying job in an automobile assembly plant or a steel mill—perhaps even without a high school education—is unlikely in today's economy to do any better than a minimum-wage job—work that is not likely to lead to a lifetime of steady employment and economic security.

These transformations also explain a lot about how work is organized within particular areas of the economy and how structures deeply built into the economy can affect even the simplest things—such as how the unconscious biases associated with people's names may shape opportunities in the labor market, a phenomenon cleverly studied by Marianne Bertrand and Sendhil Mullainathan ("Are Emily and Greg More Employable than Lakisha and Jamal? A Field Experiment on Labor Market Discrimination").

The remaining two articles in this section further detail how the move to a service-based society affects people's experiences at work. Christine L. Williams ("Racism in Toyland") details how jobs within the retail industry are organized in a division of labor that is very much pinned down by the interaction of race, class, and gender. Likewise, Sandra Weissinger's study of women workers in Wal-Mart ("Gender Matters. So Do Race and Class: Experiences of Gendered Racism on the Wal-Mart Shop Floor") explores how the inequities of race, class, and gender are reflected in the lived experiences of women workers.

FAMILIES

Families are another primary social institution profoundly influenced by intersecting systems of race, class, and gender. Dominant belief systems about the family purport that families are places for nurturing, love, and support. This ideal associates women with the private world of the family and men with the public sphere of work. Although families no longer conform to this ideal—and the extent to which they ever did is questionable.

Bonnie Thornton Dill's essay, "Our Mothers' Grief: Racial-Ethnic Women and the Maintenance of Families," shows that for African American, Chinese American, and Mexican American women, family structure is deeply affected by the relationships of families to the structures of race, class, and gender. Dill's historical analysis of racial-ethnic women and their families examines diverse patterns of family organization directly influenced by a group's placement in the larger political economy. Just as the political economy of the nineteenth century affected women's experience in families, the contemporary political economy shapes family relations for women and men of all races.

Tiffany Manuel and Ruth Enid Zambrana ("Exploring the Intersections of Race, Ethnicity, and Class on Maternity Leave Decisions") show how a specific set of social policies—namely, maternity leaves—continue to be shaped by assumptions about families and work that put low-income women at particular risk. Manuel and Zambrana show how race, gender, and class shape women's experiences of motherhood even when presumably neutral maternity leave policies are in place.

Kath Weston also explores how ideologies of the family cause us to think of some families as normative and others as deviant. In "Straight Is to Gay as Family Is to No Family," Weston shows that claiming sexual identities as gay, lesbian, or bisexual does not mean relinquishing family or, worse yet, being defined as anti-family or as being a threat to family. Increasingly, same-sex marriages are becoming recognized by state law, altering the heteronormative assumptions in beliefs about the so-called "ideal family." These changes are good examples of how group mobilization for expanded civil rights has changed social institutions, as well as public opinion.

Family diversity is also apparent in the growing number of multiracial families. Erica Chito Childs explores how mixed-race families negotiate their relationships in her study of interracial couples ("Navigating Interracial Borders: Black-White Couples and Their Social Worlds"). Her research shows how new family arrangements are constantly being constructed, even as family members have to negotiate the complex terrain of past family ideals. Re-centering one's thinking about the family by understanding the interconnections of race, class, gender, and sexuality reveals new understandings about families and the myths that have pervaded assumptions about family experience.

EDUCATION

American education has recently and frequently been described as an institution in crisis. People worry that children are not learning in school. High school dropout rates, especially among students from poor and working-class families of color, reveal deep problems in the system of education. Schools, in fact, reproduce inequalities of race, class, and gender, at the same time that they reflect the inequalities in the society at large (see Figure 2). Historically, social policies of racial and class segregation have been one important way that schools have fostered inequalities. Yet the elimination of formal segregation does not mean that desegregation as a policy yields educational equity.

Although actual desegregation policies were generated to attack schooling inequalities more than fifty years ago, racial isolation in schools is actually growing. And, as Beverly Tatum shows ("Affirming Identity in an Era of School Desegregation"), public schools are resegregating at rates that leave young African American and Latino/a youth in schools that are now described as "majority/minority" (see Table 1). How can we expect identities to be shaped

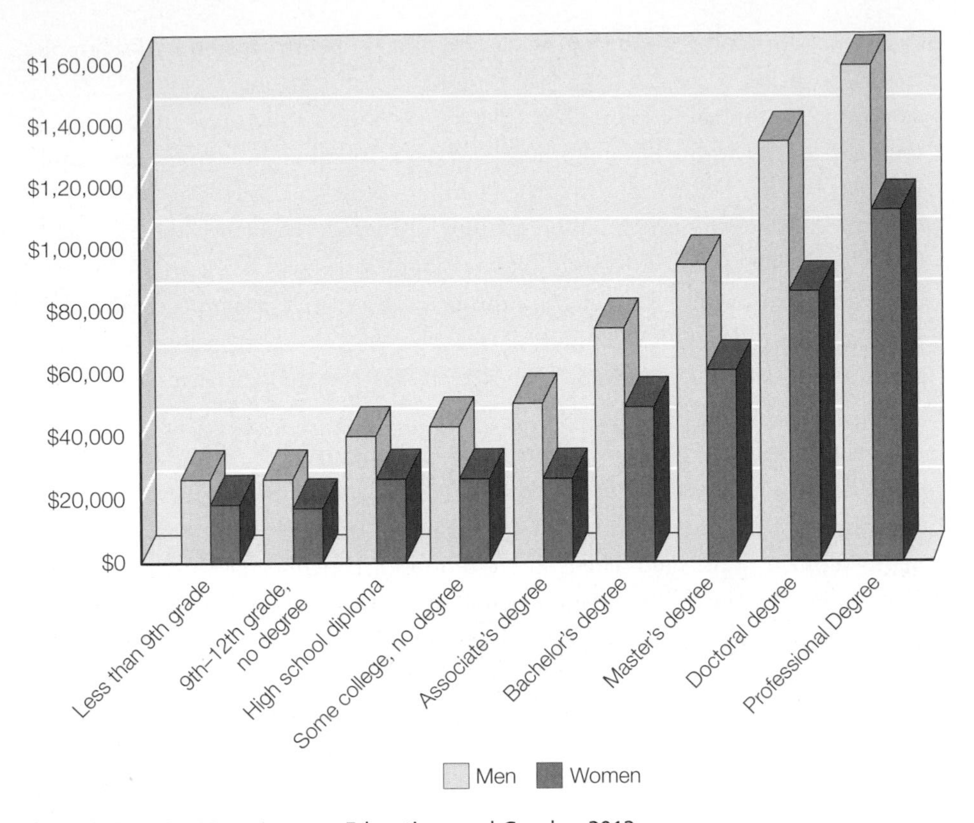

FIGURE 2 Mean Income, Education, and Gender: 2012
SOURCE: Data from U.S. Census Bureau. 2013. *Detailed Income Tables,* Table PINC-04. www.census.gov

in ways that promote racial justice when young people grow up in isolation from other groups? Tatum sees this trend toward resegregation as a fundamental threat to American democracy.

Persistent inequality in the schools has generated much attention and research on the causes and consequences of inequality in schooling—notably in the so-called achievement gap. But, as Gloria Ladson-Billings argues ("From the Achievement Gap to the Education Debt"), we should be thinking not just about the race and gender gap in achievement, but also about the debt owed to those who suffer from poor education and social neglect. Ladson-Billings, like Tatum, asks whether we can afford such bifurcated experience in a society allegedly based on democratic principles.

Segregated schools that are largely schools for poor and minority children are more often than not failing their students. Schools also serve as sites of military recruitment. In "How a Scholarship Girl Becomes a Soldier: The Militarization of Latina/o Youth in Chicago Public Schools," Gina Perez details how

TABLE 1 Percentage of Black Students in Predominantly (>50%) Minority Schools by Region, 1968–2007

Region	1968	1980	1988	1991	2005
South	81	57	57	60	72
Border	72	59	60	59	70
Northeast	67	80	77	75	78
Midwest	77	70	70	70	72
West	72	67	67	69	77
U.S. Total	77	63	63	66	73

SOURCE: U.S. Department of Education Office of Civil Rights data in reported in Gary Orfield and Chungmei Lei. 2007. (August). *Historic Reversals, Accelarating Resegregation, and the Need for New Integration Strategies.* UCLA Civil Rights Project/Proyecto Derecos Civiles. www.civilrightsproject.ucla.edu

working-class Puerto Rican and Latino youth are targets for military recruitment. The recruitment of racial-ethnic youth is not new, but Perez shows how the practices of military programs such as the Junior Reserve Officer Training Program (JROTC) illustrate intersections of race, class, gender, and sexuality. In the past, the army primarily recruited young Latino men who were considered "at risk" of falling into gangs and drug use. Today, the military also targets young Latinas, arguing that as women, they are "at risk" for unwed teenage pregnancy. Because the military seemingly offers more opportunities to escape poverty than does college, Perez shows how the new push to recruit Latina youth reinforces structures of race, class, and gender inequalities.

In a somewhat different vein, Nirmala Erevelles and Andrea Minear discuss the overrepresentation of African American children in special education. Their article ("Unspeakable Offenses: Untangling Race and Disability in Discourses of Intersectionality") uses an intersectional analysis to show how the association of race and disability creates what has been called "spirit murder." Together, the articles in this section question whether the promise of education as a path for social mobility is a fading dream.

MEDIA AND POPULAR CULTURE

How can race, class, and gender exert such a powerful influence on our everyday lives, yet remain so invisible? The answer lies in the influence of the media and popular culture in our daily lives. The products of cultural institutions, such as movies, books, television, advertisements, textbooks, and, increasingly, social media, produce highly influential images that project particular race, class, and gender images. Cultural institutions identify which ideas are valuable, which are not, and which should not be heard at all. Indeed, these images place different value on particular groups directly and indirectly influencing how we define people. Images widely distributed in the mass media produce social stereotypes, thereby limiting and distorting images that we have of others and our selves.

In her article, "Representations of Latina/o Sexuality in Popular Culture," Deborah Vargas identifies numerous ways that the mass media, especially television and film, have fused Latino/a sexual stereotypes with gender stereotypes. The "Latin lover" stereotype associates Latino masculinity with sexuality. Media representations of Latinas portray them in dualistic terms—the Madonna and the whore, juxtaposing images of sexual purity with images of sexual excess. For both Latinos and Latinas, these stereotypes distort the true complexity of Latina/o sexual and gender identities.

Because beliefs about race, class, and gender are so ordinary, they often go unnoticed—even when they can produce harm. Native American stereotypes are ubiquitous in the United States, most commonly found in the derogatory use of Native American mascots. Sports fans celebrate American Indian mascots at college and professional sporting events. But, as Dana M. Williams shows through her research ("Where's the Honor? Attitudes toward the 'Fighting Sioux' Nickname and Logo"), native people are seriously harmed by these stereotypes. Although some, especially White Americans say they are "just for fun," research shows that this is not true. As Williams documents, attitudes about these mascots reproduce the color-blind racism that dominates in today's culture.

Whereas cultural institutions can produce harm by projecting images of certain groups, other groups are simply invisible. Gregory Mantsios's essay, "Media Magic: Making Class Invisible," explores how mass media have been central in shaping ideas about social class. The invisibility of the working class (except perhaps as comedy) can perpetuate ideas that this is a classless society. His work also identifies how the working class is stigmatized through its absence in popular culture.

When you start to look at the mass media and popular culture with a critical eye, you can sometimes be amazed at how pervasive images once taken for granted are. This is well illustrated by Rebecca Hayes-Smith in her review of the popular *Twilight* series ("Gender Norms in the *Twilight* Series"). The *Twilight* series has been enormously popular, especially with adolescent girls (probably some of whom are now reading this book). How many thought, while reading these books, that the books were projecting an image of women as weak, men as strong (and sometimes violent), and thereby legitimating certain gender norms? Hayes-Smith shows how the images in these books also depict Native American men as weak and animal-like, unable to match the sophistication and presumed superiority of White, high-status men. Although many will argue that such books are sheer entertainment, they nonetheless influence (both implicitly and explicitly) our perceptions of ourselves and others in a race-, class-, and gender-framed system of inequality.

Even while cultural institutions distort and mask the experiences of diverse groups, cultural institutions can also be sites for challenging stereotypes and inequalities. Jessie Daniels explores this in her work "Rethinking Cyberfeminism(s): Race, Gender, and Embodiment." Daniels shows how women (and, presumably men) can use some of the new media technologies to experiment with new identities and to communicate in ways that can subvert dominant cultural systems.

BODIES, BEAUTY, AND SPORTS

The articles in the previous section show how the mass media produces and reproduces the ideas we hold about ourselves and others. In contemporary American culture, the body has become one of the main "canvases" on which cultural ideals are drawn. You might even say we live in a body culture—one in which images of bodies produced to perfection shape our beauty ideals. In a related vein, we also live within a culture where sports have become a major cultural institution. In these different venues—the beauty and body culture and sports institutions—images of ourselves and others are drawn.

The images so produced are *controlling images*—that is, they twist and confine our images of what is supposed to be, even while those images are produced and distributed by those with enormous economic interest in shaping our behavior. The images produced are both racialized, gendered, and sexualized, for both women and men. They have a profound influence on how we see one another—especially if we have no context other than the media for understanding what people are. Thus, such images can be especially harmful for people of color if they are segregated from other groups. Controlling images are part of the architecture of race, class, and gender oppression (Collins 2000) and they shape how we perceive, present, and judge things as basic as our bodies—and those of others.

Tressie McMillan Cottom ("Brown Body, White Wonderland") relates this by showing how Black women's bodies have been historically treated as spectacles. Portrayed as sexual objects, Black women's bodies have been violated, brutalized, and commodified, defined as objects for the pleasure and profit of others. As Cottom shows, such images are not just historical; they are rampant in the portrayals of African American women in popular culture.

Abby L. Ferber ("The Construction of Black Masculinity: White Supremacy Now and Then") provides a different analysis, showing how dominant images of Black male athletes reinforce White supremacy. By presenting Black male bodies as violent, aggressive, and hypersexual, these images buttress a system of race, class, and gender inequality, even while sports are believed to be just for entertainment and fun.

Body politics are increasingly part of the various social movements that are questioning many of the institutional practices that impact our bodies. Food movements, concerns about climate change, and sustainability efforts are more part of people's awareness about how institutions threaten the health of our bodies—indeed of our entire planet. But as Janani Balasubramanian writes ("Sustainable Food and Privilege: Why Green is Always White (and Male and Upper-Class"), these movements are often centered only in the experiences of White, middle-class people. She reminds us that people can share a vision yet must also struggle with the contradictions of race, class, and gender to bring their vision about.

Finally in this section, we look at the institutional structure of a site for bodywork—nail salons. Miliann Kang ("There's No Business Like the Nail Business") begins with a simple question: "Why are there so many nail salons where the workers are mostly Asian American immigrant women?" Her answer engages an analysis of race, class, gender, and immigration by showing the structural forces of change that have influenced the presence of Korean American women in the nail salons of New York City where Kang conducted her research. In other areas of the country, manicurists may be of other Asian ethnicity (Vietnamese, for example), but the social forces influencing these workers are the same. Moreover, the complexities of race, class, and gender shape this form of bodywork in the relationships that Kang also analyzes between the clients and the providers of body labor. Collectively, the articles in this section reveal the hidden institutional dimensions of bodywork—whether done in the music industry, on the playing field, in the beauty salon, or as part of the ordinary practices of everyday life.

THE STATE AND VIOLENCE

In the last section of this part, we examine violence and its connection to social institutions. Violence in the United States is a pressing social issue. School shootings and street criminals dominate the news and project the idea that violence is rampant. Usually violence is depicted as the action of crazed individuals who are socially maladjusted, angry, or desperate. Although this may be the case for some, deeper questions about violence are raised when we think about violence in the context of how state power organizes race, class, and gender relations. Who is perceived as violent shifts, as does who we see as victims of violence. How can we apply our knowledge of race, class, and gender to understanding violence?

Violent acts, the threat of violence, and more generalized policies based on the use of force find organizational homes in state institutions of social control—primarily the police, the military, and the criminal justice system. Intersections of race, class, and gender shape all of these institutions. In "Gender, Race, and Urban Policing: The Experiences of African American Youth," Rod S. Brunson and Jody Miller find that race alone does not influence the experiences of Black urban youth with the police. Surely race matters, but as their research finds, gender makes interactions with the police different for young, poor African American women and men: In the aftermath of the police shootings of young Black men in Ferguson and other places, their article is especially poignant and shows how race, class, and gender are all part of this pressing social problem.

Middle- and upper-class people may escape these policies and, as a result, may perceive the police and the criminal justice system as protecting society. Yet for others, the criminal justice system itself is a source of violence. High rates of incarceration, especially in African American poor communities, as shown by Michelle Alexander in "The Color of Justice," mean that huge numbers of young, African American men experience the state as a form of social control, not an institution of protection.

Moreover, the criminal justice system punishes certain forms of violence while ignoring others. Differential enforcement of state policies, even the presence of policies that define some acts as violent and others not, mean that state institutions define what counts as violence. For example, race, class, and gender have shaped the very definition of *rape* and the treatment afforded rape survivors and their rapists. In "Rape, Racism, and the Law," Jennifer Wriggins examines how the legal system's treatment of rape has disproportionately targeted Black men for punishment and made Black women especially vulnerable by denying their sexual subordination. Although violence may be experienced individually, it occurs in specific organizational and institutional contexts shaped by race, class, and gender.

Finally, who becomes a target of violence is also shaped by race, class, and gender. Hate crimes are harmful to any group or person experiencing such violence, and those disadvantaged by either race or class or gender are most likely to be victimized by such horrific acts. But, as Doug Meyer shows ("Interpreting and Experiencing Anti-Queer Violence: Race, Class, and Gender Differences among LGBT Hate Crime Victims"), LGBT people of color find it more difficult than other LGBT victims to have this crime defined as based on their sexuality. Meyer's research illustrates how the intersection of race, class, and gender with sexuality shapes both the definition and handling of this form of violence.

Altogether, the articles in this section show how embracing an intersectional analysis of race, class, gender, and sexuality is important to creating an awareness of how society can become more safe, secure, and just.

REFERENCES

U.S. Department of Labor. January 2014. *Employment and Earnings*. Washington, D.C.: U.S. Government Printing Office. www.dol.gov

Wilson, William Julius. 1997. *When Work Disappears: The New World of the Urban Poor*. Chicago, IL: University of Chicago Press.

31

Seeing in 3D

A Race, Class, and Gender Lens on the Economic Downturn

MARGARET L. ANDERSEN

If you have ever gone to a 3D movie, you know you get these funny little glasses that bring out otherwise unseen dimensions to what you see in the film. In ordinary vision, the eye sees two images—one from each eye—and the brain puts the two images together so you see it as one. When an image is projected in three-dimensions, 3D glasses filter the different colors into a three dimensional image, presenting the brain with one image perceived with depth. Thus, in 3D films and with the 3D lenses, you can see objects otherwise invisible to you, perhaps even flying at you! Without 3D lenses, the image will appear blurry, fuzzy, perhaps even difficult to see or confusing, maybe like trying to understand racism or sexism without the knowledge that helps filter what you observe.

The technology of 3D images is akin to understanding a race/class/gender framework in society. With the "lens" of a three-dimensional intersectional framework, you will see new dimensions to any single vision of race or class or gender. Indeed, in a society where race-blind, gender blind, and class-blind views dominate public understandings of inequality, a three-dimensional view that is anchored in an intersectional analysis of race, class, and gender reveals otherwise hidden aspects of structured inequality.

This essay uses an intersectional analysis to understand the recent economic downturn—a societal event where an intersectional analysis of race, class, and gender has gone largely unnoticed in the public discourse about this historic development—even though these dimensions of inequality are critical to understanding the full consequences of this economic crisis.

As the United States experienced an economic downturn beginning in 2009, daily news headlines proclaimed, "New Jobless Claims Rise Unexpectedly" (*The New York Times*, April 3, 2009); "Jobless Rate Hits 10.2% With More Underemployed" (*The New York Times*, November 11, 2009); "Job Prospects are Poor, Agency Warns" (*The New York Times*, March 20, 2009); and, "Job Market Blues" (*The New York Times*, August 8, 2009). Such headlines suggest that the economy is a neutral force, operating through abstract processes that have little to do with specific group experiences. How different might "the economy" seem were headlines to shout, "Women Disadvantaged by Economic Recovery;" or "Recession a National Disaster for African Americans and Latinos/as;" or "Race, Class, and Gender Interact to Produce Inequality"?

The reality is that, as the national unemployment rate rose to 10 percent, it reached a level that has been characteristic of Black unemployment for the past 60 years. Were African American and Latino unemployment used as a measure of the nation's economic well-being, we would have declared ourselves in an ongoing economic crisis for over half a century! Add in the factor of age and you will see that the unemployment rate for African American young people (aged 20–24)—the very point when they are entering the labor market for the first time—is a whopping 26 percent. For Latinos in the same age category, it is 17 percent (U.S. Department of Labor 2011a).

How do we then see the economic crisis that fell upon the entire nation? As *New York Times* columnist Bob Herbert put it, "The unemployment that has wrought such devastation in black communities for decades is now being experienced by a much wider swath of the population. We've been in deep denial about this" (Herbert 2009). The denial that Herbert refers to is characteristic of a society that wants to think of itself as a place where individual effort, not structural advantage and disadvantage, is the basis for anyone's economic well-being. Herbert was writing about race, but race is only one dimension of the economic crisis that has been overlooked in most reports about the economic recession. What about women?

In 2008, then Senator Edward Kennedy worried about women and the economic crisis when he wrote, "Despite their critical role in the workforce and in raising families, women and their vulnerability in economic downturns have received too little focus. [There is a] severe and disproportionate impact of this recession on women and their families. We need to act immediately to restore women's rights to fair pay, provide workers with paid sick days, and shore up programs that help workers and families endure hard times" (Kennedy 2008).

Unfortunately, few have heeded either Herbert's call for action on race or Kennedy's call for action on behalf of women, much less focusing on the intersections of race, gender, and class in analyzing the particular impact of the economic downturn on particular groups. What would we see differently were we to view the economic recession with women and people of color in mind, thus thinking about gender *and* race *and* class simultaneously?

First, you would probably notice the general absence of women in most of the reports being issued about the impact of the recession. There are exceptions and women, of course, are not the only people affected by economic crises. But, from the dominant, public narrative, you would hardly know that women existed. For example: An otherwise excellent report, *State of the Dream 2009: The Silent Depression*, provides compelling detail about the impact of the economic recession on people of color, but with no regard to the specific impact on women and men (United for a Fair Economy 2009). Likewise, in a different report on housing foreclosures, where there is great detail on the impact of housing foreclosures on African Americans, Latinos, and low-income people, there is not one mention of women—even though the report concludes that this economic crisis will result in the largest loss of Black wealth ever witnessed, wiping out a whole generation of home ownership (United for a Fair Economy 2008).

Such reports beg the further question of the recession's impact on women. What do we see when we look at women?

- Women are more worried about money than men (Newport 2009).
- Although men are more likely to be officially unemployed than women, women's unemployment has also risen, particularly when the recession hit the service sector, not just the manufacturing sector (U.S. Department of Labor 2011b).
- Women are more likely to be employed in the public sector than are men; during the recovery, there was more job growth in the private than the public sector; thus, the decline in unemployment for women has been less than that for men (U.S. Department of Labor 2011b).
- Among women heading families, the unemployment rate has grown and is higher than the national unemployment rate and twice as high as that for either married men or married women (Joint Economic Committee 2009).
- Women's wages are also more volatile than men's wages, and women face a much higher risk of seeing large drops in income than do men (Kennedy 2008).
- The number of women among the working poor continues to exceed that of men—working poor being defined as the percentage of those working 27 weeks or more but living below the poverty level. Among such workers, 6.5% of women and 5.6% of men are working poor. Black and White women are more likely to be working poor than are men in their same racial group; for Asian and Hispanic workers, men and slightly more likely to be among the working poor than are women (though the differences are negligible for these latter two groups; U.S. Bureau of Labor Statistics 2010). Thus, while men have been hardest hit by job loss, holding jobs does not necessarily lift women out of poverty.

These are facts about women as an aggregated group, but what about women of color?

- Among Black Americans, unlike other groups, women are a larger share of those employed (U.S. Department of Labor 2011c).
- As the recession unfolded, the growth for unemployment for Hispanic women nearly matched that of White men (Bureau of Labor Statistics, March 2009).
- All women are disproportionately at risk in the current foreclosure crisis, since women are 32% more likely than men to have subprime mortgages (One-third of women, compared to one-fourth of men, have subprime mortgages; and, the disparity between women and men increases in higher income brackets.) Black women earning double an area's median income

were nearly five times more likely to receive subprime home purchase mortgages than White men with similar earnings. Latino women earning twice an area's median income were about four times more likely to receive subprime mortgages than White men with similar earnings (Fishbein and Woodall 2006).

- Women of color are the most likely to hold subprime mortgages—five times more likely than men in the same income brackets. This is despite the fact that women on average have slightly higher credit scores than men. African American women are 256 percent more likely to hold a subprime loan than White men (Fishbein and Woodall 2006)!

The intent here is not to ignore men and the impact that the economic crisis is having on them. Instead, it is to be attuned to how gender and race work together to produce economic standing. As one example, when you examine gender with regard to men, while also examining race, you will learn that during the recession the growth in unemployment among Hispanic men actually exceeded that of White men. Such facts are overlooked when looking at either race or gender alone. And, in thinking about men, you might ask yourself would it not improve men's lives if women's economic well-being were improved? The point is that we need to be fully attuned to the realities of gender and race—for both women and men or we ignore the specific impacts that economic change—and thus economic and social policy—have on particular groups.

So why is there not more outrage about the persistent inequalities that women, people of color, and, especially, women of color face? In part, this is because of the persistence of a dominant narrative that remains anchored in a race, gender, and class-blind perspective. Furthermore, the dominant narrative remains focused on a one-dimensional view of individualism, not a three-dimensional view of social structural causes. Thus, the public narrative about economic impacts presumes a White, middle-class, masculinist focus. Public rhetoric about so-called "working families" is a case in point. "Working families" is a phrase that invokes images of heterosexual, married families with two potential earners present. Yet, as we know from volumes of research, such families are no longer the only, nor even, dominant form of family relationship. Moreover, the image of the "working class" is one that connotes White, male, and blue-collar, thus ignoring the vast numbers of women, and especially women of color, who now constitute the actual "working class"—that is, not just those who are employed in blue-collar, industrial work, but those working in the vast low-wage, service sector, such as fast food workers, cleaners, housekeepers, and so forth.

The individualistic perspective of the dominant narrative is also evident in how the causes of the economic recession are generally understood. Several recurrent themes appear to explain the recession:

- that the economic crisis is the result of the greedy behavior of individuals on Wall Street;
- that this crisis resulted from the irresponsible habits of Americans who overspent, did not save enough, and consumed beyond their means.

Ironically, prior to the recession, Americans were said to not be saving enough and to be running up too much debt. Now, to recover people have been encouraged to use credit cards and get the credit market going again. To whose advantage?

- that the economic crisis is about job loss. Of course, this is true, but the focus on job creation has also ignored the unpaid labor that people (mostly women and, especially women of color, do)—as if the economy were only about public, paid labor.

It is not that greed and consumer behaviors are irrelevant or that job loss in the paid public sphere is not a problem. The point is that the national discourse blinds us to the gendered, raced, and classed foundations of the economic processes of contemporary capitalism. Instead, public understanding of the economic problems assumes that individual greed and individual deviance have created the recession, thus obscuring the sociological processes that are fundamental to the workings of modern capitalism. Yet, the practices and relations that shape capitalism are fundamentally about race, class, and gender. As sociologist Joan Acker puts it, "gendering and racialization are integral to the creation and recreation of class inequalities and class divisions [and they emerge] … in complex, multifaceted, boundary-spanning capitalist activities" (Acker 2006: 7). Although the "economy" and "the market" are discussed as if they operate by abstract forces, capitalism is decidedly not neutral in terms of class, race, and gender.

Conclusion: Recovery for Whom?

Most of the public discussion about recovering from the economic crisis has focused (other than on financial institution bailouts) on job creation. But, most of the job creation plans that have stemmed from recovery plans have focused on jobs in skilled labor—the very area of the labor market where women have made the least progress. For example, women remain a mere two percent of the labor force in construction-related jobs as carpenters, electricians, sheet metal workers, highway maintenance workers, and auto mechanics. Alternative job recovery plans could focus on training women in areas such as these where women have historically been underemployed. Instead most innovation still relies on individual initiative. One example is Sarah Lateiner who graduated from college with high honors (Phi Beta Kappa) and a double degree in pre-law and women's studies. Originally planning to go to law school, she changed course and established her own automobile repair shop where she not only repairs cars but also trains women about auto mechanics. She has also started a scholarship for women to enroll in technical skills. In her one way of changing the world, she believes that the entry of women into this largely male-dominated domain is one way of empowering women (Hartman 2010).

Could not financial recovery plans provide incentives for such initiatives? Or, why could a job stimulus package not include enhancing wages in areas of the labor market where women ARE employed? Even in a recession, people still need their children cared for, hotel rooms cleaned, and offices cleaned—all areas of the labor market where women and, especially, women of color work, but at low wages and often with few or no job benefits. The federal stimulus package that was part of the nation's recovery plans did include funds for child care and housing, but to date there have been no plans for increasing wages in fields most populated by women. Although an important change has been the passage of the Lily Ledbetter Act protecting women against discrimination, for most women job segregation, not discrimination, is the depressor of wages.

The race-gender neutral way of understanding economic problems masks the reality of women's lives and blunts the political opportunity for a resurgence of a nationally visible feminist movement—one anchored in what now exists as a wide-ranging base of knowledge about the intersections of race, class, and gender in shaping people's life opportunities. Instead, the current national discourse persists in "othering" people—that is, privileging some, but blaming others for their own plight. I have to believe that, if people knew the facts about gender, race, and class, the national discourse—and national action—would be transformed. Women, for example, have particular needs during times of economic crisis. Some of those needs may complement the needs of men. But other needs are particular to their status as women—and, in the case of women of color—as racial-ethnic women. We need a more three-dimensional agenda for women and for women of color—one that is anchored in the now extensive scholarship on the opportunities allowed and denied because of the connections between race, class, and gender.

REFERENCES

Acker, Joan. 2006. *Class Questions, Feminist Answers*. Lanham, MD: Rowman and Littlefield.

Bureau of Labor Statistics. 2009. www.bls.gov

Fishbein, Allan J., and Patrick Woodall. 2006. "Women are Prime Targets for Subprime Lending: Women are Disproportionately Represented in High-Cost Mortgage Market." Washington, DC: Consumer Federation of America. www.consumerfed.org

Hartman, Steve. 2010. "Female Auto Mechanic Breaks Stereotypes." CBS Evening News, December 6. www.cbsnews.com

Herbert, Bob. 2009. "A Scary Reality." *The New York Times*, August 11, 2009.

Hirshman, Linda. 2008. "Where Are the New Jobs for Women?" *The New York Times*, December 9. www.nytimes.com

Joint Economic Committee. 2009. *Women in the Recession: Working Mothers Face High Rates of Unemployment*. Washington, DC: U.S. Senate, May 28. www.jec.senate.gov

Kennedy, Edward M. 2008. *Taking a Toll: The Effects of Recession on Women*. Washington, DC: Majority Staff of the Committee on Health, Education, Labor and Pensions, United States Senate.

Newport, Frank. 2009. "Worry about Money Peaks with Forty-Somethings." *The Gallup Poll*. New York: Gallup Organization, February 27. www.gallup.com

United for a Fair Economy. 2008. *State of the Dream: Foreclosed*. Boston, MA: United for a Fair Economy. www.faireconomy.org

United for a Fair Economy. 2009. *State of the Dream: The Silent Depression*. Boston, MA: United for a Fair Economy. www.faireconomy.org

U.S. Bureau of Labor Statistics. 2010. *A Profile of the Working Poor, 2008*. Washington, DC: U.S. Department of Labor. www.bls.gov

U.S. Department of Labor. 2011a. *Employment and Earnings*. Washington, DC: U.S. Department of Labor, January. www.dol.gov

U.S. Department of Labor. 2011b. *The Black Labor Force in the Recovery*. Washington, DC: U.S. Department of Labor, June 6. www.dol.gov

U.S. Department of Labor. 2011c. *Women's Employment during the Recovery*. Washington, DC: U.S. Department of Labor, May 3. www.dol.gov

32

Inequality and the Growth of Bad Jobs

MATT VIDAL

The Occupy movement brought the topic of economic inequality into mainstream political discourse with the simple slogan: "We are the 99 percent." The outrage of Occupy was directed at the top 1 percent of the population, an elite class consisting mainly of investment bankers, corporate executives, and lawyers who currently own 35 percent of total net wealth in the United States.

But to fully understand economic inequality in America, we need to account for the stagnation of living standards experienced by tens of millions of Americans over the last three decades. To do that, we need to examine the growth of bad jobs.

The standard explanation for rising economic inequality is that computer-based technology has generated growing demand for highly skilled workers. That demand has outpaced supply, leading to a wage premium for such skilled workers.

The cure is for the rest of the workforce to get more education. This is the perspective of mainstream economists, who focus their explanation almost exclusively on technology. The central problem with this analysis is that it ignores the fundamental issue of structural demand for low-skill occupations such as retail sales, food service, building maintenance, personal service, and healthcare support.

In 2005, fully one-quarter of the U.S. workforce was working in a low-wage job (based on the standard definition of a job paying less than two-thirds of the median wage). As Erik Wright and Rachel Dwyer showed in a 2003 article in *Socio-Economic Review,* low-wage work has accounted for an increasing proportion of total job growth in recent decades. They divided all jobs into five wage quintiles and examined periods of job expansion over four decades, finding that the lowest-wage quintile accounted for just 8 percent of total job growth in—the 1960s expansion, but 20 percent in the expansions of the 1980s and '90s....

To help explain this trend, I examined shifts in the employment structure using a typology developed by Stephen A. Herzenberg, John A. Alic, and Howard Wial in their book *New Rules for a New Economy*. They distinguished four generic job types: *high-skill autonomous* work (e.g., executives and professionals), *semiautonomous* work (e.g., supervisors and secretaries), *tightly constrained* work (e.g., clerks, cashiers, assemblers, and machine operators), and *unrationalized labor-intensive* work (e.g., waitstaff, cooks, nursing aids, and janitors).

SOURCE: Vidal, Matt. 2013. *Contexts* 12 (Fall): 70–72.

The latter two types—tightly constrained and unrationalized labor-intensive jobs—are low-skill by definition. They are the most likely to be low-wage jobs. Based on the growth of low-wage jobs, then, we would expect to see a *rise* in low-skill work. However, when I coded 840 detailed occupations from Census data into the four types for 1960 and 2005, I found a 15 percent *decrease* in low-skill jobs' share of the workforce....

There are two key findings here. First, low-wage work has risen despite a decline in low-skill jobs. This may imply that some higher-skill jobs have become low-wage, but here I want to emphasize how the economy used to provide decent pay for low-skill jobs in a way that it no longer does. Second, after 45 years of technological progress and globalization, low-skill jobs continue to account for 35 percent of total employment.

THE BREAKDOWN OF THE POSTWAR CLASS COMPROMISE

To understand the growth of low-wage work, it is helpful to analyze the underlying class dynamics of the U.S. economy over time.

The average profit rate is central to the health of the capitalist economy. However, as Karl Marx argued, there is a tendency for the profit rate to decline because labor is the source of profit, and competition spurs capitalists to substitute machines for labor. Indeed, while the corporate profit rate in the United States was high during the 1950s and '60s, it dropped from a postwar high of 26.8 percent in 1951 to a low of 9.4 percent in 1982....

Under intensified international competition and a declining profit rate, capital embarked on a wave of corporate restructuring. One outcome has been a decline in the wage share (total national income can be divided into profits and wages) from a high of 59.9 percent in 1970 to just 50.7 percent in 2011. The high profit rate of the 1950s and '60s allowed a class compromise between capital and labor. Real wages rose. As the profit rate began dropping under intensified competition and increasing capital intensity in the economy, capital abandoned the class compromise and began to recover profits by reducing wages.

... Deunionization and the declining wage share are highly correlated. This does not imply causality, but our standard understanding of what unions do strongly suggests that deunionization played a key role in the declining wage share. Deunionization, however, is best thought of [as] one indicator of a broader shift in employment relations. Specifically, in the 1950s and '60s, business operated under a dominant logic of employment internalization, in which "best practice" was understood to include internalizing activities, developing internal labor markets, and protecting workers from market forces. As a result, a large percentage of low-skill jobs provided security, opportunities for training and promotion, and decent pay through administratively-determined wages associated with positions rather than individual workers (and patterned on union contracts in the auto sector).

Employment internalization and the good wages it brought were made possible by the institutional context of a nationally-bound economy with a core of large manufacturing employers engaged in oligopolistic competition.

Under the pressure of declining profits and rising international competition, the dominant logic of employment became one of externalization. This included outsourcing, downsizing, and lean staffing strategies (which reduced opportunities for training and promotion); deunionization; part-time and temporary employment; and a return to market-determined wages even for fulltime, long-term jobs.

After the corporate restructuring of the 1970s and '80s, the core of the now-internationalized and reconfigured domestic economy was service-based—among the 10 largest employers in 2005 were Wal-Mart, UPS, McDonalds, Yum! Brands (Taco Bell, KFC, and Pizza Hut), Kroger and Home Depot. In the context of a postindustrial, internationalized economy, wage-based competition returned with a vengeance, hitting core sectors of the economy, even those not subject to international competition.

This analysis suggests that mainstream economists, and the politicians who echo them, are missing the mark. Increasing education levels across the population is a good in its own right. But with the postindustrial economy generating an occupational structure in which 35 percent of all jobs produced are low-skill jobs, education is no panacea.

We must develop institutions that allow workers who are willing to fill the demand for long-term employment in low-skill jobs to receive a living wage. As Annette Bernhardt argued in a 2012 article in *Work and Occupations*, establishing such institutions will require a multipronged, multi-level effort. Necessary interventions include reforming labor law to make it easier for workers to organize unions (based on something like the Employee Free Choice Act) and establishing a strong floor standard for wages (and health and safety). Critically, such standards need to be vigorously enforced; as Bernhardt notes, there is "a growing body of research suggesting that wage theft and other workplace violations are on the verge of becoming common business strategy in low-wage industries, impacting millions of workers—from hotel room cleaners, dishwashers, retail sales workers, and home health aides to garment factory workers, taxi drivers, janitors, and construction laborers."

There have been some notable successes at the local level, including living wage ordinances in Baltimore, Boston, Los Angeles, and San Francisco. And much of the enforcement of a higher floor of basic workplace standards must occur at the local and state level. Ultimately, though, fixing the American economy so that all workers willing and able to put in a full day's work are lifted and remain out of poverty requires progressive federal legislation on labor law and a living wage covering all workers.

33

Are Emily and Greg More Employable Than Lakisha and Jamal?

A Field Experiment on Labor Market Discrimination

MARIANNE BERTRAND AND SENDHIL MULLAINATHAN

Every measure of economic success reveals significant racial inequality in the U.S. labor market. Compared to Whites, African-Americans are twice as likely to be unemployed and earn nearly 25 percent less when they are employed (Council of Economic Advisers, 1998). This inequality has sparked a debate as to whether employers treat members of different races differentially. When faced with observably similar African-American and White applicants, do they favor the White one? Some argue yes, citing either employer prejudice or employer perception that race signals lower productivity. Others argue that differential treatment by race is a relic of the past, eliminated by some combination of employer enlightenment, affirmative action programs and the profit-maximization motive. In fact, many in this latter camp even feel that stringent enforcement of affirmative action programs has produced an environment of reverse discrimination. They would argue that faced with identical candidates, employers might favor the African-American one. Data limitations make it difficult to empirically test these views. Since researchers possess far less data than employers do, White and African-American workers that appear similar to researchers may look very different to employers. So any racial difference in labor market outcomes could just as easily be attributed to differences that are observable to employers but unobservable to researchers.

To circumvent this difficulty, we conduct a field experiment that builds on the correspondence testing methodology that has been primarily used in the past to study minority outcomes in the United Kingdom. We send resumes in response to help-wanted ads in Chicago and Boston newspapers and measure callback for interview for each sent resume. We experimentally manipulate perception of race via the name of the fictitious job applicant. We randomly assign very White-sounding names (such as Emily Walsh or Greg Baker) to half the resumes and very African-American-sounding names (such as Lakisha Washington or Jamal Jones) to the other half. Because we are also interested in how credentials affect the racial gap in callback, we experimentally vary the quality of the resumes used in response

SOURCE: Bertrand, Marianne, and Sendhil Mullainathan. 2004. "Are Emily and Greg More Employable than Lakisa and Jamal: A Field Experiment on Labor Market Discrimination." *American Economic Review* 94 (September): 991–1013.

to a given ad. Higher-quality applicants have on average a little more labor market experience and fewer holes in their employment history; they are also more likely to have an e-mail address, have completed some certification degree, possess foreign language skills, or have been awarded some honors. In practice, we typically send four resumes in response to each ad: two higher-quality and two lower-quality ones. We randomly assign to one of the higher- and one of the lower-quality resumes an African-American-sounding name. In total, we respond to over 1,300 employment ads in the sales, administrative support, clerical, and customer services job categories and send nearly 5,000 resumes. The ads we respond to cover a large spectrum of job quality, from cashier work at retail establishments and clerical work in a mail room, to office and sales management positions.

We find large racial differences in callback rates. Applicants with White names need to send about 10 resumes to get one callback whereas applicants with African-American names need to send about 15 resumes. This 50-percent gap in callback is statistically significant. A White name yields as many more callbacks as an additional eight years of experience on a resume. Since applicants' names are randomly assigned, this gap can only be attributed to the name manipulation.

Race also affects the reward to having a better resume. Whites with higher-quality resumes receive nearly 30 percent more callbacks than Whites with lower-quality resumes. On the other hand, having a higher-quality resume has a smaller effect for African-Americans. In other words, the gap between Whites and African-Americans widens with resume quality. While one may have expected improved credentials to alleviate employers' fear that African-American applicants are deficient in some unobservable skills, this is not the case in our data.

The experiment also reveals several other aspects of the differential treatment by race. First, since we randomly assign applicants' postal addresses to the resumes, we can study the effect of neighborhood of residence on the likelihood of callback. We find that living in a wealthier (or more educated or Whiter) neighborhood increases callback rates. But, interestingly, African-Americans are not helped more than Whites by living in a "better" neighborhood. Second, the racial gap we measure in different industries does not appear correlated to Census-based measures of the racial gap in wages. The same is true for the racial gap we measure in different occupations. In fact, we find that the racial gaps in callback are statistically indistinguishable across all the occupation and industry categories covered in the experiment. Federal contractors, who are thought to be more severely constrained by affirmative action laws, do not treat the African-American resumes more preferentially; neither do larger employers or employers who explicitly state that they are "Equal Opportunity Employers." In Chicago, we find a slightly smaller racial gap when employers are located in more African-American neighborhoods.

I. PREVIOUS RESEARCH

With conventional labor force and household surveys, it is difficult to study whether differential treatment occurs in the labor market. Armed only with survey data, researchers usually measure differential treatment by comparing the labor market performance of Whites and African-Americans (or men and women) for which they observe similar sets of skills. But such comparisons can be quite misleading. Standard labor force surveys do not contain all the characteristics that employers observe when hiring, promoting, or setting wages. So one can never be sure that the minority and nonminority workers being compared are truly similar from the employers' perspective. As a consequence, any measured differences in outcomes could be attributed to these unobserved (to the researcher) factors.

This difficulty with conventional data has led some authors to instead rely on pseudo-experiments. Claudia Goldin and Cecilia Rouse (2000), for example, examine the effect of blind auditioning on the hiring process of orchestras. By observing the treatment of female candidates before and after the introduction of blind auditions, they try to measure the amount of sex discrimination. When such pseudo-experiments can be found, the resulting study can be very informative; but finding such experiments has proven to be extremely challenging.

A different set of studies, known as audit studies, attempts to place comparable minority and White actors into actual social and economic settings and measure how each group fares in these settings. Labor market audit studies send comparable minority (African-American or Hispanic) and White auditors in for interviews and measure whether one is more likely to get the job than the other. While the results vary somewhat across studies, minority auditors tend to perform worse on average: they are less likely to get called back for a second interview and, conditional on getting called back, less likely to get hired.

These audit studies provide some of the cleanest nonlaboratory evidence of differential treatment by race. But they also have weaknesses.... First, these studies require that both members of the auditor pair are identical in all dimensions that might affect productivity in employers' eyes, except for race. To accomplish this, researchers typically match auditors on several characteristics (height, weight, age, dialect, dressing style, hairdo) and train them for several days to coordinate interviewing styles. Yet, critics note that this is unlikely to erase the numerous differences that exist between the auditors in a pair.

Another weakness of the audit studies is that they are not double-blind. Auditors know the purpose of the study.... This may generate conscious or subconscious motives among auditors to generate data consistent or inconsistent with their beliefs about race issues in America. As psychologists know very well, these demand effects can be quite strong. It is very difficult to insure that auditors will not want to do "a good job." Since they know the goal of the experiment, they can alter their behavior in front of employers to express (indirectly) their own views. Even a small belief by auditors that employers treat minorities differently can result in measured differences in treatment. This effect is further magnified by the fact that auditors are not in fact seeking jobs and are therefore more free to let their beliefs affect the interview process.

Finally, audit studies are extremely expensive, making it difficult to generate large enough samples to understand nuances and possible mitigating factors. Also, these budgetary constraints worsen the problem of mismatched auditor pairs. Cost considerations force the use of a limited number of pairs of auditors, meaning that any one mismatched pair can easily drive the results. In fact, these studies generally tend to find significant differences in outcomes across pairs.

Our study circumvents these problems. First, because we only rely on resumes and not people, we can be sure to generate comparability across race. In fact, since race is randomly assigned to each resume, the same resume will sometimes be associated with an African-American name and sometimes with a White name. This guarantees that any differences we find are caused solely by the race manipulation. Second, the use of paper resumes insulates us from demand effects. While the research assistants know the purpose of the study, our protocol allows little room for conscious or subconscious deviations from the set procedures. Moreover, we can objectively measure whether the randomization occurred as expected. This kind of objective measurement is impossible in the case of the previous audit studies. Finally, because of relatively low marginal cost, we can send out a large number of resumes. Besides giving us more precise estimates, this larger sample size also allows us to examine the nature of the differential treatment from many more angles.

★★★

Our results indicate that for two identical individuals engaging in an identical job search, the one with an African-American name would receive fewer interviews. Does differential treatment within our experiment imply that employers are discriminating against African-Americans (whether it is rational, prejudice-based, or other form of discrimination)? In other words, could the lower callback rate we record for African-American resumes *within our experiment* be consistent with a racially neutral review of the *entire pool* of resumes the surveyed employers receive?

In a racially neutral review process, employers would rank order resumes based on their quality and call back all applicants that are above a certain threshold. Because names are randomized, the White and African-American resumes we send should rank similarly on average. So, irrespective of the skill and racial composition of the applicant pool, a race-blind selection rule would generate equal treatment of Whites and African-Americans. So our results must imply that employers use race as a factor when reviewing resumes, which matches the legal definition of discrimination....

II. CONCLUSION

This paper suggests that African-Americans face differential treatment when searching for jobs and this may still be a factor in why they do poorly in the labor market. Job applicants with African-American names get far fewer callbacks for each resume they send out. Equally importantly, applicants with African-American names find it hard to overcome this hurdle in callbacks by improving their observable skills or credentials.

Taken at face value, our results on differential returns to skill have possibly important policy implications. They suggest that training programs alone may not be enough to alleviate the racial gap in labor market outcomes. For training to work, some general-equilibrium force outside the context of our experiment would have to be at play. In fact, if African-Americans recognize how employers reward their skills, they may rationally be less willing than Whites to even participate in these programs.

REFERENCES

Council of Economic Advisers. *Changing America: Indicators of social and economic well-being by race and Hispanic origin*. September 1998, http://www.gpoaccess.gov/eop/ca/pdfs/ca.pdf.

Goldin, Claudia and Rouse, Cecilia. "Orchestrating Impartiality: The Impact of Blind Auditions on Female Musicians." *American Economic Review*, September 2000, *90*(4), pp. 715–41.

34

Racism in Toyland

CHRISTINE L. WILLIAMS

Not long ago I had to buy a present for a six-year-old. I had at least three choices for where to shop: The Toy Warehouse, a big-box superstore with a vast array of low-cost popular toys; Diamond Toys, a high-end chain store with a more limited range of reputedly high-quality toys; or Tomatoes, a locally owned, neighborhood shop that sells a relatively small, offbeat assortment of traditional and politically correct toys. Can sociology offer any advice to consumers like me?

Unfortunately, many sociologists turn into utilitarian economists when it comes to analyzing shopping, assuming that customer behavior is determined only by price, convenience, and selection. But a number of social factors influence where we choose to shop, including the racial makeup of the store's workers and customers. In my book, *Inside Toyland: Working, Shopping, and Social Inequality* [University of California Press, 2006], I argue that racial inequality (and gender and class inequality as well) influence where we choose to shop, how we shop, and what we buy. The retail industry sustains such inequality through hiring policies that favor certain kinds of workers and advertising aimed at customers from specific racial or ethnic groups.

I noticed the connection between shopping and social inequality while working as a clerk at the Toy Warehouse and Diamond Toys for three months in 2001. These stores belonged to national chains, and both employed about 70 hourly employees. At the warehouse store, I was one of only three white women on the staff; most were African American, Hispanic, or second-generation Asian American. The "guests" (as we were required to call customers) were an amazing mix from every racial and ethnic group and social class. In contrast, only three African Americans worked at the upscale toy store; most clerks were white. Most of the customers were also white and middle to upper class. My experiences taught me to notice racial diversity (or its absence) wherever I shop.

"DON'T SHOP WHERE YOU CAN'T WORK!"

This slogan, popular during the Great Depression, rallied black protesters to demand equal access to jobs in stores, and many chains responded by hiring

SOURCE: Williams, Christine L. 2005. "Racism in Toyland." *Contexts*, 4(4): 28–32.

African Americans in predominately black neighborhoods: "We employ colored salesmen" signs appeared in Sears, Walgreens, and other stores eager for black customers.

Retail work is one of the most integrated occupations in the United States today. The proportions of whites, African Americans, Asian Americans, and Hispanics employed in retail jobs more or less match their representation in the labor force. But these statistics hide segregation at the store level. More than 15 percent of employees in shoe stores and variety stores, for instance, are black, but less than 5 percent of employees are black in stores that sell liquor, gardening equipment, or needlework supplies. Inside stores, there is further segregation by task; Whites usually have the top director and manager positions, and nonwhites have the lowest-paid, often invisible backroom jobs.

In the toy stores where I worked, the two most segregated jobs were the director positions (all white men) and the janitor jobs (all Latinas subcontracted from outside firms). The Toy Warehouse employed mostly African Americans in all the other positions, including cashiers. But over time I noticed that the managers preferred to assign younger and lighter-skinned women to this position. Older African-American women who wanted to work as cashiers had to struggle to get the assignment. Lazelle, for example, who was about 35, had been asking to work as a cashier for the two months she had been working there. She worked as a merchandiser, getting items from the storeroom, pricing items, and checking prices when bar codes were missing. Lazelle finally got her chance at the registers the same day that I started. We set up next to each other, and I noticed with a bit of envy how competent and confident on the register she was. (Later she told me she had worked registers at other stores, including fast-food restaurants.) I told her I had been hoping to get assigned as a merchandiser. I liked the idea of being free to walk around the store, talk with customers, and learn more about the toys. I had mentioned to the manager that I wanted that job, but she made it clear I was destined for cashiering and service desk (and later, to my horror, computer accounting). Lazelle looked at me like I was crazy. Most workers thought merchandising was the worst job in the store because it was so physically taxing. From her point of view, I had gotten the better job, no doubt because of my race, and it seemed to her that I wanted to throw that advantage away. (The manager may also have considered my background and educational credentials in assigning me to particular jobs.)

The preference for lighter-skinned women as cashiers reflects the importance of this job in the store's general operations. In discount stores, customers seldom talk with sales clerks. The cashier is the only person most customers deal with, giving her enormous symbolic—and economic—importance for the corporation. Transactions can break down if clerks do not treat customers as they expect. The preference for white and light-skinned women as cashiers should be understood in this light: In a racist and sexist society, managers generally believe that such women are the most friendly and solicitous, and thus most able to inspire trust and confidence in a commercial transaction.

At the upscale Diamond Toys, virtually all cashiers were white. Unlike the warehouse store, where cash registers were lined up at the front of the store, the upscale store had cash registers scattered throughout the different departments. The preference for white workers in these jobs (and throughout the store) seemed consistent with the marketing of the store's workers as "the ultimate toy experts." In retail work, professional expertise is typically associated with whiteness, much as it is in domestic service.

The purported expertise of salesclerks is one of the great deceptions of the retail industry. Here, where jobs pay little and turnover rates are high (estimated at more than 100 percent per year by the National Retail Federation), many clerks know almost nothing about the products they sell. I knew nothing about toys when I was put behind the cash register, and I received no training on the merchandise at either store. Any advice I gave I literally made up. But at the upscale store I was expected to help customers with their shopping decisions. They frequently asked questions like "What are going to be the hot toys for one-year-olds this Christmas?" or "What one item would you recommend for two sisters of different ages?" One mother asked me to help her pick out a $58 quartz watch for her seven-year-old son. A personal shopper phoned in and asked me to describe the three Britney Spears dolls we carried, help her pick out the "nicest" one, and then arrange to ship it to her employer's niece. Customers asked detailed questions about how the toys were meant to work, and they were especially curious about comparing the merits of the educational toys we offered (I was asked to compare the relative merits of the "Baby Mozart" and the "Baby Bach"). On my first day, I answered a phone call from a customer who asked me to pick out toys for a one-year-old girl and a boy who was two and a half, spend up to $100, and arrange to have the toys gift-wrapped and mailed to their recipients.

At Diamond Toys, most customers didn't mind waiting to talk with me. When the lines were long, they didn't make rude huffing noises or try to make eye contact with their fellow sufferers, as they often did at the Toy Warehouse. I couldn't help but think that the customers—mostly white—were more civil and polite at the upscale store because most of us workers were white. We were presumed to be professional, caring, and knowledgeable, even when we weren't. White customers seemed less respectful of minority service workers than white workers; they were willing to pay more and wait longer for the services of whites because they apparently assumed that whites were more refined and intelligent.

On several occasions at the warehouse store, I saw customers reveal racist attitudes toward my African-American coworkers. One night, after the store had closed, I saw Doris and Selma (fiftyish African Americans) escorting several white customers out. Getting straggling customers to leave the store after closing was always a big chore. Soon after, as I was being audited in the manager's office, Selma came in very upset because one of the women she and Doris escorted out had spit out her chewing gum at her. Doris and Selma were appalled. Doris said to them, "That is really disgusting; how could you do that?" And the woman said to Doris, "What's your name?" like she was going to report her. This got Selma so angry she said, "If you take her name, take mine too," and showed her name-tag. She told the woman that she was never welcome to come back to this store. Selma was very distressed. Talking back to customers was taboo, and she knew she could be fired for what she had said. The manager told her that some people are going to be gross and disgusting and what can you do? Clearly Selma and Doris would not get in trouble over this. But I sensed it was doubly humiliating to have to fear that she might be punished for talking back to a white woman who had spit at her.

Although I suffered from plenty of customer condescension at this store, at least putting up with racism was not part of my job. Once when Tanesha, a 23-year-old African American, was training me at the service desk, two white women elbowed up to the counter to complain to me about the long wait for service (they had waited about five minutes while we were serving another customer). I said something about being in training, and they thought I meant that I was training Tanesha. So they said, "Well, call up someone else to the register!" I said I'd have to ask Tanesha to do this. They demanded that I stop training her for a moment to call another person to the exchange desk. "No you don't understand," I said, "I'm the one who is in training; she knows what she is doing, and she is the only one who can call for backup, and she is in the middle of trying to accommodate this other customer." They seemed embarrassed at having assumed that the white woman was in charge. When they realized their mistake, they looked mortified and stepped back from the counter.

SHOPPING WHILE BLACK

During the civil rights era, equal access to stores was high on the list of demands for racial justice. Before Jim Crow laws were repealed, many stores restricted their facilities to whites only. Black customers often were not allowed to try on clothes, eat at lunch counters, or use public restroom facilities in stores.

Today the worst forms of racism have been eliminated. Gone are the "whites only" signs on restrooms and drinking fountains. But some stores build to exclude. The history of suburban malls is a history of intentional racial segregation. Even today, so-called desirable retail locations are characterized by limited access. In my city of Austin, Texas, local malls have opposed public bus service on the grounds that it would encourage undesirable (nonwhite) patrons.

But open access is not enough to ensure racial diversity. Diamond Toys was located in a racially diverse urban shopping district, next to subway and bus lines, yet nearly all its patrons were white. I didn't fully realize this until one day when Chandrika, an 18-year-old African-American gift wrapper, told me that she thought she saw one of her friends in the store. We weren't allowed to leave our section, so she asked the plainclothes security guard to look around and see if there was a black guy in the store. I asked her about this. Was it so unusual for an African-American teenager to be in the store that one black guy would be so apparent? After all, lots of young men came in to check out the new electronic toys and collectibles. Chandrika assured me that a young black man would definitely stand out.

One way that many stores show hostility to racial/ethnic minorities is through consumer racial profiling. Like racial profiling in police work, this involves detaining, searching, and harassing such people more often than is done for whites, usually because they are suspected of stealing. Some scholars have labeled this potential violation of people's rights "shopping while black." At the stores where I worked, clerks weren't allowed to pursue anyone suspected of stealing; that was the job of the plainclothes security workers. However, relations between customers and clerks sometimes broke down, and I saw double standards in the treatment of whites and minorities.

At the warehouse store, I was told to treat shoppers as if they were my mother (most shoppers at both stores were women). At the service desk, I was told to appease them by honoring all requests for returns, even if the merchandise had been used and worn out. The goal, my manager told me, was to make these shoppers so grateful that they would return to the store and spend $20,000 per child, the amount their marketers claimed was spent on an average child's toys.

In my experience, only middle-class white women could depend on this treatment. Nevertheless, I watched many white women throw fits, loudly complaining of shoddy service and merchandise with comments like "I will never shop in this store again!" Such arrogance no doubt came from being accustomed to having their demands met. On one occasion, when a white woman threw a tantrum because the bike she had ordered was not ready as scheduled, the manager offered her a $25 gift certificate for her troubles. She refused, demanded a refund, and left shouting that she would "never come to this store again." The manager then gathered the entire staff at the front of the store and chewed us out for being disorganized and incompetent.

Members of minority groups who wanted to return used merchandise or needed special consideration were rarely accommodated. The week before the bike incident, I was on a register that broke down in the middle of a credit card transaction. A middle-class black woman in her forties was buying inline skates for her ten-year-old daughter. The receipt came out of the register but not the slip for her to sign, so I had to call a manager, who came over and explained that she needed to go to another register and repeat the transaction. She refused, since it seemed to go through all right and she didn't want to be charged twice. She had to wait more than an hour to get this problem solved, and she wasn't offered any compensation. She didn't yell or make a scene; she waited stoically. I felt sorry for her and went to the service desk to tell a couple of my fellow workers what was happening while the managers tried to resolve it, and I said they should just give her the skates and let her go. My fellow workers thought that was the funniest thing they had ever heard. I said, "What about our policy of letting things go to make sure we keep loyal customers?" But they just laughed at me. Celeste said, "I want Christine to be the manager; she just lets the customers have whatever they want!"

Research has shown that African Americans suffer discrimination in public places, including stores; middle-class whites, on the other hand, are privileged. We do not recognize this precisely because it is so customary. Whites expect first-rate service; when they don't receive it, some feel victimized, even discriminated against, and some throw tantrums when they don't get what they want.

When African-American customers did shout or make a scene, the managers called or threatened to call the police. Each of these instances involved an

African-American man complaining and demanding a refund. Once a young black man, denied a cash refund, threw a toy on the service desk, and accidentally hit the telephone, which flew off the desk and hit me on the side of the head, knocking me to the floor. Within minutes, three police officers arrived and asked me if I wanted to press assault charges. I did not. After all, angry people often threw merchandise on that desk, and what happened had been an accident. But at least I was appeased. At the end of my shift, the manager gave me a "toy buck" for "taking a hit" in the line of duty, which entitled me to a free Coke.

There wasn't much shouting or throwing at the upscale store. It protected itself from conflict by catering to an upper-class clientele, much as a gated community does. This is not to say that diversity always leads to conflict, but it did at the warehouse store because race, class, and gender differences existed under a layer of power differences within the store. Clerks and customers interacted in a context where these differences had been used to shape marketing agendas, hiring practices, and labor policies—all of which benefited specific groups (especially middle-class white men and women).

WHAT ABOUT TOMATOES?

There are alternatives to shopping at large chain stores, although their numbers are dwindling. The store I'm calling Tomatoes is a small, family-owned business in an upper-middle-class neighborhood on a busy street with lots of pedestrian traffic. It's been in the neighborhood for 25 years, owned by the same family. It sells an offbeat assortment of toys, including many traditional items like kites and wooden blocks, and a variety of toys I would call "politically correct." It didn't carry Barbie, for example, but it did have "Get Real Girls," female action figures that look like G.I. Joe's sisters. I laughed when I saw a pack of plastic "multicultural" food, including spaghetti, sushi, a taco, and a bagel (all marked "made in China").

Working conditions at the store seemed very relaxed compared to what I had experienced. The owner wore shorts and a Hawaiian shirt, and the workers dressed like punk college students, including weirdly dyed hair, piercings, and tattoos. They didn't wear uniforms (as we did at the other two stores). One clerk wore her T-shirt hiked up in a knot in the front and stuffed under her bra in the back. Clerks seemed to be on a friendly, first-name basis with several of the customers, who were mostly middle-class white women.

Although I didn't get hired at Tomatoes, after several visits I noticed social patterns in the store's organization. The owner and the manager were both white men, and all the clerks were young white women. The owner was the only one who was near my age (mid-40s). You can never be sure why you aren't hired, but my impression is that I wasn't young enough or hip enough to work there. Although Tomatoes allowed more autonomy and self-expression than the stores where I worked, race, class, and gender inequality were as much a part of the social organization there as in other retail stores.

CONCLUSION

I ended up buying my gift at Tomatoes—a children's book written and autographed by Marge Piercy. My decision reflects my identity and my social relationships. But what are the implications of my choice for social inequality? My purchase supported a store that was organized around racial exclusion, gender segregation, and class distinctions.

Everyone has to shop in our consumer society, yet the way shopping is organized bolsters social divisions. The racism of shopping is reflected in labor practices, store organization, and the guidelines, explicit or unspoken, for relations between clerks and customers. When deciding where and how to shop, consumers should be aware of what their choices imply with regard to racial justice and equality. Although an individual shopper can do little to change the overall social organization of shopping, raising awareness of the inequalities that our choices support must be a first step in imagining and then creating a better alternative.

35

Gender Matters. So Do Race and Class

Experiences of Gendered Racism on the Wal-Mart Shop Floor*

SANDRA E. WEISSINGER

In this case study, inequitable access to power found within one particular institution, Wal-Mart stores, is examined. As the largest corporation in the world—with revenues larger than those accumulated by Switzerland and outselling Target, Home Depot, Sears, Kmart, Safeway, and Kroger combined—the company is quite powerful both in the United States and abroad (Litchenstein 2006). In the U.S., 1.3 million are employed (Rathke 2006) within the four thousand stores across the nation (Litchenstein 2006). Because the lives of so many converge there, Wal-Mart has the potential to mirror many of the oppressive practices observed in social relationships outside of the stores. For this reason, the statements given by Wal-Mart employees are telling of the robust and adaptive nature of discriminatory practices, regardless of geographic location. Within Wal-Mart, the multidimensional nature of inequality can be illuminated as gender, race, and class intersect and shape the experiences of the women that work at these stores.

CASE BACKGROUND

In 2004, 1.6 million plaintiffs were granted class action status, making theirs the largest sex discrimination case seen in U.S. courts.** The suit, Dukes v. Wal-Mart Stores Inc., began when current and former female Wal-Mart employees from across the United States claimed that the retailer discriminated against them in terms of access to promotions, wages similar to their male colleagues, and their job tasks. Individual employees examined their personal issues and developed a sociological imagination (Mills 2000 [1959])—the ability to see that the

*This essay is dedicated to Ms. Betty Dukes and the nearly two million women who were courageous enough to stand in the gap for other employees experiencing similar treatment.

**Editors' note: In June 2011, the U.S. Supreme Court ruled against the women who filed a class-action suit against Wal-Mart.

SOURCE: Weissinger, Sandra. 2009. "Gender Matters, So Do Race and Class: Experiences of Gendered Racism on the Wal-Mart Shop Floor." *Humanity and Society* 33 (November): 341–362.

treatment they were subject to was shared, at least in part, by those with similar biographical characteristics. For example, when her supervisor referred to her as a "Mexican princess" in front of her peers, Gina Espinoza-Price (a plaintiff) was humiliated. She also realized that others were being treated in a similar unprofessional fashion.

Wal-Mart's defense lawyers posed several challenges to the plaintiffs; the most important was whether their case qualified as a class action suit. If the case had not qualified, plaintiffs would have had to file individual suits against the corporation. At best, this could produce rulings that benefited an individual plaintiff but failed to bring about large-scale changes to Wal-Mart's employment and promotions procedures that would benefit all employees. This suit is the impetus behind changes in Wal-Mart's human resources departments nationwide. All employees can now view and apply for all positions. They are also given the opportunity to officially declare their career goals in their computerized personnel file (Featherstone 2004).

The plaintiff for whom the case is named, Betty Dukes, is still employed at Wal-Mart. Dukes has stated that her motivation to continue speaking about employment practices at Wal-Mart comes from her religious background as a Christian minister. According to Dukes, "I am participating in this case in order to insure that young women such as my nieces and other women are treated fairly at every Wal-Mart store. The time has surely come for equality for women" (Rosen 2004). Dukes's statement is unique in that she sees her actions today as connected to the lives of others who will come after her. Although the other plaintiffs did not specifically list similar motivations, all of the plaintiffs made claims concerning inequality. These claims have been bolstered by the statistical findings of Drogin (2003), who found that across geographic locations, women working for Wal-Mart earn less than male employees who occupy the same positions. In addition, Drogin found that women were promoted into management positions at lower and slower rates than male employees.

The Dukes suit rightly identifies the effects of sex-based discrimination that are generalizable across women's experiences within Wal-Mart stores. To provide another vantage point from which to understand the effects of the discrimination experienced by these employees (and perhaps those at corporations that seek to model their businesses after Wal-Mart), I contend that discrimination due to biological sex differences alone does not explain the range of the plaintiffs' experiences or properly gauge individual and social damages caused when one must navigate treacherous environments on a daily basis. Rather, I argue, those individuals who are targeted for mistreatment experience such treatment as equally raced, classed, and gendered people whose lives exist within a web of intersecting and relational inequalities. Therefore, a reexamination of what plaintiffs of the class action suit said about their employment experiences (when guided by a gender primacy framework) is necessary so that the multiple and differing daily work experiences of the plaintiffs can be illuminated, adding to sociological knowledge concerning the work lives of women across standpoints. Reexamination is needed not simply to see difference in lives, but to reveal the persistence of inequality and the multiple ways discriminatory work atmospheres are maintained.

According to the files Wal-Mart made available to the court, women worked for the company longer and received better reviews from supervisors, yet earned less money and received fewer promotions when compared to male employees.

Women's Experiences across Race and Class

... Scholarship about the social locations of women has addressed not only the lived experiences of women of color, but also those of white women across class and geographic location. Enlarging our understanding of the differences between women, sociologists have documented the ways in which white women act as oppressors as well as the oppressed—benefiting from certain race and/or class positions. As explained by Blee (2002) and Frankenberg (1993), one experience of marginalization does not automatically inspire empathy in white women toward other oppressed communities. Whiteness, as all categories of difference, is socially constructed through carrying real consequences that bear upon individuals' everyday lives. "In a white-dominated society, whiteness is invisible. Race is something that adheres to others as a mark of otherness, a stigma of difference. It is not perceived by those sheltered under the cloak of privilege that masks its own existence" (Blee 2002:56). In short, white women can be marginalized due to gender and class, but experience certain benefits because of their race.

Methodology

... In this work I identify six biographical characteristics most plaintiffs mentioned in their declarations: gender, race, class, family makeup, geographic location, and age. To be clear, I do not argue that one area of oppression or facet of one's biography is more important than another. My argument is not one that posits that there is a monolithic or authentic experience either. Certainly, lived experiences can be similar and should be juxtaposed against one another for analysis. Rather, I argue that each employee had rich and unique work experiences resulting from the intersecting characteristics that make up their biographies.

... Past studies have failed to examine women's work experiences outside of the gender lens. Questioning how the multiple features of each plaintiff's biography influence and shape the stories they tell about work life at Wal-Mart is at the heart of this work.

Findings

Insider Outsider Statuses

Very few white plaintiffs mentioned their race but when they did, it was in relation to the race of others. These statements were made when the plaintiff reflected on "backstage" (Picca and Feagin 2007), private conversations they engaged in with white male supervisors. The failure of white women plaintiffs to mention race, except when noting the difference between themselves and people of color, illustrates that "whiteness is unmarked because of the pervasive nature of white domination" (Blee 2002:56). To illuminate this argument, the following excerpts show how two white women who held positions of authority at Wal-Mart stores addressed race.

The first quote is from Ms. Lorie Williams, the youngest declarant in this sample (twenty-six years old) and a single mother who supports her child [working] as a "Front End Manager," the lowest ranking management position in the occupational hierarchy at the stores.

> In 1996, I became the front-end manager. With no training, I was single-handedly responsible for hiring door greeters, cart pushers, over sixty new cashiers, and preparing the entire front end in order to transition the Collierville Wal-Mart into a Supercenter. Almost immediately, Co-Manager Doug Ayerst and new Store Manager Robert Hayes (he replaced Wes Grab in 1996) began criticizing the way in which I was managing the front end. They repeatedly complained to me about problems in the front end but did not give me any practical management advice and often gave inconsistent instructions. Store Manager Hayes once told me that the problem was that there were "too many damn women in the front end." On another occasion, Store Manager Hayes and Co-Manager Jim Belcoff pulled me aside to tell me that I needed to "whiten up [the staff of] the front end." Both of these statements seemed to indicate that they wanted me to stop hiring women and African-American cashiers. When I asked why, they indicated that the staff was "intimidating" to the clientele. I tried to explain that I did not believe this was true and that I was hiring the individuals whom I believed were the most qualified applicants. They were unresponsive.

... In the second example, the words of Melissa Howard, a 35-year-old white woman who had risen to the rank of Store Manager while raising her biracial child in the Midwest, are examined.

> ... In July 1997, I was promoted to the position of Store Manager in Marysville, Kansas. I drove with my daughter's father, who is African-American, to find a place to live there. When we arrived in town, we were treated hostilely by a clerk in the local Wal-Mart store, the town realtor, and by a number of prospective landlords because we were a mixed-race family. At the last rental we visited, the landlord told us that my daughter was not safe and that we needed to be out of town by dark. I knew then that I could not move my family to this place. I called the Regional Personnel Manager Gary Coward, explained what had happened and asked to be placed as a store manager anywhere else. He refused and told me that I would have to go to Marysville as planned or accept a demotion and return as an assistant manager to the 86th Store in Indianapolis, one of the worst stores in the area. I took the demotion.... This was particularly humiliating because my employees in Plainfield had just thrown me a big going-away party to celebrate my promotion.

Melissa Howard was chastised by a male supervisor not just because of her gender, but because she broke a racialized stereotype of white women's chasteness (Collins 2000:132–134). By breaking this stereotype, her colleagues and supervisors saw her as a lower class white. As a result of this construction, the unequal ways individuals have access to power are produced and upheld (Brown 1995; DeVault 1999:27). As a white woman in an interracial relationship, Melissa Howard was punished and seen as an outsider by the white people with whom she interacted. It was as though she created the problems she and her family experienced in Kansas.

In both examples, these white women were treated as incompetent and at fault for the problems they experienced. In turn, this affected their work. In these examples it is clear that while these women faced oppression due to their gender, rules and relationships to power are established for individuals based on the multiple, and differently valued, statuses they hold. Melissa Howard seemingly rose above the occupational glass ceiling noted by many plaintiffs. However, her failure to adhere to dating rules held by those with power, including landlords, realtor, and the Wal-Mart Regional Manager, created experiences of discrimination. As an insider due to her race, but a subordinate due to gender and class, Lorie Williams was able to attain a lower-level management position, but was chastised and belittled because she hired men of color and women.

These stories illustrate the lived experiences of white women who work at Wal-Mart. To succeed, they were socialized to accept and replicate discrimination against people of color. Similarly, workers are trained to follow rules of "respectability" concerning their actions. Insiders like Lorie Williams who hold beliefs that make them outsiders, need support so they will continue to believe that the long-term benefits of their actions will outweigh the short-term punishments they endure for failing to replicate discriminatory practices.

The Continuing Significance of Race

... In the following statement, it is clear subordinates also draw on racial and gendered privileges to gain power, even in instances in which they occupy lower class positions. In her statement, Ms. Jennifer Johnson, a Black woman with a community college degree, describes her interpersonal relationships with store staff.

> In 1991, Mr. Pshek was promoted to District Manager in a different district. He told me that if I was willing to relocate to a store in his district, that he would promote me to Assistant Manager, a position that I had sought for some time. I agreed, although I knew that Scott Schwalback, Mark Melatesta, and a male named Rick had been promoted to Assistant Manager without having to switch districts. I was transferred to the Eustis, Florida store as an hourly employee. I worked in that store as a Department Manager for approximately five months without receiving the promotion to Assistant Manager that Mr. Pshek had promised. Finally, I spoke with Bob Hart, Regional Vice President, about the situation, and he told me I had to wait another three to fourth months to be promoted to Assistant Manager. I was finally promoted in February 1992. Both Scott Schwalback and Rick (a Department Manager in Furniture) became Department Managers after I had, since

> I had worked for Wal-Mart for longer, but they were both promoted to Assistant Manager positions before I was. Thus, I worked much longer as a Department Manager before being promoted to Assistant Manager than similarly qualified men had. Shortly after I was promoted to Co-Manager, one of the male Assistant Managers who reported to me was disrespectful and avoided doing work I assigned him. I spoke to my Store Manager Kevin Robinson about it. He told me that the man was upset that I had been promoted instead of him, and said "you have two strikes against you: 1) you're a woman; 2) you're black."

Jennifer Johnson described the years of obstacles she navigated to gain promotions. Even with her experience and dedication, others reduced her accomplishments and used them to treat her with disdain. For her and other women of color, working in these stores means that they must not only do their jobs well, but they must also navigate a workplace where others use them as scapegoats for their own frustrations, block them from training opportunities, or sabotage them by not doing work these women of color assign them. In addition, those with the power to chastise offenders and influence workplace discussions of inequality did nothing. Therefore, women of color had to find their own ways to overcome inequality.

The above example illustrates blatant, hostile racism. But racism is not always aggressively expressed. Feagin (1991) argues after a lifetime of experiencing discrimination, people of color develop a special lens through which to recognize even subtle discrimination. For example, joking can be a subtle medium through which people of color experience belittlement and struggles for dominance. Ms. Gina Espinoza-Price, an energetic, innovative woman, worked her way up to become the District Manager of Mexico. She provided an example of how a white man used joking to force others to acknowledge the labels he created for them and to show dominance. Because he knew he would not face a penalty for his actions, he was able to set a behavioral example for other men under his supervision.

> Male Photo Division Management behaved in ways that demeaned and belittled women and minorities. In fall 1996, there was a Photo District Manager meeting in Valencia, California. Wal-Mart had just hired a second female Photo District Manager for the western region, Linda Palmer. During dinner, Jeff Gwartney introduced all of the District Managers to Ms. Palmer using nicknames for the minorities and women. I was introduced as Gina, "the little Mexican princess." I was very offended by Mr. Gwartney's comment and left the dinner early. Throughout the meeting, men made sexual statements and jokes that I thought were very offensive. For example, a flyer with an offensive joke about women being stupid was left on my belongings. In February 1997, during an evaluation, I complained to One-hour Photo Divisional Manager Joe Lisuzzo about harassment based on gender at the previous Photo District Manager meeting. He replied that he would take care of it. I knew from trainings on Wal-Mart's sex harassment

> policy given by Wal-Mart Legal Department employee Canetta Ivy that company policy mandates that when someone complains of sexual harassment, an investigation must begin within twenty-four hours. Therefore, I expected to be interviewed as a part of an investigation. I was never called. A couple of weeks later, in March 1997, I saw Mr. Lisuzzo at a meeting. I asked him if he had been conducting an investigation of my sexual harassment complaint. He replied that it was being taken care of. I was never aware of any action taken in response to my complaint. Six weeks after complaining about sexual harassment, I was terminated.

Gina Espinoza-Price's statement highlights blatant examples of racism and sexism as well as subtle, institutional examples of discrimination as carried out by individuals who occupied similar class positions, but different status levels. Mr. Lisuzzo's handling of the sexual harassment claim filed by the plaintiff is an example of how the concerns of people are ignored, requiring victims of harassment and discrimination to waste emotional energy rationalizing their own reactions and creating coping tactics to help them interact with hostile colleagues. In addition, it shows how racist joking is often coupled with class and gender.

Class Matters

... Just as some women enjoy privilege and opportunities based on their race, class also shapes their experiences and actions. The following excerpt demonstrates the difficulties and frustrations of women of color who are single mothers trying to support their families with Wal-Mart wages.

... Ms. Uma Jean Minor, a single mother living in Alabama, described the need to remain gainfully employed.

> As a single mother, I could not support my family on this wage [paid by Wal-Mart] and was forced to take a second, full-time day job at Food Fair. At the time, I planned to work this second job only until I was able to move to a higher paying position at Wal-Mart. I had no idea that it would take me another seven years to obtain a management position at Wal-Mart and that I would be working two full-time jobs for this entire period.

In this example, Uma Jean Minor notes that her Wal-Mart wages were not sufficient to lift her out of poverty, an argument echoed by living wage supporters throughout the United States (see McCarthy and Ciokajlo 2006; Talbott and Dolby 2003; Warren 2005). Though she shows a tremendous amount of will power, motivation, and agency, her actions would not be sustainable for most women in her position as a single mother of five. Missing from her declaration are the stories of fear, if not hardship, she lived through raising five children and working two full time jobs. In addition, we do not know if she had a network of kin or fictive kin on which she could rely. Therefore, we can only guess at the psychological and material effects discrimination had on her and her children.

Clearly, the effects of sex-based discrimination at Wal-Mart shape the lives of women and their families in different and important ways.

... For female heads of household whose families depend on their earnings, the wages, difficult decisions about how to navigate work, and workplace stress factor into whether it is worthwhile to work for poverty wages. Studies like that of Edin and Lein (1997) have demonstrated that the benefits lost by taking a low-wage job are detrimental to these women's children, who lose state funded health care, housing assistance, and food support. For women like the defendants, work provides another burden; because they need these jobs, they must sometimes sacrifice their dignity and morals in the face of oppressive supervisors (like Mary Crawford) and glass ceilings (like Uma Jean Minor) for the sake of their children.

DISCUSSION AND CONCLUSION

... These plaintiffs did not experience the waste of their talent, energy, and potential in the same way. Rather, it can be observed that the Dukes case is a collection of varying vignettes through which social scientists can observe how some women have more access to power and resources, leaving others to struggle because of their marginalized positions in the power hierarchy. To be clear, although some of the women at Wal-Mart have suffered because of sexism, gender abuse takes multiple forms allowing certain women privilege (even in their oppression) because power is not distributed equally.

REFERENCES

Blee, Kathleen M. 2002. *Inside Organized Racism: Women in the Hate Movement*. Berkeley: University of California Press.

Brown, Elsa B. 1995. "'What has Happened Here': The Politics of Difference in Women's History and Feminist Politics." In D. C. Hine, W. King and L. Reed (eds). *We Specialize in the Wholly Impossible: A Reader in Black Women's History*, pp. 39–54. Brooklyn, NY: Carlson Publishing.

Collins, Patricia H. 2000. *Black Feminist Thought: Knowledge, Consciousness and the Politics of Empowerment*. New York: Routledge.

DeVault, Marjorie L. 1999. *Liberating Method: Feminism and Social Research*. Philadelphia: Temple University Press.

Drogin, Richard. 2003. *Statistical Analysis of Gender Patterns in Wal-Mart Workforce*. Retrieved from www.walmartclass.com on 22 Sept 2008.

Edin, Kathryn, and Lein, Laura. 1997. *Making Ends Meet: How Single Mothers Survive Welfare and Low-Wage Work*. New York: Russell Sage Foundation.

Feagin, Joe R. 1991. "The Continuing Significance of Race: Antiblack Discrimination in Public Places." *American Sociological Review* 56(1): 101–116.

Featherstone, Liza. 2004. *Selling Women Short*. New York: Basic Books.

Ferber, Abby L. 1999. "What White Supremacists Taught a Jewish Scholar about Identity." *The Chronicle of Higher Education* May 7: B6–B7.

Frankenberg, Ruth. 1993. *White Women, Race Matters: The Social Construction of Whiteness*. Minneapolis: University of Minnesota Press.

Litchenstein, Nelson. 2006. "Wal-Mart: A Template for Twenty-First Century Capitalism." Chapter 1 in Nelson Litchenstein (ed.). *Wal-Mart: The Face of Twenty-First Century Capitalism*. New York: The New Press.

McCarthy, Brendan, and Mickey Ciokajlo. 2006. "Clerics Slam Big-Box Wage Law: Ordinance Would Chase Jobs from City, They Contend." *Chicago Tribune*, July 18, 2006. Accessed from Lexis Nexis August 5, 2007.

Mills, C. Wright. 2000 [1959]. *The Sociological Imagination*. Oxford: Oxford University Press.

Picca, Leslie Houts, and Feagin, Joe R. 2007 *Two-Faced Racism: Whites in the Backstage and Frontstage*. New York: Routledge.

Rathke, Wade. 2006. "A Wal-Mart Workers Association? An Organizing Plan" in Litchenstein, Nelson (ed.). *Wal-Mart: The Face of Twenty-First Century Capitalism*. New York: The New Press.

Rosen, Ruth. 2004. *Big-Box Battle: A Review of Selling Women Short: The Landmark Battle for Workers' Rights at Wal-Mart, by Liza Featherstone*. Retrieved from http://www.longviewinstitute.org/research/rosen/walmart/sellingwomenshort on October 14, 2008.

Talbott, Madeline, and Doby, Michael. 2003. "Using the Big Box Living Wage Ordinance to Keep Wal-Mart Out of the Cities." *Social Policy* 34(2/3): 23–28.

Warren, Dorian T. 2005. "Wal-Mart Surrounded: Community Alliances and Labor Politics in Chicago." *New Labor Forum* 14(3): 17–23.

36

Our Mothers' Grief

Racial-Ethnic Women and the Maintenance of Families

BONNIE THORNTON DILL

REPRODUCTIVE LABOR[1] FOR WHITE WOMEN IN EARLY AMERICA

In eighteenth- and nineteenth-century America, the lives of white[2] women in the United States were circumscribed within a legal and social system based on patriarchal authority. This authority took two forms: public and private. The social, legal, and economic position of women in this society was controlled through the private aspects of patriarchy and defined in terms of their relationship to families headed by men. The society was structured to confine white wives to reproductive labor within the domestic sphere. At the same time, the formation, preservation, and protection of families among white settlers was seen as crucial to the growth and development of American society. Building, maintaining, and supporting families was a concern of the State and of those organizations that prefigured the State. Thus, while white women had few legal rights as women, they were protected through public forms of patriarchy that acknowledged and supported their family roles of wives, mothers, and daughters because they were vital instruments for building American society....

In colonial America, white women were seen as vital contributors to the stabilization and growth of society. They were therefore accorded some legal and economic recognition through a patriarchal family structure....

Throughout the colonial period, women's reproductive labor in the family was an integral part of the daily operation of small-scale family farms or artisan's shops. According to Kessler-Harris (1981), a gender-based division of labor was common, but not rigid. The participation of women in work that was essential to family survival reinforced the importance of their contributions to both the protection of the family and the growth of society.

Between the end of the eighteenth and mid-nineteenth century, what is labeled the "modern American family" developed. The growth of industrialization and an urban middle class, along with the accumulation of agrarian wealth among Southern planters, had two results that are particularly pertinent to this

SOURCE: Dill, Bonnie Thornton. 1988. "Our Mothers' Grief: Racial-Ethnic Women and the Maintenance of Families." *Journal of Family History* 13: 415–431.

discussion. First, class differentiation increased and sharpened, and with it, distinctions in the content and nature of women's family lives. Second, the organization of industrial labor resulted in the separation of home and family and the assignment to women of a separate sphere of activity focused on childcare and home maintenance. Whereas men's activities became increasingly focused upon the industrial competitive sphere of work, "women's activities were increasingly confined to the care of children, the nurturing of the husband, and the physical maintenance of the home" (Degler 1980, p. 26).

This separate sphere of domesticity and piety became both an ideal for all white women as well as a source of important distinctions between them. As Matthaei (1982) points out, tied to the notion of wife as homemaker is a definition of masculinity in which the husband's successful role performance was measured by his ability to keep his wife in the homemaker role. The entry of white women into the labor force came to be linked with the husband's assumed inability to fulfill his provider role.

For wealthy and middle-class women, the growth of the domestic sphere offered a potential for creative development as homemakers and mothers. Given ample financial support from their husband's earnings, some of these women were able to concentrate their energies on the development and elaboration of the more intangible elements of this separate sphere. They were also able to hire other women to perform the daily tasks such as cleaning, laundry, cooking, and ironing. Kessler-Harris cautions, however, that the separation of productive labor from the home did not seriously diminish the amount of physical drudgery associated with housework, even for middle-class women.... In effect, household labor was transformed from economic productivity done by members of the family group to home maintenance; childcare and moral uplift done by an isolated woman who perhaps supervised some servants.

Working-class white women experienced this same transformation but their families' acceptance of the domestic code meant that their labor in the home intensified. Given the meager earnings of working-class men, working-class families had to develop alternative strategies to both survive and keep the wives at home. The result was that working-class women's reproductive labor increased to fill the gap between family need and family income. Women increased their own production of household goods through things such as canning and sewing; and by developing other sources of income, including boarders and homework. A final and very important source of other income was wages earned by the participation of sons and daughters in the labor force. In fact, Matthaei argues that "the domestic homemaking of married women was supported by the labors of their daughters" (1982, p. 130)....

Another way in which white women's family roles were socially acknowledged and protected was through the existence of a separate sphere for women. The code of domesticity, attainable for affluent women, became an ideal toward which nonaffluent women aspired. Notwithstanding the personal constraints placed on women's development, the notion of separate spheres promoted the growth and stability of family life among the white middle class and became the basis for working-class men's efforts to achieve a family wage, so that they could keep their wives at home. Also, women gained a distinct sphere of authority and expertise that yielded them special recognition.

During the eighteenth and nineteenth centuries, American society accorded considerable importance to the development and sustenance of European immigrant families. As primary laborers in the reproduction and maintenance of family life, women were acknowledged and accorded the privileges and protections deemed socially appropriate to their family roles. This argument acknowledges the fact that the family structure denied these women many rights and privileges and seriously constrained their individual growth and development. Because women gained social recognition primarily through their membership in families, their personal rights were few and privileges were subject to the will of the male head of the household. Nevertheless, the recognition of women's reproductive labor as an essential building block of the family, combined with a view of the family as the cornerstone of the nation, distinguished the experiences of the white, dominant culture from those of racial ethnics.

Thus, in its founding, American society initiated legal, economic, and social practices designed to promote the growth of family life among European colonists. The reception colonial families found in the United States contrasts sharply with the lack of attention given to the families of racial-ethnics. Although the presence of racial-ethnics was equally as important for the growth of the nation, their political, economic, legal, and social status was quite different.

REPRODUCTIVE LABOR AMONG RACIAL-ETHNICS IN EARLY AMERICA

Unlike white women, racial-ethnic women experienced the oppressions of a patriarchal society but were denied the protections and buffering of a patriarchal family. Their families suffered as a direct result of the organization of the labor systems in which they participated.

Racial-ethnics were brought to this country to meet the need for a cheap and exploitable labor force. Little attention was given to their family and community life except as it related to their economic productivity. Labor, and not the existence or maintenance of families, was the critical aspect of their role in building the nation. Thus they were denied the social structural supports necessary to make *their* families a vital element in the social order. Family membership was not a key means of access to participation in the wider society. The lack of social, legal, and economic support for racial-ethnic families intensified and extended women's reproductive labor, created tensions and strains in family relationships, and set the stage for a variety of creative and adaptive forms of resistance.

African-American Slaves

Among students of slavery, there has been considerable debate over the relative "harshness" of American slavery, and the degree to which slaves were permitted or encouraged to form families. It is generally acknowledged that many slaveowners found it economically advantageous to encourage family formation as a way of reproducing and perpetuating the slave labor force. This became increasingly true after 1807 when the importation of African slaves was explicitly prohibited. The existence of these families and many aspects of their functioning, however, were directly controlled by the master. In other words, slaves married and formed families but these groupings were completely subject to the master's decision to let them remain intact. One study has estimated that about 32 percent of all recorded slave marriages were disrupted by sale, about 45 percent by death of a spouse, about 10 percent by choice, with the remaining 13 percent not disrupted at all (Blassingame 1972, pp. 90–92). African slaves thus quickly learned that they had a limited degree of control over the formation and maintenance of their marriages and could not be assured of keeping their children with them. The threat of disruption was perhaps the most direct and pervasive cultural assault[3] on families that slaves encountered. Yet there were a number of other aspects of the slave system which reinforced the precariousness of slave family life.

In contrast to some African traditions and the Euro-American patterns of the period, slave men were not the main provider or authority figure in the family. The mother-child tie was basic and of greatest interest to the slaveowner because it was critical in the reproduction of the labor force.

In addition to the lack of authority and economic autonomy experienced by the husband-father in the slave family, use of the rape of women slaves as a weapon of terror and control further undermined the integrity of the slave family.... The slave family, therefore, was at the heart of a peculiar tension in the master-slave relationship. On the one hand, slaveowners sought to encourage familial ties among slaves because, as Matthaei (1982) states: "... these provided the basis of the development of the slave into a self-conscious socialized human being" (p. 81). They also hoped and believed that this socialization process would help children learn to accept their place in society as slaves. Yet the master's need to control and intervene in the familial life of the slaves is indicative of the other side of this tension. Family ties had the potential for becoming a competing and more potent source of allegiance than the slavemaster himself. Also, kin were as likely to socialize children in forms of resistance as in acts of compliance.

It was within this context of surveillance, assault, and ambivalence that slave women's reproductive labor took place. She and her menfolk had the task of preserving the human and family ties that could ultimately give them a reason for living. They had to socialize their children to believe in the possibility of a life in which they were not enslaved. The slave woman's labor on behalf of the family was, as Davis (1971) has pointed out, the only labor the slave engaged in that could not be directly appropriated by the slaveowner for his own profit. Yet, its indirect appropriation, as labor crucial to the reproduction of the slaveowner's labor force, was the source of strong ambivalence for many slave women. Whereas some mothers murdered their babies to keep them from being slaves, many sought within the family sphere a degree of autonomy and creativity denied them in other realms of the society. The maintenance of a distinct African-American culture is testimony to the ways in which slaves maintained a degree of cultural autonomy and resisted the creation of a slave family that only served the needs of the master.

Gutman (1976) provides evidence of the ways in which slaves expressed a unique Afro-American culture through their family practices. He provides data on naming patterns and kinship ties among slaves that flies in the face of the dominant ideology of the period. That ideology argued that slaves were immoral and had little concern for or appreciation of family life.

Yet Gutman demonstrated that within a system which denied the father authority over his family, slave boys were frequently named after their fathers, and many children were named after blood relatives as a way of maintaining family ties. Gutman also suggested that after emancipation a number of slaves took the names of former owners in order to reestablish family ties that had been disrupted earlier. On plantation after plantation, Gutman found considerable evidence of the building and maintenance of extensive kinship ties among slaves. In instances where slave families had been disrupted, slaves in new communities reconstituted the kinds of family and kin ties that came to characterize Black family life throughout the South. These patterns included, but were not limited to, a belief in the importance of marriage as a long-term commitment, rules of exogamy that included marriage between first cousins, and acceptance of women who had children outside of marriage. Kinship networks were an important source of resistance to the organization of labor that treated the individual slave, and not the family, as the unit of labor (Caulfield 1974).

Another interesting indicator of the slaves' maintenance of some degree of cultural autonomy has been pointed out by Wright (1981) in her discussion of slave housing. Until the early 1800s, slaves were often permitted to build their housing according to their own design and taste. During that period, housing built in an African style was quite common in the slave quarters. By 1830, however, slaveowners had begun to control the design and arrangement of slave housing and had introduced a degree of conformity and regularity to it that left little room for the slave's personalization of the home. Nevertheless, slaves did use some of their own techniques in construction and often hid it from their masters....

Housing is important in discussions of family because its design reflects sociocultural attitudes about family life. The housing that slaveowners provided for their slaves reflected a view of Black family life consistent with the stereotypes of the period. While the existence of slave families was acknowledged, it certainly was not nurtured. Thus, cabins were crowded, often containing more than one family, and there were no provisions for privacy. Slaves had to create their own....

Perhaps most critical in developing an understanding of slave women's reproductive labor is the gender-based division of labor in the domestic sphere. The organization of slave labor enforced considerable equality among men and women. The ways in which equality in the labor force was translated into the family sphere is somewhat speculative....

We know, for example, that slave women experienced what has recently been called the "double day" before most other women in this society. Slave narratives (Jones 1985; White 1985; Blassingame 1977) reveal that women had primary responsibility for their family's domestic chores. They cooked (although on some plantations meals were prepared for all of the slaves), sewed, cared for their children, and cleaned house, all after completing a full day of labor for the master. Blassingame (1972) and others have pointed out that slave men engaged in hunting, trapping, perhaps some gardening, and furniture making as ways of contributing to the maintenance of their families. Clearly, a gender-based division of labor did exist within the family and it appears that women bore the larger share of the burden for housekeeping and child care....

Black men were denied the male resources of a patriarchal society and therefore were unable to turn gender distinctions into female subordination, even if that had been their desire. Black women, on the other hand, were denied support and protection for their roles as mothers and wives and thus had to modify and structure those roles around the demands of their labor. Thus, reproductive labor for slave women was intensified in several ways: by the demands of slave labor that forced them into the double-day of work; by the desire and need to maintain family ties in the face of a system that gave them only limited recognition; by the stresses of building a family with men who were denied the standard social privileges of manhood; and by the struggle to raise children who could survive in a hostile environment.

This intensification of reproductive labor made networks of kin and quasi-kin important instruments in carrying out the reproductive tasks of the slave community. Given an African cultural heritage where kinship ties formed the basis of social relations, it is not at all surprising that African American slaves developed an extensive system of kinship ties and obligations (Gutman 1976; Sudarkasa 1981). Research on Black families in slavery provides considerable documentation of participation of extended kin in child rearing, childbirth, and other domestic, social, and economic activities (Gutman 1976; Blassingame 1972; Genovese and Miller 1974)....

With individual households, the gender-based division of labor experienced some important shifts during emancipation. In their first real opportunity to establish family life beyond the controls and constraints imposed by a slavemaster, family life among Black sharecroppers changed radically. Most women, at least those who were wives and daughters of able-bodied men, withdrew from field labor and concentrated on their domestic duties in the home. Husbands took primary responsibility for the fieldwork and for relations with the owners, such as signing contracts on behalf of the family. Black women were severely criticized by whites for removing themselves from field labor because they were seen to be aspiring to a model of womanhood that was considered inappropriate for them. This reorganization of female labor, however, represented an attempt on the part of Blacks to protect women from some of the abuses of the slave system and to thus secure their family life. It was more likely a response to the particular set of circumstances that the newly freed slaves faced than a reaction to the lives of their former masters. Jones (1985) argues that these patterns were "particularly significant" because at a time when industrial development was introducing a labor system that divided male and female labor, the freed Black family was establishing a pattern of joint work and complementary tasks between males and females that was reminiscent of the preindustrial American families. Unfortunately, these former slaves had to do this without the institutional supports that white farm families had in the midst of a sharecropping system that deprived them of economic independence.

Chinese Sojourners

An increase in the African slave population was a desired goal. Therefore, Africans were permitted and even encouraged at times to form families subject to the authority and whim of the master. By sharp contrast, Chinese people were explicitly denied the right to form families in the United States through both law and social practice. Although male laborers began coming to the United States in sizable numbers in the middle of the nineteenth century, it was more than a century before an appreciable number of children of Chinese parents were born in America. Tom, a respondent in Nee and Nee's (1973) book, *Longtime Californ'*, says: "One thing about Chinese men in America was you had to be either a merchant or a big gambler, have lot of side money to have a family here. A working man, an ordinary man, just can't!" (p. 80).

Working in the United States was a means of gaining support for one's family with an end of obtaining sufficient capital to return to China and purchase land. The practice of sojourning was reinforced by laws preventing Chinese laborers from becoming citizens, and by restrictions on their entry into this country. Chinese laborers who arrived before 1882 could not bring their wives and were prevented by law from marrying whites. Thus, it is likely that the number of Chinese-American families might have been negligible had it not been for two things: the San Francisco earthquake and fire in 1906, which destroyed all municipal records; and the ingenuity and persistence of the Chinese people who used the opportunity created by the earthquake to increase their numbers in the United States. Since relatives of citizens were permitted entry, American-born Chinese (real and claimed) would visit China, report the birth of a son, and thus create an entry slot. Years later the slot could be used by a relative or purchased. The purchasers were called "paper sons." Paper sons became a major mechanism for increasing the Chinese population, but it was a slow process and the sojourner community remained predominantly male for decades.

The high concentration of males in the Chinese community before 1920 resulted in a split-household form of family....

The women who were in the United States during this period consisted of a small number who were wives and daughters of merchants and a larger percentage who were prostitutes. Hirata (1979) has suggested that Chinese prostitution was an important element in helping to maintain the split-household family. In conjunction with laws prohibiting intermarriage, Chinese prostitution helped men avoid long-term relationships with women in the United States and ensured that the bulk of their meager earnings would continue to support the family at home.

The reproductive labor of Chinese women, therefore, took on two dimensions primarily because of the split-household family form. Wives who remained in China were forced to raise children and care for in-laws on the meager remittances of their sojourning husband. Although we know few details about their lives, it is clear that the everyday work of bearing and maintaining children and a household fell entirely on their shoulders. Those women who immigrated and worked as prostitutes performed the more nurturant aspects of reproductive labor, that is, providing emotional and sexual companionship for men who were far from home. Yet their role as prostitute was more likely a means of supporting their families at home in China than a chosen vocation.

The Chinese family system during the nineteenth century was a patriarchal one wherein girls had little value. In fact, they were considered only temporary members of their father's family because when they married, they became members of their husband's families. They also had little social value: girls were sold by some poor parents to work as prostitutes, concubines, or servants. This saved the family the expense of raising them, and their earnings also became a source of family income. For most girls, however, marriages were arranged and families sought useful connections through this process.

With the development of a sojourning pattern in the United States, some Chinese women in those regions of China where this pattern was more prevalent would be sold to become prostitutes in the United States. Most, however, were married off to men whom they saw only once or twice in the 20- or 30-year period during which he was sojourning in the United States. Her status as wife ensured that a portion of the meager wages he earned would be returned to his family in China. This arrangement required considerable sacrifice and adjustment on the part of wives who remained in China and those who joined their husbands after a long separation....

Despite these handicaps, Chinese people collaborated to establish the opportunity to form families and settle in the United States. In some cases it took as long as three generations for a child to be born on United States soil....

Chicanos

Africans were uprooted from their native lands and encouraged to have families in order to increase the slave labor force. Chinese people were immigrant laborers whose "permanent" presence in the country was denied. By contrast, Mexican-Americans were colonized and their traditional family life was disrupted by war and the imposition of a new set of laws and conditions of labor. The hardships faced by Chicano families, therefore, were the result of the United States colonization of the indigenous Mexican population, accompanied by the beginnings of industrial development in the region. The treaty of Guadalupe Hidalgo, signed in 1848, granted American citizenship to Mexicans living in what is now called the Southwest. The American takeover, however, resulted in the gradual displacement of Mexicans from the land and their incorporation into a colonial labor force (Barrera 1979). In addition, Mexicans who immigrated into the United States after 1848 were also absorbed into the labor force.

Whether natives of Northern Mexico (which became the United States after 1848) or immigrants from Southern Mexico, Chicanos were a largely peasant population whose lives were defined by a feudal economy and a daily struggle on the land for economic survival. Patriarchal families were important instruments of community life and nuclear family units were linked together through an elaborate system of kinship and godparenting. Traditional life was characterized by hard work and a fairly distinct pattern of sex-role segregation....

As the primary caretakers of hearth and home in a rural environment, *Las Chicanas* labor made a vital and important contribution to family survival....

Although some scholars have argued that family rituals and community life showed little change before World War I (Saragoza 1983), the American conquest of Mexican lands, the introduction of a new system of labor, the loss of Mexican-owned land through the inability to document ownership, plus the transient nature of most of the jobs in which Chicanos were employed, resulted in the gradual erosion of this pastoral way of life. Families were uprooted as the economic basis for family life changed. Some immigrated from Mexico in search of a better standard of living and worked in the mines and railroads. Others who were native to the Southwest faced a job market that no longer required their skills and moved into mining, railroad, and agricultural labor in search of a means of earning a living. According to Camarillo (1979), the influx of Anglo[4] capital into the pastoral economy of Santa Barbara rendered obsolete the skills of many Chicano males who had worked as ranchhands and farmers prior to the urbanization of that economy. While some women and children accompanied their husbands to the railroad and mine camps, they often did so despite prohibitions against it. Initially many of these camps discouraged or prohibited family settlement.

The American period (post-1848) was characterized by considerable transiency for the Chicano population. Its impact on families is seen in the growth of female-headed households, which was reflected in the data as early as 1860. Griswold del Castillo (1979) found a sharp increase in female-headed households in Los Angeles, from a low of 13 percent in 1844 to 31 percent in 1880. Camarillo (1979, p. 120) documents a similar increase in Santa Barbara from 15 percent in 1844 to 30 percent by 1880. These increases appear to be due not so much to divorce, which was infrequent in this Catholic population, but to widowhood and temporary abandonment in search of work. Given the hazardous nature of work in the mines and railroad camps, the death of a husband, father or son who was laboring in these sites was not uncommon. Griswold del Castillo (1979) reports a higher death rate among men than women in Los Angeles. The rise in female-headed households, therefore, reflects the instabilities and insecurities introduced into women's lives as a result of the changing social organization of work.

One outcome, the increasing participation of women and children in the labor force was primarily a response to economic factors that required the modification of traditional values....

Slowly, entire families were encouraged to go to railroad workcamps and were eventually incorporated into the agricultural labor market. This was a response both to the extremely low wages paid to Chicano laborers and to the preferences of employers who saw family labor as a way of stabilizing the work-force. For Chicanos, engaging all family members in agricultural work was a means of increasing their earnings to a level close to subsistence for the entire group and of keeping the family unit together....

While the extended family has remained an important element of Chicano life, it was eroded in the American period in several ways. Griswold del Castillo (1979), for example, points out that in 1845 about 71 percent of Angelenos lived in extended families and that by 1880, fewer than half did. This decrease in extended families appears to be a response to the changed economic conditions and to the instabilities generated by the new sociopolitical structure. Additionally, the imposition of American law and custom ignored and ultimately undermined some aspects of the extended family. The extended family in traditional Mexican life consisted of an important set of familial, religious, and community obligations. Women, while valued primarily for their domesticity, had certain legal and property rights that acknowledged the importance of their work, their families of origin and their children....

In the face of the legal, social, and economic changes that occurred during the American period, Chicanas were forced to cope with a series of dislocations in traditional life. They were caught between conflicting pressures to maintain traditional women's roles and family customs and the need to participate in the economic support of their families by working outside the home. During this period the preservation of some traditional customs became an important force for resisting complete disarray....

Of vital importance to the integrity of traditional culture was the perpetuation of the Spanish language. Factors that aided in the maintenance of other aspects of Mexican culture also helped in sustaining the language. However, entry into English-language public schools introduced the children and their families to systematic efforts to erase their native tongue....

Another key factor in conserving Chicano culture was the extended family network, particularly the system of *compadrazgo* or godparenting. Although the full extent of the impact of the American period on the Chicano extended family is not known, it is generally acknowledged that this family system, though lacking many legal and social sanctions, played an important role in the preservation of the Mexican community (Camarillo 1979, p. 13). In Mexican society, godparents were an important way of linking family and community through respected friends or authorities. Named at the important rites of passage in a child's life, such as birth, confirmation, first communion, and marriage, *compadrazgo* created a moral obligation for godparents to act as guardians, to provide financial assistance in times of need, and to substitute in case of the death of a parent. Camarillo (1979) points out that in traditional society these bonds cut across class and racial lines....

The extended family network—which included godparents—expanded the support groups for women who were widowed or temporarily abandoned and for those who were in seasonal, part-, or full-time work. It suggests, therefore, the potential for an exchange of services among poor people whose income did not provide the basis for family subsistence.... This family form is important to the continued cultural autonomy of the Chicano community.

CONCLUSION: OUR MOTHERS' GRIEF

Reproductive labor for Afro-American, Chinese-American, and Mexican-American women in the nineteenth century centered on the struggle to maintain family units in the face of a variety of cultural assaults. Treated primarily as individual units of labor rather than as members of family groups, these women labored to maintain, sustain, stabilize, and reproduce their families while working in both the public (productive) and private (reproductive) spheres. Thus, the concept of reproductive labor, when applied to women of color, must be modified to account for the fact that labor in the productive sphere was required to achieve even minimal levels of family subsistence. Long after industrialization had begun to reshape family roles among middle-class white families, driving white women into a cult of domesticity, women of color were coping with an extended day. This day included subsistence labor outside the family and domestic labor within the family. For slaves, domestics, migrant farm laborers, seasonal factory-workers, and prostitutes, the distinctions between labor that reproduced family life and which economically sustained it were minimized. The expanded workday was one of the primary ways in which reproductive labor increased.

Racial-ethnic families were sustained and maintained in the face of various forms of disruption. Yet racial-ethnic women and their families paid a high price in the process. High rates of infant mortality, a shortened life span, the early onset of crippling and debilitating disease provided some insight into the costs of survival.

The poor quality of housing and the neglect of communities further increased reproductive labor. Not only did racial-ethnic women work hard outside the home for a mere subsistence, they worked very hard inside the home to achieve even minimal standards of privacy and cleanliness. They were continually faced with disease and illness that directly resulted from the absence of basic sanitation. The fact that some African women murdered their children to prevent them from becoming slaves is an indication of the emotional strain associated with bearing and raising children while participating in the colonial labor system.

We have uncovered little information about the use of birth control, the prevalence of infanticide, or the motivations that may have generated these or other behaviors. We can surmise, however, that no matter how much children were accepted, loved, or valued among any of these groups of people, their futures in a colonial labor system were a source of grief for their mothers. For those children who were born, the task of keeping them alive, of helping them to understand and participate in a system that exploited them, and the challenge of maintaining a measure—no matter how small—of cultural integrity, intensified reproductive labor.

Being a racial-ethnic woman in nineteenth-century American society meant having extra work both inside and outside the home. It meant having a contradictory relationship to the norms and values about women that were being generated in the dominant white culture. As pointed out earlier, the notion of separate spheres of male and female labor had contradictory outcomes for the nineteenth-century whites. It was the basis for the confinement of women to the household and for much of the protective legislation that subsequently developed. At the same time, it sustained white families by providing social acknowledgment and support to women in the performance of their family roles. For racial-ethnic women, however, the notion of separate spheres served to reinforce their subordinate status and became, in effect, another assault. As they increased their work outside the home, they were forced into a productive labor sphere that was organized for men and "desperate" women who were so unfortunate or immoral that they could not confine their work to the domestic sphere. In the productive sphere, racial-ethnic women faced exploitative jobs and depressed wages. In the reproductive sphere, however, they were denied the opportunity to embrace the dominant ideological definition of "good" wife or mother. In essence, they were faced with a double-bind situation, one that required their participation in the labor force to sustain family life but damned them as women, wives, and mothers because they did not confine their labor to the home. Thus, the conflict between ideology and reality in the lives of racial-ethnic women during the nineteenth century sets the stage for stereotypes, issues of self-esteem, and conflicts around gender-role prescriptions that surface more fully in the twentieth century. Further, the tensions and conflicts that characterized their lives during this period provided the impulse for community activism to jointly address the inequities, which they and their children and families faced.

NOTES

1. The term *reproductive labor* is used to refer to all of the work of women in the home. This includes but is not limited to: the buying and preparation of food and clothing, provision of emotional support and nurturance for all family members, bearing children, and planning, organizing, and carrying out a wide variety of tasks associated with their socialization. All of these activities are necessary for the growth of patriarchal capitalism because they maintain, sustain, stabilize, and *reproduce* (both biologically and socially) the labor force.
2. The term *white* is a global construct used to characterize peoples of European descent who migrated to and helped colonize America. In the seventeenth century, most of these immigrants were from the British Isles. However, during the time period covered by this article, European immigrants became increasingly diverse. It is a limitation of this article that time and space does not permit a fuller discussion of the variations in the white European immigrant experience. For the purposes of the argument made herein and of the contrast it seeks to draw between the experiences of mainstream (European) cultural groups and that of racial/ethnic minorities, the differences among European settlers are joined and the broad similarities emphasized.
3. Cultural assaults, according to Caulfield (1974), are benign and systematic attacks on the institutions and forms of social organization that are fundamental to the maintenance and flourishing of a group's culture.
4. This term is used to refer to white Americans of European ancestry.

REFERENCES

Barrera, Mario. 1979. *Race and Class in the Southwest.* South Bend, IN: Notre Dame University Press.

Blassingame, John. 1972. *The Slave Community: Plantation Life in the Antebellum South.* New York: Oxford University Press.

Blassingame, John. 1977. *Slave Testimony: Two Centuries of Letters, Speeches, Interviews, and Autobiographies.* Baton Rouge, LA: Louisiana State University Press.

Camarillo, Albert. 1979. *Chicanos in a Changing Society.* Cambridge, MA: Harvard University Press.

Caulfield, Mina Davis. 1974. "Imperialism, the Family, and Cultures of Resistance." *Socialist Review* 4(2)(October): 67–85.

Davis, Angela. 1971. "The Black Woman's Role in the Community of Slaves." *Black Scholar* 3(4)(December): 2–15.

Degler, Carl. 1980. *At Odds.* New York: Oxford University Press.

Genovese, Eugene D., and Elinor Miller, eds. 1974. *Plantation, Town, and County: Essays on the Local History of American Slave Society.* Urbana: University of Illinois Press.

Griswold del Castillo, Richard. 1979. *The Los Angeles Barrio: 1850–1890.* Los Angeles: The University of California Press.

Gutman, Herbert. 1976. *The Black Family in Slavery and Freedom: 1750–1925.* New York: Pantheon.

Hirata, Lucie Cheng. 1979. "Free, Indentured, Enslaved: Chinese Prostitutes in Nineteenth-Century America." *Signs* 5 (Autumn): 3–29.

Jones, Jacqueline. 1985. *Labor of Love, Labor of Sorrow.* New York: Basic Books.

Kessler-Harris, Alice. 1981. *Women Have Always Worked.* Old Westbury: The Feminist Press.

Matthaei, Julie. 1982. *An Economic History of Women in America.* New York: Schocken Books.

Nee, Victor G., and Brett de Bary Nee. 1973. *Longtime Californ'.* New York: Pantheon Books.

Saragoza, Alex M. 1983. "The Conceptualization of the History of the Chicano Family: Work, Family, and Migration in Chicanos." Research Proceedings of the Symposium on Chicano Research and Public Policy. Stanford, CA: Stanford University, Center for Chicano Research.

Sudarkasa, Niara. 1981. "Interpreting the African Heritage in Afro-American Family Organization." Pp. 37–53 in *Black Families*, edited by Harriette Pipes McAdoo. Beverly Hills, CA: Sage Publications.

White, Deborah Gray. 1985. *Ar'n't I a Woman?: Female Slaves in the Plantation South.* New York: W. W. Norton.

Wright, Gwendolyn. 1981. *Building the Dream: A Social History of Housing in America.* New York: Pantheon Books.

37

Exploring the Intersections of Race, Ethnicity, and Class on Maternity Leave Decisions

Implications for Public Policy

TIFFANY MANUEL AND RUTH ENID ZAMBRANA

... Today, just beyond the threshold of the twenty-first century, all industrialized nations provide some form of maternity leave; almost all do so at fairly high levels of wage replacement; and most have provisions for paternity leave as well. In this context, the United States is an outlier among other industrialized nations in that it offers families very little in the way of job protection and direct financial support to working caregivers while they are needed to provide direct care for their families. Rather, U.S. policymakers have created institutional arrangements for family leave that are primarily privately financed (mostly through employer fringe benefit programs and employee earnings), cover a small percentage of firms, and disproportionately benefit workers in the shrinking primary labor market rather than universally cover American families (Garrett, Wenk, & Lubeck, 1990). Moreover these programs operate with virtually no coordination across federal, state, and local levels nor do they provide a coherent set of policies to help families address work and family conflicts (Vogel, 1993).

Most family leave policies in the United States were passed in the last three decades, raising the question: why is the United States only now experiencing a "big boom" with regard to the creation of such policies? The primary catalyst for the dramatic turnaround seems to be the significant shift in the gender composition of the U.S. workforce. Women in prime working years moved from just under 40 percent employment in 1950, to nearly 80 percent by the turn of the last century. For the first time in the nation's history, women represent nearly half of the U.S. workforce and women's employment trajectories now closely follow that of men in the same age categories. Interestingly enough, the widest variation in labor force participation today is not between men and women (or the gender disparity) but rather between workers with and without substantial care responsibilities for young children. This is an important trend as 72 percent of all mothers

SOURCE: Manuel, Tiffany, and Ruth Enid Zambrana. 2009. "Leave Decisions: Implications for Public Policy." Pp. 123–149 in *Emerging Intersections: Race, Class, and Gender in Theory, Policy, and Practice*, edited by Bonnie Thornton Dill and Ruth Enid Zambrana. New Brunswick, NJ: Rutgers University Press. Reprinted by permission.

are employed in the United States and 90 percent of all American parents with children under the age of eighteen (more than sixty-nine million Americans) are employed.…

Recent legislative attempts to extend work supports to working caregivers, in conjunction with the emergence of family-leave policies represent the opening of an important policy window. Generally family-leave policies guarantee the ability of workers to return to their jobs following short-term absences related to their family care-giving responsibilities. The term "maternity leave" … is one form of family-leave policy and can be defined simply as a family leave taken by a mother to care for her newborn. Some leave policies provide wage replacement while leave-takers are away from their jobs and others simply include job protection provisions. These policies have been passed at all levels of government (particularly at the state level), as well as by private employers, although there is great diversity across the workforce in terms of eligibility, access, and usage of such benefits.

The specific policy origins of maternity leave policies in the workplace were largely the result of the Equal Pay Act of 1963 and the Civil Rights Act of 1964 (predecessors to the FMLA and the Pregnancy Disability Act). These civil rights laws outlawed discrimination with respect to "compensation, terms, condition, or privileges of employment" (Title VII of the Civil Rights Act). Although neither piece of legislation specifically mentioned pregnancy, these laws opened the door for later court battles that challenged a multitude of sex discriminating employment practices. After several court battles and pressure from coalition groups, the Pregnancy Discrimination Act (PDA) was passed in 1978 as an amendment to the Civil Rights Act of 1964. Essentially the PDA made it unlawful for employers to terminate women because they were pregnant or to otherwise force them out of their jobs and mandated that, when a woman was "disabled" because of a pregnancy, she receive the same sick pay, insurance coverage, and job protection as an employer's other employees with short-term disabilities.

This early legislation paved the way for more aggressive leave policies later—first at the state and local levels and eventually in 1993, the Family and Medical Leave Act passed becoming the most sweeping federal family leave law in the United States (albeit more than one hundred years after the first maternity leave law was enacted in Europe).… The FMLA covers about half of the U.S. workforce (about 6 percent of the nation's firms) and requires that employers with fifty or more employees allow eligible employees up to twelve weeks of unpaid, job protected leave to care for newborn or newly adopted children, for seriously ill spouses, children, parents, or to take care of their own illnesses, including pregnancy.… Although the FMLA was the first piece of federal legislation to specifically address working caregivers, other federal policies, such as public assistance programs, help some mothers defray the cost of taking family leave.

In addition to the federal FMLA policy, there are other sources of family leave assistance in the United States. Although only half of the nation's workers are covered by the FMLA, more workers are covered by state and employer leave policies. Most states have maternity leave policies that actually predate the FMLA and thirty-five states have legislation that is more generous than the federal family-leave provisions. Additionally five states and Puerto Rico have Temporary Disability Insurance programs (TDI) that provide financial assistance to mothers on disability due to pregnancy related illnesses. Finally employers also provide maternity leave benefits and many allow their employees to use sick and vacation time to replace lost wages during the leave.

Although maternity leave policies predate the FMLA, most analysts agree that the FMLA has been an important catalyst for increasing maternity leave coverage and dramatically changing maternal employment behavior around childbirth. For example, the majority of American women no longer routinely quit their jobs when they become pregnant. Between 1961 and 1965, more than 60 percent of all new mothers quit their jobs when they became pregnant with their first child, whereas during 1991–1995, less than 30 percent of new mothers did.... Generally women with family leave benefits are more likely to return to jobs and to return faster than those without such benefits (Baum, 2003; Hofferth, 1996). Taken together, these trends have contributed to the formulation of a wide variety of work and family policies implemented by government agencies and private employers....

Intersectionality theory argues that the socially constructed ways of identifying and categorizing individuals that arise in societies are transformed by and transform history themselves, and that those categories must be studied as interconnected rather than as separate systems (Amott & Matthaei, 1991), In recent years, there has been a growing interest in the notion of intersectionality for many reasons; not the least of which has been scholarly attempts to better conceptualize policy outcomes. There are enormous possibilities for how an intersectional research approach might offer important policy directions that have been largely ignored....

FAMILY LEAVE AND LOW-INCOME MOTHERS

... Three particular trends in access to family leave benefits have emerged as important catalysts in the legislative debate of leave policies as a work support for low-income mothers. First, low-income mothers are substantially less likely to have access to work supports and particularly lack those that address work and family conflicts. One reason is that the eligibility requirements associated with leave policies and other work supports target full-time workers in the primary labor market excluding low-income mothers who are more likely to be part-time, seasonal, or intermittent workers. Furthermore low-income mothers are more likely to work in jobs like those in the retail industry that do not typically offer fringe benefits or job-protected time off (such as sick or vacation leave)....

... Second, low-income mothers are more likely to be the sole financial support of their families and without some form of wage-replacement are unable to fully take advantage of leave benefits when they are offered. The case of low-income mothers with access to paid leave (likely obtained because of sick or vacation leave) is not much better: most do not have enough accrued time off to cover an entire maternity leave and as a result, they go without any income from their jobs for some portion of their leave.... Moreover the use of paid time off for sick and vacation days is still at the discretion of supervisors and is mediated by informal institutional practices that make it difficult for many workers (especially low-wage workers, men, and upper-level managers) to take advantage of it when they need to do so (Fried, 1998).

In response, low-income mothers have devised a mix of strategies to ease the financial burdens of caring for their children.... If they have sick and vacation benefits, the vast majority simply try to save up as much sick and vacation time as they can, work as far into their pregnancies as possible and use whatever paid time off they have. Thereafter they exercise one of three options: (1) use up whatever financial savings/resources they have at their disposal to stay home; (2) return immediately to work because they cannot afford to stay home without pay; or (3) turn to their social networks (such as family, friends) and/or public assistance programs to help defray the costs of leave-taking....

The turn to public assistance by low-income mothers is related to a third trend: the shift in the racial/ethnic composition of public assistance recipients. Beginning in the late 1960s, the racial and ethnic composition of public assistance programs began to change such that by the early 1990s, minority women were overrepresented among recipients. As this shift occurred, welfare became socially constructed in the public consciousness as a "Black program" (Neuback & Cazenave, 2001). It was the political currency of the association between race (actually, with being Black) and welfare that helped enact the PRWORA....

As a result, employment behavior—seeking, getting, and keeping jobs—became the goal of social welfare policymaking operationalized in the case of the PRWORA as promoting "self-sufficiency" by encouraging the "transition" to work, limiting reproduction among low-income mothers, and, consequently, reducing public financial support.

Although very little data on the racial dynamics of welfare programs in the Untied States following the passage of the PRWORA exist, initial data on welfare reform has not been particularly encouraging for African American and Latino women as they have had much more difficulty finding employment and leaving the rolls compared to White women. "As the welfare rolls continue to plunge, White recipients are leaving the system much faster than Black and Hispanic recipients, pushing the minority share of the caseload to the highest level on record" (DeParle, 1998, A1). Racial imbalances in welfare rolls characterize at least half of the states and welfare rolls continue to "darken," leaving the stereotypical images of women of color and of welfare programs largely intact....

Existing literature clearly links the period around childbirth with increased economic vulnerability, likelihood of being in poverty, and public assistance use....

FINDINGS

... Finally, given our particular interest in the leave-taking behavior of low-income mothers, we summarize the findings that are noteworthy for this group of mothers:

1. Low-income mothers take shorter periods of maternity leave than mothers in all other socioeconomic groups.

2. Low-income mothers are most dramatically affected by the experience and work histories of their parents and the work history of their own mothers is shown to be particularly important.
3. Low-income mothers with less than one year on the job (a likely circumstance for many low-income women "transitioning" off of public assistance programs) take the shortest maternity leave of all socioeconomic groups.
4. Low-income mothers with one to two years tenure with their current employers take longer maternity leaves when compared with low-income mothers who have been with their current employers for longer periods.
5. Flexible hour benefits have a more dramatic impact in terms of increasing the length of leave taken by low-income mothers than maternity-leave benefits.

DISCUSSION

Each year in the United States about four million families welcome the birth of a new child and make critical decisions to balance the care needs and financial responsibilities of their growing families. Our findings show that race, ethnicity, and class characteristics significantly affect those decisions; that a variety of additional factors at the individual, family, employment, and government policy levels evaluated separately and intersectionally also affect those decisions. When these variables are tested (intersectionally), as nested phenomenon, dramatically different outcomes for different groups of mothers are observed. Finally, we find that leave length is determined most powerfully by variables that operate at the family and employment levels....

... [S]triking differences were observed in leave length by race, ethnicity, and class. For example, higher income Black women with maternity leave benefits from their employers seem to make very different maternal employment decisions than similarly situated Hispanic women, and we argue that those decisions are directly related to the unique social location of each woman. Understanding maternity leave outcomes as a function of social location provides an alternative analytical discourse for researchers interested in these issues....

Finally the social location of low-income mothers with regard to maternity leave represents a particularly important consideration. Low-income mothers are generally assumed to have the strongest economic incentive to return to work quickly both because they do not earn enough to forego wages in the labor market for extended periods and also because of recent changes in public assistance programs that require employment as a condition for financial assistance. Additionally the weight of these work "incentives" are likely to be felt more deeply and experienced differently by low-income women of color, who are already disproportionately affected by reductions in public assistance programs. With little capacity to purchase their own work supports (such as the labor of third parties like nannies, babysitters, etc.) nor to replace lost wages when their labor is needed at home and away from work, the increased work demands for low-income mothers can conflict powerfully with their roles and responsibilities as family and community members.

In the midst of these already difficult transitions, low-income mothers have critical choices to make about how to respond to periods of heightened family-care needs (such as the birth of a new child, a short-term illness, etc.). Low-income mothers use multiple survival strategies to cope with these circumstances and our findings suggest that the strategies they use do not follow neat or consistent patterns across economic categories....

Our findings also have enormous implications for public policymakers and are especially germane given the level of recent policy attention aimed at shaping the fertility and employment decisions of low-income women. The implications ... are that: race, ethnicity, and class matter a great deal in terms of how women sort out their employment options and that social policies that do not address the centrality of family and employment variables may very well miss what seems to be key resources women use to make employment decisions. Most importantly, these findings give more credence to the notion that "one-size" policies do not "fit all." That is, policies that seek to help mothers maintain employment by embracing only one set of policy tools are likely to have the intended effect for some but are likely to engender a host of unintended and/or deleterious effects for others. Flexible leave benefits are a good example of how one policy tool used to help employees manage work and family conflicts results in very different outcomes for different groups of new mothers....

REFERENCES

Amott, T., & Matthaei, J. (1991). *Race, Gender & Work: A Multicultural Economic History of Women in the United States*. Boston: South End Press.

Baum, C. (2003). The Effects of Maternity Leave Legislation on Mothers' Labor Supply after Childbirth. *Southern Economic Journal*, *69*(4), 772–799.

DeParle, J. (1998, July 27). Shrinking Welfare Rolls Leave Record High Share of Minorities. *New York Times*, p. A1.

Fried, M. (1998). *Taking Time*. Philadelphia: Temple University Press.

Garrett, P., Wenk, D., & Lubeck, S. (1990). Working around Childbirth: Comparative and Empirical Perspectives on Parental-Leave Policy. *Child Welfare*, *69*(5), 401–413.

Gruber, J. (1994). The Incidence of Mandated Maternity Benefits. *American Economic Review*, *84*, 622–641.

Hattingadi, A. (2000). *Paid Family Leave: At What Cost?* Washington, DC: Employment Policy Foundation.

Hofferth, S. (1996). The Effects of Public and Private Policies on Working after Childbirth. *Work and Occupations*, *23*(4), 378–404.

Neuback, K., & Cazenave, N. (2001). *Welfare Racism: Playing the Race Card against America's Poor*. New York: Routledge Press.

Trzcinski, E., & Alpert, W. (2000). The Economics of Family and Medical Leave in Canada and in the United States. In W. Alpert & S. Woodbury (Eds.), *Employee*

Benefits and Labor Markets in Canada and the United States (pp. 129–180). Kalamazoo, MI: W. E. Upjohn Institute for Employment Research.

Vogel, L. (1993). *Mothers on the job: Maternity Policy in the U.S. Workplace*. New Brunswick, NJ: Rutgers University Press.

Waldfogel, J. (1997). The Effect of Children on Women's Wages. *American Sociological Review*, *62*, 209–217.

Waldfogel, J., Higuchi, Y., & Abe, M. (1999). Family Leave Policies and Women's Retention after Childbirth: Evidence from the United States, Britain, and Japan. *Journal of Population Economics*, *12*, 523–545.

38

Straight Is to Gay as Family Is to No Family

KATH WESTON

IS "STRAIGHT" TO "GAY" AS "FAMILY" IS TO "NO FAMILY"?

For years, and in an amazing variety of contexts, claiming a lesbian or gay identity has been portrayed as a rejection of "the family" and a departure from kinship. In media portrayals of AIDS, Simon Watney ... observes that "we are invited to imagine some absolute divide between the two domains of 'gay life' and 'the family,' as if gay men grew up, were educated, worked and lived our lives in total isolation from the rest of society." Two presuppositions lend a dubious credence to such imagery: the belief that gay men and lesbians do not have children or establish lasting relationships, and the belief that they invariably alienate adoptive and blood kin once their sexual identities become known. By presenting "the family" as a unitary object, these depictions also imply that everyone participates in identical sorts of kinship relations and subscribes to one universally agreed-upon definition of family.

Representations that exclude lesbians and gay men from "the family" invoke what Blanche Wiesen Cook ... has called "the assumption that gay people do not love and do not work," the reduction of lesbians and gay men to sexual identity, and sexual identity to sex alone. In the United States, sex apart from heterosexual marriage tends to introduce a wild card into social relations, signifying unbridled lust and the limits of individualism. If heterosexual intercourse can bring people into enduring association via the creation of kinship ties, lesbian and gay sexuality in these depictions isolates individuals from one another rather than weaving them into a social fabric. To assert that straight people "naturally" have access to family, while gay people are destined to move toward a future of solitude and loneliness, is not only to tie kinship closely to procreation, but also to treat gay men and lesbians as members of a nonprocreative species set apart from the rest of humanity....

It is but a short step from positioning lesbians and gay men somewhere beyond "the family"—unencumbered by relations of kinship, responsibility, or

SOURCE: Weston, Kath. 1991. *Families We Choose: Lesbians, Gays, Kinship.* New York: Columbia University Press, pp. 22–29. Reprinted by permission of Columbia University Press.

affection—to portraying them as a menace to family and society. A person or group must first be outside and other in order to invade, endanger, and threaten. My own impression from fieldwork corroborates Frances Fitzgerald's … observation that many heterosexuals believe not only that gay people have gained considerable political power, but also that the absolute number of lesbians and gay men (rather than their visibility) has increased in recent years. Inflammatory rhetoric that plays on fears about the "spread" of gay identity and of AIDS finds a disturbing parallel in the imagery used by fascists to describe syphilis at mid-century, when "the healthy" confronted "the degenerate" while the fate of civilization hung in the balance.…

A long sociological tradition in the United States of studying "the family" under siege or in various states of dissolution lent credibility to charges that this institution required protection from "the homosexual threat."…

… By shifting without signal between reproduction's meaning of physical procreation and its sense as the perpetuation of society as a whole, the characterization of lesbians and gay men as nonreproductive beings links their supposed attacks on "the family" to attacks on society in the broadest sense. Speaking of parents who had refused to accept her lesbian identity, a Jewish woman explained, "They feel like I'm finishing off Hitler's job." The plausibility of the contention that gay people pose a threat to "the family" (and, through the family, to ethnicity) depends upon a view of family grounded in heterosexual relations, combined with the conviction that gay men and lesbians are incapable of procreation, parenting, and establishing kinship ties.

Some lesbians and gay men … had embraced the popular equation of their sexual identities with the renunciation of access to kinship, particularly when first coming out. "My image of gay life was very lonely, very weird, no family," Rafael Ortiz recollected. "I assumed that my family was gone now—that's it." After Bob Korkowski began to call himself gay, he wrote a series of poems in which an orphan was the central character. Bob said the poetry expressed his fear of "having to give up my family because I was queer." When I spoke with Rona Bren after she had been home with the flu, she told me that whenever she was sick, she relived old fears. That day she had remembered her mother's grim prediction: "You'll be a lesbian and you'll be alone the rest of your life. Even a dog shouldn't be alone."

Looking backward and forward across the life cycle, people who equated their adoption of a lesbian or gay identity with a renunciation of family did so in the double-sided sense of fearing rejection by the families in which they had grown up, and not expecting to marry or have children as adults. Although few in numbers, there were still those who had considered "going straight" or getting married specifically in order to "have a family." Vic Kochifos thought he understood why:

> It's a whole lot easier being straight in the world than it is being gay.… You have built-in loved ones: wife, husband, kids, extended family. It just works easier. And when you want to do something that requires children, and you want to have a feeling of knowing that there's gonna be someone around who cares about you when you're 85 years old, there are thoughts that go through your head, sure. There must be. There's a way of doing it gay, but it's a whole lot harder, and it's less secure.

Bernie Margolis had been sexually involved with men since he was in his teens, but for years had been married to a woman with whom he had several children. At age 67 he regretted having grown to adulthood before the current discussion of gay families, with its focus on redefining kinship and constructing new sorts of parenting arrangements.

> I didn't want to give up the possibility of becoming a family person. Of having kids of my own to carry on whatever I built up.... My mother was always talking about [how] she's looking forward to the day when she would bring her children under the canopy to get married. It never occurred to her that I wouldn't be married. It probably never occurred to me either.

The very categories "good family person" and "good family man" had seemed to Bernie intrinsically opposed to a gay identity. In his fifties at the time I interviewed him, Stephen Richter attributed never having become a father to "not having the relationship with the woman." Because he had envisioned parenting and procreation only in the context of a heterosexual relationship, regarding the two as completely bound up with one another, Stephen had never considered children an option.

Older gay men and lesbians were not the only ones whose adult lives had been shaped by ideologies that banish gay people from the domain of kinship. Explaining why he felt uncomfortable participating in "family occasions," a young man who had no particular interest in raising a child commented, "When families get together, what do they talk about? Who's getting married, who's having children. And who's not, okay? Well, look who's not." Very few of the lesbians and gay men I met believed that claiming a gay identity automatically requires leaving kinship behind. In some cases people described this equation as an outmoded view that contrasted sharply with revised notions of what constitutes a family.

Well-meaning defenders of lesbian and gay identity sometimes assert that gays are not inherently "anti-family," in ways that perpetuate the association of heterosexual identity with exclusive access to kinship. Charles Silverstein ..., for instance, contends that lesbians and gay men may place more importance on maintaining family ties than heterosexuals do because gay people do not marry and raise children. Here the affirmation that gays and lesbians are capable of fostering enduring kinship ties ends up reinforcing the implication that they cannot establish "families of their own," presumably because the author regards kinship as unshakably rooted in heterosexual alliance and procreation. In contrast, discourse on gay families cuts across the politically loaded couplet of "pro-family" and "anti-family" that places gay men and lesbians in an inherently antagonistic relation to kinship solely on the basis of their nonprocreative sexualities. "Homosexuality is not what is breaking up the Black family," declared Barbara Smith ..., a black lesbian writer, activist, and speaker at the 1987 Gay and Lesbian March on Washington. "Homophobia is. My Black gay brothers and my Black lesbian sisters are members of Black families, both the ones we were born into and the ones we create."

At the height of gay liberation, activists had attempted to develop alternatives to "the family," whereas by the 1980s many lesbians and gay men were struggling to legitimate gay families as a form of kinship.... Gay or chosen families might incorporate friends, lovers, or children, in any combination. Organized through ideologies of love, choice, and creation, gay families have been defined through a contrast with what many gay men and lesbians ... called "straight," "biological," or "blood" family. If families we choose were the families lesbians and gay men created for themselves, straight family represented the families in which most had grown to adulthood.

What does it mean to say that these two categories of family have been defined through contrast? One thing it emphatically does *not* mean is that heterosexuals share a single coherent form of family (although some of the lesbians and gay men doing the defining believed this to be the case). I am not arguing here for the existence of some central, unified kinship system vis-à-vis which gay people have distinguished their own practice and understanding of family. In the United States, race, class, gender, ethnicity, regional origin, and context all inform differences in household organization, as well as differences in notions of family and what it means to call someone kin.

In any relational definition, the juxtaposition of two terms gives meaning to both. Just as light would not be meaningful without some notion of darkness, so gay or chosen families cannot be understood apart from the families lesbians and gay men call "biological," "blood," or "straight." Like others in their society, most gay people ... considered biology a matter of "natural fact." When they applied the terms "blood" and "biology" to kinship, however, they tended to depict families more consistently organized by procreation, more rigidly grounded in genealogy, and more uniform in their conceptualization than anthropologists know most families to be. For many lesbians and gay men, blood family represented not some naturally given unit that provided a base for all forms of kinship, but rather a procreative principle that organized only one possible *type* of kinship. In their descriptions they situated gay families at the opposite end of a spectrum of determination, subject to no constraints beyond a logic of "free" choice that ordered membership. To the extent that gay men and lesbians mapped "biology" and "choice" onto identities already opposed to one another (straight and gay, respectively), they polarized these two types of family along an axis of sexual identity.

The following chart recapitulates the ideological transformation generated as lesbians and gay men began to inscribe themselves within the domain of kinship.

What this chart presents is not some static substitution set, but a historically motivated succession. To move across or down the chart is to move through

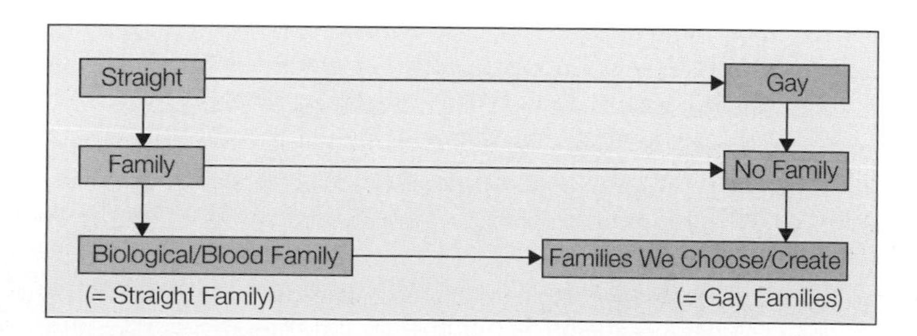

time. Following along from left to right, time appears as process, periodized with reference to the experience of coming out. In the first opposition, coming out defines the transition from a straight to a gay identity. For the person who maintains an exclusively biogenetic notion of kinship, coming out can mark the renunciation of kinship, the shift from "family" to "no family" portrayed in the second opposition. In the third line, individuals who accepted the possibility of gay families after coming out could experience themselves making a transition from the biological or blood families in which they had grown up to the establishment of their own chosen families.

Moving from top to bottom, the chart depicts the historical time that inaugurated contemporary discourse on gay kinship. "Straight" changes from a category with an exclusive claim on kinship to an identity allied with a specific kind of family symbolized by biology or blood. Lesbians and gay men, originally relegated to the status of people without family, later lay claim to a distinctive type of family characterized as families we choose or create. While dominant cultural representations have asserted that straight is to gay as family is to no family (lines 1 and 2), at a certain point in history gay people began to contend that straight is to gay as blood family is to chosen families (lines 1 and 3).

What provided the impetus for this ideological shift? Transformations in the relation of lesbians and gay men to kinship are inseparable from sociohistorical developments: changes in the context for disclosing a lesbian or gay identity to others, attempts to build urban gay "community," cultural inferences about relationships between "same-gender" partners, and the lesbian baby boom associated with alternative (artificial) insemination.... If ... kinship is something people use to act as well as to think, then its transformations should have unfolded not only on the "big screen" of history, but also on the more modest stage of day-to-day life, where individuals have actively engaged novel ideological distinctions and contested representations that would exclude them from kinship.

39

Navigating Interracial Borders

Black-White Couples and Their Social Worlds

ERICA CHITO CHILDS

The 1967 Academy Award-winning movie *Guess Who's Coming to Dinner* concluded with a warning from a white father to his daughter and her "Negro" fiancé. That same year, the Supreme Court overturned any laws against interracial marriage as unconstitutional. Yet how does the contemporary U.S. racial landscape compare? In this ever-changing world of race and color, where do black-white couples fit, and has this unimaginable opposition disappeared?

While significant changes have occurred in the realm of race relations largely from the civil rights struggle of the 1960s, U.S. society still has racial borders. Most citizens live, work, and socialize with others of the same race—as if living within borders, so to speak—even though there are no longer legal barriers such as separate facilities or laws against intermarriage. Yet if these largely separate racial worlds exist, what social world(s) do black-white couples live in and how do they navigate these racial borders? Even more important, how do white communities and black communities view and respond to black-white couples? In other words, do they navigate the racial borders by enforcing, ignoring, or actively trying to dismantle them? My goal is to explore these issues to better understand the contemporary beliefs and practices surrounding black-white couples.... My data come from varied sources, including Web sites, black-white couples, Hollywood films, white communities, and black communities....

My own story also brought me to this research. The social world of black-white couples is the world I navigate. From my own experiences, I have seen the ways most whites respond to an interracial relationship. Growing up white, second-generation Portuguese in a predominantly white Rhode Island suburb, race never was an issue, or at least not one I heard about. After moving to Los Angeles during high school and beginning college, I entered into a relationship with an African American man (who I eventually married), never imagining what it would bring. My family did not disown me or hurl racial slurs. Still, in many ways I learned what it meant to be an "interracial couple" and how this was not what my family, community, or countless unknown individuals had scripted for me. Not many whites ever said outright that they were opposed to the relationship, yet their words and actions signaled otherwise.

SOURCE: Childs, Erica Chito. 2005. *Navigating Interracial Borders: Black-White Couples and Their Social World.* New Brunswick, NJ: Rutgers University Press.

One of the most telling examples occurred a few years into our relationship. An issue arose when my oldest sister's daughter wanted to attend her prom with an African American schoolmate she was dating. My sister and her husband refused to let him in the house the night of the prom or any other time because, they said, he was "not right" for her. It was clear to everyone, however, that skin color was the problem. To this day, my niece will tell you that her parents would never have accepted her with a black man. Yet my sister and her family never expressed any opposition to my relationship and even seemed supportive, in terms of inviting us over to their house, giving wedding and holiday gifts, and so forth. Although my sister never openly objected to my relationship, she drew the line with her daughter—quite literally enforcing a racial boundary to protect her daughter and family from blackness. For me, this personal story and the countless stories of other interracial couples point to the necessity of examining societal attitudes, beliefs, images, and practices regarding race and, more specifically, black-white relations. Interracial couples—because of their location on the line between white and black—often witness or bring forth racialized responses from both whites and blacks. As with my sister, opposition may exist yet is not visible until a close family member or friend becomes involved or wants to become involved interracially....

INTERRACIAL RELATIONSHIPS AS A MINER'S CANARY

It is these community and societal responses, as well as the images and beliefs produced and reproduced about these unions that provide the framework within which to understand the issue of interracial couplings. Underlying these responses and images is a racial ideology, or, in other words, a dominant discourse, that posits interracial couples and relationships as deviant. Still, the significance of these discourses and what exactly they reveal about race in society can be hard to see. For some of us the effects of race are all too clear, while for others race—and the accompanying advantages and disadvantages—remain invisible. As a white woman, it was only through my relationship, and raising my two children, that I came to see how race permeates everything in society. Being white yet now part of a multiracial family, I experienced, heard, and even thought things much differently than before, primarily because whites and blacks responded to me differently than before. I think of the metaphor of the "miner's canary"—the canaries miners use to alert them to a poisonous atmosphere. In *The Miner's Canary: Enlisting Race, Resisting Power, Transforming Power,* Lani Guinier and Gerald Torres argue that the experiences of racial minorities, like the miner's canary, can expose the underlying problems in society that ultimately affect everyone, not just minorities. In many ways, the experiences of black-white couples are a miner's canary, revealing problems of race that otherwise can remain hidden, especially to whites. The issues surrounding interracial couples—racialized/sexualized stereotypes, perceptions of difference, familial opposition, lack of community acceptance—should not be looked at as individual problems, but rather as a reflection of the larger racial issues that divide the races. Since interracial couples exist on the color-line in society—a "borderland" between white and black—their experiences and the ways communities respond to these relationships can be used as a lens through which we can understand contemporary race relations....

LIFE ON THE BORDER: NARRATIVES OF BLACK-WHITE COUPLES

From 1999 to 2001, I interviewed fifteen black-white heterosexual couples who were referred to me through personal and professional contacts, and some of whom I encountered randomly in public. They ranged in age from twenty to sixty-nine and all were in committed relationships of two to twenty-five years. Nine were married. The couples' education levels varied. All respondents had finished high school or its equivalent; twenty-one respondents had attended some college and/or had received a bachelor's degree; and four respondents had advanced degrees. The socioeconomic status of the couples ranged from working class to upper middle class. The respondents included a college student, waitress, manager, factory worker, university professor, social worker, salesperson, and postal worker. All couples lived in the northeastern United States, from Maine to Pennsylvania, yet many of the couples had traveled extensively and had lived in other parts of the country, including California, Florida, and the South.

I interviewed the couples together, since I was interested in their experiences, accounts, narratives, and the ways they construct their lives and create their "selves" and their identities as "interracial couples." The interviews lasted for two to three hours, and I ended up with more than forty hours of interview data.... These accounts are seen not only as "descriptions, opinions, images, or attitudes about race relations but also as 'systems of knowledge' and 'systems of values' in their own right, used for the discovery and organization of reality."

THE SEPARATE WORLDS OF WHITES AND BLACKS

To explore the larger cultural and sociopolitical meanings that black-white couplings have for both the white and black communities in which they occur, a significant portion of this work is based on original qualitative research in white communities and black communities about their ideas, beliefs, and views on interracial sexuality and marriage. Community research was conducted to further explore the responses to interracial couples that are found in social groups and communities—family, friends, neighbors, religious groups, schools, etc. The ways that these couples provide the occasion for groups to express and play out their ideas and prejudices about race and sex are integral to understanding the social construction of interracial couples....

A black person and a white person coming together has been given many names—miscegenation, amalgamation, race mixing, and jungle fever—conjuring up multiple images of sex, race, and taboo. Black-white relationships and marriages have long been viewed as a sign of improving race relations and assimilation, yet these unions have also been met with opposition from both white and black communities. Overall, there is an inherent assumption that interracial couples are somehow different from same-race couples. Within the United States, the responses to black-white couplings have ranged from disgust to curiosity to endorsement, with the couples being portrayed as many things—among them, deviant, unnatural, pathological, exotic, but always sexual. Even the way that couples are labeled or defined as "interracial" tells us something about societal expectations. We name what is different. For example, a male couple is more likely to be called a "gay couple" than a gender-mixed couple is to be called a "heterosexual couple."

Encompassed by the history of race relations and existing interracial images, how do black-white couples view themselves, their relationships, and the responses of their families and communities? And how do they interpret these familial and community responses? Black-white couples, like all of us, make meaning out of their experiences in the available interpretive frameworks and often inescapable rules of race relations in this country....

Black-white couples come together across the boundaries of race and perceived racial difference seemingly against the opposition of their communities. This is not to say, however, that the couples are free from racialized thinking, whether it be in their use of color-blind discourse or their own racial preferences, such as to date only interracially or to live in all-white neighborhoods. Nonetheless, these couples create multiracial families, not only creating multiracial families of their own but also changing the racial dynamic of the families from which they come. What significance does this have for the institution of family, and how does this play out for the white and black families to whom it occurs?

It might be expected that the family is the source of the greatest hostility toward interracial relationships. It is in families that the meanings and attachments to racial categories are constructed and learned; one's family is often "the most critical site for the generation and reproduction of racial formations." This includes who is and is not an acceptable marriage partner. In white and black families, certain discourses are used when discussing black-white relationships that reproduce the image of these unions as different, deviant, even dangerous. Interracial relationships and marriage often bring forth certain racialized attitudes and beliefs about family and identity that otherwise may have remained hidden. The ways that white and black families understand and respond to black-white interracial couples and the racialized discourses they use are inextricably tied to ideas of family, community, and identity. White and black families' (and communities') interpretations and responses to interracial couples are part of these available discourses on race and race relations in our society. Many times, black-white couples provide the occasion for families to express and play out their ideas and prejudices about race and sex, which is integral to understanding the social construction of "interracial couples" within America today.

ALL IN THE FAMILY: WHITE FAMILIES

Among whites, the issue of interracial marriage is often a controversial topic, and even more so when they are asked to discuss their own views of their family's views. The white community respondents in this study were hesitant to discuss their personal views on family members becoming involved interracially. One strategy used by the respondents was to discuss other families they knew rather than their own views. For example, during the white focus group interviews, the first responses came from two individuals who had some experience or knowledge of interracial couples or families. Sara discussed a friend who adopted two "very dark" black children and how white people would stare at her and the children in public. Anne mentioned her niece who married a black man and said "the family is definitely against it," which causes them problems.

In group interviews with white college students, a number of the students also used stories about other interracial families they knew to explain why their family would prefer they marry someone of the same race. One college student said her family would have a problem with her marrying interracially, explaining that their opinion is based on their experiences with an interracial family in their neighborhood. She had babysat for this family and, according to her, they had "social issues because the dad was *real dark* and the mom was white, and the kids just had major issues." Her choice of words reveals the importance of color and the use of a discursive strategy such as referring to the children as a problem and not the relationship itself....

White "Concern" and Preference

For the white college students interviewed, the role of family is key and certainly influences their decisions not to date interracially. The majority of white students expressed verbally or by raising their hand that their parents, white parents in general, would have a difficult time with an interracial partner for a number of reasons. Yet their parents' opposition was described in nonracial terms, much like their own views. For example, one white male student said, "All parents find something wrong [with the person their child chooses] if they're not exactly as they imagined. My parents would be *surprised* because [an interracial relationship] is not what they are expecting, so it would be difficult in that sense." Many students simply stated something to the effect of "my parents aren't prejudiced; they just wouldn't want me to marry a black guy."

Other students described their families' views against interracial dating as based on the meanings of family and marriage in general. For example, one male student stated that his parents would have a problem with an interracial relationship, because "they brought me up to date within [my] race. They're not like racist, but [they say] just keep with your own culture, be proud of who you are and carry that on to your kids." Some students cited the difference between dating and marriage. As one female college student said, "It becomes more of an issue when you get to be juniors or seniors, because you start thinking long-term, about what your parents will say, and dating a black person just isn't an option." Another student described dating interracially in similar terms: "Dating's not an issue, they always encouraged [me] to interact with all people. For the future and who you're gonna spend the rest of your life with, there's a difference between being with someone of the same race and someone of a different race, [interracial marriage] is not like it's wrong but ... just that it would be too difficult."

In one of the discussions of family with the white college students, an interesting incident occurred. One of the male students, who was vocal throughout the focus group, sat listening to the other students and looking around the room. Finally he interrupted the discussion: "Are you kidding me? ... parents would shit, they'd have a freaking heart attack. [*He dramatically grabs the front of his shirt and imitates a growling father's voice*]. "Uhh, son, how could you do this to us, the family?" Everyone laughed at his performance, but this student's use of humor to depict his father (or a white father in general) having "a heart attack" is an interesting discursive strategy. Although he presented it in a comical way, his statement does not seem unrealistic to the group. Often, jocular speech is used to convey a serious message in order to avoid being labeled racist or prejudiced.

When asked why their families would respond in these ways to an interracial relationship, the group largely cited the "opposition of the larger society" as the reason why they and/or their family personally would prefer that their family not become involved interracially. Among the white college students, parents were described as "concerned about how difficult it would be."...

IT'S A FAMILY AFFAIR: BLACK FAMILIES

Black families, like white families, can operate as a deterrent to interracial relationships. A family member becoming involved with a white individual is seen as problematic on a number of levels, yet the black families often raised different issues than the white families, such as the importance of "marrying black" and the negative meanings attached to becoming involved interracially. These issues figured prominently among the black partners, the black community respondents, and even black popular culture.

Among the black college students and community respondents, a main issue was the emphasis on marrying within their race, explicitly identifying race as the issue, unlike the white communities. Most students discussed how their parents would have a problem. One black female student stated, "My family would outright disown me" (which received a number of affirmations from the group). Another college student commented on the beliefs her family instilled in her: "My family raised me to be very proud of who I am, a black woman, and they instilled in me the belief that I would never want to be with anyone but a strong black man." Other students described incidents or comments they had heard that let them know they were expected to marry black. Similarly, Leslie, a black community respondent, stated she was raised by her parents to date anyone, but they were "adamant about me marrying black." She added that they told her not to "even think about marrying interracially." Allen also stated that his parents told him that "high school dating is fine but not marriage, ohhh nooo!" Couples like Gwen and Bill, Chris and Victoria, and others also recounted how the black partner's family had more difficulty accepting the relationship when they realized that the couple was getting married as opposed to just dating.

A significant piece of black familial opposition involves the perceived racism of whites in the larger society. Black families were described as having a hard time accepting a family member getting involved with a white person because of lingering racism and a distrust in whites in general. For example, one black college student stated that her family would have a problem if she brought home a white man, "because they would always be wondering what his family was saying, you know, do they talk about me behind my back?" Other students' responses echoed these views, such as one student who remarked, "My mom would have a problem with it. She just doesn't trust white people."

Also, black community members such as Alice and Jean argued that their opposition to a family member getting involved with a white person was based on the belief that the white individual (or their white family/neighborhood) would mistreat their family member. All but two of the black community members expressed the concern that since white society is racist there is no reason to become involved with a white person....

The black families related to this study objected to having a white person in their family and intimate social circles because they viewed whites as the "enemy" and their presence as a sign that the black partner is not committed to his or her community or family. Not surprisingly, black individuals opposed interracial marriage much more than interracial dating, primarily because marriage represents a legitimation of the union and formally brings the white partner into the family....

Despite such opposition, the black college students argued that black families are still more accepting than white families....

While white families discourage their family members from engaging in interracial relationships to maintain white privilege, black families discourage these unions to maintain the strength and solidarity of black communities. Black families view interracial relationships as a loss in many ways—the loss of individuals to white society, the weakening of families and communities, and the devaluing of blackness....

FAMILY MATTERS

... The familial responses discussed throughout the research clearly demonstrate how images of oneself and one's family [are ...] linked to the concept of race and otherness, especially within the construction of families, both white and black.

White families often objected to the idea of a member dating interracially, not because they met the black individual and were confronted with overwhelming "racial" differences, but because they were merely responding based on their ideas and beliefs about interracial relationships and blacks in general....

Being in an interracial relationship often brings forth problems within families, since white and black families overwhelmingly want to remain monoracial....

This ... highlights how difficult it is to negotiate issues of race and family. Discussing race and, more important, views on interracial relationships in nonracial terms often makes it even more difficult for individuals to challenge and confront their family's views. By using phrases such as "It's not my personal preference" or "I just worry about the problems you will face in society," families and individuals are able to oppose interracial relationships without appearing prejudiced or racist. A family member's opposition, however, is also tied to the issue of identity—the identity of the family and the individuals involved. Families express concern over the identities of the biracial children who will be produced, and in many ways there is a tendency for white families to worry that the children will be "too black," and the black families worry about the children being "more white." The white individuals who enter into a relationship with a black person are seen as "less white" and as tainting the white family, while the black individuals who get involved are seen as "not black enough" and as leaving their blackness behind. All of these fears and beliefs demonstrate the centrality of race to the constructions of families and identities and, more important, the socially constructed nature of race, if one's relationship can change one's "race" in this society still divided by racial boundaries. Ultimately, black-white couples and biracial children are forced to exist somewhere in between, with or without their families.

40

Affirming Identity in an Era of School Desegregation

BEVERLY TATUM

"Identities are the stories we tell ourselves and the world about who we are, and our attempt to act in accordance with these stories."[1] I love this quote, because it captures so vividly the meaning of identity. Yet before we can tell the stories ourselves, they are told to us. Our sense of identity—of self-definition—is very much shaped in childhood by what is reflected back to us by those around us. If you were asked to describe yourself using a set of adjectives, and you replied, "I am tall," "I am smart," "I am attractive," "I am outgoing," or "I am shy," whatever those descriptors might be, one might ask, "Why do you think so?" And the answer to that question might easily be, "Because that is what people have said about me. That's the feedback that I have received." Identity is shaped by the social context in which we learn about ourselves over time. Group identities—gender, race, social class, to name a few—are part of that developmental process.

When we think about identity as it is shaped in schools, one of the questions we must ask is, How do students see themselves reflected in that environment? What stories are being told about who they are? What messages are being transmitted to them in their daily interactions in classrooms and in the school hallways, and by whom? The answer today is different than it was for my parents' generation. During the school segregation of the pre-*Brown* era, Black students typically attended schools staffed by people who looked like them—educators who shared their racial and ethnic background and knew firsthand the identity stories that were being told at home and in the neighborhood. Even with inadequate school resources in impoverished communities, the shared efforts of the teachers, administrators, and families created stories of success.

One consequence of the desegregation process in the South was the dismissal of thousands of Black teachers. When predominantly Black schools were closed and the busing of Black children began in southern school districts, Black teachers and administrators were displaced, replaced by White teachers and administrators. Active discrimination on the part of White school officials kept Black teachers out of racially mixed classrooms, particularly in the South. Even very experienced teachers who had earned advanced degrees in education at such prestigious northern institutions as Teachers College at Columbia University, the

SOURCE: Tatum, Beverly Daniel. 2007. *Can We Talk about Race? And Other Conversations in an Era of School Resegregation*. Boston: Beacon Press.

University of Wisconsin, and other leading education programs in the North (which allowed Black students to enroll when southern universities did not) found themselves demoted or unemployed.

Their displacement represented the rapid loss of role models, models of academic achievement, for young Black students. As doors were closing on Black teachers in the 1960s and 1970s, young African American college students interested in teaching were surely discouraged by what appeared to be declining employment opportunities. Meanwhile, doors were beginning to open in business, law, medicine, and other professions during the affirmative action years of the civil rights era. Not surprisingly, Black enrollment in teacher-education programs declined as enrollments in business administration increased. The ranks of Black educators still remain well below the pre-*Brown* levels.

Indeed, of the more than 3 million teachers in the United States, only 15.6 percent are teachers of color, 7.5 percent African American, specifically. Most students of color today are being taught by a teaching force that is predominantly White and female, particularly at the elementary school level. Nowhere is the current cultural mismatch between students and teachers more visible than in urban school districts, where White women make up 65 to 76 percent (depending on grade level) of the teaching population and students of color represent 76 percent of the urban student population.

Can White teachers—male or female—affirm the identities of the students of color in their classrooms? An ahistorical and idealistic response to this question might be, yes, of course. But in his essay "White Women's Work: On the Front Lines in Urban Education," Stephen Hancock reminds us that, "instead of providing students, schools and communities with better learning environments, *Brown* created (and continues to create) environments where African American and other minority students and White women teachers share dysfunctional relationships built on fear, ignorance, mistrust and resentment."[2] His description might seem harsh to some, but we cannot wish away the history of hostility that greeted Black students at school in the era of school desegregation, hostility that represented an assault on one's personhood rather than an affirmation of it. This generation of students and teachers may seem far away from that past, but its legacy lingers in the form of misinformation and stereotypes to which we are continuously exposed. If, for example, your knowledge of African American or Latino communities was based only on watching the real-life courtroom dramas so common on television today—where the frequency of people of color as plaintiffs and defendants is high—or perhaps based on a steady diet of popular music videos, what images would you hold in your mind? We carry a lifetime of these and other images with us as we interact across racial lines. How do those images shape the stories we tell students about who they are and who they will be?

Can any teacher transcend our shared history to affirm rather than assault student identities? Yes, but not without considerable effort and intention. Teachers of all backgrounds must be willing to engage in significant self-reflection about their own racial and cultural identities … to understand the assaulting stories they tell without conscious awareness. They also need to be willing to learn deeply about the lives of their students in their full cultural, socioeconomic, and sociopolitical contexts in order to affirm their identities authentically—with identity stories of hope and empowerment.

In her book *The Dreamkeepers: Successful Teachers of African American Children,* Gloria Ladson-Billings documents her classroom observations of both Black and White teachers who told such stories—teachers who worked effectively with their urban African American students, communicating high expectations and inspiring their students' best efforts. While the teachers differed in style, what they shared in common was a clear and demonstrable respect for the students and their families, and knowledge of the community from which the child came. In return they held the trust of the children and their parents. Such community knowledge takes time for an outsider to acquire, and trusting relationships in a school community take time to build. One critical challenge that urban school districts face is that the teacher turnover rate in racially isolated schools with concentrated poverty is high, limiting the opportunity to gain the local knowledge needed to truly understand and then affirm the identities important to the students.

We must also consider that it is not just the teachers that changed in the post-*Brown* era. The curriculum in 1954, particularly in segregated Black schools, often included some cultural dimension specific to the African American experience. Ask somebody who went to school in 1954 to recite the lines of a poem by Langston Hughes or to sing a verse of "Lift Every Voice and Sing," the James Weldon Johnson song once referred to as the "Negro National Anthem," and it is very likely that you will get a positive response....

...If we think about our school environments as an illustrated book in which students look to see themselves, we have to ask, what story is being told, and who is included in the illustrations? As the environment becomes increasingly segregated, what pictures are they seeing? For young people of color in largely segregated schools, are they seeing themselves in the story, and how? They may be seeing themselves among their classmates, but they may not be seeing themselves in the curriculum in meaningful and substantive ways. In all likelihood, they are not seeing themselves among the teachers and they are not seeing themselves in the administration.

What does that mean for their own view about their possibilities, their future? Is there a relationship between invisibility in the curriculum and the underachievement of Black and Latino students? Certainly we know that motivation to learn is related to one's sense of connection to both the content and the teacher. We know that "how learners feel about the setting they are in, the respect they receive from the people around them, and their ability to trust their own thinking and experience powerfully influence their concentration, their imagination, their effort, and their willingness to continue."[3]

This point is clearly illustrated in Herbert Kohl's essay, "I Won't Learn from You."[4] He describes observing a history class being taught in a public junior high school in San Antonio, a school that served low-income Latino students but had very few Latino teachers and no Latino administrators. As the White male teacher began the day's lesson on "the first people to settle Texas" by asking students to read from the history textbook, the students demonstrated their disengagement by slumping in their seats, rolling their eyes, grimacing, and refusing to volunteer. The teacher began to read aloud the history text's account of the first settlers of Texas—pioneers from New England and the South—when one student interrupted. Knowing full well that Mexicans (his ancestors) lived in what is now known as Texas long before any New Englanders arrived, the student blurted, "What are we, animals or something?" The teacher, ignoring his student's point completely, replied, "What does that have to do with the text?" In apparent frustration, the teacher left the room, leaving his visitor, Herbert Kohl, in charge of the class. Kohl reread the passage from the text and asked the students whether they believed what they had just heard. His question captured their attention, and he continued, saying, "This is lies, nonsense. In fact, I think the textbook is racist and an insult to everyone in this room." Kohl's response to the text opened the door for an important dialogue. He writes:

> The class launched into a serious and sophisticated discussion of the way in which racism manifests itself in their everyday lives at school. And they described the stance they took in order to resist that racism and yet not be thrown out of school. It amounted to nothing less than full-blown and cooperative not-learning. They accepted the failing grades it produced in exchange for the passive defense of their personal and cultural integrity. This was a class of school failures, and perhaps, I believed then and still believe, the repository for the positive leadership and intelligence of their generation.[5]

Kohl captures the essence of their resistance in this conclusion: "To agree to learn from a stranger who does not respect your integrity"—or as I would say, your identity—"causes a major loss of self. The only alternative is to not-learn and reject their world." As the noted theorist Jean Baker Miller once said, we all want to feel "seen, heard, and understood."[6] At its core, that is what affirming identity means. It is not just about what pictures are hanging on the wall, or what content is included in the curriculum, though these things are important. It is about recognizing students' lives—and helping them make connections to them. In Kohl's example, the State of Texas or the local school district may have required that the teacher use that particular history text, but the conversation was not scripted. It was Kohl's willingness to acknowledge the contradiction between the students' lives and the text that affirmed them and engaged them.

Affirming identity is not just about being nice—it is about being knowledgeable about who our students are, and reflecting a story that resonates with their best hope for themselves....

Affirming identity is about asking who they are, and where they want to go, and conveying a fundamental belief that they can get there—through the development of their intellect and their critical capacity to think. Any teacher—White or of color—willing to work at affirming identity will have engaged students.

However, the task of creating identity stories is not that of the school alone. Of course the messages we receive at home from family and friends from the time of our birth are powerful parts of our narrative as well—for better or worse. But as educators we must acknowledge the impact of the many hours spent in school and the influence even one teacher can have on the story a student tells him or herself—also for better or worse. We cannot control the stories others are telling—but we must take responsibility for the identity stories we tell....

What about affirming the identities of White children? White children in a largely White school environment typically see themselves in the curriculum. They learn about White authors, scientists, inventors, artists, and explorers—most often male, but not exclusively so. The opportunity to envision oneself in similar roles is regularly offered to White children through the example of White adults. While certainly there is ethnic variation, socioeconomic variation, and religious variation that mediates the ease with which a child might identify with such examples, it is still likely that there will be places where White students see themselves reflected, at least in the faces of their teachers and their administrators—the adults they arguably observe most closely doing their jobs in the larger world for the most extended period of time—a privilege that students of color cannot take for granted. While the individual narratives they are constructing in childhood will vary with family circumstance and personal characteristics, as they do for all of us, the group story of what it means to be White is a story of achievement, success, and of being in charge.

But how do White children see *others* reflected? Are they learning about people of color as equals or does the curriculum continue to reinforce old notions of assumed White superiority as the result of unchallenged stereotypes and unrecognized omissions of information about the societal contributions of people of color? Are they receiving information that will help them navigate a global society, information that will help them engage with people who are different from themselves in that environment? In the absence of such information, the story is incomplete and they are not well served by their education....

In a race-conscious society, the development of a positive sense of racial or ethnic identity, not based on assumed superiority or inferiority, is an important task for everyone. It is an important task for people of color. It is an important task for White people. Sometimes when people hear the phrase "White identity," what comes to mind are connotations of White supremacy, as embodied by the Ku Klux Klan, perhaps. But of course the notion of White identity relevant here is not one based on a sense of assumed superiority. What is necessary, rather, is recognition of the meaning of Whiteness in our society. As many scholars and writers have explored in recent years, Whiteness is not an identity without meaning. Some White people who haven't thought much about these issues will say, "Well, you know, I'm an individual. I want you to see me as an individual." And of course, each of us is an individual, and we want our individuality recognized. But we each also have a social identity, with a social history, a social meaning. Recognition of the meaning of Whiteness in our society is recognition of the meaning of *privilege* in the context of a society that advantages being White.

Now, urging White teachers and students to recognize the meaning of their Whiteness is *not* equivalent to asking them to feel guilty about their privilege, although sometimes guilt *is* part of that exploration of identity for many people. Feeling badly about one's own Whiteness is a stage that many people experience. It's certainly not the goal of the educational process nor should it be the end point.

Ideally, we should each be able to embrace all of who we are, and to recognize that in a society where race is still meaningful and where Whiteness is still a source of power and privilege, that it is possible to resist being in the role of dominator, or "oppressor," and to become genuinely antiracist in one's White identity, and to actively work against systems of injustice and unearned privilege. It is possible to claim both one's Whiteness as a part of who one is and of one's daily experience, and the identity of being what I like to call a "White ally": namely, a White person who understands that it is possible to use one's privilege to create more equitable systems; that there are White people throughout history who have done exactly that; and that one can align oneself with that history. That is the identity story that we have to reflect to White-children, and help them see themselves in it in order to continue the racial progress in our society.

When White adults have not thought about their own racial identity, it is difficult for them to respond to the identity-development needs of either White children or children of color. Consequently, it becomes very important to engage teachers around these issues in pre-service preparation and in ongoing professional development. The intergenerational transmission of incomplete and distorted identity stories is a problem that we must address at the level of teacher preparation—and for the thousands of teachers already in the classroom, as part of ongoing professional development.... The need is particularly pressing for White teachers, who represent the vast majority of the public school educators in the United States, but it applies to all teachers. We cannot assume that teachers of color are confident in their abilities to talk about these issues as well. None of us can teach what we haven't learned ourselves. The good news is that those who have engaged in a process of examining their own racial or ethnic identity, and who feel affirmed in it, are more likely to be respectful of the self-definition that others claim, and are much more effective working in multiracial settings. It is these members of our society who can help us move beyond the regressive state of our current educational system, and move us forward into the twenty-first century with hope.

NOTES

1. Theresa Perry, "Freedom for Literacy," in *Young, Gifted, and Black: Promoting High Achievement among African-American students*, ed. Theresa Perry, Claude Steele, and Asa Hilliard III (Boston: Beacon Press, 2003), 50.
2. Stephen D. Hancock, "White Women's Work: On the Front Lines in Urban Education," in *White Teachers, Diverse Classrooms,* ed. Julie Landsman and Chance W. Lewis (Sterling, Va.: Stylus, 2006), 55.
3. Raymond J. Wlodkowski and Margery B. Ginsberg, *Diversity & Motivation: Culturally Responsive Teaching* (San Francisco: Jossey-Bass, 1995), 2.
4. Herbert Kohl, "I Won't Learn from You: Confronting Student Resistance," in *Rethinking Our Classrooms: Teaching for Equity and Justice* (Milwaukee: Rethinking Schools, 1994), 134–35.
5. Ibid., 135.
6. Jean Baker Miller, "Connections, Disconnections, and Violations," Work in Progress no. 33 (Wellesley, Mass.: Stone Center Working Paper Series, 1988).

41

From the Achievement Gap to the Education Debt

Understanding Achievement in U.S. Schools

GLORIA LADSON-BILLINGS

One of the most common phrases in today's education literature is "the . . . achievement gap." The term produces more than 11 million citations on Google. "Achievement gap," much like certain popular culture music stars, has become a crossover hit. It has made its way into common parlance and everyday usage. The term is invoked by people on both ends of the political spectrum, and few argue over its meaning or its import. According to the National Governors' Association, the achievement gap is "a matter of race and class. Across the U.S., a gap in academic achievement persists between minority and disadvantaged students and their white counterparts." It further states: "This is one of the most pressing education-policy challenges that states currently face" (2005). The story of the achievement gap is a familiar one. The numbers speak for themselves. In the 2005 National Assessment of Educational Progress results, the gap between Black and Latina/o fourth graders and their White counterparts in reading scaled scores was more than 26 points. In fourth-grade mathematics the gap was more than 20 points (Education Commission of the States, 2005). In eighth-grade reading, the gap was more than 23 points, and in eighth-grade mathematics the gap was more than 26 points. We can also see that these gaps persist over time (Education Commission of the States).

Even when we compare African Americans and Latina/os with incomes comparable to those of Whites, there is still an achievement gap as measured by standardized testing (National Center for Education Statistics, 2001). While I have focused primarily on showing this gap by means of standardized test scores, it also exists when we compare dropout rates and relative numbers of students who take advanced placement examinations; enroll in honors, advanced placement, and "gifted" classes; and are admitted to colleges and graduate and professional programs.

Scholars have offered a variety of explanations for the existence of the gap. In the 1960s, scholars identified cultural deficit theories to suggest that children of color were victims of pathological lifestyles that hindered their ability to

SOURCE: Ladson-Billings, Gloria. 2006. "From the Achievement Gap to the Education Debt: Understanding Achievement in U.S. Schools," *Educational Researcher* 35: 3–12.

benefit from schooling (Hess & Shipman, 1965; Bereiter & Engleman, 1966; Deutsch, 1963). The 1966 Coleman Report, *Equality of Educational Opportunity* (Coleman et al.), touted the importance of placing students in racially integrated classrooms. Some scholars took that report to further endorse the cultural deficit theories and to suggest that there was not much that could be done by schools to improve the achievement of African American children. But Coleman et al. were subtler than that. They argued that, more than material resources alone, a combination of factors was heavily correlated with academic achievement. Their work indicated that the composition of a school (who attends it), the students' sense of control of the environments and their futures, the teachers' verbal skills, and their students' family background all contribute to student achievement. Unfortunately, it was the last factor—family background—that became the primary point of interest for many school and social policies.

Social psychologist Claude Steele (1999) argues that a "stereotype threat" contributes to the gap. Sociolinguists such as Kathryn Au (1980), Lisa Delpit (1995), Michèle Foster (1996), and Shirley Brice Heath (1983), and education researchers such as Jacqueline Jordan Irvine (2003) and Carol Lee (2004), have focused on the culture mismatch that contributes to the gap. Multicultural education researchers such as James Banks (2004), Geneva Gay (2004), and Carl Grant (2003), and curriculum theorists such as Michael Apple (1990), Catherine Cornbleth (and Dexter Waugh; 1995), and Thomas Popkewitz (1998) have focused on the nature of the curriculum and the school as sources of the gap. And teacher educators such as Christine Sleeter (2001), Marilyn Cochran-Smith (2004), Kenneth Zeichner (2002), and I (1994) have focused on the pedagogical practices of teachers as contributing to either the exacerbation or the narrowing of the gap.

But I want to use this opportunity to call into question the wisdom of focusing on the achievement gap as a way of explaining and understanding the persistent inequality that exists (and has always existed) in our nation's schools. I want to argue that this all-out focus on the "Achievement Gap" moves us toward short-term solutions that are unlikely to address the long-term underlying problem.

DOWN THE RABBIT-HOLE

Let me begin the next section of this discussion with a strange transition from a familiar piece of children's literature:

> *Alice started to her feet, for it flashed across her mind that she had never before seen a rabbit with either a waistcoat-pocket, or a watch to take out of it, and burning with curiosity, she ran across the field after it, and fortunately was just in time to see it pop down a large rabbit-hole under the hedge. In another moment down went Alice after it, never once considering how in the world she was to get out again.*
>
> Lewis Carroll, *Alice's Adventures in Wonderland*

The relevance of this passage is that I, like Alice, saw a rabbit with a watch and waistcoat-pocket when I came across a book by economist Robert Margo entitled *Race and Schooling in the American South, 1880–1950* (1990). And, like Alice, I chased the rabbit called "economics" down a rabbit-hole, where the world looked very different to me. Fortunately, I traveled with my trusty copy of Lakoff and Johnson's (1980) *Metaphors We Live By* as away to make sense of my sojourn there. So, before making my way back to the challenge of school inequality, I must beg your indulgence as I give you a brief tour of my time down there.

NATIONAL DEBT VERSUS NATIONAL DEFICIT

Most people hear or read news of the economy every day and rarely give it a second thought. We hear that the Federal Reserve Bank is raising interest rates, or that the unemployment numbers look good. Our ears may perk up when we hear the latest gasoline prices or that we can get a good rate on a mortgage refinance loan. But busy professionals rarely have time to delve deeply into all things economic. Two economic terms—"national deficit" and "national debt"—seem to befuddle us. A deficit is the amount by which a government's, company's, or individual's spending exceeds income over a particular period of time. Thus, for each budget cycle, the government must determine whether it has a balanced budget, a budget surplus, or a deficit. The debt, however is the sum of all previously incurred annual federal deficits. Since the deficits are financed by government borrowing, national debt is equal to all government debt.

Most fiscal conservatives warn against deficit budgets and urge the government to decrease spending to balance the budget. Fiscal liberals do not necessarily embrace deficits but would rather see the budget balanced by increasing tax revenues from those most able to pay. The debt is a sum that has been accumulating since 1791, when the U.S. Treasury recorded it as $75,463,476.52 (Gordon, 1998). Thomas Jefferson (1816) said, "I ... place economy among the first and most important virtues, and public debt as the greatest of dangers to be feared. To preserve our independence, we must not let our rulers load us with perpetual debt."...

But the debt has not merely been going up.... Even in those years when the United States has had a balanced budget, that is, no deficits, the national debt continued to grow. It may have grown at a slower rate, but it did continue to grow....

THE DEBT AND EDUCATION DISPARITY

By now, readers might assume that I have made myself firmly at home at the Mad Hatter's Tea Party. What does a discussion about national deficits and national debt have to do with education, education research, and continued education disparities? It is here where I began to see some metaphorical concurrences between our national fiscal situation and our education situation. I am arguing that our focus on the achievement gap is akin to a focus on the budget deficit, but what is actually happening to African American and Latina/o students is really more like the national debt. We do not have an achievement gap; we have an education debt....

... I am arguing that the historical, economic, sociopolitical, and moral decisions and policies that characterize our society have created an education debt. So, at this point, I want to briefly describe each of those aspects of the debt.

THE HISTORICAL DEBT

Scholars in the history of education, such as James Anderson (1989), Michael Fultz (1995), and David Tyack (2004), have documented the legacy of educational inequities in the United States. Those inequities initially were formed around race, class, and gender. Gradually, some of the inequities began to recede, but clearly they persist in the realm of race. In the case of African Americans, education was initially forbidden during the period of enslavement. After emancipation we saw the development of freedmen's schools whose purpose was the maintenance of a servant class. During the long period of legal apartheid, African Americans attended schools where they received cast-off textbooks and materials from White schools. In the South, the need for farm labor meant that the typical school year for rural Black students was about 4 months long. Indeed, Black students in the South did not experience universal secondary schooling until 1968 (Anderson, 2002). Why, then, would we not expect there to be an achievement gap?

The history of American Indian education is equally egregious. It began with mission schools to convert and use Indian labor to further the cause of the church. Later, boarding schools were developed as General George Pratt asserted the need "to kill the Indian in order to save the man." This strategy of deliberate and forced assimilation created a group of people, according to Pulitzer Prize writer N. Scott Momaday, who belonged nowhere (Lesiak, 1992). The assimilated Indian could not fit comfortably into reservation life or the stratified mainstream. No predominately White colleges welcomed the few Indians who successfully completed the early boarding schools. Only historically Black colleges, such as Hampton Institute, opened their doors to them. There, the Indians studied vocational and trade curricula.

Latina/o students also experienced huge disparities in their education. In Ferg-Cadima's report *Black, White, and Brown: Latino School Desegregation Efforts in the Pre– and Post–*Brown v. Board of Education *Era* (2004), we discover the longstanding practice of denial experienced by Latina/os dating back to 1848. Historic desegregation cases such as *Mendez v. Westminster* (1946) and the Lemon Grove Incident detail the ways that Brown children were (and continue to be) excluded from equitable and high-quality education.

It is important to point out that the historical debt was not merely imposed by ignorant masses that were xenophobic and virulently racist. The major leaders of the nation endorsed ideas about the inferiority of Black, Latina/o, and Native peoples. Thomas Jefferson (1816), who advocated for the education of the American citizen, simultaneously decried the notion that Blacks were capable of education. George Washington, while deeply conflicted about slavery, maintained a substantial number of slaves on his Mount Vernon Plantation and gave no thought to educating enslaved children.

A brief perusal of some of the history of public schooling in the United States documents the way that we have accumulated an education debt over time. In 1827 Massachusetts passed a law making all grades of public school open to all pupils free of charge. At about the same time, most Southern states already had laws forbidding the teaching of enslaved Africans to read. By 1837, when Horace Mann had become head of the newly formed Massachusetts State Board of Education, Edmund Dwight, a wealthy Boston industrialist, felt that the state board was crucial to factory owners and offered to supplement the state salary with his own money. What is omitted from this history is that the major raw material of those textile factories, which drove the economy of the East, was cotton—the crop that depended primarily on the labor of enslaved Africans (Farrow, Lang, & Frank, 2005). Thus one of the ironies of the historical debt is that while African Americans were enslaved and prohibited from schooling, the product of their labor was used to profit Northern industrialists who already had the benefits of education....

This pattern of debt affected other groups as well. In 1864 the U.S. Congress made it illegal for Native Americans to be taught in their native languages. After the Civil War, African Americans worked with Republicans to rewrite state constitutions to guarantee free public education for all students. Unfortunately, their efforts benefited White children more than Black children. The landmark *Plessy v. Ferguson* (1896) decision meant that the segregation that the South had been practicing was officially recognized as legal by the federal government.

Although the historical debt is a heavy one, it is important not to overlook the ways that communities of color always have worked to educate themselves. Between 1865 and 1877, African Americans mobilized to bring public education to the South for the first time. Carter G. Woodson (1933/1972) was a primary critic of the kind of education that African Americans received, and he challenged African Americans to develop schools and curricula that met the unique needs of a population only a few generations out of chattel slavery.

THE ECONOMIC DEBT

As is often true in social research, the numbers present a startling picture of reality. The economics of the education debt are sobering. The funding disparities that currently exist between schools serving White students and those serving students of color are not recent phenomena. Separate schooling always allows for differential funding. In present-day dollars, the funding disparities between urban schools and their suburban counterparts present a telling story about the value we place on the education of different groups of students.

The Chicago public schools spend about $8,482 annually per pupil, while nearby Highland Park spends $17,291 per pupil. The Chicago public schools have an 87% Black and Latina/o population, while Highland Park has a 90% White population. Per pupil expenditures in Philadelphia are $9,299 per pupil for the city's 79% Black and Latina/o population, while across City Line Avenue in Lower Merion, the per pupil expenditure is $17,261 for a 91% White population. The New York City public schools spend $11,627 per pupil for a student population that is 72% Black and Latina/o, while suburban Manhasset spends $22,311 for a student population that is 91% White (figures from Kozol, 2005).

One of the earliest things one learns in statistics is that correlation does not prove causation, but we must ask ourselves why the funding inequities map so neatly and regularly onto the racial and ethnic realities of our schools. Even if we cannot prove that schools are poorly funded *because* Black and Latina/o students attend them, we can demonstrate that the amount of funding rises with the rise in White students. This pattern of inequitable funding has occurred over centuries. For many of these populations, schooling was nonexistent during the early history of the nation; and, clearly, Whites were not prepared to invest their fiscal resources in these strange "others."

Another important part of the economic component of the education debt is the earning ratios related to years of schooling. The empirical data suggest that more schooling is associated with higher earnings; that is, high school graduates earn more money than high school dropouts, and college graduates earn more than high school graduates. Margo (1990) pointed out that in 1940 the average annual earnings of Black men were about 48% of those of White men, but by 1980 the earning ratio had risen to 61%. By 1993, the median Black male earned 74% as much as the median White male....

THE SOCIOPOLITICAL DEBT

The sociopolitical debt reflects the degree to which communities of color are excluded from the civic process. Black, Latina/o, and Native communities had little or no access to the franchise, so they had no true legislative representation. According to the Civil Rights Division of the U.S. Department of Justice, African Americans and other persons of color were substantially disenfranchised in many Southern states despite the enactment of the Fifteenth Amendment in 1870 (U.S. Department of Justice, Civil Rights Division, 2006).

The Voting Rights Act of 1965 is touted as the most successful piece of civil rights legislation ever adopted by the U.S. Congress (Grofman, Handley, & Niemi, 1992). This act represents a proactive attempt to eradicate the sociopolitical debt that had been accumulating since the founding of the nation.

... The dramatic changes in voter registration are a result of Congress's bold action. In upholding the constitutionality of the act, the Supreme Court ruled as follows:

> Congress has found that case-by-case litigation was inadequate to combat wide-spread and persistent discrimination in voting, because of the inordinate amount of time and energy required to overcome the obstructionist tactics invariably encountered in these lawsuits. After enduring nearly a century of systematic resistance to the Fifteenth Amendment, Congress might well decide to shift the advantage of time and inertia from the perpetrators of the evil to its victims. (*South Carolina v. Katzenbach*, 1966; U.S. Department of Justice, Civil Rights Division, 2006)

It is hard to imagine such a similarly drastic action on behalf of African American, Latina/o, and Native American children in schools. For example, imagine that an examination of the achievement performance of children of color provoked an immediate reassignment of the nation's best teachers to the schools serving the most needy students. Imagine that those same students were guaranteed places in state and regional colleges and universities. Imagine that within one generation we lift those students out of poverty.

The closest example that we have of such a dramatic policy move is that of affirmative action. Rather than wait for students of color to meet predetermined standards, the society decided to recognize that historically denied groups should be given a preference in admission to schools and colleges. Ultimately, the major beneficiaries of this policy were White women. However, Bowen and Bok (1999) found that in the case of African Americans this proactive policy helped create what we now know as the Black middle class.

As a result of the sociopolitical component of the education debt, families of color have regularly been excluded from the decision-making mechanisms that should ensure that their children receive quality education. The parent—teacher organizations, school site councils, and other possibilities for democratic participation have not been available for many of these families. However, for a brief moment in 1968, Black parents in the Ocean Hill-Brownsville section of New York exercised community control over the public schools (Podair, 2003). African American, Latina/o, Native American, and Asian American parents have often advocated for improvements in schooling, but their advocacy often has been muted and marginalized. This quest for control of schools was powerfully captured in the voice of an African American mother during the fight for school desegregation in Boston. She declared: "When we fight about schools, we're fighting for our lives" (Hampton, 1986).

Indeed, a major aspect of the modern civil rights movement was the quest for quality schooling. From the activism of Benjamin Rushing in 1849 to the struggles of parents in rural South Carolina in 1999, families of color have been fighting for quality education for their children (Ladson-Billings, 2004). Their more limited access to lawyers and legislators has kept them from accumulating the kinds of political capital that their White, middle-class counterparts have.

THE MORAL DEBT

A final component of the education debt is what I term the "moral debt." I find this concept difficult to explain because social science rarely talks in these terms....

... [A] moral dent reflects the disparity between what we know is right and what we actually do. Saint Thomas Aquinas saw the moral debt as what human beings owe to each other in the giving of, or failure to give, honor to another when honor is due. This honor comes as a result of people's excellence or because of what they have done for another. We have no trouble recognizing that we have a moral debt to Rosa Parks, Martin Luther King, Cesar Chavez, Elie Wiesel, or Mahatma Gandhi. But how do we recognize the moral debt that we owe to entire groups of people? How do we calculate such a debt?...

What is that we might owe to citizens who historically have been excluded from social benefits and opportunities? Randall Robinson (2000) states:

> No nation can enslave a race of people for hundreds of years, set them free bedraggled and penniless, pit them, without assistance in a hostile environment, against privileged victimizers, and then reasonably expect the gap between the heirs of the two groups to narrow. Lines, begun parallel and left alone, can never touch, (p. 74)

Robinson's sentiments were not unlike those of President Lyndon B. Johnson, who stated in a 1965 address at Howard University: "You cannot take a man who has been in chains for 300 years, remove the chains, take him to the starting line and tell him to run the race, and think that you are being fair"' (Miller, 2005)....

... Taken together, the historic, economic, sociopolitical, and moral debt that we have amassed toward Black, Brown, Yellow, and Red children seems insurmountable, and attempts at addressing it seem futile. Indeed, it appears like a task for Sisyphus. But as legal scholar Derrick Bell (1994) indicated, just because something is impossible does not mean it is not worth doing.

WHY WE MUST ADDRESS THE DEBT

... On the face of it, we must address it because it is the equitable and just thing to do. As Americans we pride ourselves on maintaining those ideal qualities as hallmarks of our democracy. That represents the highest motivation for paying this debt. But we do not always work from our highest motivations.

Most of us live in the world of the pragmatic and practical. So we must address the education debt because it has implications for the kinds of lives we can live and the kind of education the society can expect for most of its children. I want to suggest that there are three primary reasons for addressing the debt—(a) the impact the debt has on present education progress, (b) the value of understanding the debt in relation to past education research findings, and (c) the potential for forging a better educational future.

The Impact of the Debt on Present Education Progress

... As I was attempting to make sense of the deficit/debt metaphor, educational economist Doug Harris (personal communication, November 19, 2005) reminded me that when nations operate with a large debt, some part of their current budget goes to service that debt. I mentioned earlier that interest payments on our national debt represent the third largest expenditure of our national budget. In the case of education, each effort we make toward improving education is counterbalanced by the ongoing and mounting debt that we have accumulated. That debt service manifests itself in the distrust and suspicion about what schools can and will do in communities serving the poor and children of color. Bryk and Schneider (2002) identified "relational trust" as a key component in school reform. I argue that the magnitude of the education debt erodes that trust and represents a portion of the debt service that teachers and administrators pay each year against what they might rightfully invest in helping students advance academically.

The Value of Understanding the Debt in Relation to Past Research Findings

The second reason that we must address the debt is somewhat selfish from an education research perspective. Much of our scholarly effort has gone into looking at educational inequality and how we might mitigate it. Despite how hard we try, there are two interventions that have never received full and sustained hypothesis testing—school desegregation and funding equity. Orfield and Lee (2006) point out that not only has school segregation persisted, but it has been transformed by the changing demographics of the nation. They also point out that "there has not been a serious discussion of the costs of segregation or the advantages of integration for our most segregated population, white students" (p. 5). So, although we may have recently celebrated the 50th anniversary of the *Brown* decision, we can point to little evidence that we really gave *Brown* a chance. According to Frankenberg, Lee, and Orfield (2003) and Orfield and Lee (2004), America's public schools are more than a decade into a process of resegregation. Almost three-fourths of Black and Latina/o students attend schools that are predominately non-White. More than 2 million Black and Latina/o students—a quarter of the Black students in the Northeast and Midwest—attend what the researchers call apartheid schools. The four most segregated states for Black students are New York, Michigan, Illinois, and California.

The funding equity problem, as I illustrated earlier in this discussion, also has been intractable. In its report entitled *The Funding Gap 2005*, the Education Trust tells us that "in 27 of the 49 states studied, the highest-poverty school districts receive fewer resources than the lowest-poverty districts.... Even more states shortchange their highest minority districts. In 30 states, high minority districts receive less money for each child than low minority districts" (p. 2). If we are unwilling to desegregate our schools *and* unwilling to fund them equitably, we find ourselves not only backing away from the promise of the *Brown* decision but literally refusing even to take *Plessy* seriously. At least a serious consideration of *Plessy* would make us look at funding inequities....

The Potential for Forging a Better Educational Future

Finally, we need to address what implications this mounting debt has for our future. In one scenario, we might determine that our debt is so high that the only thing we can do is declare bankruptcy. Perhaps, like our airline industry, we could use the protection of the bankruptcy laws to reorganize and design more streamlined, more efficient schooling options. Or perhaps we could be like developing nations that owe huge sums to the IMF and apply for 100% debt relief. But what would such a catastrophic collapse of our education system look like? Where could we go to begin from the ground up to build the kind of education system that would aggressively address the debt? Might we find a setting where a catastrophic occurrence, perhaps a natural disaster—a hurricane—has completely obliterated the schools? Of course, it would need to be a place where the schools weren't very good to begin with. It would have to be a place where our Institutional Review Board and human subject concerns would not keep us from proposing aggressive and cutting-edge research. It would have to be a place where people were so desperate for the expertise of education researchers that we could conduct multiple projects using multiple approaches. It would be a place so hungry for solutions that it would not matter if some projects were quantitative and others were qualitative. It would not matter if some were large-scale and some were small-scale. It would not matter if some paradigms were psychological, some were social, some were economic, and some were cultural. The only thing that would matter in an environment like this would be that education researchers were bringing their expertise to bear on education problems that spoke to pressing concerns of the public. I wonder where we might find such a place?...

... [T]he cumulative effect of poor education, poor housing, poor health care, and poor government services creates a bifurcated society that leaves more than its children behind. The images should compel us to deploy our knowledge, skills, and expertise to alleviate the suffering of the least of these. They are the images that compelled our attention during Hurricane Katrina. Here, for the first time in a very long time, the nation—indeed the world—was confronted with the magnitude of poverty that exists in America.

In a recent book, Michael Apple and Kristen Buras (2006) suggest that the subaltern can and do speak. In this country they speak from the barrios of Los Angeles and the ghettos of New York. They speak from the reservations of New Mexico and the Chinatown of San Francisco. They speak from the levee breaks of New Orleans where they remind us, as education researchers, that we do not merely have an achievement gap—we have an education debt.

REFERENCES

Anderson, J. D. (1989). *The education of Blacks in the South, 1860–1935.* Chapel Hill, NC: University of North Carolina Press.

Anderson, J. D. (2002, February 28). *Historical perspectives on Black academic achievement.* Paper presented for the Visiting Minority Scholars Series Lecture. Wisconsin Center for Educational Research, University of Wisconsin, Madison.

Apple, M. (1990). *Ideology and curriculum* (2nd ed.). New York: Routledge.

Apple, M., & Buras, K. (Eds.). (2006). *The subaltern speak: Curriculum, power and education struggles.* New York: Routledge.

Au, K. (1980). Participation structures in a reading lesson with Hawaiian children. *Anthropology and Education Quarterly, 11*(2), 91–115.

Banks, J. A. (2004). Multicultural education: Historical development, dimensions, and practices. In J. A. Banks & C. M. Banks (Eds.), *Handbook of research in multicultural education* (2nd ed., pp. 3–29). San Francisco: Jossey-Bass.

Bell, D. (1994). *Confronting authority: Reflections of an ardent protester.* Boston: Beacon Press.

Bereiter, C., & Engleman, S. (1966). *Teaching disadvantaged children in preschool.* Englewood Cliffs, NJ: Prentice Hall.

Bowen, W., & Bok, D. (1999). *The shape of the river.* Princeton, NJ: Princeton University Press.

Brice Heath, S. (1983). *Ways with words: Language, life and work in communities and classrooms.* Cambridge, UK: Cambridge University Press.

Brown v. Board of Education 347 U.S. 483 (1954).

Bryk, A., & Schneider, S. (2002). *Trust in schools: A core resource for improvement.* New York: Russell Sage Foundation.

Cochran-Smith, M. (2004). Multicultural teacher education: Research, practice and policy. In J. A. Banks & C. M. Banks (Eds.), *Handbook of research in multicultural education* (2nd ed., pp. 931–975). San Francisco: Jossey-Bass.

Coleman, J., Campbell, E., Hobson, C., McPartland, J., Mood, A., Weinfeld, F. D., et al. (1966). *Equality of educational opportunity.* Washington, DC: Department of Health, Education and Welfare.

Cornbleth, C., & Waugh, D. (1995). *The great speckled bird: Multicultural politics and education.* Mahwah, NJ: Lawrence Erlbaum.

Delpit, L. (1995). *Other people's children: Cultural conflict in the classroom.* New York: Free Press.

Deutsch, M. (1963). The disadvantaged child and the learning process. In A. H. Passow (Ed.), *Education in depressed areas* (pp. 163–179). New York: New York Bureau of Publications, Teachers College, Columbia University.

Education Commission of the States. (2005). *The nation's report card.* Retrieved January 2, 2006, from http://nces.ed.gov/nationsreportcard

Education Trust. (2005). *The funding gap 2005.* Washington, DC: Author.

Farrow, A., Lang, J., & Frank, J. (2005). *Complicity: How the North promoted, prolonged and profited from slavery.* New York: Ballantine Books.

Ferg-Cadima, J. (2004, May). *Black, White, and Brown: Latino school desegregation efforts in the pre– and post–* Brown v. Board of Education *era.* Washington, DC: Mexican-American Legal Defense and Education Fund.

Foster, M. (1996). *Black teachers on teaching*. New York: New Press.

Frankenberg, E., Lee, C., & Orfield, G. (2003, January). *A multiracial society with segregated schools: Are we losing the dream?* Cambridge, MA: The Civil Rights Project, Harvard University.

Fultz, M. (1995). African American teachers in the South, 1890–1940: Powerlessness and the ironies of expectations and protests. *History of Education Quarterly, 35*(4), 401–422.

Gay, G. (2004). Multicultural curriculum theory and multicultural education. In J. A. Banks & C. M. Banks (Eds.), *Handbook of research in multicultural education* (2nd ed., pp. 30–49). San Francisco: Jossey-Bass.

Gordon, J. S. (1998). *Hamilton's blessing: The extraordinary life and times of our national debt.* New York: Penguin Books.

Grant, C. A. (2003). *An education guide to diversity in the classroom.* Boston: Houghton Mifflin.

Grofman, B., Handley, L., & Niemi, R. G. (1992). *Minority representation and the quest for voting equality.* New York: Cambridge University Press.

Hampton, H. (Director). (1986). *Eyes on the prize* [Television video series], Blackside Productions (Producer). New York: Public Broadcasting Service.

Hess, R. D., & Shipman, V. C. (1965). Early experience and socialization of cognitive modes in children. *Child Development, 36*, 869–886.

Irvine, J. J. (2003). *Educating teachers for diversity: Seeing with a cultural eye.* New York: Teachers College Press.

Jefferson, T. (1816, July 21). *Letter to William Plumer. The Thomas Jefferson Paper Series. 1. General correspondence*, 1651–1827. Retrieved September 11, 2006, from http://rs6.loc.gov/cgi-bin/ampage

Kozol, J. (2005). *The shame of the nation: The restoration of apartheid schooling in America.* New York: Crown Publishing.

Ladson-Billings, G. (1994). *The dreamkeepers: Successful teachers of African American children.* San Francisco: Jossey-Bass.

Ladson-Billings, G. (2004). Landing on the wrong note: The price we paid for *Brown. Educational Researcher, 33*(7), 3–13.

Lakoff, G., & Johnson, M. (1980). *Metaphors we live by.* Chicago: University of Chicago Press.

Lee, C. D. (2004). African American students and literacy. In D. Alvermann & D. Strickland (Eds.), *Bridging the gap: Improving literacy learning for pre-adolescent and adolescent learners, Grades 4–12.* New York: Teachers College Press.

Lesiak, C. (Director). (1992). *In the White man's image* [Television broadcast]. New York: Public Broadcasting Corporation.

Margo, R. (1990). *Race and schooling in the American South, 1880–1950.* Chicago: University of Chicago Press.

Mendez v. Westminster 64F. Supp. 544 (1946).

Miller, J. (2005, September 22). New Orleans unmasks apartheid American style [Electronic version]. *Black Commentator, 151.* Retrieved September 11, 2006, from http://www.blackcommentator.com/151/151_miller_new_orleans.html

National Center for Education Statistics. (2001). *Education achievement and Black-White inequality*. Washington, DC: Department of Education.

National Governors' Association. (2005). *Closing the achievement gap*. Retrieved October 27, 2005, from http://www.subnet.nga.org/educlear/achievement!

Orfield, G., & Lee, C. (2004, January). *Brown at 50: King's dream or Plessy's nightmare?* Cambridge, MA: The Civil Rights Project, Harvard University.

Orfield, G., & Lee, C. (2006, January). *Racial transformation and the changing nature of segregation*. Cambridge, MA: The Civil Rights Project, Harvard University.

Plessy v. Ferguson 163 U.S. 537 (1896).

Podair, J. (2003). *The strike that changed New York: Blacks, Whites and the Ocean Hill-Brownsville Crisis*. New Haven, CT: Yale University Press.

Popkewitz, T. S. (1998). *Struggling for the soul: The politics of schooling and the construction of the teacher*. New York: Teachers College Press.

Robinson, R. (2000). *The debt: What America owes to Blacks*. New York: Dutton Books.

Sleeter, C. (2001). *Culture, difference and power*. New York: Teachers College Press.

South Carolina v. Katzenbach 383 U.S. 301, 327–328 (1966).

Steele, C. M. (1999, August). Thin ice: "Stereotype threat" and Black college students. *Atlantic Monthly, 284*, 44–47, 50–54.

Tyack, D. (2004). *Seeking common ground: Public schools in a diverse society*. Cambridge, MA: Harvard University Press.

U.S. Department of Justice, Civil Rights Division. (2006, September 7). *Introduction to federal voting rights laws*. Retrieved September 11, 2006, from http://www.usdoj.gov/crt/voting/intro/intro.htm

Woodson, C. G. (1972). *The mis-education of the Negro*. Trenton, NJ: Africa World Press. (Original work published 1933).

Zeichner, K. M. (2002). The adequacies and inadequacies of three current strategies to recruit, prepare, and retain the best teachers for all students. *Teachers College Record, 105*(3), 490–511.

42

How a Scholarship Girl Becomes a Soldier

The Militarization of Latina/o Youth in Chicago Public Schools

GINA M. PÉREZ

Like many working-class communities, Chicago Puerto Ricans have had a complicated relationship with the United States military, an institution that represents, in a particularly visible way, Puerto Rico's unresolved political status, one that is contested by community activists. Recent broad-based political mobilization in both Chicago and New York City demanding the end of naval operations on the island of Vieques, as well as a smaller, less visible presence of island residents supporting the navy's role on the island, is just one example of the complex responses Puerto Ricans have to the military. The military also is an institution that is understood to be (and is often successfully used as) an important avenue of social mobility for families and households with limited resources and opportunities available to them, particularly for young women who often are sources of productive and reproductive labor critical to a household's survival....

In what follows, I draw on ethnographic and interview data to discuss how Puerto Rican and Latina/o youth in Chicago are implicated in an increasingly militarized world. Military programs like the Junior Reserve Officer Training Program (JROTC) are not new (it was founded in 1916) and have long targeted Puerto Rican youth. In recent years, however, these programs have operated with new intensity, appearing more frequently in urban schools with an explicit goal of targeting populations deemed "at risk" in "less affluent large urban schools." It is not surprising that this "at risk" group includes young men who are allegedly in danger of falling into gangs or drug use. What is alarming, however, is the extent to which these programs are now targeting young women who are deemed "at risk" because of the possibility that they might become unwed teenage mothers. The gendered dynamics through which militarization is occurring on the homefront are in need of critical attention if we are to understand how poor and working-class youth and their families struggle and strategize to make ends meet....

SOURCE: Pérez, Gina M. 2006. "How a Scholarship Girl Becomes a Soldier: The Militarization of Latina/o Youth in Chicago Public Schools." Pp. 53–72 in *Identities: Global Studies in Culture and Power*, edited by Jonathan Hill. New York: Routledge. Reprinted by permission.

CHICAGO: A GLOBAL LATINO CITY

With more than three-quarters of a million Latina/o residents, Chicago is home to the third largest and one of the most diverse Latina/o populations in the country. It also is the only place where large numbers of Mexicans and Puerto Ricans of several generations live, marry, and work side by side....

Chicago Latinas/os, however, are also some of the city's most impoverished residents, with nearly twenty-four percent of Latinas/os living in poverty. Chicago Puerto Ricans are the poorest of all Chicago Latinos. With limited employment opportunities, primarily in the low-wage service sector, Puerto Ricans in Chicago index the highest poverty rates in the city at 33.8 percent, compared to thirty-three percent for the city's African-American families.[1] In contrast with Cubans and South Americans, Puerto Ricans and Mexicans have lower educational and average income levels and are concentrated in the low-wage service sector and as operatives.[2] Moreover, studies by the Latino Institute, The Chicago Urban League, and Northern Illinois University demonstrate that none of the jobs requiring only a high school diploma—and even many of those demanding some post-secondary education—pay a living wage for a family with dependent children.[3] According to the Latino Institute (1994), more than seventy-five percent of Puerto Ricans are employed in those sectors of the economy. In short, while transnational investment and the loss of manufacturing jobs in the Chicago area have rendered Latinos in general much poorer than a decade earlier, Puerto Ricans specifically remain the most economically disadvantaged group in all of Chicago. It is no wonder that many Latina/o youth consider military service as one of the few opportunities that will guarantee employment, provide them with marketable skills, and serve as a vehicle for achieving economic stability.

JROTC, MILITARY SERVICE, AND MAKING ENDS MEET

Beginning at a very early age, military service becomes one of the most appealing and seemingly sure options for many poor and working-class Latina/o families. Bombarded by television advertisements, visits by recruiters in junior high and high schools, and the prospect of receiving JROTC programs' financial incentives (both immediate and promises of money for the future), poor children and particularly children of color quickly come to consider the military—rather than college—as an employment venue after high school. Money, promises of free education, training, discipline, honor, and respect seduce young men and women into the military ranks.

Chicago schools provide particularly fertile ground for cultivating military careers after high school. Nearly eighty-five percent of Chicago public school students come from low-income families and are either African American or Latina/o.[4] They also are increasingly enmeshed in what Pauline Lipman has identified as "stratified academic programs," whereby African-American and

Latina/o high school students attend schools with "limited offerings of advanced courses and new vocational academies, basic skills transitional high schools" or public military academies rather than the "[n]ew academically selective magnet schools and programs, mainly located in largely white upper-income and/or gentrifying neighborhoods" (Lipman 2003: 81). The Chicago public schools, for example, lead the nation with more than 10,000 students participating in a wide range of expanding public school military programs. Chicago is home to two school-wide military academies—Chicago Military Academy and Carver Military Academy—both affiliated with the United States Army and located on Chicago's South Side; eight military academies within regular high schools (four affiliated with the Army, two with the Marines, and one each with the Navy and the Air Force); a middle school military academy in the predominately Mexican neighborhood of Little Village; and part-time JROTC in forty-three Chicago high schools and after-school programs at seventeen middle schools.[5] …

In Chicago, JROTC is part of the Education to Careers (ETC) program, whose mission is to "equip students to successfully transition into postsecondary education, advanced training, and the workplace." … High school students may participate in JROTC to satisfy the Chicago public high school career education requirement or, in some schools, to satisfy the physical education requirement for graduation. JROTC also offers many extracurricular activities, including honor guard, competitions, service, field trips, and opportunities to visit military installations. Many students explain that it is precisely these curricular and extra-curricular offerings that make JROTC appealing. Almost all of the young Latina/o and African-American JROTC participants I interviewed in one Chicago high school also cited "being treated with respect" (especially while wearing the cadet uniform) both as a reason for joining, as well as one of the greatest advantages of participating in, JROTC.

Students' concern with respect is no small matter, since most reside in poor and working-class (and, often, slowly gentrifying) neighborhoods regarded in local media as dangerous and are enmeshed in racialized policing practices aimed at containing suspect youth.[6] Latina/o and African-American youth are painfully aware of how their bodies are read. Thus, wearing a military uniform is, perhaps, one way of negotiating the racialized systems of surveillance that not only operate within their neighborhoods, but also within their own schools.…

Even when young people are not in formal high schools, the military continues to shape many of their educational and employment opportunities in other ways. For example, when nineteen-year-old Marvin Polanco[7] entered a G.E.D. class I taught at a Puerto Rican cultural center in Chicago, he regaled me with stories about his experience in Lincoln's Challenge, a military boarding school affiliated with the Illinois National Guard.[8] Shortly after dropping out of Clemente High School, Marvin's mother signed him up for the five-month program, which promised to instill discipline, prepare him for the G.E.D., and pay him $2,200 after successfully graduating from the school. "But they tricked my mom," Marvin told me solemnly one day after class. He never passed the G.E.D., a failure he attributes to the fact that they required him to exercise, run, and engage in other physical training immediately before the exam, leaving him exhausted and unable to finish the test. Marvin received less than $500 when he finished the program. "They try to brainwash you there," he explained another day, referring to the deeply militarized atmosphere and training each cadet receives while in residence. Between the physical discipline imposed by the officers and the danger of getting jumped by gang members also participating in the program, Marvin said he often was afraid he wouldn't get out alive. After a few months in the program, he ran away from the school and returned home, begging his mother not to send him back to the school and promising to get his G.E.D. somewhere in Chicago.

Many of my male G.E.D. students wanted to go into the military once they completed their G.E.D. When they would approach me about this, I encouraged them to consider enrolling in one of Chicago's city colleges and eventually transferring to either University of Illinois–Chicago (UIC) or Northeastern University, which both had programs specifically targeting and supporting incoming students with their G.E.D. With the help of the Centro's counselors and staff, we put many students in touch with representatives from these universities, invited them on a tour of one of the city colleges, and helped organize a campus visit to a small liberal arts college to whet their educational appetites. Michael, one of the Centro's counselors, was extremely invested in pushing our students to consider college. As a G.E.D. graduate himself and current student at Northeastern, he knew firsthand the strong push toward military service and vocational training, and he worked tirelessly with me to provide the students with alternatives.

Considering the military as an option, however, infected even the brightest of students. Nineteen-year-old Eddie Vélez, for example, a smart student who delighted in being a good student and a math wiz after years of being told he was not, arrived late to G.E.D. class one day, announcing that he met with an army recruiter who explained to him how the G.I. Bill would pay for his college education. Eddie was seriously considering this option and his enthusiasm spread to Frankie, a sixteen-year-old student who was also interested in joining the military. I was furious and gathered the class together and asked them to think about why the military aggressively recruited them and other poor people of color and not the mostly white affluent Northwestern University students I also taught on the city's northern suburbs. Initially stunned, my students responded passionately about the "honor of serving your country," telling me about their fathers, grandfathers, uncles, and brothers who had all served time in the armed forces. They were angry with me for being so upset and critical of what for many of them had been small steps toward social and economic mobility. "My father and *grandfather* were in the army," Eddie insisted. "It's a good job. And that's what I need. I need to get out of my house and *away* from the drugs and get into a good place. I can't concentrate when I'm at home, you know. They're smokin', drinkin', and I can't concentrate. Even if I go to college, I still have to go *home*. Why do you think I work so many hours and *like* to come to school? I need to get out of that environment." The class was silent, and a few students nodded their heads.

The idea of using the military as a haven or as a way to avoid troubled households and communities is not lost on those who run JROTC, as well as some of the programs' staunchest advocates.... The idea of the benefits JROTC programs provide for "inner-city" and "at-risk" youth is so entrenched that the program's Pentagon funding [was] expected to increase more than fifty percent of its current budget of $215 million in 2001 to $326 million by 2004.[9] Much of this money goes to local school districts eager to implement JROTC programs, since doing so guarantees the infusion of federal funds that many proponents argue can be a strategy for broadening school curriculum and securing more money for already financially strapped schools.

GENDER, SEXUALITY, AND THE QUEST FOR AUTONOMY

Young men, of course, are not alone in considering the military as a viable educational and employment option. Increasingly, young girls swell the JROTC ranks, a phenomenon that allegedly demonstrates the military's egalitarian gender roles.... In Chicago public schools, young women outnumber young men in Army JROTC programs by a slim margin; they also comprise nearly half of all JROTC program participants, with Latinas and African-American women constituting the overwhelming majority.

Perhaps the appeal of more egalitarian gender relations partially explains young women's interest in JROTC. Ethnographic evidence, however, suggests that young girls' decisions to enter JROTC programs are slightly more complicated. Latina adolescents, for example, emphasize how household economic imperatives, as well as parental pressure and encouragement from friends and kin, influence their ideas about JRTOC and military service. Participating in JROTC's extracurricular programs may also offer freedom and autonomy otherwise unavailable to some young girls who are expected to abide by culturally prescribed norms of behavior requiring them to be in the home. Some parents support their daughter's participation in JROTC because of the program's promise to cultivate in its participants the values of structure, discipline, and honor. As parents grow increasingly concerned with their adolescent daughter's sexuality, JROTC participation can be regarded as another way of preserving their daughter' s chastity and honor, while simultaneously reinforcing their critical role in their household's economic survival.

Thus, while military involvement allegedly keeps young men out of gangs, for young women it allegedly functions as a way to mitigate unwed teen pregnancy, providing necessary discipline for "unruly," "unpredictable" women's sexuality. Chicana and Puerto Rican scholars have demonstrated how culturally prescribed notions of female chastity and virginity among Latinas frequently result in family policing of young unmarried women's bodies by observing them, being vigilant of their movements and activities, and designating the household as the appropriate place for young women (Souza 2002; Trujillo 1991; Zavella 1997). These concerns with female chastity and purity also characterize poor and working-class *puertorriqueñas* whose families (usually mothers) not only monitor young women's sexual behavior, but who also depend on them for the economic survival of their households.

As Caridad Souza (2002) has shown in her research among working class Puerto Rican women in Queens, young women's reproductive work is critical to households whose survival depends on family and community solidarity through the exchange of goods, services, and the pooling of resources across multiple kin and non-kin households. In this context, young women are expected to be *en la casa* (inside the home or family) for important material and cultural reasons: Without their reproductive work, adult women—mothers, grandmothers, and aunts—are easily overwhelmed by (and perhaps unable to meet) the reproductive demands within their own households and kin networks. Being *una muchacha de la casa* also means abiding by the cultural norms of respectability, chastity, and family honor valued by the community. To be *de la calle* (outside the home or literally "on the streets") is to be transgressive, sexually promiscuous, and dangerous (Souza 2002: 35).

For young Latinas, therefore, participating in JROTC programs offers both a way for them to contribute to their household economies as well as the possibility of creating a space for them to exercise some autonomy, giving them "legitimate" and "respectable" reasons to be outside of the home in ways that conform to cultural expectations of young women. In this way, young Puerto Rican women creatively construct ways of fulfilling their culturally prescribed gendered responsibilities, although they do so by participating in a hyper-masculinized social context that regards female sexuality as a problem in need of discipline and control (Enloe 2000). Seventeen-year-old Jasmín Rodríguez's story provides a glimpse into these competing demands and dreams that shape young women's thinking of JROTC and their sense of obligation to their families

Throughout middle and high school, Jasmín has been a model student, consistently making honor roll, being ranked at the top of her high school class, and actively participating in school-sponsored community projects, writing contests, and youth-leadership programs. Jasmín has also been the recipient of summer scholarship programs targeting young women of color to attend summer-long programs at prestigious universities like Princeton and Cornell. She is an avid reader with a keen sense of social justice and inequality, characteristics that led her to announce at the age of ten that she was going to go to Harvard Law School one day so she could help make the world a better place. As a junior at one of Chicago's public high schools, she scored well on her PSATs and began receiving letters from four-year colleges and universities, whetting her desire to go away to college, to be the first woman in her family to do so. Certainly, how she is going to pay for college weighs heavily on her mind, but with the assurance of her high school counselors and mentors, Jasmín was confident that she would be able to get enough money from scholarships and financial aid to pay for college without having to rely on her mother for help.

Jasmín's enthusiasm for college has spread to her younger sister, Myrna, who also talks excitedly about college. Together with their college degrees, they talk confidently that they will soon be financially secure enough to help support their mother, Aida Rodríguez, and the rest of their family [who] barely survive month to month on the pooled resources of Aida's salary as a high school clerk and their older sister's and brother's part-time jobs as a cashier and a security guard at O'Hare Airport. Both Jasmín and Myrna believe college is a sure way to economic security and to helping their mother achieve her dream of finally owning a modest home for their family one day. Jasmín's determination to go to college has been such a defining feature of her personality that I was amazed when, in the Spring of 2002, she told me she had signed up for the military. When she made the announcement at her nephew's first birthday party, she glowed with excitement and even seemed to take pleasure in the shocked response her decision provoked. When I asked her why she had changed her mind and why she was now considering the military, she repeated several times, "I just had to do it. It was the right decision. I'm really going to do this *and* go to college. And JROTC will help pay." When I assured her that based on her grades and family's financial need she would be able to pay for college without relying on money from military service, she replied, "Yeah, I know, but this is just another way to be sure I can pay for college. And they also give you money now. And we really need it."

Aida was equally confused by her daughter's decision. Aida takes enormous pride in Jasmín's accomplishments, and the walls and shelves in their small second story apartment, covered with mostly Jasmín's diplomas, awards, and trophies recognizing her academic achievements, are a clear sign of Aida's pride in and admiration for her daughter. Jasmín's decision, therefore, was tremendously distressing to Aida, who could not explain why Jasmín wanted to join JROTC and serve in the military after graduation. Aida had encouraged her oldest daughter, Milly, to participate in JROTC in high school and Aida even hoped Milly would go into the military after graduation, assuring her that doing so would benefit her financially and educationally. Aida feared that Milly wouldn't have the same kinds of options Jasmín had and military service, she assured me on many occasions, was also a way to keep Milly "out of trouble" and to prevent her from making the same mistakes Aida made as a young girl, namely becoming a teenaged single mother. Like many Puerto Rican mothers, Aida worried that Milly and her other daughters would *meterse las patas* (to mess up, and in this case, to get pregnant) as teenagers and she continuously strategized ways of protecting her daughters, including contemplating sending Milly to Puerto Rico when she was fourteen years old. Jasmín, however, was different and carried the hope and responsibility of being the first woman in her family to go away and graduate from college.

Jasmín understood her mother's hopes for her. But she was also painfully aware of her family's precarious economic situation that became even more tenuous with the birth of Milly's son, who brought an incredible amount of joy to the family. When I interviewed Jasmín about her decision later that summer, she provided a narrative filled with concerns with money, the need for security, and a desire to protect and serve both her nation and her family. She explained how one morning she was approached by the JROTC sergeant at her high school to take the ASVAB. His praise of her high scores, his relationship with her older sister, and his promise of $2,000 if she participated in a summer boot camp that would prepare her for her service in the army reserves all highlight a critical feature of successful JROTC programs: Their strength derives from a trusting and respectful relationship between the school commanding officer and the cadets. In June 2004 I observed, for example, the ease with which young men and women cadets interacted with their commanding officer, and his sensitive understanding of each cadet's struggles and hopes for the future. Many of the cadets I interviewed had siblings or cousins who also participated in JROTC, and one young man, Johnny, explained how he convinced his younger sister to join. When I asked him why, Johnny jokingly offered that since he is older and has "rank" in his unit, his younger sister would have to "show me respect" and defer to him in ways that she refused to do when at home. While Johnny admitted this was only one reason for persuading his sister to participate, his comments speak to the role of kinship and social networks in expanding the program's ranks. Over time, JROTC commanders meet these family members and friends and are able to draw on those networks not only to support his cadets, but also to approach young people who otherwise may not consider JROTC or military service after high school. It was precisely these networks that facilitated Jasmín's conversations with one of the JROTC officers in the school. She explains, "He was pretty convincing. He said he knew my family since he recruited my sister ... and I saw it as a precaution just in case I didn't get any scholarship or financial help for college. I could do a lot with this money. My family has never been rich, and that money seems like so much.... I had my senior fees coming up and I can't expect my mother to pay for it. My senior trip, my prom. And it feels really good to have the money."

When I asked her if there were other reasons for her decision, she admitted that the events of 9/11 were also important.

> I have family in New York ... and I sure didn't feel protected and I felt like [by joining the military] I would be protecting those I love. And even those I don't know, I felt like I could be doing something to help.... The women in my family have had to be protected by bad men [all their lives] and I don't want to have to do that. I could protect myself and not have to depend on others, but [my friends and family] can depend on me. Protecting people and standing up for what you believe in and being a good example of what the U.S. can be. I think our armed forces embody that.

The desire to feel protected and to equip oneself with important skills and financial resources to ensure independence fills Jasmín's narrative of education, her future, and her family indebtedness. They are also themes emerging from many Chicago Puerto Rican women's life histories and explain, in part, Aida's ambivalent embrace of military service for her daughters. For both Aida and Jasmín, the military's promise of money is clearly a means to ultimately achieve a better education and financial security....

Jasmín's decision, however, is also informed by her quest for autonomy and power that will allow the women in her family to depend on her rather than have to be "protected by bad men." Military service provides some possibilities not only for her, but specifically the women in her family whose options are often circumscribed because of limited education and low-wage employment.

Young women's participation in JROTC, therefore, is informed by social and economic need, although public officials are often quick to seize upon the deeply gendered concern with sexuality to promote the benefits of military service for young women. Public officials, for example, deploy the language of security and protection to advance the idea that military service protects "at risk" youth from danger and instills critical values allegedly absent in poor families, namely discipline, honor, and the value of hard work. Some JROTC officers and elected officials share parental concern with young women's sexuality and suggest that JROTC might be a new way to battle teen pregnancy. In Jackson, Mississippi, for example, state senator Robert Johnson recently applauded the increasing number of women entering into JROTC programs saying, "Jackson has more unwed mothers than just about any city of its size in the nation. We're talking about second- and third-generation single parents. The people criticizing JROTC are not the people living in these communities, because if they were, they would know that the people making the biggest difference, doing the most grass-roots work, are people with military backgrounds.

Regardless of whether the senator's statement is true, what is striking about his analysis is the explicit way in which he and others connect women's military service with sexual discipline, and how poor and working families also see these benefits of military service as well as the possibility it provides for sustaining precarious household economies and leading the way out of poverty. There are clearly a number of success stories of how this has happened, and these narratives are part of a larger American dream ideology that continues to animate impoverished Latina/o families.

CONCLUSION

With high poverty rates, low levels of educational attainment, and high drop-out rates among Latinas/os, military service becomes an appealing option for many Latinas/os whose life chances and economic options are increasingly circumscribed. These realities are not lost on military recruiters and the Department of Defense and the Pentagon, who have increased spending for JROTC programs in urban schools and target Latinos for military enlistment with the goal of boosting "the Latino numbers in the military." ... Like their male counterparts, young Latinas are particularly vulnerable to military appeals to their sense of family obligation as well as notions of pride, discipline, loyalty, honor, and citizenship as they consider military service as one of many economic strategies to make ends meet. This discussion, therefore, is an attempt to analyze Latinas/os' choice to participate in military programs within a broader cultural and political-economic context. What is needed, however, is more historically informed ethnographic research that engages with these complex issues of military service, poverty, race, gender, and notions of citizenship and patriotism in a way that honors the real lived experiences and struggles of our informants, but also challenges conventional notions of community based on militarized (and masculinist) notions of home and nation, to build, instead, communities of justice, equal opportunity, and solidarity.

NOTES

1. Latino Institute (1995).
2. John Betancur et al. (1993) point out that despite Cubans' and South Americans' economic success, their wages according to the 1980 census still approximated those of Puerto Ricans and Mexicans rather than that of whites. The 1990 census, however, paints a very different picture, emphasizing the growing gap between Puerto Ricans and other Latino groups in terms of average incomes, employment rates, and poverty levels.
3. See Chicago Urban League et al. (1994, 1995a, 1995b).
4. Chicago Public Schools, CPS at a Glance, February 2005. Available at www.cps .k.12.il.us/AtAGlance.html. The three full-time military academies, Chicago Military Academy, Carver Area High School, and Austin Community Academy, all opened since 1999 (www.cps.k12.il.us/AtAGlance.html).
5. Ana Beatriz Cholo, Military Marches into Middle Schools, *Chicago Tribune*, 26 July 2002; Chicago Public Schools, JROTC Program Book, nd; Education to Careers (ETC) FY 2003–2004.
6. Elsewhere (Pérez 2002), I have documented how Latina/o youth (especially those in rapidly gentrifying neighborhoods) are implicated in the policing of urban space aimed at curbing, for example, gang activity. Although Chicago's anti-loitering ordinance was declared unconstitutional in 1999, some scholars and activists have highlighted the "ongoing attempts to legalize harassment and street sweeps of youth," particularly youth of color who are regarded as dangerous and who allegedly "need to be locked up or removed from public space" (Lipman 2003: 95).
7. All names are pseudonyms.
8. Lincoln's Challenge is a federally funded youth program for at-risk youth between the ages of sixteen and eighteen whose quasi-military training in "discipline, esprit-de-corps, leadership, and teamwork is producing rapid and effective change in individual behaviors and attitude." Lincoln's Challenge is the largest National Guard Youth Challenge program in the nation (www.lincolnschallengeacademy.org/challenge/challenge.htm).
9. Racism and Conscription in the JROTC, *Peace Review*, September 2002; Class Warfare, *Time*, 4 March 2002, p. 50.

REFERENCES

American Friends Service Committee. 1999. *Trading Books for Soldiers: The True Cost of JROTC*. Philadelphia, PA: AFSC.

Betancur, John J., Teresa Cordova, and Maria de los Angeles Torres. 1993. Economic Restructuring and the Process of Incorporation of Latinos into the Chicago

Economy. In *Latinos in a Changing U.S. Economy*. Rebecca Morales and Frank Bonilla, eds. New York: Sage. Pp. 109–132.

Chicago Urban League, Latino Institute, and Northern Illinois University. 1994. The Changing Economic Standing of Minorities and Women in the Chicago Metropolitan Area, 1970-1990. Chicago, IL: Chicago. Urban League, Final Report.

Chicago Urban League, Latino Institute, and Northern Illinois University. 1995a. When the Job Doesn't Pay: Contingent Workers in the Chicago Metropolitan Area. In *The Working Poor Project*. Chicago, IL: Chicago Urban League.

Chicago Urban League, Latino Institute, and Northern Illinois University. 1995b. Jobs That Pay: Are Enough Jobs Available in Metropolitan Chicago? In *The Working Poor Project*. Chicago, IL: Chicago, Urban League.

Enloe, Cynthia. 2000. *Maneuvers: The International Politics of Militarizing Women's Lives*. Berkeley: University of California Press.

Latino Institute. 1994. *A Profile of Nine Latino Groups in Chicago*. Chicago IL: Latino Institute.

Latino Institute. 1995. *Facts on Chicago's Puerto Rican Population*. Chicago IL: Latino Institute.

Lipman, Pauline. 2003. Cracking Down: Chicago School Policy and the Regulation of Black and Latino Youth. In *Education as Enforcement: The Militarization and Corporatization of Schools*. Kenneth J. Saltman and David A. Gabbard, eds. New York and London: Routledge Falmer. Pp. 81–101.

Pérez, Gina. 2002. The other "Real World": Gentrification and the social construction of place in Chicago. *Urban Anthropology* 3 (1): 37–67.

Souza, Caridad. 2002. Sexual identities of young Puerto Rican mothers. *Dialogo* 6 Winter/Spring: 33–39.

Trujillo, Carla. 1991. Chicana Lesbians: Fear and Loathing in the Chicano Community. In *Chicana Critical Issues*. Norma Alarcon, Rafaela Castro, Emma Perez, Beatriz Pesquera, Adaljiza Sosa Riddel, and Patricia Zavella, eds. Berkeley: Third World Woman Press. Pp. 117–126.

Zavella, Patricia. 1997. "Playing with Fire": The Gendered Constructions of Chicana/Mexicana Sexuality. In *The Gender/Sexuality Reader: Culture, History, Political Economy*. Micaela di Leonardo and Roger Lancaster, eds. New York: Routledge. Pp. 392–408.

43

Unspeakable Offenses

Untangling Race and Disability in Discourses of Intersectionality

NIRMALA EREVELLES AND ANDREA MINEAR

In her essay, "Spirit Murdering the Messenger," Critical Race Feminist … (CRF) Patricia Williams describes the brutal murder of a poor, elderly, overweight, disabled, black woman by several heavily armed police officers. Trapped at the intersections of multiple oppressive contexts, Eleanor Bumpurs's tattered body was quite literally torn apart by her multiple selves—being raced, classed, gendered, and disabled. In the essay, Williams reads this murder as an unambiguous example of "racism [experienced] as … an offense so painful and assaultive as to constitute … 'spirit murder'" (230). Toward the end of the essay Williams struggles to fathom why the officer who fired the fatal shots saw such an '"immediate threat and endangerment to life' … [that he] could not allay his need to kill a sick old lady fighting off hallucinations with a knife" (234). In this quote, Williams recognizes Eleanor Bumpurs's disability when invoking her arthritis and possible mental illness. However, Williams deploys disability merely as a descriptor, a difference that is a matter of "magnitude" or "context," what another … scholar, Angela Harris, has described as "nuance theory" (14). According to Harris, "nuance theory constitutes black women's oppression as only an intensified example of (white) women's oppression" and is therefore used as the "ultimate example of how bad things [really] are" for all women (15).

While we agree with the critique of nuance theory in feminist analyses that ignore the real experiences of black women, we argue that CRF [critical race feminist] scholars deploy a similar analytical tactic through their unconscious non-analysis of disability as it intersects with race, class, and gender oppression. Disability, like race, offers not just a "nuance" to any analysis of difference. For example, one could argue that the outrage emanating from a heaving, black body wielding a knife sent a nervous (and racist) police officer into panic when confronted by his own racialized terror of otherness. But what about the other ideological terrors that loomed large in this encounter? Could the perception of Eleanor Bumpurs as a dangerous, obese, irrational, black woman also have

SOURCE: Erevelles, Nirmala, and Andrea Minear. 2013. "Unspeakable Offenses: Untangling Race and Disability in Discourses of Intersectionality." Pp. 354-368 in *The Disability Studies Reader*, 4th ed., edited by Leonard J. Davis. Liverpool, UK: Liverpool University Press. Republished with permission of Liverpool University Press. Permission conveyed through Copyright Clearance Center, Inc.

contributed to her construction as criminally "insane" (disability) because her reaction to a "mere" legal matter of eviction (class) was murderous rage? And did our socially sanctioned fears of the mentally ill and our social devaluation of disabled (arthritic and elderly) bodies of color justify the volley of shots fired almost instinctively to protect the public from the deviant, the dangerous, and the disposable? We, therefore, argue that in the violent annihilation of Eleanor Bumpurs's being, disability as it intersects with race, class, and gender served more than just a "context" or "magnifier" to analyze the oppressive conditions that caused this murder.

In this article, we demonstrate how the omission of disability as a critical category in discussions of intersectionality has disastrous and sometimes deadly consequences for disabled people of color caught at the violent interstices of multiple differences....

INTERSECTIONALITY AT THE CROSSROADS: THEORIZING MULTIPLICATIVE DIFFERENCES

With the deconstruction of essentialism, the challenge of how to theorize identity in all its complex multiplicity has preoccupied feminist scholars of color. Kimberle Crenshaw, one of the key proponents of the theory of intersectionality, has argued that "many of the experiences black women face are not subsumed within the traditional boundaries of race or gender oppression as these boundaries are currently understood" (358). Part of the problem of "relying on a static and singular notion of being or of identity" (Pastrana, 75) is that the single characteristic that is foregrounded (e.g. female or black) is expected to explain all of the other life experiences of the individual or the group. Additionally, Crenshaw points out that social movements based on a single identity politics (e.g. the Feminist Movement, Black Power Movement, GLBT and the Disability Rights Movement) have historically conflated or ignored intra-group differences and this has sometimes resulted in growing tensions between the social movements themselves.

Feminists of color have, therefore, had the difficult task of attempting to theorize oppression faced at the multiple fronts of race, class, gender, sexuality, and disability. Thus, if one is poor, black, elderly, disabled, and lesbian, must these differences be organized into a hierarchy such that some differences gain prominence over others? What if some differences coalesce to create a more abject form of oppression (e.g. being poor, black, and disabled) or if some differences support both privilege/invisibility within the same oppressed community (e.g. being black, wealthy, and gay)? What happens if we use "race" as a stable register of oppression against which other discriminations gain validity through their similarity and difference from that register? (Arondekar).

In the face of this theoretical challenge, intersectionality has been set up as the most appropriate analytical intervention expected to accomplish the formidable task of mediating multiple differences. For example, Patricia Hill Collins writes that "[a]s opposed to examining gender, race, class, and nation as separate systems of oppression, intersectionality explores how these systems mutually construct one another ..." (63)....

But this is all much easier said than done. Attempts to deploy intersectionality as an analytical tool in academic research have taken on different forms with varying analytical outcomes—some more useful than others....

POINTS OF CONTACT: AT THE INTERSECTION OF CRT AND DISABILITY STUDIES

In educational contexts, the association of race with disability has resulted in large numbers of students of color (particularly African American and Latino males) being subjected to segregation in so-called special-education classrooms through sorting practices such as tracking and/or through labels such as mild mental retardation and/or emotional disturbance. The PBS film, *Beyond Brown: Pursuing the Promise* (Haddad, Readdean, & Valadez) substantiates these claims with the following statistics.

- Black children constitute 17 percent of the total school enrollment, but 33 percent of those labeled "mentally retarded."
- During the 1998–1999 school year more than 2.2 million children of color in U.S. schools were served by special education. Post-high school outcomes for these students were striking. Among high school youth with disabilities, about 75 percent of African Americans compared to 39 percent of whites, are still not employed three to five years out of school. In this same time period, the arrest rate for African Americans with disabilities is 40 percent, compared to 27 percent for whites.
- States with a history of legal school segregation account for five of the seven states with the highest overrepresentation of African Americans labeled mentally retarded. They are Mississippi, South Carolina, North Carolina, Florida, and Alabama.
- Among Latino students, identification for special education varies significantly from state to state. Large urban school districts in California exhibit disproportionately large numbers of Latino English-language learners represented in special education classes in secondary schools.
- Some 20 percent of Latino students in grades 7 through 12 had been suspended from school according to statistics from 1999 compared with 15 percent of white students and 35 percent of African American students.

The association of race with disability has been extremely detrimental to people of color in the U.S.—not just in education, but also historically where associations of race with disability have been used to justify the brutality of slavery, colonialism, and neo-colonialism. Unfortunately, rather than nurturing an alliance between race and disability, CRT scholars (like other radical scholars) have mistakenly conceived of disability as a biological category, as an immutable and pathological abnormality rooted in the "medical language of symptoms and diagnostic categories" (Linton, 8). Disability studies scholars, on the other hand, have critiqued this "deficit" model of disability and have described disability as a socially constructed category that derives meaning and social (in) significance from the historical, cultural, political, and economic structures that frame social life.

Thus, at the first point of contact, both CRT and disability scholars begin with the critical assumption that race and disability are, in fact, social constructs. Thus, Haney Lopez explains "Biological race is an illusion.... Social race, however, is not.... Race has its genesis and maintains its vigorous strength in the realm of social beliefs" (172). Similarly, Garland-Thomson describes disability as "the attribution of corporeal deviance—not so much a property of bodies [but rather] ... a product of cultural rules about what bodies should be or do" (6). At their second point of contact, race and disability are both theorized as relational concepts. Thus, CRT scholars argue that "[r]aces are constructed relationally against one another, rather than in isolation" (Haney Lopez, 168) such that the privileges that Whites enjoy are linked to the subordination of people of color (Harris). Similarly, Lennard Davis points out that "our construction of the normal world is based on a radical repression of disability" (22) because "without the monstrous body to demarcate the borders of the generic ... and without the pathological to give form to the normal, the taxonomies of bodily value that underlie political, social and economic arrangements would collapse" (Garland-Thomson, 20). Finally, at the third point of contact, both perspectives use stories and first-person accounts to foreground the perspectives of those who have experienced victimization by racism and ableism first-hand....

In building on these alliance possibilities, disability-studies scholars have argued that disability is, in fact, constitutive of most social differences, particularly race.... One example to support the above claim lies in the historical narrative of eugenics as a program of selective breeding to prevent the degeneration of the human species. Colonial ideologies conceiving of the colonized races as intrinsically degenerate sought to bring these "bodies" under control via segregation and/or destruction. Such control was regarded as necessary for the public good. The association of degeneracy and disease with racial difference also translated into an attribution of diminished cognitive and rational capacities of non-white populations. Disability related labels such as feeble-mindedness and mental illness were often seen as synonymous with bodies marked oppressively by race (Baynton; Gould). Fearing that such characteristics could be passed down from generation to generation and further pose a threat to the dominant white race, "protective" practices such as forced sterilizations, rigid miscegenation laws, residential segregation in ghettoes, barrios, reservations and other state institutions and sometimes even genocide (e.g. the Holocaust) were brought to bear on non-white populations under the protected guise of eugenics. However, constructing the degenerate "other" was not just an ideological intervention to support colonialism....

It is easy to dismiss eugenics as a relic of a bygone era, but the continued association of race and disability in debilitating ways necessitates that we examine how eugenic practices continue to reconstitute social hierarchies in contemporary contexts via the deployment of a hegemonic ideology of disability that have real material effects on people located at the intersections of difference....

"SPIRIT MURDER" AND THE "NEW" EUGENICS: CRITICAL RACE THEORY MEETS DISABILITY STUDIES

... Police brutality, false imprisonment, and educational negligence are commonplace in the lives of people of color—especially those who are located at the margins of multiple identity categories. So common are these practices that CRF scholar Patricia Williams has argued that these kinds of assaults should not be dismissed as the "odd mistake" but rather be given a name that associates them with criminality. Her term for such assaults on an individual's personhood is "spirit murder," which she describes as the equivalent of body murder....

... Clearly, in our educational institutions there are millions of students of color, mostly economically disadvantaged and disabled, for whom spirit murder is the most significant experience in their educational lives. In fact, it is this recognition of spirit murder in the everyday lives of disabled students of color that forges a critical link between disability studies and CRT/F through the intercategorical analysis of intersectionality. In other words, utilizing an intercategorical analysis from the critical standpoint of disability studies will foreground the structural forces in place that constitute certain students as a surplus population that is of little value in both social and economic terms. That most of these students are poor, disabled, and of color is critical to recognize from within a CRT/F perspective. By failing to undertake such an analysis, we could miss several political opportunities for transformative action.

WORKS CITED

Arondekar, Anjali. "Border/Line Sex: Queer Positionalities, or How Race Matters Outside the United States." *Interventions* 7.2 (2005): 236–50.

Baynton, Douglas. "Disability in History." *Perspectives* 44.9 (2006): 5–7.

Crenshaw, Kimberlé. "Mapping the Margins: Intersectionality, Identity Politics, and Violence against Women." *Critical Race Theory: The Key Writings that Formed the Movement.* Ed. Kimberlé Crenshaw, Neil Gotanda, Gary Pellar, and Kendall Thomas. New York: New Press, 1996. 357–83.

Davis, Lennard J. *Enforcing Normalcy: Disability, Deafness, and the Body.* New York: Verso, 1995.

Garland-Thomson, Rosemarie. *Extraordinary Bodies: Figuring Physical Disability in American Culture and Literature*, New York: Columbia UP, 1997.

Gould, Stephen. J. *The Mismeasure of Man.* New York: Norton, 1981.

Haney Lopez, Ian. F. "The Social Construction of Race." *Critical Race Theory: The Cutting Edge*. Ed. Richard Delgado and Jean Stefancic. Philadelphia: Temple UP, 2007. 163–75.

Linton, Simi. *Claiming Disability: Knowledge and Identity*. New York: New York UP, 1998.

Pastrana, Antonio. "Black Identity Constructions: Inserting Intersectionality, Bisexuality, and (Afro-) Latinidad into Black Studies." *Journal of African American Studies* 8. 1–2 (2004): 74–89.

Williams, Patricia. J. "Spirit Murdering the Messenger: The Discourse of Fingerpointing as the Law's Response to Racism." *Critical Race Feminism: A Class Reader*. Ed. Adrien. K. Wing. New York: New York UP, 1997. 229–42.

44

Representations of Latina/o Sexuality in Popular Culture

DEBORAH R. VARGAS

CRITICAL ANALYSIS OF THE ISSUES: REPRESENTATION AS STEREOTYPES

As the twentieth century gave way to the new millennium, popular culture representations of Latina/o sexuality seemed omnipresent. From late-night talk shows, music videos, and movies to the popularity of sports stars utilizing media to become popular icons, our visual and sonic senses have been met with the booty-shaking moves of Selena, Ricky Martin, and Jennifer Lopez while even boxer Oscar De La Hoya took advantage of his suave sexy image and heavily Latina/o fan base to enter the popular arena of *ranchero* music....

One inescapable representation of Latino masculinity and sexuality is the "Latin lover," most notably and problematically associated with Rudolph Valentino. The fact that Valentino, an Italian immigrant, was represented as "Latin" serves to remind us of the range of ethnicities—from the Mexican actor Gilbert Roland to the Italian Valentino—that fell within the racially charged Latin lover category in early cinema. In the early twentieth century, the term Latin was not synonymous with Latin American but encompassed both Europeans who spoke Latinate languages, such as Italians and Spaniards, and Latin Americans. As Clara E. Rodriguez argues, a key racialized shift occurs in the U.S. context when non-Hispanic European ethnics, such as Italians, assimilate into whiteness, leaving non-European ethnics to bear the brunt of the racist stereotypes of excess and deviant sexuality.[1]

For the most part, the Latin lover trope remained a dominant pathologization of Latino masculinity in order to maintain the masculinist hegemony of the great white hero as emblematized, for instance, in films starring John Wayne or Clint Eastwood.... From Ramón Novarro in the 1920s to Cesar Romero in the 1940s and Ricardo Montalbán and Fernando Lamas in the 1950s, Latino actors

SOURCE: Vargas, Deborah R. 2010. "Representations of Latina/o Sexuality in Popular Culture." Pp. 117–136 in *Latina/o Sexualities: Probing Powers, Passions, Practices, and Policies*, edited by Marysol Asencio. New Brunswick, NJ: Rutgers University Press. Reprinted by permission.

were typecast as Latin lovers, and found it difficult to establish their careers beyond the predictable suave sexuality hinting at danger and perhaps contagion. The Latin lover became the "possessor of a primal sexuality that made him capable of making a sensuous but dangerous—and clearly non-WASP—brand of love.[2]

The Latin lover stereotype has extended to other constructions of masculinity on television and in cinema as well. Including figures like that of the boxer and *pachuco* or *cholo*. Boxing in particular has provided one of the most visible stages for performing Latino hypermasculinity as virile and physically dominant. Actual sports figures and fictional narratives of boxers exemplify the extension of the Latin lover figure....

For Latinas, representations of their sexuality have been dichotomized as virgin/whore or mother/prostitute—what Chicana feminist scholars refer to as "La Virgen/La Malinche" dichotomy. These polarized representations have their origins in the cinematic figures of the demure señoritas and spitfires of the 1920s and 1930s, respectively.... Once a stock figure in the genre of Western films, the equivalent of the Latin lover for Latinas his been the lusty, hot-tempered harlot, a character most widely associated with Lupe Vélez Having starred in eight movies as the "Mexican spitfire" Carmelita Woods, Vélez herself was often described as "the hot baby of Hollywood" or "just a Mexican wild kitten" in publicity photo captions. In the following assertion.... Katy Jurado best exemplifies this stereotype in her role as the home-wrecking other woman. Helen Ramirez, opposite Gary Cooper's Will Kane in *High Noon* (1952).

Historically for Latinas, those extreme representations as either hot-blooded spitfire or dutiful mother have reified dominant notions of ostensibly American subjectivity, citizenship, and family, especially in politically tense contexts. As feminist analyses of nationalism argue, notions of country, homeland, region, locality, and ethnicity are constructed through the racialization, sexualization, and genderization of female corporeality. In a U.S. context, then, representations of Latina sexuality as hot-blooded and excessive become the markers of what is morally wrong, set against the good morals and hegemonic U.S. citizenship values of non-Latino whites....

More contemporary reproductions of Latinas have desexualized their gender constructions into that of mother or domestic worker. This strategy has been the result of what I contend is a resistance strategy to overly correct hegemonic hypersexual representations. As feminist analyses of Latina/o cultural representations demonstrate, there is a dual pressure based on U.S. nationalist and Chicano nationalist projects to represent Latina sexuality as domestic, virginal and asexual.... Almost always in the background of movies that focus on heroic Latino characters looms the sacrificing mother....

... [T]he notion of positive Latina characters often resulted in a reification of Latina femininity as heteronormative mother or daughter....

... Extremes of hyper- or asexuality have left Latinas in roles that represent little agency over their sexuality.... A combination of "spitfire" and comedic release, Carmen Miranda-born Maria do Carmen Miranda da Cunha in Portugal and raised in Brazil, and nicknamed the "lady in the tutti-frutti hat"—best exemplifies the representation of Latinas as sensually playful and sexually primitive.... Miranda was a key figure of Latina representation during the era of the Good Neighbor policy initiated by President Franklin Roosevelt that emphasized cooperation and trade with Latin America in order to maintain U.S. influence and deter any possible political, economic, and cultural invasion by Germany and its allies before and during World War II. Miranda's playful sexuality came to embody Latin America as a cultural and sexual playground. Miranda's excessive yet clownish femininity—marked by her platform shoes, energetic rhythmic dancing, and massive exotic headdresses—represented a sexual subjectivity not threatening to white hegemonic femininity. In this historical context, Latinas and Latinos were represented as infantalized characters or by cheerful musical performances, such as those by Xavier Cugat and Desi Amaz....

COLOR MATTERS

A key issue with regard to the representations of Latina/o sexuality in popular culture is that of race, or more precisely, color. Following the popular mantra that social structural systems of power matter, color matters when it comes to the configuration of demonized, pathologized, and desexualized Latina/o representations.... [C]ontemporary figures of hyper-sexual Latino masculinity like the gangster, drug runner, and inner-city urban homeboy—ruthlessly in pursuit of vulgar cravings for money, power, and sexual pleasure—have their origins in the *bandido* figure dating back to early "greaser" films such as *Broncho Billy and the Greaser* (1914).... The Mexican bandit—most often of dark skin—whose outlaw characteristics include deviant sexuality, made numerous appearances in films throughout the twentieth century....

Color matters with regard to representations of Latina/o sexuality in the arena of popular music as well....

In our contemporary moment of popular culture, particularly in the arena of music, representations of Latino/o sexuality direct attention to the varied manifestations of racialized subjectivities within the panethnic formation of *Latinidad* in the U.S. public sphere....

The politics of color and representation is thus at the center of much feminist analysis of Latina/o pop icons. For instance, analyses of Jennifer Lopez, the highest-paid musical and cinematic Latina artist in the United States, call attention to the ways in which her lighter skin tone has imparted her with access to a range of sexual subjectivities in ways a darker woman can not without risking negative consequences. Similarly and insightfully. Maria Elena Cepeda notes how Shakira (who once declared in a press release her "desire to seduce the U.S."), Lopez, and Christina Aguilera have grown thinner and blonder as they have moved up the ladder of success....

IMPLICATIONS FOR POLICY

At the turn of the new century, popular culture remains, to paraphrase Stuart Hall, a highly charged arena of consent and resistance for minority populations in the United States. In particular, public cultural production, especially of movies, music, television programs, public performance, and art—because of technological reach—is increasingly the sphere in which representations of Latina/o sexualities provide the main ideological frame by which to make sense of policies related to immigration, education, criminology, and health care among others. Certainly, popular culture has had a close relationship to policy agendas in the United States, yet most often in terms of the media regulating offensive language, nudity, and copyright issues. For the most part, popular culture has not been the arena policy-makers have turned to for political insight or for models addressing social inequality....

Representations of family, criminality, and sexual deviance are not mere sideshows: they are central to discourses that frame policy arguments such as those justifying more prisons and fewer social welfare programs for single mothers. Equally as problematic are public discourses forging "positive' representations of familia narratives in movies or untenable icons such as Oscar De La Hoya's "Golden Boy" masculinity. Such normative representations are powerful, if ill-conceived, mechanisms for perpetuating Latina/os as assimilatable citizens; they do little to contest racist and sexist hierarchies of masculinity and femininity among Latina/os. Heterogeneous representations of Latina/o sexualities in popular culture need to be more seriously integrated in policy agendas addressing Latina/o poverty, illiteracy, educational attainment and incarceration. In this way, more funding for independent cultural productions is critical as this arena often has presented the public with alternative ideologies and discourses of gender and sexuality not found in commercial venues....

NOTES

1. C. E. Rodrigues, *Herces, Lovers, and Others: The Story of Latinos in Hollywood* (Washington, D.C.: Smithsonion Books, 2004).
2. Fregoso, Mexicana Encounters C. Ramirez-Berg, *Latino Images in Film* (Austin: University of Texas Press, 2002): quotations on 67.76. Rodriguez, *Heroes, Lovers and Others*. Berumen, *Chicana/Hispanic Image in American Film*.

45

Where's the Honor? Attitudes toward the "Fighting Sioux" Nickname and Logo

DANA M. WILLIAMS

The practice of American sports teams using racial nicknames and mascots is … and has been controversial. By far, the most common racial nicknames are of Native Americans (Nuessel, 1994).... This practice has been criticized for fostering inaccurate stereotypes of a discriminated-against racial minority group. Indeed, there are many historical, political, economic, and sociological factors that can be used to explain the existence of Native American sport nicknames and why they have prevailed over time. There are strong views both in favor of retaining and eliminating this practice.

What do Native Americans themselves think about these images? Native American tribes and organizations that have condemned the practice as racist stereotyping have provoked heated reaction and controversy, particularly from sports fans. Many authors and commentators have argued that Native Americans have a wide variety of views about their likenesses being employed by sports teams....

Recent actions by the National Collegiate Athletics Association (NCAA) to prohibit the use of Native American nicknames and mascots in college sports because such names are "hostile or abusive" (NCAA, 2005) and overwhelming opposition to the practice from Native American organizations and most tribal leaders (AISTM, 2006) have led to further contention, because some schools have been granted exceptions due to local tribal support for names. Yet there have been few quantitative studies comparing the attitudes of individual Natives and Whites. In addition, no research has comprehensively shown the extent of opinion and feeling by the most directly affected constituency at a school that uses Native nicknames, Native Americans students themselves.

The purpose of this study was to explore the present-day attitudes of students towards a specific instance of this practice, specifically UND's "Fighting Sioux" nickname and logo. I seek to explain the attitudes of Native, White, and non-Native minority UND students regarding the Fighting Sioux, and how these attitudes may differ across demographics and within UND. This school presents a

SOURCE: Williams, Dana M. 2007. *Sociology of Sport Journal* 24: 437–456. © Human Kinetics, Inc. Abridged and republished with permission.

unique test case; UND is close to many Native reservations in the Northern Great Plains and has a sizable Native student enrollment including actual Lakotan (or "Sioux") peoples. Thus, these conditions represent an ideal situation in which to explore the opinions of a substantial Native student population regarding a Native American nickname at a predominantly White university....

RESEARCH ON NATIVE AMERICAN NICKNAMES

Academic research has begun to cover a wide variety of issues related to the practice of American sports teams using Native American imagery. Evidence indicates that even well intentioned acts, however historically inaccurate that intent is deployed, can have negative consequences for Native Americans. The American Psychological Association (APA, 2005) has determined that "the continued use of American Indian mascots, symbols, images, and personalities by school systems appear to have a negative impact on the self-esteem of American Indian children" (p. 1), an impact that may negatively affect life chances. Social-psychological research has demonstrated that Native logos harm Native youth in a variety of ways (Fryberg, 2003). When presented with such imagery, Native youth exhibited decreased self-esteem, lowered self-efficacy, and diminished perception of their potential achievement. Ironically, the use of same imagery increased the self-esteem of White youth.

LaRocque (2004) studied the reactions of a sample of UND students to images related to the school's nickname and logo ("neutral" slides) and related controversial elements ("controversial" slides). The results indicated that White and Native American students had significantly different reactions to the images they were shown. Native students experienced higher levels of dysphoria, anxiety, and depression after viewing the "neutral" slide show, whereas White students experienced no distress viewing standard images of the Fighting Sioux. White students did, however, exhibit a major increase in hostility after watching the "controversial" slide show. LaRocque concluded that White students at UND were less likely than Native Americans to see racist imagery regarding the Fighting Sioux nickname as disrespectful.

Not only do Native people suffer from Native American nicknames and Whites benefit from them, but sports fans also tend to be strong supporters of keeping such team nicknames (Sigelman, 1998). King and Springwood (2000) argue that sports reinforce stereotypes about Native Americans as fans try to "play Indian" with the misperceived "fighting spirit" of Native peoples in the combat of sport. The visceral violence of some sports is linked to a particularly masculine expression of school pride, which in turn can provoke a combative relationship toward any critic of the Fighting Sioux nickname—particularly for fans of the most violent sports like football and hockey (Williams, 2006). Sigelman (1998) found that football fans were significantly less likely to want the Washington Redskins nickname changed than nonfans.

Whites, who are the most militant defenders of these nickname practices, view nonsupportive Native Americans as disorderly interruptions in the acceptable discourse of sports (Farnell, 2004). Consequently, Native criticisms of nicknames are positioned as "politically correct" or of having no appreciation for school spirit....

Scholars have suggested that the use of Native American imagery as mascots warps not only the view of Native Americans held by the dominant White culture but also harms the relationship between the two groups. The common representations of Native people by sports teams—performances and imagery—are merely tokens of a still-existing people who are not permitted to speak for themselves and are consequentially fetishized (Slowikowski, 1993). Thus, the "mascotting" of Native Americans perpetuates White hegemony over Natives and forces false "unity" between Native Americans and Whiles (Black 2002)....

DATA AND METHODS

... Data for this study come from a four-part survey conducted in 2000 by UNO's Social Science Research Institute (SSRI) at the behest of the UND Presidential Nickname Commission. The Commission was charged with the goal of determining any possible directions for change at UND regarding its nickname. SSRI did a random sample of four constituent populations: alumni, employees, students, and minority students at UND....

DISCUSSION

... In the present study Whites demonstrated the greatest support for the Fighting Sioux nickname, more than non-Native minorities who in turn support the nickname to a greater extent than do Native Americans. Consequently, the data strongly demonstrated that Native students at UND did *not* share the positive sentiments expressed by their White classmates. Thus, claims that Native Americans find the UND nickname honorable were not supported by these data.

Driscoll and Schieve's (1987) survey of Native American students at UND in the 1980s showed support *for* the nickname while the results of the current study have demonstrated the opposite. There are a number of possible reasons for this, some of which are statistical. First, the current measures are more valid for reasons previously discussed. The present study used two highly reliable measures, each composed of three questions, that measured the notions of respect and change in a more thorough fashion. Bias in how the questions were asked is a possibility, because people will respond differently depending on how questions are asked. Second, weaknesses in the earlier sampling methodology could have resulted in a nonrepresentative sample of Native Americans compared with the more thorough and comprehensive sampling done for the more recent 2000 SSRI survey used in this study. The SSRI sampled more than half of the UND students who were officially registered as Native American, resulting in a more representative and accurate pool of respondents. Third, a shift in the sentiments of Native American students at UND from 1987 to 2000, or in the type of Native students who attend the school, could have occurred.

Sigelman's (1998) finding that sports fans were less favorable to changing the Washington Redskins nickname is echoed in the descriptive findings of the SSRI data. UND students who attended less than five sporting events were significantly more likely to want the Fighting Sioux nickname changed than students who had attended more than five events. Football and men's basketball fans in particular had significantly greater resistance to change than those who did not attend these sporting events. Clearly, sport attendance is associated with stronger demands to keep UND's nickname. This seems sensible on the face of it: sports fans have a vested interest in retaining the team identities they support. Additionally, those students less inclined to sports attendance are bound to be less committed to such imagery and less concerned with the possibility of seeing the nickname removed. Incidentally, men were significantly more likely than women to attend football and men's basketball events. Since women were more inclined to change at a bivariate level (and when considering respect at the multivariate level), this gender difference could help explain why the attendees of these two sports are less inclined to change. Yet, none of these findings appear to bolster the color-blind racism framework, since there are no expressed claims that "race does not matter" here. In fact, race is always present at sporting events—on jerseys, signs, loudspeaker announcements, in concession stands, and audience cheers....

Although the current study did not indicate whether the Fighting Sioux nickname is specifically "hostile or abusive" as charged by the NCAA (2005), Native American students did not view the nickname as respectful. In light of these findings and psychological research that Native youth and students are harmed by these practices (Fryberg, 2003; LaRocque, 2004), the NCAA case against the use of Native American nicknames is strongly bolstered.

This research has demonstrated that White students at UND tend to adhere to a racial and racist view of a nickname that Native students themselves reject. The strong negative correlation between respect and change—the greater the perception of disrespect, the greater the support for change—is a finding that should not be ignored by nickname-change advocates. With an increased understanding that most Native American students see the nickname as disrespectful, support for a nickname change is likely to increase....

REFERENCES

American Indian Sports Team Mascots. (2006). *Lists of organizations endorsing retirement of "Indian" sports team tokens*. Retrieved September 26, 2006, from http://aistm.org/fr.groups.htm.

American Psychological Association. (2005, October 18). *APA resolution recommending the immediate retirement of American Indian mascots, symbols, images, and personalities by schools, colleges, universities, athletic teams, and organizations*. Retrieved January 5, 2006, from http://www.apa.org/releases/ResAmIndianMascots.pdf.

Black, J.E. (2002). The 'mascotting' of Native America. *American Indian Quarterly, 26*, 605–622.

Driscoll, T., & Schieve, D. (1987). *UND Native American logo survey 1987* (RR-SA-070787). Grand Forks, ND: University of North Dakota.

Farnell, B. (2004). The fancy dance of racializing discourse. *Journal of Sport & Social Issues, 28*, 30–55.

Fryberg, S.A. (2003). *Really? You don't look like an American Indian: Social representations and social group identities* (Doctoral dissertation, Stanford University).

King, C.R., & Springwood, C.F. (2000). Fighting spirits: The racial politics of sports mascots. *Journal of Sport & Social Issues, 24*, 282–304.

LaRocque, A. (2004), *Psychological distress between American Indian and majority culture college students regarding the use of the Fighting Sioux nickname and logo* (Doctoral dissertation, University of North Dakota).

National Collegiate Athletic Association. (2005, August 5). *NCAA Executive Committee issues guidelines for use of Native American mascots at championship events.* Retrieved August 24, 2005, from http://www.ncaasports.com/story/8706763.

Nuessel, F. (1994). Objectionable sport team designations. *Names, 42*, 101–119.

Sigelman, L. (1998). Hail to the Redskins? Public reactions to a racially insensitive team name. *Sociology of Sport Journal, 15*, 317–325.

Slowikowski, S.S. (1993). Cultural performance and sport mascots. *Journal of Sport & Social Issues, 17*, 23–33.

Williams, D.M. (2006). Patriarchy and 'the Fighting Sioux': A gendered look at racial college sports nicknames. *Race, Ethnicity & Education, 9*, 325–340.

46

Media Magic

Making Class Invisible

GREGORY MANTSIOS

Of the various social and cultural forces in our society, the mass media is arguably the most influential in molding public consciousness. Americans spend an average twenty-eight hours per week watching television. They also spend an undetermined number of hours reading periodicals, listening to the radio, and going to the movies. Unlike other cultural and socializing institutions, ownership and control of the mass media is highly concentrated. Twenty-three corporations own more than one-half of all the daily newspapers, magazines, movie studios, and radio and television outlets in the United States.[1] The number of media companies is shrinking and their control of the industry is expanding. And a relatively small number of media outlets is producing and packaging the majority of news and entertainment programs. For the most part, our media is national in nature and single-minded (profit-oriented) in purpose. This media plays a key role in defining our cultural tastes, helping us locate ourselves in history, establishing our national identity, and ascertaining the range of national and social possibilities. In this essay, we will examine the way the mass media shapes how people think about each other and about the nature of our society.

The United States is the most highly stratified society in the industrialized world. Class distinctions operate in virtually every aspect of our lives, determining the nature of our work, the quality of our schooling, and the health and safety of our loved ones. Yet remarkably, we, as a nation, retain illusions about living in an egalitarian society. We maintain these illusions, in large part, because the media hides gross inequities from public view. In those instances when inequities are revealed, we are provided with messages that obscure the nature of class realities and blame the victims of class-dominated society for their own plight. Let's briefly examine what the news media, in particular, tells us about class.

ABOUT THE POOR

The news media provides meager coverage of poor people and poverty. The coverage it does provide is often distorted and misleading.

SOURCE: Mantsios, Gregory. 1998. "Media Magic: Making Class Invisible." In *Race, Class, and Gender in the United States: An Integrated Study*, 4th ed., edited by Paula Rothenberg. New York: St. Martin's Press. Reprinted with permission of the author.

The Poor Do Not Exist

For the most part, the news media ignores the poor. Unnoticed are forty million poor people in the nation—a number that equals the entire population of Maine, Vermont, New Hampshire, Connecticut, Rhode Island, New Jersey, and New York combined. Perhaps even more alarming is that the rate of poverty is increasing twice as fast as the population growth in the United States. Ordinarily, even a calamity of much smaller proportion (e.g., flooding in the Midwest) would garner a great deal of coverage and hype from a media usually eager to declare a crisis, yet less than one in five hundred articles in the *New York Times* and one in one thousand articles listed in the *Readers Guide to Periodic Literature* are on poverty. With remarkably little attention to them, the poor and their problems are hidden from most Americans.

When the media does turn its attention to the poor, it offers a series of contradictory messages and portrayals.

The Poor Are Faceless

Each year the Census Bureau releases a new report on poverty in our society and its results are duly reported in the media. At best, however, this coverage emphasizes annual fluctuations (showing how the numbers differ from previous years) and ongoing debates over the validity of the numbers (some argue the number should be lower, most that the number should be higher). Coverage like this desensitizes us to the poor by reducing poverty to a number. It ignores the human tragedy of poverty—the suffering, indignities, and misery endured by millions of children and adults. Instead, the poor become statistics rather than people.

The Poor Are Undeserving

When the media does put a face on the poor, it is not likely to be a pretty one. The media will provide us with sensational stories about welfare cheats, drug addicts, and greedy panhandlers (almost always urban and Black). Compare these images and the emotions evoked by them with the media's treatment of middle class (usually white) "tax evaders," celebrities who have a "chemical dependency," or wealthy businesspeople who use unscrupulous means to "make a profit." While the behavior of the more affluent offenders is considered an "impropriety" and a deviation from the norm, the behavior of the poor is considered repugnant, indicative of the poor in general, and worthy of our indignation and resentment.

The Poor Are an Eyesore

When the media does cover the poor, they are often presented through the eyes of the middle class. For example, sometimes the media includes a story about community resistance to a homeless shelter or storekeeper annoyance with panhandlers. Rather than focusing on the plight of the poor, these stories are about middle-class opposition to the poor. Such stories tell us that the poor are an inconvenience and an irritation.

exp.) minimum wage?

The Poor Have Only Themselves to Blame

In another example of media coverage, we are told that the poor live in a personal and cultural cycle of poverty that hopelessly imprisons them. They routinely center on the Black urban population and focus on perceived personality or cultural traits that doom the poor. While the women in these stories typically exhibit an "attitude" that leads to trouble or a promiscuity that leads to single motherhood, the men possess a need for immediate gratification that leads to drug abuse or an unquenchable greed that leads to the pursuit of fast money. The images that are seared into our mind are sexist, racist, and classist. Census figures reveal that most of the poor are white not Black or Hispanic, that they live in rural or suburban areas not urban centers, and hold jobs at least part of the year.[2] Yet, in a fashion that is often framed in an understanding and sympathetic tone, we are told that the poor have inflicted poverty on themselves.

The Poor Are Down on Their Luck

During the Christmas season, the news media sometimes provides us with accounts of poor individuals or families (usually white) who are down on their luck. These stories are often linked to stories about soup kitchens or other charitable activities and sometimes call for charitable contributions. These "Yule time" stories are as much about the affluent as they are about the poor: they tell us that the affluent in our society are a kind, understanding, giving people—which we are not. The series of unfortunate circumstances that have led to impoverishment are presumed to be a temporary condition that will improve with time and a change in luck.

Despite appearances, the messages provided by the media are not entirely disparate. With each variation, the media informs us what poverty is not (i.e., systemic and indicative of American society) by informing us what it is. The media tells us that poverty is either an aberration of the American way of life (it doesn't exist, it's just another number, it's unfortunate but temporary) or an end product of the poor themselves (they are a nuisance, do not deserve better, and have brought their predicament upon themselves).

By suggesting that the poor have brought poverty upon themselves, the media is engaging in what William Ryan has called "blaming the victim."[3] The media identifies in what ways the poor are different as a consequence of deprivation, then defines those differences as the cause of poverty itself. Whether blatantly hostile or cloaked in sympathy, the message is that there is something fundamentally wrong with the victims—their hormones, psychological make up, family environment, community, race, or some combination of these—that accounts for their plight and their failure to lift themselves out of poverty.

But poverty in the United States is systemic. It is a direct result of economic and political policies that deprive people of jobs, adequate wages, or legitimate support. It is neither natural nor inevitable: there is enough wealth in our nation to eliminate poverty if we chose to redistribute existing wealth or income. The plight of the poor is reason enough to make the elimination of poverty the nation's first priority. But poverty also impacts dramatically on the non-poor. It has a dampening effect on wages in general (by maintaining a reserve army of unemployed and underemployed anxious for any job at any wage) and breeds crime and violence (by maintaining conditions that invite private gain by illegal means and rebellion-like behavior, not entirely unlike the urban riots of the 1960s). Given the extent of poverty in the nation and the impact it has on us all, the media must spin considerable magic to keep the poor and the issue of poverty and its root causes out of the public consciousness.

ABOUT EVERYONE ELSE

Both the broadcast and the print news media strive to develop a strong sense of "we-ness" in their audience. They seek to speak to and for an audience that is both affluent and like-minded. The media's solidarity with affluence, that is, with the middle and upper class, varies little from one medium to another. Benjamin DeMott points out, for example, that the *New York Times* understands affluence to be intelligence, taste, public spirit, responsibility, and a readiness to rule and "conceives itself as spokesperson for a readership awash in these qualities."[4] Of course, the flip side to creating a sense of "we," or "us," is establishing a perception of the "other." The other relates back to the faceless, amoral, undeserving, and inferior "underclass." Thus, the world according to the news media is divided between the "underclass" and everyone else. Again the messages are often contradictory.

The Wealthy Are Us

Much of the information provided to us by the news media focuses attention on the concerns of a very wealthy and privileged class of people. Although the concerns of a small fraction of the populace, they are presented as though they were the concerns of everyone. For example, while relatively few people actually own stock, the news media devotes an inordinate amount of broadcast time and print space to business news and stock market quotations. Not only do business reports cater to a particular narrow clientele, so do the fashion pages (with $2,000 dresses), wedding announcements, and the obituaries. Even weather and sports news often have a class bias. An all news radio station in New York City, for example, provides regular national ski reports. International news, trade agreements, and domestic policies issues are also reported in terms of their impact on business climate and the business community. Besides being of practical value to the wealthy, such coverage has considerable ideological value. Its message: the concerns of the wealthy are the concerns of us all.

The Wealthy (as a Class) Do Not Exist

While preoccupied with the concerns of the wealthy, the media fails to notice the way in which the rich as a class of people create and shape domestic and foreign policy. Presented as an aggregate of individuals, the wealthy appear without special interests, interconnections, or unity in purpose. Out of public view are the class interests of the wealthy, the interlocking business links, the concerted actions to preserve their class privileges and business interests (by running for public office, supporting political candidates, lobbying, etc.). Corporate lobbying is ignored, taken for granted, or assumed to be in the public interest. (Compare this with the media's portrayal of the "strong arm of labor" in attempting to defeat trade legislation that is harmful to the interests of working people.) It is estimated that two-thirds of the U.S. Senate is composed of millionaires. Having such a preponderance of millionaires in the Senate, however, is perceived to be neither unusual nor antidemocratic; these millionaire senators are assumed to be serving "our" collective interests in governing.

The Wealthy Are Fascinating and Benevolent

The broadcast and print media regularly provide hype for individuals who have achieved "super" success. These stories are usually about celebrities and superstars from the sports and entertainment world. Society pages and gossip columns serve to keep the social elite informed of each others' doings, allow the rest of us to gawk at their excesses, and help to keep the American dream alive. The print media is also fond of feature stories on corporate empire builders. These stories provide an occasional "insider's" view of the private and corporate life of industrialists by suggesting a rags to riches account of corporate success. These stories tell us that corporate success is a series of smart moves, shrewd acquisitions, timely mergers, and well thought out executive suite shuffles. By painting the upper class in a positive light, innocent of any wrongdoing (labor leaders and union organizations usually get the opposite treatment), the media assures us that wealth and power are benevolent. One person's capital accumulation is presumed to be good for all. The elite, then, are portrayed as investment wizards, people of special talent and skill, who even their victims (workers and consumers) can admire.

The Wealthy Include a Few Bad Apples

On rare occasions, the media will mock selected individuals for their personality flaws. Real estate investor Donald Trump and New York Yankees owner George Steinbrenner, for example, are admonished by the media for deliberately seeking publicity (a very un-upper class thing to do); hotel owner Leona Helmsley was caricatured for her personal cruelties; and junk bond broker Michael Milkin was condemned because he had the audacity to rob the rich. Michael Parenti points out that by treating business wrongdoings as isolated deviations from the socially beneficial system of "responsible capitalism," the media overlooks the features of the system that produce such abuses and the regularity with which they occur. Rather than portraying them as predictable and frequent outcomes of corporate power and the business system, the media treats abuses as if they were isolated and atypical. Presented as an occasional aberration, these incidents serve not to challenge, but to legitimate, the system.[6]

The Middle Class Is Us

By ignoring the poor and blurring the lines between the working people and the upper class, the news media creates a universal middle class. From this perspective, the size of one's income becomes largely irrelevant: what matters is that most of "us" share an intellectual and moral superiority over the disadvantaged. As *Time* magazine once concluded, "Middle America is a state of mind."[7] "We are all middle class," we are told, "and we all share the same concerns": job security, inflation, tax burdens, world peace, the cost of food and housing, health care, clean air and water, and the safety of our streets. While the concerns of the wealthy are quite distinct from those of the middle class (e.g., the wealthy worry about investments, not jobs), the media convinces us that "we [the affluent] are all in this together."

The Middle Class Is a Victim

For the media, "we" the affluent not only stand apart from the "other"—the poor, the working class, the minorities, and their problems—"we" are also victimized by the poor (who drive up the costs of maintaining the welfare roles), minorities (who commit crimes against us), and by workers (who are greedy and drive companies out and prices up). Ignored are the subsidies to the rich, the crimes of corporate America, and the policies that wreak havoc on the economic well-being of middle America. Media magic convinces us to fear, more than anything else, being victimized by those less affluent than ourselves.

The Middle Class Is Not a Working Class

The news media clearly distinguishes the middle class (employees) from the working class (i.e., blue collar workers) who are portrayed, at best, as irrelevant, outmoded, and a dying breed. Furthermore, the media will tell us that the hardships faced by blue collar workers are inevitable (due to progress), a result of bad luck (chance circumstances in a particular industry), or a product of their own doing (they priced themselves out of a job). Given the media's presentation of reality, it is hard to believe that manual, supervised, unskilled, and semiskilled workers actually represent more than 50 percent of the adult working population.[8] The working class, instead, is relegated by the media to "the other."

In short, the news media either lionizes the wealthy or treats their interests and those of the middle class as one in the same. But the upper class and the middle class do not share the same interests or worries. Members of the upper class worry about stock dividends (not employment), they profit from inflation and global militarism, their children attend exclusive private schools, they eat and live in a royal fashion, they call on (or are called upon by) personal physicians, they have few consumer problems, they can escape whenever they want from environmental pollution, and they live on street and travel to other areas under the protection of private police forces.*[9]

The wealthy are not only a class with distinct life-styles and interests, they are a ruling class. They receive a disproportionate share of the country's yearly income, own a disproportionate amount of the country's wealth, and contribute a disproportionate number of their members to governmental bodies and decision-making groups—all traits that William Domhoff, in his classic work *Who Rules America,* defined as characteristic of a governing class.[10]

This governing class maintains and manages our political and economic structures in such a way that these structures continue to yield an amazing proportion of our wealth to a minuscule upper class. While the media is not above referring to ruling classes in other countries (we hear, for example, references to Japan's ruling elite),[11] its treatment of the news proceeds as though there were no such ruling class in the United States.

Furthermore, the news media inverts reality so that those who are working class and middle class learn to fear, resent, and blame those below, rather than those above them in the class structure. We learn to resent welfare, which accounts for only two cents out of every dollar in the federal budget (approximately $10 billion) and provides financial relief for the needy,** but learn little about the $11 billion the federal government spends on individuals with incomes in excess of $100,000 (not needy),[12] or the $17 billion in farm subsidies, or the $214 billion (twenty times the cost of welfare) in interest payments to financial institutions.

Middle-class whites learn to fear African Americans and Latinos, but most violent crime occurs within poor and minority communities and is neither inter-racial† nor interclass. As horrid as such crime is, it should not mask the destruction and violence perpetrated by corporate America. In spite of the fact that 14,000 innocent people are killed on the job each year, 100,000 die prematurely, 400,000 become seriously ill, and 6 million are injured from work-related accidents and diseases, most Americans fear government regulation more than they do unsafe working conditions.

Through the media, middle-class—and even working-class—Americans learn to blame blue collar workers and their unions for declining purchasing power and economic security. But while workers who managed to keep their jobs and their unions struggled to keep up with inflation, the top 1 percent of American families saw their average incomes soar 80 percent in the last decade.[13]

---

*The number of private security guards in the United States now exceeds the number of public police officers. (Robert Reich, "Secession of the Successful," *New York Times Magazine,* February 17, 1991, p. 42.)

**A total of $20 billion is spent on welfare when you include all state funding. But the average state funding also comes to only two cents per state dollar.

†In 92 percent of the murders nationwide the assailant and the victim are of the same race (46 percent are white/white, 46 percent are black/black), 5.6 percent are black on white, and 2.4 percent are white on black. (FBI and Bureau of Justice Statistics, 1985–1986, quoted in Raymond S. Franklin, *Shadows of Race and Class,* University of Minnesota Press, Minneapolis, 1991, p. 108.)

Much of the wealth at the top was accumulated as stockholders and corporate executives moved their companies abroad to employ cheaper labor (56 cents per hour in El Salvador) and avoid paying taxes in the United States. Corporate America is a world made up of ruthless bosses, massive layoffs, favoritism and nepotism, health and safety violations, pension plan losses, union busting, tax evasions, unfair competition, and price gouging, as well as fast buck deals, financial speculation, and corporate wheeling and dealing that serve the interests of the corporate elite, but are generally wasteful and destructive to workers and the economy in general.

It is no wonder Americans cannot think straight about class. The mass media is neither objective, balanced, independent, nor neutral. Those who own and direct the mass media are themselves part of the upper class, and neither they nor the ruling class in general have to conspire to manipulate public opinion. Their interest is in preserving the status quo, and their view of society as fair and equitable comes naturally to them. But their ideology dominates our society and justifies what is in reality a perverse social order—one that perpetuates unprecedented elite privilege and power on the one hand and widespread deprivation on the other. A mass media that did not have its own class interests in preserving the status quo would acknowledge that inordinate wealth and power undermines democracy and that a "free market" economy can ravage a people and their communities.

NOTES

1. Martin Lee and Norman Solomon, *Unreliable Sources*, Lyle Stuart (New York, 1990), p. 71. See also Ben Bagdikian, *The Media Monopoly*, Beacon Press (Boston, 1990).
2. Department of Commerce, Bureau of the Census, "Poverty in the United States: 92," *Current Population Reports, Consumer Income*, Series *P60–185*, pp. xi, xv, 1.
3. William Ryan, *Blaming the Victim*, Vintage (New York, 1971).
4. Benjamin Demott, *The Imperial Middle*, William Morrow (New York, 1990), p. 123.
5. Fred Barnes, "The Zillionaires Club," *The New Republic*, January 29, 1990, p. 24.
6. Michael Parenti, *Inventing Reality*, St. Martin's Press (New York, 1986), p. 109.
7. *Time*, January 5, 1979, p. 10.
8. Vincent Navarro, "The Middle Class—A Useful Myth," *The Nation*, March 23, 1992, p. 1.
9. Charles Anderson, *The Political Economy of Social Class*, Prentice Hall (Englewood Cliffs, N.J., 1974), p. 137.
10. William Domhoff, *Who Rules America*, Prentice Hall (Englewood Cliffs, N.J., 1967), p. 5.
11. Lee and Solomon, *Unreliable Sources*, p. 179.
12. *Newsweek*, August 10, 1992, p. 57.
13. *Business Week*, June 8, 1992, p. 86.

47

Gender Norms in the *Twilight* Series

REBECCA HAYES-SMITH

Chances are, you've at least heard of the *Twilight* series by Stephanie Meyer (if only because of the film adaptations). For those new to the books, the series revolves around a love story between vampire Edward and human Bella. To liven up the plot, Meyer creates a triangle in which a werewolf, Jacob, is also in love with Bella, but the series essentially describes Edward and Bella's "true" and mostly requited love and the many troubles they face due to their different worlds.

Among tween girls and their moms, these books have achieved tremendous popularity. On the one hand, I understand it. The books are entertaining (I read all four in a little over a week), and the romantic vampire story is somewhat beautiful and mysterious. The vampire theme is hardly new, but Meyer's accessible, simplistic style, revamps familiar imagery of what it means to be a vampire. Her vamps are out during the day (where they can show off their beautiful sparkling skin), and she even makes some of them "vegetarian" (her vampires can "just say no" to human blood). By way of contrast, another famous novelist, Anne Rice, has vampires who are seductive but still retain an edge of evil that keeps them from becoming even remotely incorporated into the human world.

What nags me is how, as a sociologist, I find it difficult to ignore the underlying message of gender conformity in Meyer's books. As a society we have multiple ways to communicate how men and women should act in order to be happy. These messages play an important role in the marginalization of women and girls, and *Twilight* reinforces these messages in several ways: through traditional notions of femininity and masculinity, and intersecting stereotypes of race, class, and gender.

Throughout the series, women are weak, passive, and in need of protection, while men are strong and violent. In the first volume, our heroine Bella narrates her story of how she came to fall in love with a vampire. Instead of a strong female lead who assists young women readers in the building of confidence and self esteem, we are presented with a whiny, unassertive, timid young lady. Bella complains, before the first day of school, that she doesn't fit the stereotype of a girl from Phoenix: "I should be tan, sporty, blond—a volley-ball player, or a cheerleader, perhaps—all the things that go with living in the valley of the sun." Her insecurity is what makes her easy to identify with, but how can we change women's (and society's) views without challenging the romanticization of victimhood?

SOURCE: Hayes-Smith, Rebecca. 2011. *Contexts* 10 (Spring): 78–79.

Continuing the stereotype, Bella selflessly takes care of everyone in her life, although constantly complaining about them all (except for Edward). She tells us from the start that she's moved to her father's home only now that her "loving, erratic, harebrained mother" has a new husband to guarantee "bills would get paid, there would be food in the refrigerator, and someone to call when she got lost," Only with a man to take care of her, can Bella's mother do without her daughter. This fragility of female relationships is repeated in Bella's interactions with her peers. A young woman, Jessica, attempts to befriend Bella, but throughout the books Bella complains about how Jessica's jealous and talks too much. Reading how Bella and Jessica's relationship never becomes a full-on friendship led me to think about how often I hear young women say, "I just get along better with guys." I never hear men say that about other men. But it's when women see one another mostly as competitors, that their common oppression becomes less visible.

Edward and Jacob also uphold their gender stereotypes—they're strong, reckless, violent men. They act out their masculine dominance by treating women as property. Edward is constantly speeding in his car, "saving" Bella, stalking Bella (because after all she is "his"), and getting into fights. He rescues Bella from strange men, from a runaway van, and from various vampires (including his own family). Basically, Edward rescues Bella from herself and her bad choices, such as daring to walk alone at night. Then, there's Jacob, who can't control his temper, also gets into fights, and literally forces himself onto Bella. Inevitably, Jacob and Edward fight over the "prize." They play out their masculinity and indicate that in the end, it is not Bella's choice as to which man (so to speak) will win her love.

In the real world, most female victimization occurs between intimate partners, and this theme is prevalent in *Twilight.* When Bella has "sex" with Edward (if that's what you can call it), she walks away from the event literally injured: "[L]arge purplish bruises were beginning to blossom across the pale skin of my arm. My eyes followed the trail they made up to my shoulder, and then down across my ribs, I pulled my hand free to poke at a discoloration on my left forearm, watching it fade where I touched and then reappear. It throbbed a little." The expression of dangerous love-making in other vampire-themed novels is likely similar, but at least in those, the vampires are considered evil.

Stephanie Meyer has been quite clear about her message of abstinence, but at what cost? All right, I get it: a teenager becoming pregnant isn't the best idea. Is that why the description of Bella and Edward's consummation sounds more like a domestic violence incident? This isn't so far from the music videos, films, video games, and other forms of popular culture that sexualize violence among intimate partners. In the end, Edward maintains his masculinity by acting aggressively even in a situation that is supposed to be about intimacy and love. The subtle message is that as long as you're in love or at home, violence is not objectionable. The resulting pregnancy almost kills Bella, but here too, she ignores personal costs as she selflessly chooses the life of her baby.

In *New Moon* and *Eclipse*, Jacob, the werewolf, becomes a more central character. He eventually plants a kiss on Bella, and the description sounds like sexual assault: "His lips crushed mine, stopping my protest. He kissed me angrily, roughly, his other hand gripping tight around the back of my neck, making escape impossible. I shoved against his chest with all my strength, but he didn't even seem to notice." Bella initially attempts to say "No" but does not get it out. Afterward, she punches Jacob, but ends up hurting herself rather than him. Later in the book, she concedes, deciding she does love Jacob after all. This suggests women are not only physically weaker than men, but that women like Bella don't even know what they want from men. Indeed, this repeats a common romance novel theme where women are accepting of their own subordination, all in the name of love.

Last, but certainly not least, are assumptions throughout the entire story that confirm existing systems of race, gender, and class. Bella—as a low-income, white woman—is weak, subservient, and deserves to be protected. Edward—also white, but of a high socio-economic status and, of course, male—is strong, powerful, and dominant. Jacob, as a low-income, Native American male, attempts to be powerful and dominant, but alas, he isn't enough of either to beat out Edward. Indeed the device whereby Native Americans are werewolves (animals) and the white male is the "sophisticated" vampire upholds notions of racial superiority. A sly, but consistent reminder? Edward and the other vampires constantly complain about how the werewolves (dogs, they call them) smell bad.

Yes, these are novels: should feminists and scholars lighten up? These books might simply be innocent entertainment, or potentially harmful to young women. Plenty of research describes how people construct their social reality depending on the cultural messages around them. This is especially important when considering the sheer volume of media images young people now constantly absorb (both actively and passively). It occurs at all ages, but is especially important among teenagers who are navigating the in-between years of not quite being an adult, yet not a kid anymore. My concern is that the most popular young adult books of the last few years are repeating stereotypes sociologists have been debunking for decades. The reading we get is: "Young women, if you dress nice, go to school, and don't have sex, you'll find a nice boy to take care of you. Don't worry if he's a bit violent, that's just how boys are." We could be wrong—perhaps all young women will read *Twilight* purely as fiction without being touched, consciously or otherwise, by the underlying messages. Considering the messages, let's hope so.

48

Rethinking Cyberfeminism(s)

Race, Gender, and Embodiment

JESSIE DANIELS

"[C]yberfeminism" refers to a range of theories, debates, and practices ... about the relationship between gender and digital culture (Flanagan and Booth 2002, 12), so it is perhaps more accurate to refer to the plural, "cyberfeminism(s)." Within and among cyberfeminism(s) there are a number of distinct theoretical and political stances in relation to Internet technology and gender as well as a noticeable ambivalence about a unified feminist political project (Chatterjee 2002, 199). Further, some distinguish between the "old" cyberfeminism, characterized by a utopian vision of a postcorporeal woman corrupting patriarchy, and a "new" cyberfeminism, which is more about "confronting the top-down from the bottom-up" (Fernandez, Wilding, and Wright 2003, 22–23). Thus, any attempt to write about cyberfeminism as if it were a monolith inevitably results in a narrative that is inaccurately totalizing. However, what provides common ground among these variants of cyberfeminism(s) is the sustained focus on gender and digital technologies and on cyberfeminist practices (Flanagan and Booth 2002, 12; Chatterjee 2002, 199; Fernandez, Wilding, and Wright 2003, 9–13).

Cyberfeminist practices involve experimentation and engagement with various Internet technologies by self-identified women across several domains, including work..., education ... domestic life..., civic engagement, feminist political organizing ... art ... and play. While there is no consistent feminist political project associated with cyberfeminist practices, within a culture in which Internet technology is so pervasively coded as "masculine" (Adam 2004; Kendall 2000), there is something at least potentially transgressive in such practices (Fernandez, Wilding, and Wright 2003)....

In the following two sections, I explore the evidence for the view that the Internet is a technology that facilitates gender and racial equality. First, I focus on questions related to political economy and internetworked global feminism. Then, I turn to debates about "identity tourism" and the allure of disembodiment by contrasting examples of the way girls and women are using the Internet to transform their bodies.

SOURCE: Daniels, Jessie. 2009. "Rethinking Cyberfeminism(s): Race, Gender, and Embodiment," *Women's Studies Quarterly* 37, nos. 1 & 2 (Spring–Summer): 101–124.

"A LIBERATING TERRITORY OF ONE'S OWN": POLITICAL ECONOMY AND INTERNETWORKED GLOBAL FEMINISM

A central debate within cyberfeminism has to do with the tension between the political economy required to mass produce the infrastructure of the Internet and its reliance on the exploited labor, on the one hand, and, on the other, claims for the subversive potential of those same technologies....

POLITICAL ECONOMY

To take a global perspective, it is clear that those in industrialized nations are more likely to own computers and have Internet access than are those in developing societies (Norris 2001). The material reality of the global political economy is that women remain the poorest global citizens; the digital era has not shifted this in significant ways (Eisenstein 1998). However, aggregate-level country-specific data show that women have increasing rates of participation online, often at faster rates than men (Sassen 2002, 376). It is not surprising that women lag behind men globally in computer use and Internet access, given that these are so clearly linked to economic resources (Bimber 2000; Leggon 2006; Norris 2001). What is intriguing is that despite women's place at the bottom of the global economic hierarchy, their Internet participation is rapidly increasing.

In the United States, the empirical research indicates that most of the apparent "digital divide" in computer ownership and Internet access, has been the effect of class (or socioeconomic status) more than of gender and race (Norris 2001). In the United States, the rate of Internet access has converged for men and women who are white (Leggon 2006, 100). There remain some small differences in access and kinds of usage between Hispanic women and men and between African American women and men; these differences, however, are negligible (Leggon 2006,100).Yet despite the convergence and negligible differences across gender and race, public intellectuals such as Henry Louis Gates Jr. and Anthony Walton do not hesitate to assert that Black culture is "the problem" when it comes to the digital divide (Wright 2002, 2005). Discourse of "the digital divide" that configures "women" or "Blacks and Hispanics" or "the poor" living in the global South as information "have-nots" is a disabling rhetoric (Everett 2004,1280) that fails to recognize the agency and technological contributions of African Americans, Asians, Chicanos, Latinos, and working-class whites (Wright 2002, 57). What we need is a more multidimensional view of inequality of access that allows for individual agency....

INTERNETWORKED GLOBAL FEMINISM

Within the context of a global political economy, internetworked global feminism can and does bypass national states, local opposition, mass media indifference, and major national economic actors, thus opening a whole new terrain for activism that addresses gender and racial inequality....

For women of color who want to connect globally across diasporas—what Chela Sandoval refers to as "U.S. third world feminism" (2000)—the cyberfeminist practice of online organizing and discursive space takes on added significance....

Many women in and out of global feminist political organizations view Internet technology as a crucial medium for movement toward gender equality (Cherny and Weise 1996; Harcourt 1999, 2000, 2004; Purweal 2004; Merithew 2004; Jacobs 2004). Wendy Harcourt, an Australian feminist researcher with the Society for International Development, a nongovernmental organization (NGO) based in Rome and the author of *Women@Internet: Creating New Cultures in Cyberspace*, is a leading proponent of this view. She summarizes this stance when she writes that there is "convincing evidence that the Internet is a tool for creating a communicative space that when embedded in a political reality can be an empowering mechanism for women" (1999, 219)....

Many individual women outside any formal political organization experience the Internet as a "safe space" for resisting the gender oppression that they encounter in their day-to-day lives offline. In her edited volume *On Shifting Ground; Muslim Women in the Global Era*, Fereshteh Nouraie-Simone (2005a) includes essays about the importance of global information technology for women living in and resisting repressive gender regimes. Nouraie-Simone's description of the importance of the Internet is noteworthy: "For educated young Iranian women, cyberspace is a liberating territory of one's own—a place to resist a traditionally imposed subordinate identity while providing a break from pervasive Islamic restrictions in public physical space. The virtual nature of the Internet—the structure of interconnection in cyberspace that draws participants into ongoing discourses on issues of feminism, patriarchy, and gender politics, and the textual process of self-expression without the prohibition or limitation of physical space—offers new possibilities for women's agency and empowerment" (2005b, 61–62).

Here, Nouraie-Simone evokes Virginia Woolf's call for a "room of one's own" as a prerequisite for feminist consciousness when she describes her experience online as a "liberating territory of one's own." Rather than the "tool" imagery invoked by so many of the global feminist organizations when describing information technology, Nouraie-Simone chooses the term "cyberspace" to suggest that she goes to a "place to resist," where she participates in discussions of "feminism, patriarchy, and gender politics." For her, cyberspace makes global feminism possible in her life offline on an intimate, immediate, and personal level....

THE ALLURE OF IDENTITY TOURISM AND DISEMBODIMENT

... [T]he two ideas that hold the most allure for cyberfeminists interested in the subversive potential of the Internet are identity tourism and disembodiment. Lisa Nakamura in her book *Cybertypes* coins the term "identity tourism" to describe "the process by which members of one group try on for size the descriptors generally applied to persons of another race or gender" (2002, 8). The allure of changing identities online has been part of the sociological writing about the Internet since Sherry Turkle's *Life on the Screen*. Turkle contends that assuming alternate identities online can have positive psychological and social effects by loosening repressive boundaries (1997, 12)....

However, changing identities online may not be as subversive an experience as Turkle and others suggest. Jodi O'Brien notes that gender-switching online is only acceptable within very narrow boundaries and that there is an "earnestness with which gender-policing is conducted" when gender-switching occurs (1999, 82). O'Brien interprets the earnest "gender-policing" to mean that when it is intended as play or performance, switching identities is tolerated as long as there is agreement that a "natural" (read physical/biological) referent remains "intact, embodied and immutable" (O'Brien 1999, 82). Switching identities online seems much less prevalent than the kinds of online experiences that Pitts describes in her research on women with breast cancer who seek and find real community and create new forms of knowledge via sites such as Women.com's BreastFest (Pitts 2004, 55).

Additional research into actual online practices suggests that rather than going online to "switch" gender or racial identities, people actively seek out online spaces that affirm and solidify social identities along axes of race, gender, and sexuality. For example, young girls and teens who have access to the Internet increasingly form their identities, at least in part, through their online interactions (Mazzarella 2005), often via social networking sites such as MySpace or Facebook (boyd 2004); people of color affirm racial identities online through BlackPlanet.com, MiGente, and AsianAvenue.com (Byrne 2007; Lee and Wong 2003); and self-identified QLBT (queer, lesbian, bisexual, and transgender) women go online to "learn to be queer" (Bryson 2004, 251) by using sites such as QueerSisters (Nip 2004; see also Alexander 2002). In large measure, the notion of "identity tourism," in which people switch gender and racial identities, functions as a heuristic device for thinking about gender and race rather than this activity being a commonplace online practice....

... While some cyberfeminists are wildly enthusiastic about the subversive potential of a cyborg future, identity tourism, and disembodiment that is offered by digital technologies, evidence from cyberfeminist practices and empirical research on what people are actually doing online points to a more complicated reality. For some, the Internet economy reproduces oppressive workplace hierarchies that are rooted in a global political economy. For others, the Internet represents a "tool" for global feminist organizing and an opportunity to be protagonists in their own revolution. For still others, the Internet offers a "safe space" and a way to not just survive, but also resist, repressive sex/gender regimes. Girls and self-identified women are engaging with Internet technologies in ways that enable them to transform their embodied selves, not escape embodiment. Girls involved in pro-ana communities deploy Internet technologies that include text and images in order to control their bodies in ways that are both disturbing for others and deeply meaningful for them. Self-identified queer and transgendered women engage with digital technologies in order to transform their bodies, not to play at switching gender identities online.

Scholar-activists who wish to challenge the status quo of racial and gender domination have also been slow to seize the opportunity of engaged public discourse offered by the Internet. Risman (2004) urges feminist sociologists to find means to transform as well as inform society, and the Internet offers such an opportunity. Yet, curiously, most academic sociologists do not have an Internet presence beyond their college or university-sponsored faculty webpage, they do not create content for the Internet, and they do not participate in online communities or social networks. I echo Michelle Wright's call for scholar-activists to engage with the Internet "beyond email" (Wright 2005, 57). It is critically important for those of us who hope that our work can and should speak to audiences beyond the academy to follow the lead of critical cyberfeminists and "hollaback" by engaging the Internet as a discursive space and a site of political struggle....

WORKS CITED

Adam, Alison E. 2004. Hacking into Hacking: Gender and the Hacker Phenomenon. *ACM SIGCAS Computers and Society* 32 (7):0095–2737.

Alexander, Jonathan. 2002. "Queer Webs: Representations of LGBT People and Communities on the World Wide Web." *International Journal of Sexuality and Gender Quarterly* 81:868–76.

Bimber, Bruce. 2000. Measuring the Gender Gap on the Internet. *Social Science Quarterly* 81 (3):868–876.

boyd, danah. 2004. "Friendster and Publicly Articulated Social Networks." Paper presented at the Conference on Human Factors and Computing Systems (CHI 2004). Vienna: ACM, April 24–29.

Bryson, Mary. 2004. "When Jill Jacks In: Queer Women and the Net." *Feminist Media Studies* 4:239–54.

Byrne, Dara. 2007. "The Future of (the) Race: Identity and the Rise of Computer-Mediated Public Spheres." In *Learning Race and Ethnicity*, ed. Anna Everett. MacArthur Series, Digital Media and Learning. Cambridge, Mass.: MIT Press.

Chatterjee, Bela Bonita. 2002. "Razorgirls and Cyberdykes: Tracing Cyberfeminism and Thoughts on its Use in a Legal Context." *International Journal of Gender and Sexuality Studies* 7(2/3):197–213.

Cherny, Lynn, and Elizabeth Reba Weise, eds. 1996. *Wired Women: Gender and New Realities in Cyberspace*. Seattle: Seal Press.

Eisenstein, Zillah. 1998. *Global Obscenities: Patriarchy, Capitalism, and the Lure of Cyberfantasy*. New York: New York University Press.

Everett, Anna. 2004. "On Cyberfeminism and Cyberwomanism: High-Tech Mediations of Feminism's Discontents." *Signs* 30:1278–86.

Fernandez, Maria, Faith Wilding, and Michelle M. Wright, eds. 2003. *Domain Errors! Cyberfeminist Practices*. Brooklyn, N.Y.: Autonomedia.

Flanagan, Mary, and Austin Booth, eds. 2002. *Reload: Rethinking Women + Cyberculture*. Cambridge, Mass.: MIT Press.

Harcourt, Wendy. 1999. *Women@Internet: Creating New Cultures in Cyberspace*. New York: Zed Books.

———. 2000. "The Personal and the Political: Women Using the Internet." *CyberPsychology and Behavior* 3:693–97.

———. 2004. "Women's Networking for Change: New Regional and Global Configurations." *Journal of Interdisciplinary Gender Studies* 8(1/2). http://www.newcastle.edu.au/centre/jigs/issues/index.html (accessed October 22, 2008).

Jacobs, Susie. 2004. "Introduction: Women's Organizations and Networks; Some Debates and Directions." *Journal of Interdisciplinary Gender Studies* 8(1/2). http://www.newcastle.edu.au/centre/jigs/issues/index.html (accessed October 22, 2008).

Kendall, Lori. 2000. " 'OH NO! I'M A NERD!' Hegemonic Masculinity in an Online Forum." *Gender and Society* 14(2): 256–74.

Lee, Rachel C., and Sau-Ling Cynthia Wong. 2003. *AsianAmerican.Net: Ethnicity, Nationalism, and Cyberspace*. New York: Routledge.

Leggon, Cheryl B. 2006. "Gender, Race/Ethnicity, and the Digital Divide." In *Women, Gender, and Technology*, eds. M. F. Fox, D. G. Johnson, and S. V. Rosser. Urbana: University of Illinois.

Mazzarella, Sharon A., ed. 2005. *Girl Wide Web: Girls, the Internet, and the Negotiation of Identity*. New York: Peter Lang.

Merithew, Charlene. 2004. Women of the (Cyber)world: The Case of Mexican Feminist NGOs. *Journal of Interdisciplinary Gender Studies* 8(1/2). http://www.newcastle.edu.au/centre/jigs/issucs/index.html (accessed October 22, 2008).

Nakamura, Lisa. 2002. *Cybertypes: Race, Ethnicity, and Identity on the Internet*. New York: Routledge.

Nip, Joyce Y. M. 2004. "The Relationship Between Online and Offline Communities: The Case of the Queer Sisters." *Media, Culture, and Society* 26(3):409–28.

Norris, Pippa. 2001. *Digital Divide: Civic Engagement, Information Poverty, and the Internet Worldwide*. New York: Cambridge University Press.

Nouraie-Simone, Fereshteh, ed. 2005a. *On Shifting Ground: Muslim Women in the Global Era*. New York: The Feminist Press.

———. 2005b. "Wings of Freedom: Iranian Women, Identity, and Cyberspace." In *On Shifting Ground*, ed. Fereshteh Nouraie-Simone. New York: The Feminist Press.

O'Brien, Jodi. 1999. "Writing in the Body: Gender (Re)Production in Online Interaction." In *Communities in Cyberspace*, eds. Marc Smith and Peter Kollock. New York: Routledge.

Pitts, Victoria. 2004. "Illness and Internet Empowerment: Writing and Reading Breast Cancer in Cyberspace." *Health: An Interdisciplinary Journal for the Social Study of Health, Illness, and Medicine* 8:33–59.

Purweal, Navtej. 2004. "Sex Selection and Internet-Works." *Journal of Interdisciplinary Gender Studies* 8(1/2). http://www.newcastle.edu.au/centre/jigs/issues/index.html (accessed October 22, 2008).

Risman, Barbara. 2004. "Gender as a Social Structure: Theory Wrestling with Activism." *Gender and Society* 18 (4):429–50.

Sandoval, Chela. 2000. "New Sciences: Cyborg Meminism and the Methodology of the Oppressed." In *The Cybercultures Reader*, eds. David Bell and Barbara M. Kennedy. New York: Routledge.

Sassen, Saskia. 2002. "Towards a Sociology of Information Technology." *Current Sociology* 50(3):365–88.

Turkle, Sherry. 1997. *Life on the Screen: Identity in the Age of the Internet.* Cambridge, Mass.: MIT Press.

Wright, Michelle M. 2002. "Racism, Technology, and the Limits of Western Knowledge." In *Domain Errors!* eds. Maria Fernandez, Faith Wilding, and Michelle M. Wright. Brooklyn, N.Y.: Autonomedia.

———. 2005. "Finding a Place in Cyberspace: Black Women, Technology, and Identity." *Frontiers* 26(1):48–59.

49

Brown Body, White Wonderland

TRESSIE MCMILLAN COTTOM

Miley Cyrus made news this week with a carnival-like stage performance at the MTV Video Music Awards that included life-size teddy bears, flesh-colored underwear, and plenty of quivering brown buttocks. Almost immediately after the performance, many black women challenged Cyrus' appropriation of black dance—"twerking." Many white feminists defended Cyrus' right to be a sexual woman without being slut-shamed. Yet many others wondered why Cyrus' sad attempt at twerking was news when the U.S. is planning military action in Syria.

I immediately thought of a summer I spent at University of North Carolina-Chapel Hill. My partner at the time fancied himself a revolutionary born too late for all the good protests. At a Franklin Street pub one night—one of those college-town places where bottom-shelf liquor is served in fishbowls for pennies on the dollar—we were the only black couple at a happy hour. I saw a few white couples imbibing and beginning some version of bodily grooving to the DJ. I told my partner that one of them would be offering me free liquor and trying to feel my breasts within the hour.

He balked, thinking I was joking. I then explained my history of being accosted by drunk white men and women in atmospheres just like this one. Women asking to feel my breasts in the ladies' restroom. Men asking me for a threesome as a drunk girlfriend or wife looks on smiling. Frat boys offering me cash to "motorboat" my cleavage. Country boys in cowboy hats attempting to impress their buddies by grinding on my ass to an Outkast music set. My friends have witnessed it countless times.

Not 30 minutes later, with half the fishbowl gone, the white woman bumped and grinded up to our table and, laughing, told me that her boyfriend would love to see us dance. "C'mon girl! I know you can daaaaannnce," she said. To sweeten the pot, they bought us our own fishbowl.

That summer we visited lots of similar happy hours. By the third time this scene played out, my partner had taken to stonily staring down every white couple that looked my way. We were kicked out of a few bars when he challenged some white guy to a fight about it. I hate such scenes, but I gave him a break. As a man, he did not have the vocabulary borne of black breasts that sprouted before bodies have cleared statutory rape guidelines. He did what he could to tell me he was sorry: He tried to kick every white guy's ass in Chapel Hill.

SOURCE: Cottom, Tressie McMillan. August 2013. "Brown Body, White Wonderland." www.slate.com. Reprinted by permission of the author.

I am not beautiful. I phenotypically exist in a space where I am not usually offensive-looking enough to have it be an issue for my mobility, but neither am I a threat to anyone's beauty market. There is no reason for me to assume this pattern of behavior is a compliment. What I saw in Cyrus' performance was not just a clueless, culturally insensitive attempt to assert her sexuality or a simple act of cultural appropriation at the expense of black bodies. Instead I saw what kinds of black bodies were on that stage with Cyrus.

Cyrus' dancers look more like me than they do Rihanna or Beyoncé or Halle Berry. The difference is instructive.

Fat, non-normative black female bodies are kith and kin with historical caricatures of black women as worksites, production units, subjects of victimless sexual crimes, and embodied deviance. As I wrote in an analysis of hip-hop and country music crossovers, playing the desirability of black female bodies as a wink-wink joke is a way of lifting up our deviant sexuality without lifting up black women as equally desirable to white women. Cyrus did not just have black women gyrating behind her. She had particularly rotund black women. She gleefully slaps the ass of one dancer like she intends to eat it on a cracker. She is playing a type of black female body as a joke to challenge her audience's perceptions of herself, while leaving their perceptions of black women's bodies firmly intact. It's a dance between performing sexual freedom and maintaining a hierarchy of female bodies from which white women benefit materially.

That hierarchy explains why background dancers generally conform to dominant beauty norms. The performance works as spectacle precisely because the background dancers embody a specific kind of black female body. That spectacle unfolds against a long history of how capitalism is a gendered enterprise and subsequently how gendered beauty norms are resisted and embraced to protect the dominant beauty ideal of a certain type of white female beauty. So, when I saw the type of black dancers chosen to juxtapose Cyrus' performance of sexual power, I was given pause. Whether Cyrus is aware of this history or not, her performance is remarkable for how clearly it is situated in the history of racialized, gendered capitalism that makes my body a public playground for the sexual dalliances of white men and women.

Being desirable is a commodity. Capital and capitalism are gendered systems. The very form that money takes is rooted in a historical enterprise of controlling an economic sphere where women might amass wealth. As wealth is a means of power in a capitalist society, controlling this commodity is a way of controlling the accumulation, distribution, and ownership of capital and indirectly controlling women.

50

The Construction of Black Masculinity

White Supremacy Now and Then

ABBY L. FERBER

Black male bodies are increasingly admired and commodified in rap, hip ... hop, and certain sports, but at the same time they continue to be used to invoke fear. Black men are both held in contempt and valued as entertainment (Collins, 2005; Leonard, 2004). Yet this is really nothing new. Black men have been defined as a threat throughout American history while being accepted in roles that serve and entertain White people, where they can ostensibly be controlled and made to appear nonthreatening. Furthermore, within the contemporary context of color-blind ideology, the embrace of Black athletes helps White fans to assure themselves that America really is not racist after all. In this article, I will provide a reading of sports popular culture through the lens of historical and contemporary White supremacist ideology. Although seemingly harmless entertainment, mainstream sports culture reiterates the common themes evident in White supremacist constructions of Black masculinity....

THE HISTORICAL CONTEXT

Contemporary White supremacy and the new racism remain bolstered by the historical constructions of race and gender on which this nation was founded. The ongoing discrimination African Americans experience today is often rationalized or justified in the minds of Whites by deeply rooted stereotypes of Black men and women. Historically, African Americans were defined as animals, as property to be owned by White men. Racist imagery took gender-specific forms. "Because black men did hard manual labor, justifying the harsh conditions forced upon them required objectifying their bodies as big, strong, and stupid" (Collins, 2005, p. 56).

This imagery possessed a sexual component as well. Fear of Black men's sexuality remained pervasive, and they were constantly depicted as a threat to White womanhood if not controlled. "White elites reduced Black men to their bodies, and identified their muscles and their penises as their most important sites" (Collins, 2005, p. 57). This narrative, which defines Black males as hypersexual, animalistic, and savage, is central to White American identity, and, as Houston

SOURCE: Ferber, Abby L. 2007. "The Construction of Black Masculinity: White Supremacy Now and Then." *Journal of Sport and Social Issues* vol. 31 no. 1 (February): 11–24.

A. Baker (1993) reminds us, "this scene plays itself out ... with infinite variation in American history" (p. 38).

Black men were also constructed as inherently violent....

Black men were defined as beasts who had to be controlled and tamed to be put into service.

Black women were also defined as hypersexual and denied any rights to control their own bodies. They were defined as unrapable (Joseph, 2006; see also Leonard, 2007, for a fuller discussion). In the White imagination, both Black men and women have been reduced to their physical bodies. These stereotypes were relied on to justify scores of rapes and lynchings and remain entrenched still. Examining news coverage of contemporary rape cases, Susan Fraiman (1994) argues that narratives about race, gender, and sexuality inform battles between racialized men over the bodies of women. She identifies this "paradigm of American racism, available during slavery but crystallized in the period following Reconstruction and still influential today, in which White men's control of Black men is mediated by the always-about-to-be-violated bodies of White women" (p. 71).

In the past, these stereotypes were attributed to biology. Today, consistent with the new racism, they are instead attributed to Black culture. In both the old and new variations, they are used to justify inequality as a result of inherent characteristics of Black people themselves, "[pointing] to the damaged values and relationships among Black people as the root cause of Black social disadvantage" (Collins, 2005, p. 180). Think more recently about the images of Willie Horton, thug-like rapsters, or welfare queens. The images have changed very little. Collins (2005) argues that these "controlling images" of Black men and women are so entrenched they have "become common-sense 'truths'" (p. 151) in many people's minds.

AN INTERSECTIONAL APPROACH

As this brief historical synopsis reveals, gender is central to the workings of racism. In this article, I use an intersectional approach that sees race and gender as interacting and inseparable. Gender is constructed through race, and race is constructed through gender; they are intersectional and mutually constitutive.

In response to the advances of women during the past half century, many scholars have observed a "crisis of masculinity." Because gender identities are relational, masculinity is defined in opposition to femininity. As definitions of femininity have been changing, many men have been left wondering what precisely it means to be a man. It is within this cultural context of struggles over racial and gender meanings that a discussion of contemporary cultural constructs of Black masculinity must be situated....

... I will examine these common themes:

1. A continued emphasis on Black bodies and essential racial differences. African Americans continue to be defined as aggressive, hypersexual, threatening, and potentially violent.

2. A concern with taming and controlling Black males.
3. Inequality is depicted as a product of a deficient Black culture.
4. White supremacy, and White male superiority, are naturalized.

These four themes permeate the contemporary construction of Black masculinity and work to justify color-blind racism and inequality.

Although more covert, coded, and cultural, the new racism continues to uphold the same White supremacist suppositions of the past, rearticulating and churning out anew the very same constructions of Black masculinity so prevalent throughout American history and cultural expression....

MAINSTREAM CONSTRUCTIONS OF BLACK MASCULINITY

Sport is a particularly powerful institution, "a cultural text" central to American identity (Leonard, 2004, p. 285). According to Mary Jo Kane (1996), "Sport consists of a set of ideological beliefs and practices that are closely tied to traditional power structures.... Sport has become such a bedrock of our national psyche that sport figures often come to symbolize larger pressing social concerns" (pp. 95, 97). Sporting events are more than simply entertainment; "they're also sites where racial and ethnic relations happen and change" (Coakley, 2006, p. 282).

Successful women and Black athletes may be seen as potentially threatening to the notion of White male superiority (Duncan & Messner, 1998; Kane, 1996). However, depictions of African American athletes may also reinforce the traditional hierarchy by reifying stereotypes of their animal-like nature, emphasizing their sexuality, aggressiveness, and physical power. Just as we observed in far-Right racist ideology, there is a similar naturalization of racial difference in sports discourse, where Black men are often assumed to be naturally more athletic. This assumption follows from the historical stereotype of physically aggressive Black male bodies. Despite much evidence to the contrary, the myth that Blacks have more natural athletic ability is hard to dispel (Coakley, 2006; Graves, 2004). To understand how widespread this view is, just recall the recent claims of the Air Force Academy coach who rationalized his team's loss as a result of the disproportionate number of Black men on the opposing team. As Collins (2005) argues, the actual work of Black male athletes is made invisible—so that they are constructed as naturally athletic.

Although one might hope that the success of Black male athletes might help to undermine racism, sports represents an arena where Black men have historically been allowed to succeed—in the entertainment and service industries (Coakley, 2006). Although African American men have been very successful in certain sports, they are rarely found in positions of power and control—as coaches or owners. Within the industry, they are largely under the control of White men. Success in the field of athletics also does nothing to undermine the historical propensity to reduce Black men to their bodies. As Jay Coakley (2006) argues, Black men's talent is often attributed to nature, whereas the accomplishments of White athletes are instead characterized as "fortitude, intelligence, moral character, strategic preparation, coachability, and good organization" (p. 288). Thus, the success of Black men in sports is entirely consistent with White supremacist ideology.

Collins (2005) argues that there is a traditional family script in place in sports that works to minimize the threat of Black masculinity. The coach is similar to the White male father figure, whereas Black male athletes are like the children, under the father's control and subject to his rule. It is only when they accept and play this role that they are fully embraced and accepted and seen as nonthreatening. Their bodies can be admired as long as they are perceived as controlled by White males. These athletes are then defined as the "good Blacks."

At the same time, the demonization of certain Black male athletes as "bad boys" is used as a tool to exert control over those men who do not so easily submit to White male authority. According to Collins (2005), "The contested images of Black male athletes, especially 'bad boy' Black athletes who mark the boundary between admiration and fear, speak to the tensions linking Western efforts to control Black men" (p. 153). The negative depiction of bad boys works to reinforce efforts to tame their "out of control" nature.

When Latrell Sprewell choked his coach in 1997, Collins (2005) argues that Sprewell's media coverage symbolized the larger depictions of Black masculinity as overly physical, out of control, prone to violence, driven by instinct, and hypersexual. The disproportionate media coverage focused on violent or sexual assault charges brought against Black male athletes, compared with similar charges against White male athletes, reifies this stereotype of Black men as inherently dangerous and in need of civilizing. The message is that all Black men are essentially bad boys but that some can become "good guys" if tamed and controlled by White men.

Collins (2005) notes that fans display a certain amount of ambivalence toward Black male athletes, whom many fans seem to "love to hate" (p. 155). Sprewell, Allen Iverson, Charles Barkley, Dennis Rodman, Barry Bonds, and Terrell Owens are all seen as unruly and disrespectful, but at the same time, this bad boy image may enhance their reputation and media coverage....

Collins (2005) argues that athletes like Sprewell and Iverson are examples of Black males who refuse to assimilate and play by the rules, unsettling "prevailing norms of race and gender" (p. 156). At the same time, however, they reinforce the stereotype of Black men out of control and feed into racist White supremacist definitions of Black masculinity.

The stereotype of Black men as sexual predators, especially as threats to White women, is central to the good-bad dichotomy (Leonard, 2004). The White supremacist obsession with the dangers of interracial sexuality is relied on and reinforced by the mainstream media as well. This historical narrative informed perceptions of the O. J. Simpson arrest and trial. The darkening of Simpson's face on the cover of a popular magazine reinforced the correlation between blackness and danger, and a Gallup poll found that 39% of White respondents and 43 % of African American respondents claimed that they would be less interested in the Simpson case if it did not involve an interracial relationship (or, we might surmise, if the man were White and the victim an African American woman)....

This division between the good guys who have been tamed and know their place versus the bad boys who refuse to submit to control reflects the historical and ongoing construction of Black masculinity in White supremacist culture and limits the ways in which Black men are seen in our culture. It reinforces the old presumption, widespread as slavery declined, that Black men are safe and acceptable only when under the control and civilizing influence of Whites. However, they have an inherently violent, aggressive nature lying just beneath the surface, threatening to spring forth at any time, At the same time, the good guy space reinforces color-blind racism. By embracing the successful good guys, Whites can tell themselves they are not racist, and they can blame African Americans for their own failures (Leonard, 2004).

CONCLUSION

The four racist themes evident in both cultural sites produce the illusion that White male supremacy is the natural result of Black men's inherently inferior, violent, aggressive, and hypersexual natures. Black men are defined as responsible for their own failure to succeed, and they must be controlled for their own good and that of society. These four themes directly support and reinforce the new racism. They underscore the assumption that we now live in a color-blind nation and that racism is a thing of the past. Any inequality is now seen as the result of natural and cultural differences or African Americans' own poor choices. Although the construction of Black masculinity has remained virtually unchanged from slavery through the present, it has been malleable enough to reinforce both the old and new racisms.

These steadfast images of Black men naturalize and reinforce racial inequality. They reinforce the message that Black men are naturally aggressive, are violent, cannot succeed on their own, are not suited for professional careers, are not good fathers, and need to be controlled by White men. This imagery justifies in many people's minds the disproportionate imprisonment of Black men today. Black men continue to be reduced to their physical bodies and defined as inferior to White men.

Images of successful Black athletes also provide a "bootstraps" story, sending the message that these Black men have succeeded; therefore, there is no reason other Black men can't. This story allows White folks to see themselves as non-racist and imagine that we now live in a color-blind nation. As Leonard (2004) argues, "One of the most powerful discursive spaces in which colorblindness is employed and deployed is the arena of sports" (p. 287). The vast reality of discrimination and institutionalized racism is erased from view. Athletics and entertainment are the two primary realms in which we actually see Black men presented as successful in our culture, and they are consistent with the historical stereotypes and limited opportunities available to Black men. Furthermore, these images continue to limit the aspirations and role models for young Black boys. According to Collins (2005), "Most Black American boys will never achieve the wealth and fame of their athletic role models through sports. Keeping them mesmerized with sports heroes may actually weaken their ability to pursue other avenues to success" (p. 157). As Yousman (2003) argues, White adoration of Black entertainers "allows Whites to contain their fears and animosities toward Blacks through rituals not of ridicule, as in previous eras, but of adoration … [nevertheless] the act is still a manifestation of White supremacy" (p. 369).

As poet Hemphill argues, the fact that White Americans accept Black men on the court has not led to a similar acceptance off the court. Although White folks may be willing to embrace Black men as athletes, they still do not embrace them as neighbors. Segregated housing and schools result in many White people having very little opportunity to get to know people of color in their daily lives. Only 2% of White people have a Black neighbor (Williams, 1997).

Clearly, the success of Black men as athletes does little to challenge the systematic and institutionalized system of White supremacy. Instead, within a White supremacist culture, even this success is manipulated and rearticulated to support White supremacy and hegemonic White masculine privilege.

The sheer pervasive nature of this imagery means that more extremist forms of White supremacy will be more likely to resonate when encountered by White folks. In today's high-tech world, where children and adults are likely to stumble on White supremacist Web sites at some point, it is more important than ever that we interrogate our more mainstream discourses of race and present a conscious antiracist agenda. Instead, the range of racist imagery to which we are all exposed normalizes racism and naturalizes inequality.

REFERENCES

Baker, H. A. (1993). Scene … not heard. In R. Gooding-Williams (Ed.), *Reading Rodney King/reading urban uprising* (pp. 38–50). New York: Routledge.

Coakley, J. (2006). *Sports in society: Issues and controversies.* New York: McGraw-Hill.

Collins, P. H. (2005). *Black sexual politics: African Americans, gender, and the new racism.* New York: Routledge.

Duncan, M. C., & Messner, M. A. (1998). The media image of sport and gender. In L. A. Wenner (Ed.), *Media sport* (pp. 170–185). New York: Routledge.

Fraiman, S. (1994, Spring). Geometries of race and gender: Eve Sedgewick, Spike Lee, Charlayne Hunter-Gault. *Feminist Studies, 20*, 67–84.

Graves, J. L. (2004). *The race myth: Why we pretend race exists in America.* New York: Dutton.

Hemphill, E. (1992). American hero. In *Ceremonies: Prose and poetry* (p. 3). San Francisco: Cleis Press.

Joseph, J. (2006). Intersectionality of race/ethnicity, class, and justice: Women of color. In A. V. Merio & J. M. Pollock (Eds.). *Women, law, and social control* (pp. 292–312). New York: Pearson.

Kane, M. J. (Spring, 1996). Media coverage of the post Title IX female athlete: A feminist analysis of sport, gender, and power, *Duke Journal of Gender Law and Policy*, *3*(1), 95–127.

Leonard, D. J. (2004). The next M. J. or the next O. J.? Kobe Bryant, race, and the absurdity of colorblind rhetoric. *Journal of Sport and Social Issues*, *28*, 284–313.

Leonard, D. J. (2007). Innocent until proven innocent: In defense of Duke lacrosse and White power. *Journal of Sport and Social Issues*, *31*, 25–44.

Williams, P. J. (1997). *Seeing a color-blind future: The paradox of race*. New York: Noonday.

Yousman, B. (2003). Blackophilia and Blackophobia: White youth, the consumption of rap music, and White supremacy. *Communication Theory*, *13*, 366–391.

51

Sustainable Food and Privilege

Why Green Is Always White (and Male and Upper-Class)

JANANI BALASUBRAMANIAN

When asked to name the heroes of food reform and sustainable agriculture, who comes to mind? Michael Pollan, Joel Salatin, Eric Schlosser, Peter Singer, Alice Waters maybe? Notice any patterns? The food reform movement is predicated on rather shaky foundations with regard to how it deals with race and other issues of identity, with its focus on a largely white and privileged American dream.

Still, what could be better than a return to family farms and home-cooking, which many of these gurus champion? The images are powerfully nostalgic and idyllic: cows grazing on sweet alfalfa, kids' mouths stained red with fresh heirloom tomato juice, and mom in the kitchen rolling out dough for homegrown-apple pie. But this is not an equal-access trip down memory lane. While we would like to think the American dream of social communion around food is a universal one, this assumption glosses over the very real differentials in gender, class, race, ethnicity, and nationality that were enabled and exacerbated by specific communities (white plantation owners, for example) through the use of food.

This is not to say that activists in the sustainable food movement are unconcerned with issues of identity, but that their rhetoric tends to disallow discussions on race, history, and food in a number of ways. First, Pollan and others situate the current state of American consumption in a patriarchal paradigm. These writers speak about a disappearance of food culture that for the most part accompanies male privilege. For example, Pollan, in an article for *The New York Times* on cooking and entertainment aptly titled "Out of the Kitchens, onto the Couch," explores the relationship between second-wave feminism and the gender politics of cooking. He argues that Betty Friedan's *The Feminine Mystique* convinced women to regard their housework, specifically cooking, as drudgery. Friedan did not, in fact, construct this sentiment herself; she merely observed the existent trends in white women's attitudes about food and housewifery. Pollan goes on to describe how Julia Child inspired his mother and other women like her, empowering them to channel their creativity into the kitchen. This is apt praise for the lively and engaging cook, but can Pollan not drive home the point that Americans need to cook more often without guilting American feminists?

Second, the emphasis on the local food economy, though admirable, has certain anti-global and overly nationalist undertones. Let us take the example of Joel Salatin, owner of Polyface Farms, featured in many of Pollan's books, as well as the movies *Food Inc.* and *Fresh!* Salatin is an ex-lawyer, of considerable means, who moves to the countryside, establishes a dynamic, organic, solar-powered farm, and sells top-quality animal products at top-quality dollar. If the nation is truly to scale up sustainable foods, we cannot fixate on the early image of the American farmer as white, male, and conservative. Instead, we must acknowledge (as USDA statistics tell us) that the face of farming is changing, and women and people of color will continue to grow in number as stewards of sustainable agriculture. Furthermore, we need to consider the real impact of foods we purchase, rather than mindlessly buying produce labeled "local" and "organic." The United States supports a lot of global agriculture through its food purchases, and this is a relationship we should not break off entirely. True, we can do more to support efficient, environmentally friendly purchasing, but we should also not be too hasty to reject globalization.

Finally, the major voices in food are not talking about race and class as often as they should. Food justice is fundamentally a race and class issue. Schlosser's *Fast Food Nation* elucidates labor practices that disproportionately affect people of color, but does not engage the issue of race specifically. Partly, this stagnancy is a matter of perception: after all, activists of color like Bryant Terry and Winona La Duke do brilliant work in their communities with regards to food justice. For some reason, however, their work goes largely underappreciated.

All social movements need a variety of voices, but I argue that food reform requires this diversity even more urgently because it is so universal in its reach. And if we can reach all those voices, then think of all the activists we will have as allies—feminists, anti-racists, interfaith leaders, and so on—interested and involved because food justice speaks to the needs of their communities and their call for action (activists: this is on you too—get on board!). As consumers of this kind of liberal rhetoric, we need to demand that the powers and big hitters in the food world diversify their representations. The food movement can only grow more powerful for it.

52

There's No Business Like the Nail Business

MILIANN KANG

A manicure is no longer a purely private ritual that a woman gives herself, her daughter, or a girlfriend in the quiet of her own bathroom. Instead, it is something she increasingly purchases in a nail salon and from an Asian manicurist. In the buying and selling of manicuring services, women both implicate their own bodies in intimate commercialized exchanges and expand the boundaries of the service economy to encompass regimens of hygiene and physical adornment that were once private. In so doing, they also encounter at close range women whom they would normally regard only from a safe social distance.

Why have nail salons cropped up on city blocks and in suburban strip malls across the United States? Why do so many women get manicures? Why do so many Asian women, and Korean women in particular, own and work in these salons? These questions recognize the simultaneous ways that supply, demand, and location shape the growth of the global service economy and the development of a specific ethnic-dominated niche like nail salons within it. While these are distinct questions, they are also closely interrelated. The question of why so many Asian women work in nail salons can be answered only in relation to the question of why so many women in the United States desire and purchase these services.... [I]t is important to examine not only the individual consumers who purchase these services and the providers who offer them but also the economic, political, and cultural contexts in which these exchanges occur.

In this chapter I respond to these questions—why nails ... and why Asian women?—by examining the appeal of nail services to diverse customers, the growth of the nail salon industry and the clustering of Asian women, particularly Koreans, in this employment sector. The politics of race, class, and immigration in the United States and the shifting dynamics of the global service economy provide the context that shapes the relations that women forge around the manicuring table. The growth in manicuring services reflects a general expansion of capitalist markets, the specifically gendered processes relating to the commercialization of women's bodies, and the positions of women in the labor market. The influx of women into the paid labor force has increased the demand for such services, because more women can now afford them. However, another

SOURCE: Kang, Miliann. 2010. *The Managed Hand: Race, Gender, and the Body in Beauty Service Work*, pp. 32–50. Copyright © 2010. Berkeley, CA: University of California Press.

important factor is heightened desires for beauty as a commodity. The purchase of body-related services is fueled by ramped-up social standards for women's appearance, as well as by women's own longing for the accoutrements of beauty, including the pampering services associated with it.

At the same time women's desires for beauty services would simply be longings rather than daily enactments if it were not for the presence of a ready fleet of immigrant women workers to provide these services. The lifestyle that many urbanites take for granted in cities such as New York is possible only because of the influx of new immigrants and their willingness to work long arduous hours for minimal pay in jobs that many native-born Americans view as beneath them. While immigrant women from specific ethnic groups are not the sole creators of these jobs or the terms under which they perform them, they contribute to creation of these specialized niches by capitalizing on the limited choices available to them.

The formation of New York City's nail salon industry and its domination by Asian women simultaneously draws on and contests two competing racial discourses. On the one hand, representations of Asian success in this industry exemplify praise for the innovation and diligence of Asian Americans and their independence from government "special treatment."... On the other hand, their success fuels the anti-Asian and anti-immigrant sentiments held by those who blame these groups for downgrading U.S. working conditions. Neither of these discourses ... adequately accounts for the composite factors that drive nail salon growth in New York and other cities. While Asian immigrant women are indeed hardworking and resourceful, these characteristics alone cannot account for their domination of the nail salon industry.

Rather, it is the context of the "global city" that shapes the consistent demand for inexpensive and convenient beauty services and the terms of who does this work.... In other words, while customers' desire for manicured nails and manicurists' need to earn a living are certainly important factors in the increase in nail salons, these factors are driven by larger processes pertaining to the postindustrial transformation of cities and city life. Through a fortuitous convergence of global and local factors, rather than through their unique cultural traits, Korean women have successfully mobilized individual and community resources to sustain entrepreneurship and employment in this service niche.

WHY NAILS? MANICURES AND THE COMMERCIALIZATION OF THE BODY

Why do more and more women now pay for manicures instead of doing their nails themselves? The answer to this question, far from a simple story of women's innate longing for physical beauty, instead pulls back the curtain on the surreptitious but revolutionary reorganization of social life in the late twentieth century. Arlie Hochschild describes this sea change as "the commercialization of intimate life," in which more and more human activity that was formerly engaged in by family, friends, and community members has been subsumed into the global capitalist economy.... In short, capitalism has expanded geographically to hinterlands previously untouched by commodity markets as well as into areas of human life once viewed as private and even sacred. Other scholars have similarly explored how formerly unpaid activities, such as raising children, caring for the elderly, preparing food, doing laundry, and mowing the lawn, are now routinely farmed out to paid service workers.... In addition, market capitalism has spawned new occupations, ranging from on-line matchmaker to personal assistant, home organizer, party escort, and life coach. These services are designed to meet the emotional and social needs formerly met by friends, churches, bowling leagues, neighborhood associations and community groups and reflect an overall decline in civic ties.... Most important for this study, a new range of service providers—personal trainers, massage therapists, plastic surgeons, and manicurists, to name a few—have staked out the body, its appearance, comfort, and health, as a profit-making venue.

Whereas Hochschild's study focuses on the commercialization of human feelings, the interactions in nail salons illuminate the complexities of buying and selling services that cater to both human emotions and bodies. These interactions are not unique to nail salons but reveal how the routine upkeep of the body and its appearance have spawned an array of purchasable services. Indeed, the growth in nail salons has been impressive—and the overall growth in beauty services has been staggering. Like the manicure, a child's first haircut is no longer a ritual performed on a stool in a family kitchen but is farmed out to hair-cutting chains around the country.... Rather than giving each other backrubs at the end of a long day, two tired spouses can opt for a fifteen-minute chair massage on the way home. A child's diaper or an elderly parent's bedpan are increasingly changed by paid caregivers rather than family members. The process of assigning market value to bodies—their appearance, functions, and the forms of contact between them—generates new forms of work, which I refer to as *body labor.* By examining how body labor exchanges occur, and what the participants gain and lose through them, the study of manicures can illuminate similar patterns in other embodied services.

While the levels and kinds of consumption of these complex embodied and emotional services increase, the means to purchase them has been eroded for most segments of the population. However, rather than decreasing the demand for body services, these economic pressures can fuel the market for them. Just like going to the movies, indulging in body services can be the ideal escape in a recession. With unemployment and downsizing looming as constant threats, workday concerns take over more and more of individual and collective life. Rather than turning away from market solutions, people increasingly turn to the power of consumption, not only to obtain material goods but also to purchase services that provide care and connection, especially for tired and stressed bodies.

This process of seeking commercialized solutions for intimate needs, however, is far from fluid and care free. As Viviana Zelizer writes in *The Purchase of Intimacy,* while the arenas of intimacy and economics have long intermingled in various forms, the negotiation of the terms of this intermingling is often confusing and fraught, and increasingly so in new venues where social conventions are not fully worked out.... Thus social actors must negotiate new forms of intimate relations within existing frameworks, even when these frameworks are not fully up to the task of making sense of how commerce and intimacy mix in both uncharted and ubiquitous ways.

Into this brave new world of commercialized intimacy enters the nail salon. In a day devoid of touch and beauty, the nail salon provides a taste of both. For $15 or less customers can brush their cares away while their manicurists dote on them, massaging the day's tensions out of their hands and putting polish on their cracked or lackluster nails.... A manicure thus can serve as a quick fix to a host of problems, ranging from demanding children, nagging spouses, critical bosses, and needy friends to larger anxieties about seemingly unsolvable personal and social problems.

While the avenues for relief or distraction from the cares and demands of domestic, work, and social life are many and varied, the combination of emotional and physical attention offered in a haven of women who cater to the needs of other women carries a particular appeal. The changing dynamics of women's place in the home and workplace, as well as their substantial increase in earnings, are thus important pieces in understanding the growth of the nail salon industry and, more broadly, the commercialization of the body and body-related services. Since the late 1960s women have entered into the paid labor force in historically unprecedented numbers.... While their paychecks are usually central to maintaining basic economic survival for themselves and their dependents, women's paid labor also gives them greater control over discretionary spending. Work provides them with income to purchase nail services but also fuels their consumption of these services, as many women feel that professionally manicured nails are an expected part of their work attire.... Even in workplaces where such appearance standards are not explicit, well-manicured nails can augment professional appearance and give some women a confidence boost.

Manicures can also be a way of reassuring women and those around them that they do indeed conform to norms of traditional femininity, even as they challenge and redefine these standards in various areas of their lives. On the other hand, some women use original nail designs to express an identity that is distinct from or in opposition to mainstream feminine norms. Given the wide-ranging needs that a weekly manicure can fulfill, is it any wonder that nail salons have become a major growth industry?

NAIL SALONS AS A GROWTH INDUSTRY

"There's No Business Like the Nail Business" trumpeted the headline in a Vietnamese community newspaper, *Nguoi Viet,* reporting that revenues from nail salons in the United States topped $6 billion in 2004 and the number of nail salons in the United States grew from 32,674 in 1993 to 53,615 in 2003, an increase of more than 60 percent in a decade.... In 2006–2007 *Nails* magazine estimated the United States had 58,330 nail salons and 347,898 nail technicians in a $6.16 billion industry.... Even the much more conservative figures from the U.S. Bureau of Labor Statistics projected 28 percent growth between 2006 and 2016.... Despite this tremendous growth, the job of manicurist and the workplace of nail salon easily slip below the radar of both official statistics and everyday perceptions.

Nail salon growth has also been fueled by two technological innovations—the electric file and acrylic nail products—often to mixed reviews. By adding speed and versatility to the manicuring process, while also fostering dependence on regular salon visits, these innovations have substantially increased the volume and kinds of services the salons offer, enabling them to reach out to a much wider consumer base.... These technological break-throughs have revolutionized the techniques and products available for nail care and design. Unlike old-fashioned press-on nail tips, acrylic compounds form a durable, thin, and natural-looking surface that holds various colors and applications.... Acrylics can be used to repair broken nails, smooth out uneven or damaged nail surfaces, discourage nail biting, and to create long and thick nail extensions. However, they require particular skills to apply and maintain, necessitating frequent return visits to the nail salon.

While greatly expanding the overall industry, advances in technology and products have not necessarily translated into greater profits for individual nail salons or technicians. Mass marketing means more customers who pay less. By speeding up and slashing the price of a manicure, the electric file allows a faster turnover in the salon chair, thereby expanding the market for potential nail care customers. At the same time the number of dissatisfied customers increases with this assembly-line style of service provision....

Furthermore, the increase in customers also draws more competitors into the industry. Whereas many major cities once had one nail salon for every three or four blocks, now it is not uncommon to see several salons on a single block. Long-time nail salon owner Jean Hwang lamented, "There are too many nail salons today. When I first started there weren't as many. However, the number of people getting their nails done have increased tremendously. Everyone, from little children to grandmothers, are getting their nails done. But the prices have gone down and my income remained pretty much the same. The only way it could go is down." In aspiring to keep up with competitors, many salon owners have upgraded their services to include massages, eyebrow and leg waxing, name brand products, and high-tech equipment....

In this climate of intense competition, salons must adopt different strategies to stay competitive. They can cut their prices, along with wages and other costs, to the bare minimum; they can justify higher prices by offering higher quality products and services; or they can create new products and services. In other words, nail salons must not only tap into existing demand but must generate new demand by reaching out to customers who otherwise would not purchase these services on a regular basis. In attempting to expand their consumer base, nail salons have targeted different kinds of customers with different kinds of services. The three main types of salons on which this study focuses are: nail spas catering largely to middle- and upper-class white women; nail art salons serving mostly African American working-class customers; and discount salons targeting a mixed racial and socioeconomic clientele. These different kinds of salons ... reflect the innovation and resourcefulness of individual salon owners and the fluidity of the industry as a whole as well as the shifting dynamics of consumption in the global service economy. In understanding the importance of gender, race, and immigration in shaping the divergent patterns of consumption and provision of beauty services, it is illuminating to compare contrasting patterns in nail salons versus hair salons.

HAIR VERSUS NAIL SALONS: "ONLY BLACK WOMEN CAN DO BRAIDS"

What is different about doing nails and hair? Or, to put it another way, how do ethnic-owned nail salons differ from ethnic-owned hair salons? One glaring difference is that nail salons attract customers from diverse racial and ethnic groups, whereas many beauty salons, especially those that are ethnically owned, cater overwhelmingly to coethnic customers. For example, Julie Willett describes how beauty salons in African American communities have fulfilled the needs of clientele who historically have been excluded from mainstream beauty salons or whose hair requires care that mainstream salons cannot provide. As one African American customer whom I interviewed commented, "Only black women can do braids." Furthermore, black beauty salons have also served as centers for social networking and, in some cases, political organizing. Similar dynamics emerge among Latinas, as Ginetta Candelario shows that Dominican-owned salons in Washington Heights, a neighborhood on the Upper West Side of Manhattan, import specific styles and procedures from the home country while also providing newcomers with social contacts and information that is crucial to negotiating jobs, housing, and schools.... While these establishments serve important community functions, they are also largely dependent on patronage by members of their own ethnic group. This limits their customer base and their location to ethnic enclaves and, hence, their overall profit-making ability.

But why do hair salons cater mostly to members of the same racial and ethnic group, while nail salons have been able to attract diverse clients? The answer lies partly in the differences between hair and nails and their racial meanings and partly in the differences between the representations and resources of the groups that operate these enterprises. Interestingly, in stark contrast to Asian-owned nail salons, which proliferate in a wide range of urban, suburban, and even small-town settings, Asian-owned beauty salons are a very circumscribed phenomenon, visible mainly in Asian ethnic enclaves in large cities. The contrast in the patterns of clients who support Asian-owned beauty salons (same race or ethnic group) versus nail salons (diverse customers) suggests that hair, even more so than nails, emerges as a primary signifier of racial identity, both in appearance and care.

Hairstyles, textures, and treatments are markers of racial and ethnic identity, and the experience and skills to attend to particular kinds of hair emerge as forms of racial differentiation. Thus hair stands out in both practical and symbolic terms as a racial marker....

Nails also carry racial meanings ... with pastel French manicures carrying associations with white womanhood and air-brushed acrylic nails signaling black and Latina femininities. However, the knowledge necessary to create certain kinds of nails is not seen as limited to members of the same racial and ethnic group as that of the customer. Instead,... diverse customers intentionally value and seek out Asian women *because* of distinct racialized characteristics that they regard as desirable in their manicurists.

These racializations of Asian women as preferred nail service providers emerge in the specific context of large urban centers, such as New York City, that receive many immigrants. Furthermore, customers' willingness to invest regularly in a range of nail services is notably higher in large cities, where appearances carry a particular kind of cachet....

53

Gender, Race, and Urban Policing

The Experience of African American Youths

ROD K. BRUNSON AND JODY MILLER

Law enforcement strategies in poor urban communities produce a range of harms to African American residents. This includes disproportionate experiences with surveillance and stops, disrespectful treatment, excessive force, police deviance, and fewer police protections (Fagan and Davies 2000; Mastrofski, Reisig, and McCluskey 2002; Smith and Holmes 2003). Attempts to explain these patterns examine how young Black men come to symbolize the stereotypical offender (Skolnick 1994).... However, few studies have considered how gender intersects with race and neighborhood context in determining how police behaviors are experienced. It is taken for granted that young minority *men* are the primary targets of negative police experiences.

Feminist scholars suggest that young Black women are far from immune from negative experiences with the justice system. Girls are more likely than boys to experience juvenile justice interventions for relatively minor offenses (MacDonald and Chesney-Lind 2001), and African American women and girls receive more punitive treatment than their white counterparts (Bush-Baskette 1998; Miller 1999; Visher 1983). Moreover, research suggests that Black women crime victims are less likely than white women to receive police assistance (Robinson and Chandek 2000)....

GENDER, RACE, AND URBAN POLICING

Police actions in poor urban communities are different from those in middle- and upper-class neighborhoods. Areas characterized by concentrated poverty and minority racial segregation are subject to aggressive policing strategies, including drug and gang suppression efforts, higher levels of police misconduct, and under-responsive policing (Bass 2001; Kane 2002; Klinger 1997). Aggressive policing disproportionately targets African Americans (Bass 2001). Even when such strategies result in temporary crime reductions, they undermine relations

SOURCE: Brunson, Rod K., and Jody Miller. 2006. "Gender, Race, and Urban Policing: The Experience of African American Youths." *Gender & Society* 20: 531–552.

between police and minority communities and expose large numbers of law-abiding citizens to unwelcome police contacts.

A consistent finding in policing research is that legal cynicism is more prevalent among African Americans than whites. Distrust of the police is correlated with both concentrated neighborhood disadvantage (Sampson and Bartusch 1998) and personal experiences with negative and involuntary police contacts (Weitzer and Tuch 2002). Although "juveniles make up a disproportionately large segment of the population subject to police contacts and arrests" (Leiber, Nalla, and Farnworth 1998, 152), most research on race and policing has focused on adults. The few studies to examine adolescents suggest they have less favorable attitudes toward the police than adults; African American youths experience more police contacts than white youths; and they also have greater distrust of the police than their white counterparts (Hurst, Frank, and Browning 2000; Leiber, Nalla, and Farnworth 1998).

Feminist scholarship demonstrates that race, class, and gender inequalities cannot be understood in isolation to one another....

Visher's (1983) groundbreaking study was the first to demonstrate how gender and race intersect to shape police/citizen interactions. It was long assumed that the police treat women in a "chivalrous" manner, providing preferential treatment in arrest decisions. Visher (1983, 5) challenged this assumption, suggesting instead that "chivalry exists ... for those women who display appropriate gender behaviors and characteristics." Drawing from data on police/citizen encounters, she found that older, white, and deferential women received more leniency than other women. Younger women received harsher treatment, and African American women were significantly more likely to be arrested than white women or men. In fact, they faced arrest at rates comparable to those of African American men....

Although few studies have specifically compared how African American young women and men experience discretionary police practices, the research reviewed here suggests this is an important area of inquiry. While young Black men experience higher rates of involuntary police contact and police violence, young Black women are far from immune from harmful encounters with the police. Our goal in this investigation is to examine how gender influences youths' experiences with the police in their neighborhoods and to consider these issues from the perspectives of minority youths themselves.

METHOD

This investigation is based on survey and in-depth interviews with 75 African American youths living in St. Louis, Missouri. The sample includes 35 young women and 40 young men. They range in age from 12 to 19, with a mean age of approximately 16 for both genders. Interviewing began in spring 1999 and was completed in the spring of 2000. Interviews were voluntary and typically lasted about one and a half hours. Youths were paid $20 for their participation and promised strict confidentiality. Pseudonyms are used throughout for research participants and the streets they occasionally reference.

Youths were recruited into the project with the cooperation of several organizations working with "at-risk" and delinquent youths, including a local community agency and two alternative public high schools....

Data collection began with the survey, and youths were then asked to participate in an audiotaped in-depth interview, typically completed the same day. Surveys provided baseline information about youths' participation in delinquency, perceptions of the police in their neighborhoods, personal experiences with police harassment and mistreatment, and knowledge of incidents involving others in their communities. Survey responses were used as a reference point during the in-depth interviews.

Our goal was to collect data that could provide a relatively holistic assessment of youths' experiences with the police and their perceptions of policing in their communities. The in-depth interviews were semistructured with open-ended questions that allowed for considerable probing. Youths were reminded of their survey responses about experiences with the police and were asked to provide detailed descriptions of the circumstances surrounding these events, their consequences, and the youths' interpretation of what happened and why. They also were asked their perceptions of policing in their neighborhoods, and they were encouraged to discuss problematic police incidents they had witnessed or heard about....

STUDY SETTING

St. Louis is a highly distressed U.S. city with large concentrations of extreme poverty that result in social isolation, limited resources, and high crime rates. Table 1 compares youths' neighborhoods, St. Louis city and county. Study participants were drawn from neighborhoods characterized by racial segregation and high rates of poverty, unemployment, and female-headed families.

Youths provided a stark account of how these statistics translate into lived experience. Asked to describe her neighborhood, Cleshay explained,

TABLE 1 Select Neighborhood Characteristics (in Percentages)

	Youths' Neighborhoods	*St. Louis City*	*St. Louis County*
African American	82.6	51.2	18.9
Poverty	33.8	24.6	6.9
Unemployment	18.0	11.3	4.6
Female-Headed Families with Children	43.1	28.8	10.7

SOURCE: U.S. Census (2000; see http://www.census.gov/).

> [It's] terrible. Every man for theyself. Ghetto, in the sense of raggedy, people uncool to people, just outside, street light never come on, police don't come in after four o'clock.... Heavy drug dealing. They loud, they don't care about, you know, the old people in the neighborhood or nuttin'.

Likewise, Maurice explained,

> [There's] a lot of gangs, lot of drugs, dirt. Dirty, like the streets are polluted. A lot of abandoned houses, lot of burned up houses. 'Cause of the drugs and the gangs I guess.... Vandalism. They get into a lot of fights. Bring property value down, you know, people don't take care of they houses. And you know, don't nobody really wanna live there.

Many youths said their neighborhoods were physically run-down, and most described drug dealing, street gangs, and associated violence as commonplace. Moreover, youths saw their own neighborhoods as typical of the surrounding community. Asked how her neighborhood compared to others nearby, Tisha surmised, "It's not different at all. They all do the same thing." Raymond noted, "In every neighborhood there's drug activity and gang activity." And Tami explained, "It's mostly every neighborhood got drug dealers in they neighborhood. Or people that be shootin' and stuff."

These descriptions are consistent with scholarly research, which demonstrates that poor African American neighborhoods tend to be ecologically clustered and lacking in institutional resources necessary to insulate them from crime (Krivo and Peterson 1996). In fact, even when youths described their immediate blocks as relatively problem free (which they attributed to having primarily older adults or young children present), they nonetheless described gangs, drugs, and violence nearby.

These are precisely the ecological contexts associated with aggressive policing, police deviance, and underpolicing. Policing strategies in respondents' neighborhoods relied heavily on proactive encounters to address problems such as drugs and gangs. This involved frequent pedestrian and vehicle stops by patrol officers, detectives, and members of specialized units. To examine these issues further, we now turn to youths' narrative accounts of their experiences with the police, focusing on similarities and differences across gender.

FINDINGS

The survey reports offer evidence of the gendered nature of policing in urban Black neighborhoods. Table 2 shows that most youths knew someone who had been harassed by the police, including 37 young men and 33 young women.

However, young men more often reported being mistreated themselves. In all, 33 young men, versus 16 young women, described experiencing police harassment. In addition, young men reported harassment regardless of their participation in delinquency, while more of the young women we interviewed reported harassment when they were involved in delinquency. Specifically,

TABLE 2 Perceptions of Neighborhood Policing

	Young Men (n = 40)	*Young Women (n = 35)*
Harassed or mistreated by the police	33	16
Knows someone who has been harassed or mistreated by the police	37	30
"The police are often easy to talk to"	5	5
"The police are almost never easy to talk to"	26	19
"The police are often polite to people in the neighborhood"	4	2
"The police often harass or mistreat people in the neighborhood"	19	18

19 of the 24 young men who did not engage in serious delinquency nonetheless said they had been harassed by the police, compared to 6 of 20 young women. The majority of young men (14 of 16) and young women (10 of 15) who reported participation in serious delinquency also noted experiences with police harassment.

Youths' responses to survey items about how the police behave in their neighborhoods were consistent across gender. As Table 2 shows, similar numbers of young men and women described the police as impolite and difficult to talk to, and about half said the police often harass and mistreat people in their neighborhoods. In fact, only about one in five youths said this almost never occurs. These survey findings corroborate youths' in-depth interview accounts of their treatment by the police, the extent and nature of police harassment in their neighborhoods, and their beliefs about how gender shapes police/youth interactions....

YOUNG MEN AND THE POLICE

Young men's discussions focused on their frequent involuntary contact with the police. They believed that the police besieged their neighborhoods because officers believed that many of the people living there, particularly young Black men, were criminals. Ricky described how hanging out on the street attracted police attention, regardless of whether anyone was involved in crime:

> Two blocks up from me it's like a lil' heroin area. The police is *real* hot on this area now.... One day I was walkin' from over there and the police ... seen me standing out over there. I mean, they know I don't sell dope, they know I go to school every day.... Everybody don't got to be doing the same thing. This what I tried to explain to them: "Just 'cause I'm out here don't mean I sell dope, man. I mean, every time you check me

> [I'm clean]." I mean I don't even carry money no more.... How am I sellin' dope and I don't never have no money in my pocket? Check my shoes, make me take my socks off. Man it's cold outside. "Pull your pants down."... All types of stuff, man, and it's freezin' outside. Make you lay on the ground. To me, that's police brutality.

Such searches were also described as physically intrusive, with numerous complaints about the police "trying to put they hands all in your mouth."

Young men believed the police sought to limit their use of public space by designating neighborhood locations as crime hot spots. Shaun noted, "They a trip, we be sitting on the front [porch] or something, they'll pull up just 'cause we sitting there. Or we be chillin' in front of the store, [they] get out checking everybody."...

Young men believed officers failed to consider that designated crime hot spots might also be places where law-abiding youths gather. Because pedestrian stops seemed arbitrary, they questioned whether the police were really concerned with addressing neighborhood crime or were merely interested in harassing them....

While they understood that certain contexts might subject them to increased suspicion, young men nonetheless found it prejudicial. They believed officers viewed them as criminals because they were young Black men living in poor neighborhoods....

Finally, young men were critical of officers' routine use of antagonistic language, derogatory remarks, and racial epithets. Cooper commented, "They need to change the way they talk to people.... They show us no respect ... [call us] niggers and all that." Ricky concurred: "They'll talk bad, call you all types of punks and sissies, and say you don't wanna be nothing and you ain't gon' be nothing." Tony described whistling to a friend when a passing officer "told me, 'Shut up whistling, you Black monkey!'" Thus, young men's complaints about police harassment were not just about routinely being stopped and treated as suspects but were tied to their sense that officers refused to treat them with dignity and respect.

YOUNG WOMEN AND THE POLICE

Young women's descriptions of police harassment differed from young men's. Their most common complaint, particularly when alone or in the company of other girls, was being stopped for curfew violations. Katie explained:

> On weekends we can be out 'til twelve, but long as you sittin' on your front or close by your house. But if you like outside our gate or in the street or something the police, they be like, ... If we ride past again then we're gonna lock you all up ... or whatever.... We just sittin' outside the house there, I don't know why they always messin' with us....

LaSondra, who had been sexually assaulted in her neighborhood, illustrates the tension for young women between concern over how the police treat people and the desire for protection from neighborhood dangers:

> [The police] just pull people over for no reason at all. Sometimes they just check people, see if they got some drugs. It could be the innocent person on the block they just pull 'em over and just start checking 'em.... It's messed up. They'll yell at 'em too for no reason you know.... Sometimes the police, they just sit on the block and they just watch the block and stuff.... But over there you have to watch. Especially when you walking or something 'cause you never know who might be behind you.

Although the police came to the hospital after LaSondra's sexual assault, she felt they were not responsive. Asked why, she explained, "Police don't do nothing. Some of 'em just give up. They could care less."

In the survey, only four girls (vs. nine boys) described the police as responding quickly to calls for service. Several girls described calling the police when a woman was victimized but reported they did not come. Jamellah saw a woman robbed and beaten near her home and brought the victim in her house to call the police: "[They] ain't never come.... We waited like 30 minutes and they ain't never show. So I waited at the bus stop with her ... and she caught the bus home." Rennesha's parents called the police after a neighboring child asked for help because her stepfather was beating her mother. Rennesha explained, "The lil' girl looked so sorry. But the police never did come."

Complaints about police responsiveness emerged primarily in interviews with young women and centered on incidents of violence against women. This may be because such events are more likely to result in calls for service; thus, negative outcomes are more apparent. Or it may be that the police are less responsive to such calls because they do not fit neatly into the urban crime-fighting mission (Robinson and Chandek 2000). Either way, the young women who were interviewed expected the police to come to the aid of crime victims, particularly women. Their frustration with the police in their neighborhoods stemmed both from negative interactions and from the lack of responsiveness toward crime victims. Jamelle surmised, "I feel the police are there to protect and serve you, not to harass you for no reason."...

POLICE VIOLENCE

In addition to harassment, 21 young men and 14 young women reported knowledge of or experience with more serious forms of police misconduct. Most incidents involved police violence toward family or friends.... In addition, 10 young men and 3 young women recounted personal experiences with police violence. The level of violence that boys reported was more severe. On the other hand, several girls noted concerns about police sexual violence.

Many young men regarded police violence as an expected consequence of being young Black men in their neighborhoods. Ricky noted, "I been thrown on the ground, I been kicked [laughs], I been choked, man I could go on forever." Tyrell concurred: "I know people getting beat up by the police all the time." Travis described an undercover officer's choking him in an attempt to recover drugs:

> They thought I had some dope in my mouth. So this one cop grabbed me and just started squeezing [my throat]. I was coughing and spitting up stuff and I'm like, 'What you all doing this for?' and they kept on like, 'Don't swallow it son.' I'm like 'Swallow? I ain't got no dope!' I opened up my mouth after they let go. I was showing them and everything. I mean that's they job to make sure dope isn't on the street but I mean I don't think it is their job to literally squeeze someone's Adam's apple....

While officers' encounters with young men often began with aggressive physical contact (being pushed against walls or the ground, having pockets rifled through or mouths probed), such incidents were at the extreme end of girls' contacts with the police....

More commonly, youths described incidents of police violence against young men, which they had witnessed or heard about. These events fostered anger toward the police. Frank recounted an incident in which

> my friend had got harassed by the police in front of his big brother and his little sister: I wasn't there, but they said—make it bad—it was two Black cops who did it. They had grabbed him, threw the dude on the car, they maced him. He couldn't see nuttin'. Got to punching him, slamming his head against the car. Dude's brother like, 'Ya'll leave my little brother alone!' Little sister like, 'Call momma! call momma! call momma!' Got to slamming his head against the thing, boom, boom, boom. They had took the handcuffs off and scurried off. Man, that's wrong. How they gonna do it front of they little sister?...

Several girls described attempts to intervene when officers mistreated family or friends. Nicole saw two officers "grab [my cousin], take him all back there ... in the alley," and "was like, "Scuse me, what y'all doin'?' And they was like, 'Mind yo' business lil' girl.' I said, 'I'm not little,' I was like, 'You messin' with family, you need to get yo' hands off him 'fore I call in y'all [car] number....

... Youths said neighborhood residents watch closely when male relatives are detained by the police. However, only young women described intervening directly. Perhaps girls have greater confidence in challenging officers. Since they are not the typical targets of police violence, they have less fear that the police will turn their aggression toward them.

In fact, as earlier examples illustrate, more of the young women than young men in our sample said that they challenged officers when they were stopped. While young men disapproved of police behaviors, they typically described complying with officers, because, as Cooper explained, officers "sometimes try to rough you up [if you're uncooperative]." In fact, research shows that the highest rates of police compliance are found in encounters between white officers and Black men (Mastrofski, Snipes, and Supina 1996). Our research suggests this is because police interactions with young men are seen to pose danger (Anderson 1990).

DISCUSSION

... In keeping with previous research, we found that young men were the disproportionate recipients of aggressive policing tactics such as stops and searches. Youths characterized such incidents as harassment because of their intrusive and antagonistic nature. While both young men and young women said such incidents were routine in their neighborhoods, they also distinguished them as gendered. They believed young Black men were burdened by a presumption of guilt that served as justification for aggressive police behavior.

Young men emphasized their frustration with the unilateral suspicion against them. They described being stopped on a regular basis and treated as suspects. They said officers routinely used disrespectful language, engaged in physically intrusive actions such as strip searches and cavity probes, and assumed young men merely "got lucky" rather than were innocent when no evidence of criminal wrongdoing was discovered. There was an important temporal dimension as well: Young men described being harassed at all hours, including in the mornings as they walked to school. Most of the young men who were interviewed recounted incidents of police harassment, including those who were not involved in serious delinquency. Proactive policing in urban communities targets activities—street-level drug sales and gang participation—that disproportionately involve adolescents. While this contextualizes police stops and searches, it is insufficient for explaining why so many young men are treated uniformly as suspects, even when their behavior belies this interpretation (Bass 2001).

Previous research has explained such patterns by drawing on minority group threat theories and social ecological models. Our research suggests that gender plays a significant role as well. It is not simply their status as minority youths living in poor urban communities that exposes young men to aggressive policing strategies but also that they are young African American *men* (Quillian and Pager 2001). The controlling image of young Black men as "symbolic assailants," whereby they are defined and responded to as criminals, is deeply entrenched in American culture but also deeply gendered. The young men in our sample illustrate that these messages are powerfully conveyed in adolescence. In fact, research demonstrates that such responses to African American boys begin in early childhood and has reverberating consequences (Ferguson 2001).

This is not to say young women were immune from negative police encounters. Gender shaped the kinds of treatment they experienced as well. Previous research suggests that young women are more likely than young men to face juvenile justice interventions for relatively minor offenses (MacDonald and

Chesney-Lind 2001) and that African American girls face more punitive interventions than white girls (Miller 1999)....

In addition, many young women expressed specific concern about the lack of police responsiveness to crime victims in their communities. They displayed deep pessimism about police efforts to protect community members—especially women—from crime. Direct knowledge of both police inaction and sexual misconduct by the police served to heighten their distrust further. On the other hand, we did not receive systematic reports from young women of police sexual misconduct. It may be that officers who engage in such behaviors are more reticent to sexually mistreat adolescent girls precisely because they are underage and the penalties for such action could be much more severe. Certainly there is other evidence that police sexual misconduct can be a widespread problem (Kraska and Kappeler 1995) and is reported by adult women in urban criminal networks (Maher 1997)....

Finally, youths' experiences with police violence were deeply gendered as well. With few exceptions, young men faced more severe violence at the hands of the police, and youths were deeply troubled by the frequency of such incidents in their neighborhoods. Perhaps as a consequence, young women often challenged officers who they believed were conducting themselves improperly. Our findings suggest that young women may do so more than men because they believe such challenges will not be met with physical aggression....

African American youths provide vital knowledge about how they experience policing in their neighborhoods and its effects on police/community relations. In addition, they reveal police practices as deeply gendered. Our study contributes to research on race, gender, and policing by offering further evidence of the differential harms experienced by African American girls and boys within the juvenile justice system—in this case, with a focus on events beyond the scope of formal intervention. Future research on discriminatory policing will benefit from more systematic attention to the intersections of race, place, and gender....

REFERENCES

Anderson, E. 1990. *Streetwise: Race, class, and change in an urban community*. Chicago: University of Chicago Press.

Bass, S. 2001. Policing space, policing race: Social control imperatives and police discretionary decisions. *Social Justice* 28:156–76

Bush-Baskette, S. R. 1998. The war on drugs as a war against Black women. In *Crime control and women*, edited by S. Miller. Thousand Oaks, CA: Sage.

Fagan, J., and G. Davies. 2000. Street stops and broken windows: Terry, race and disorder in New York City. *Fordham Urban Law Journal* 28:457–504.

Ferguson, A. A. 2001. *Bad boys: Public schools in the making of Black masculinity*. Ann Arbor: University of Michigan Press.

Hurst, Y. G., J. Frank, and S. L. Browning. 2000. The attitudes of juveniles toward the police: A comparison of Black and white youth. *Policing* 23:37–53.

Kane, R. J. 2002. The social ecology of police misconduct. *Criminology* 40:867–96.

Klinger, D. A. 1997. Negotiating order in patrol work. *Criminology* 35:277–306.

Kraska, P. B., and V. E. Kappeler. 1995. To serve and pursue: Exploring police sexual violence against women. *Justice Quarterly* 12:85–112.

Krivo, L. L., and R. D. Peterson. 1996. Extremely disadvantaged neighborhoods and urban crime. *Social Forces* 75:619–50.

Leiber, M. J., M. K. Nalla, and M. Farnworth. 1998. Explaining juveniles' attitudes toward the police. *Justice Quarterly* 15:151–74.

MacDonald, J. M., and M. Chesney-Lind. 2001. Gender bias and juvenile justice revisited. *Crime & Delinquency* 47:173–95.

Maher, L. 1997. *Sexed work: Gender, race, and resistance in a Brooklyn drug market.* Oxford, UK: Clarendon Press.

Mastrofski, S. D., M. D. Reisig, and J. D. McCluskey. 2002. Police disrespect toward the public: An encounter based analysis. *Criminology* 40:515–51.

Mastrofski, S. D., J. B. Snipes, and A. E. Supina. 1996. Compliance on demand. *Journal of Research in Crime and Delinquency* 33:269–305.

Miller, J. 1999. An examination of disposition decision-making for delinquent girls. In *The intersection of race, gender and class in criminology*, edited by M. D. Schwartz and D. Milovanovic. New York: Garland.

Quillian, L., and D. Pager. 2001. Black neighbors, higher crime? The role of racial stereotypes in evaluations of neighborhood crime. *American Journal of Sociology* 106:717–67.

Robinson, A. L., and M. S. Chandek. 2000. Differential police response to Black battered women. *Women & Criminal Justice* 12:29–61.

Sampson, R. J., and D. J. Bartusch. 1998. Legal cynicism and (subcultural?) tolerance of deviance: The neighborhood context of racial differences. *Law and Society Review* 32:777–804.

Skolnick, J. H. 1994. *Justice without trial.* 3d ed. New York: Macmillan.

Smith, B. W., and M. D. Holmes. 2003. Community accountability, minority threat, and police brutality, *Criminology* 41:1035–64.

Visher, C. A. 1983. Gender, police arrest decisions, and notions of chivalry. *Criminology* 21:5–28.

Weitzer, R., and S. A. Tuch. 2002. Perceptions of racial profiling: Race, class and personal experience. *Criminology* 40:435–57.

54

The Color of Justice

MICHELLE ALEXANDER

Imagine you are Emma Faye Stewart, a thirty-year-old, single African American mother of two who was arrested as part of a drug sweep in Hearne, Texas. All but one of the people arrested were African American. You are innocent. After a week in jail, you have no one to care for your two small children and are eager to get home. Your court-appointed attorney urges you to plead guilty to a drug distribution charge, saying the prosecutor has offered probation. You refuse, steadfastly proclaiming your innocence. Finally, after almost a month in jail, you decide to plead guilty so you can return home to your children. Unwilling to risk a trial and years of imprisonment, you are sentenced to ten years probation and ordered to pay $1,000 in fines, as well as court and probation costs. You are also now branded a drug felon. You are no longer eligible for food stamps; you may be discriminated against in employment; you cannot vote for at least twelve years; and you are about to be evicted from public housing. Once homeless, your children will be taken from you and put in foster care.

A judge eventually dismisses all cases against the defendants who did not plead guilty. At trial, the judge finds that the entire sweep was based on the testimony of a single informant who lied to the prosecution. You, however, are still a drug felon, homeless, and desperate to regain custody of your children.

Now place yourself in the shoes of Clifford Runoalds, another African American victim of the Hearne drug bust. You returned home to Bryan, Texas, to attend the funeral of your eighteen-month-old daughter. Before the funeral services begin, the police show up and handcuff you. You beg the officers to let you take one last look at your daughter before she is buried. The police refuse. You are told by prosecutors that you are needed to testify against one of the defendants in a recent drug bust. You deny witnessing any drug transaction; you don't know what they are talking about. Because of your refusal to cooperate, you are indicted on felony charges. After a month of being held in jail, the charges against you are dropped. You are technically free, but as a result of your arrest and period of incarceration, you lose your job, your apartment, your furniture, and your car. Not to mention the chance to say good-bye to your baby girl.

This is the War on Drugs. The brutal stories described above are not isolated incidents, nor are the racial identities of Emma Faye Stewart and Clifford

SOURCE: Alexander, Michelle. 2010. *The New Jim Crow*. New York: The New Press.

Runoalds random or accidental. In every state across our nation, African Americans—particularly in the poorest neighborhoods—are subjected to tactics and practices that would result in public outrage and scandal if committed in middle-class white neighborhoods. In the drug war, the enemy is racially defined. The law enforcement methods described [here] have been employed almost exclusively in poor communities of color, resulting in jaw-dropping numbers of African Americans and Latinos filling our nation's prisons and jails every year. We are told by drug warriors that the enemy in this war is a thing—drugs—not a group of people, but the facts prove otherwise.

Human Rights Watch reported in 2000 that, in seven states, African Americans constitute 80 to 90 percent of all drug offenders sent to prison. In at least fifteen states, blacks are admitted to prison on drug charges at a rate from twenty to fifty-seven times greater than that of white men. In fact, nationwide, the rate of incarceration for African American drug offenders dwarfs the rate of whites. When the War on Drugs gained full steam in the mid-1980s, prison admissions for African Americans skyrocketed, nearly quadrupling in three years, and then increasing steadily until it reached in 2000 a level *more than twenty-six times* the level in 1983. The number of 2000 drug admissions for Latinos was twenty-two times the number of 1983 admissions. Whites have been admitted to prison for drug offenses at increased rates as well—the number of whites admitted for drug offenses in 2000 was eight times the number admitted in 1983—but their relative numbers are small compared to blacks' and Latinos. Although the majority of illegal drug users and dealers nationwide are white, three-fourths of all people imprisoned for drug offenses have been black or Latino. In recent years, rates of black imprisonment for drug offenses have dipped somewhat—declining approximately 25 percent from their zenith in the mid-1990s—but it remains the case that African Americans are incarcerated at grossly disproportionate rates throughout the United States.

There is, of course, an official explanation for all of this: crime rates. This explanation has tremendous appeal—before you know the facts—for it is consistent with, and reinforces, dominant racial narratives about crime and criminality dating back to slavery. The truth, however, is that rates and patterns of drug crime do not explain the glaring racial disparities in our criminal justice system. People of all races use and sell illegal drugs at remarkably similar rates. If there are significant differences in the surveys to be found, they frequently suggest that whites, particularly white youth, are more likely to engage in illegal drug dealing than people of color. One study, for example, published in 2000 by the National Institute on Drug Abuse reported that white students use cocaine at seven times the rate of black students, use crack cocaine at eight times the rate of black students, and use heroin at seven times the rate of black students. That same survey revealed that nearly identical percentages of white and black high school seniors use marijuana. The National Household Survey on Drug Abuse reported in 2000 that white youth aged 12–17 are more than a third more likely to have sold illegal drugs than African American youth. Thus the very same year Human Rights Watch was reporting that African Americans were being arrested and imprisoned at unprecedented rates, government data revealed that blacks were no more likely to be guilty of drug crimes than whites and that white youth were actually the *most likely* of any racial or ethnic group to be guilty of illegal drug possession and sales. Any notion that drug use among blacks is more severe or dangerous is belied by the data; white youth have about three times the number of drug-related emergency room visits as their African American counterparts.

The notion that whites comprise the vast majority of drug users and dealers—and may well be more likely than other racial groups to commit drug crimes—may seem implausible to some, given the media imagery we are fed on a daily basis and the racial composition of our prisons and jails. Upon reflection, however, the prevalence of white drug crime—including drug dealing—should not be surprising. After all, where do whites get their illegal drugs? Do they all drive to the ghetto to purchase them from somebody standing on a street corner? No. Studies consistently indicate that drug markets, like American society generally, reflect our nation's racial and socioeconomic boundaries. Whites tend to sell to whites; blacks to blacks. University students tend to sell to each other. Rural whites, for their part, don't make a special trip to the 'hood to purchase marijuana. They buy it from somebody down the road. White high school students typically buy drugs from white classmates, friends, or older relatives. Even Barry McCaffrey, former director of the White House Office of National Drug Control Policy, once remarked, if your child bought drugs, "it was from a student of their own race generally." The notion that most illegal drug use and sales happens in the ghetto is pure fiction. Drug trafficking occurs there, but it occurs everywhere else in America as well. Nevertheless, black men have been admitted to state prison on drug charges at a rate that is more than thirteen times higher than white men. The racial bias inherent in the drug war is a major reason that 1 in every 14 black men was behind bars in 2006, compared with 1 in 106 white men. For young black men, the statistics are even worse. One in 9 black men between the ages of twenty and thirty-five was behind bars in 2006, and far more were under some form of penal control—such as probation or parole. These gross racial disparities simply cannot be explained by rates of illegal drug activity among African Americans.

What, then, does explain the extraordinary racial disparities in our criminal justice system? Old-fashioned racism seems out of the question. Politicians and law enforcement officials today rarely endorse racially biased practices, and most of them fiercely condemn racial discrimination of any kind. When accused of racial bias, police and prosecutors—like most Americans—express horror and outrage. Forms of race discrimination that were open and notorious for centuries were transformed in the 1960s and 1970s into something un-American—an affront to our newly conceived ethic of colorblindness. By the early 1980s, survey data indicated that 90 percent of whites thought black and white children should attend the same schools, 71 percent disagreed with the idea that whites have a right to keep blacks out of their neighborhoods, 80 percent indicated they would support a black candidate for president, and 66 percent opposed laws prohibiting intermarriage. Although far fewer supported specific policies designed to achieve racial equality or integration (such as busing), the mere fact that large majorities of whites were, by the early 1980s, supporting the antidiscrimination principle reflected a profound shift in racial attitudes. The margin of support for colorblind norms has only increased since then.

This dramatically changed racial climate has led defenders of mass incarceration to insist that our criminal justice system, whatever its past sins, is now largely fair and nondiscriminatory. They point to violent crime rates in the African American community as a justification for the staggering number of black men who find themselves behind bars. Black men, they say, have much higher rates of violent crime; that's why so many of them are locked in prisons.

Typically, this is where the discussion ends.

The problem with this abbreviated analysis is that violent crime is *not* responsible for the prison boom. As numerous researchers have shown, violent crime rates have fluctuated over the years and bear little relationship to incarceration rates—which have soared during the past three decades regardless of whether violent crime was going up or down. Today violent crime rates are at historically low levels, yet incarceration rates continue to climb.

Murder convictions tend to receive a tremendous amount of media attention, which feeds the public's sense that violent crime is rampant and forever on the rise. But like violent crime in general, the murder rate cannot explain the prison boom. Homicide convictions account for a tiny fraction of the growth in the prison population. In the federal system, for example, homicide offenders account for 0.4 percent of the past decade's growth in the federal prison population, while drug offenders account for nearly 61 percent of that expansion. In the state system, less than 3 percent of new court commitments to state prison typically involve people convicted of homicide. As much as a third of state prisoners are violent offenders, but that statistic can easily be misinterpreted. Violent offenders tend to get longer prison sentences than nonviolent offenders, and therefore comprise a much larger share of the prison population than they would if they had earlier release dates. The uncomfortable reality is that convictions for drug offenses—not violent crime—are the single most important cause of the prison boom in the United States, and people of color are convicted of drug offenses at rates out of all proportion to their drug crimes.

These facts may still leave some readers unsatisfied. The idea that the criminal justice system discriminates in such a terrific fashion when few people openly express or endorse racial discrimination may seem far-fetched, if not absurd. How could the War on Drugs operate in a discriminatory manner, on such a large scale, when hardly anyone advocates or engages in explicit race discrimination?… Despite the colorblind rhetoric and fanfare of recent years, the design of the drug war effectively guarantees that those who are swept into the nation's new undercaste are largely black and brown.

This sort of claim invites skepticism. Nonracial explanations and excuses for the systematic mass incarceration of people of color are plentiful. It is the genius of the new system of control that it can always be defended on nonracial grounds, given the rarity of a noose or a racial slur in connection with any particular criminal case. Moreover, because blacks and whites are almost never similarly situated (given extreme racial segregation in housing and disparate life experiences), trying to "control for race" in an effort to evaluate whether the mass incarceration of people of color is really about race or something else—anything else—is difficult. But it is not impossible.

A bit of common sense is overdue in public discussions about racial bias in the criminal justice system. The great debate over whether black men have been targeted by the criminal justice system or unfairly treated in the War on Drugs often overlooks the obvious. What is painfully obvious when one steps back from individual cases and specific policies is that the system of mass incarceration operates with stunning efficiency to sweep people of color off the streets, lock them in cages, and then release them into an inferior second-class status. Nowhere is this more true than in the War on Drugs.

The central question, then, is *how* exactly does a formally colorblind criminal justice system achieve such racially discriminatory results? Rather easily, it turns out. The process occurs in two stages. The first step is to grant law enforcement officials extraordinary discretion regarding whom to stop, search, arrest, and charge for drug offenses, thus ensuring that conscious and unconscious racial beliefs and stereotypes will be given free reign. Unbridled discretion inevitably creates huge racial disparities. Then, the damning step: Close the courthouse doors to all claims by defendants and private litigants that the criminal justice system operates in racially discriminatory fashion. Demand that anyone who wants to challenge racial bias in the system offer, in advance, clear proof that the racial disparities are the product of intentional racial discrimination—i.e., the work of a bigot. This evidence will almost never be available in the era of colorblindness, because everyone knows—but does not say—that the enemy in the War on Drugs can be identified by race. This simple design has helped to produce one of the most extraordinary systems of racialized social control the world has ever seen.

55

Rape, Racism, and the Law

JENNIFER WRIGGINS

The history of rape in this country has focused on the rape of white women by Black men. From a feminist perspective, two of the most damaging consequences of this selective blindness are the denials that Black women are raped and that all women are subject to pervasive and harmful sexual coercion of all kinds....

THE NARROW FOCUS ON BLACK OFFENDER/ WHITE VICTIM RAPE

There are many different kinds of rape. Its victims are of all races, and its perpetrators are of all races. Yet the kind of rape that has been treated most seriously throughout this nation's history has been the illegal forcible rape of a white woman by a Black man. The selective acknowledgement of Black accused/ white victim rape was especially pronounced during slavery and through the first half of the twentieth century. Today a powerful legacy remains that permeates thought about rape and race.

During the slavery period, statutes in many jurisdictions provided the death penalty or castration for rape when the convicted man was Black or mulatto and the victim white. These extremely harsh penalties were frequently imposed. In addition, mobs occasionally broke into jails and courtrooms and lynched slaves alleged to have raped white women, prefiguring Reconstruction mob behavior.

In contrast to the harsh penalties imposed on Black offenders, courts occasionally released a defendant accused of raping a white woman when the evidence was inconclusive as to whether he was Black or mulatto. The rape of Black women by white or Black men, on the other hand, was legal; indictments were sometimes dismissed for failing to allege that the victim was white. In those states where it was illegal for white men to rape white women, statutes provided less severe penalties for the convicted white rapist than for the convicted Black one. In addition, common-law rules both defined rape narrowly and made it a difficult crime to prove....

SOURCE: Wriggins, Jennifer. 1983. "Rape, Racism, and the Law." *Harvard Women's Law Journal* 6 (Spring): 103–141.

After the Civil War, state legislatures made their rape statutes race-neutral, but the legal system treated rape in much the same way as it had before the war. Black women raped by white or Black men had no hope of recourse through the legal system. White women raped by white men faced traditional common-law barriers that protected most rapists from prosecution.

Allegations of rape involving Black offenders and white victims were treated with heightened virulence. This was manifested in two ways. The first response was lynching, which peaked near the end of the nineteenth century. The second, from the early twentieth century on, was the use of the legal system as a functional equivalent of lynching, as illustrated by mob coercion of judicial proceedings, special doctrinal rules, the language of opinions, and the markedly disparate numbers of executions for rape between white and Black defendants.

Between 1882 and 1946 at least 4,715 persons were lynched, about three-quarters of whom were Black. Although lynching tapered off after the early 1950s, occasional lynch-like killings persist to this day. The influence of lynching extended far beyond the numbers of Black people murdered because accounts of massive white crowds torturing, burning alive, and dismembering their victims created a widespread sense of terror in the Black community.

The most common justification for lynching was the claim that a Black man had raped a white woman. The thought of this particular crime aroused in many white people an extremely high level of mania and panic. One white woman, the wife of an ex-Congressman, stated in 1898: "If it needs lynching to protect woman's dearest possession from human beasts, then I say lynch a thousand times a week if necessary." The quote resonates with common stereotypes that Black male sexuality is wanton and bestial, and that Black men are wild, criminal rapists of white women.

Many whites accepted lynching as an appropriate punishment for a Black man accused of raping a white woman. The following argument made to the jury by defense counsel in a 1907 Louisiana case illustrates this acceptance:

> Gentlemen of the jury, this man, a nigger, is charged with breaking into the house of a white man in the nighttime and assaulting his wife, with the intent to rape her. Now, don't you know that, if this nigger had committed such a crime, he never would have been brought here and tried; that he would have been lynched, and if I were there I would help pull on the rope.[1]

It is doubtful whether the legal system better protected the rights of a Black man accused of raping a white woman than did the mob. Contemporary legal literature used the term "legal lynching" to describe the legal system's treatment of Black men. Well past the first third of the twentieth century, courts were often coerced by violent mobs, which threatened to execute the defendant themselves unless the court convicted him. Such mobs often did lynch the defendant if the judicial proceedings were not acceptable to them. A contemporary authority on lynching commented in 1934 that "the local sentiment which would make a lynching possible would insure a conviction in the courts." Even if the mob was not overtly pressuring for execution, a Black defendant accused of raping a white woman faced a hostile, racist legal system. State court submission to mob pressure is well illustrated by the most famous series of cases about interracial rape, the Scottsboro cases of the 1930s. Eight young Black men were convicted of what the Alabama Supreme Court called "a most foul and revolting crime," which was the rape of "two defenseless white girls." The defendants were summarily sentenced to death based on minimal and dubious evidence, having been denied effective assistance of counsel. The Alabama Supreme Court upheld the convictions in opinions demonstrating relentless determination to hold the defendants guilty regardless of strong evidence that mob pressure had influenced the verdicts and the weak evidence presented against the defendants. In one decision, that court affirmed the trial court's denial of a change of venue on the grounds that the mobs' threats of harm were not imminent enough although the National Guard had been called out to protect the defendants from mob executions. The U.S. Supreme Court later recognized that the proceedings had in fact taken place in an atmosphere of "tense, hostile, and excited public sentiment." After a lengthy appellate process, including three favorable Supreme Court rulings, all of the Scottsboro defendants were released, having spent a total of 104 years in prison.

In addition, courts applied special doctrinal rules to Black defendants accused of the rape or attempted rape of white women. One such rule allowed juries to consider the race of the defendant and victim in drawing factual conclusions as to the defendant's intent in attempted rape cases. If the accused was Black and the victim white, the jury was entitled to draw the inference, based on race alone, that he intended to rape her. One court wrote, "In determining the question of intention, the jury may consider social conditions and customs founded upon racial differences, such as that the prosecutrix was a white woman and defendant was a Negro man."[2] The "social conditions and customs founded upon racial differences" which the jury was to consider included the assumption that Black men always and only want to rape white women, and that a white woman would never consent to sex with a Black man.

The Georgia Supreme Court of 1899 was even more explicit about the significance of race in the context of attempted rape, and particularly about the motivations of Black men. It held that race may properly be considered "to rebut any presumption that might otherwise arise in favor of the accused that his intention was to obtain the consent of the female, upon failure of which he would abandon his purpose to have sexual intercourse with her."[3] Such a rebuttal denied to Black defendants procedural protection that was accorded white defendants....

The outcome of this disparate treatment of Black men by the legal system was often the same as lynching—death. Between 1930 and 1967, thirty-six percent of the Black men who were convicted of raping a white woman were executed. In stark contrast, only two percent of all defendants convicted of rape involving other racial combinations were executed. As a result of such disparate treatment, eighty-nine percent of the men executed for rape in this country were Black. While execution rates for all crimes were much higher for Black men than for white men, the differential was most dramatic when the crime was the rape of a white woman.

The patterns that began in slavery and continued long afterwards have left a powerful legacy that manifests itself today in several ways. Although the death penalty for rape has been declared unconstitutional, the severe statutory penalties for rape continue to be applied in a discriminatory manner. A recent study concluded that Black men convicted of raping white women receive more serious sanctions than all other sexual assault defendants. A recent attitudinal study found that white potential jurors treated Black and white defendants similarly when the victim was Black. However, Black defendants received more severe punishment than white defendants when the victim was white.

The rape of white women by Black men is also used to justify harsh rape penalties. One of the few law review articles written before 1970 that takes a firm position in favor of strong rape laws to secure convictions begins with a long quote from a newspaper article describing rapes by three Black men, who at 3 a.m. on Palm Sunday "broke into a West Philadelphia home occupied by an eighty-year-old widow, her forty-four-year-old daughter and fourteen-year-old granddaughter," brutally beat and raped the white women, and left the grandmother unconscious "lying in a pool of blood." This introduction presents rape as a crime committed by violent Black men against helpless white women. It is an image of a highly atypical rape—the defendants are Black and the victims white, the defendants and victims are strangers to each other, extreme violence is used, and it is a group rape. Contemporaneous statistical data on forcible rapes reported to the Philadelphia police department reveals that this rape case was virtually unique.[4] Use of this highly unrepresentative image of rape to justify strict rape laws is consistent with recent research showing that it is a prevalent, although false, belief about rape that the most common racial combination is Black offender and white victim.[5]

Charges of rapes committed by Black men against white women are still surrounded by sensationalism and public pressure for prosecution. Black men seem to face a special threat of being unjustly prosecuted or convicted. One example is Willie Sanders.[6] Sanders is a Black Boston man who was arrested and charged with the rapes of four young white women after a sensational media campaign and intense pressure on the police to apprehend the rapist. Although the rapes continued after Sanders was incarcerated, and the evidence against him was extremely weak, the state subjected him to a vigorous twenty-month prosecution. After a lengthy and expensive trial, and an active public defense, he was eventually acquitted. Although Sanders was clearly innocent, he could have been convicted; he and his family suffered incalculable damage despite his acquittal....

From slavery to the present day, the legal system has consistently treated the rape of white women by Black men with more harshness than any other kind of rape....

This selective focus is significant in several ways. First, since tolerance of coerced sex has been the rule rather than the exception, it is clear that the rape of white women by Black men has been treated seriously not because it is coerced sex and thus damaging to women, but because it is threatening to white men's power over both "their" women and Black men. Second, in treating Black offender/white victim illegal rape much more harshly than all coerced sex experienced by Black women and most coerced sex experienced by white women, the legal system has implicitly condoned the latter forms of rape. Third, this treatment has contributed to a paradigmatic but false concept of rape as being primarily a violent crime between strangers where the perpetrator is Black and the victim white. Finally, this pattern is perverse and discriminatory because rape is painful and degrading to both Black and white victims regardless of the attacker's race.

THE DENIAL OF THE RAPE OF BLACK WOMEN

The selective acknowledgement of the existence and seriousness of the rape of white women by Black men has been accompanied by a denial of the rape of Black women that began in slavery and continues today. Because of racism and sexism, very little has been written about this denial. Mainstream American history has ignored the role of Black people to a large extent; systematic research into Black history has been published only recently. The experiences of Black women have yet to be fully recognized in those histories, although this is beginning to change. Indeed, very little has been written about rape from the perspective of the victim, Black or white, until quite recently. Research about Black women rape victims encounters all these obstacles.

The rape of Black women by white men during slavery was commonplace and was used as a crucial weapon of white supremacy. White men had what one commentator called "institutionalized access" to Black women. The rape of Black women by white men cannot be attributed to unique Southern pathology, however, for numerous accounts exist of northern armies raping Black women while they were "liberating" the South.

The legal system rendered the rape of Black women by any man, white or Black, invisible. The rape of a Black woman was not a crime. In 1859 the Mississippi Supreme Court dismissed the indictment of a male slave for the rape of a female slave less than 10 years old, saying:

> [T]his indictment can not be sustained, either at common law or under our statutes. It charges no offense known to either system. [Slavery] was unknown to the common law … and hence its provisions are inapplicable.… There is no act (of our legislature on this subject) which embraces either the attempted or actual commission of a rape by a slave on a female slave.… Masters and slaves can not be governed by the same a slave on a female slave.… Masters and slaves can not be governed by the same system or laws; so different are their positions, rights and duties.[7]

This decision is illuminating in several respects. First, Black men are held to lesser standards of sexual restraint with Black women than are white men with white women. Second, white men are held to lesser standards of restraint with

Black women than are Black men with white women. Neither white nor Black men were expected to show sexual restraint with Black women.

After the Civil War, the widespread rape of Black women by white men persisted. Black women were vulnerable to rape in several ways that white women were not. First, the rape of Black women was used as a weapon of group terror by white mobs and by the Ku Klux Klan during Reconstruction. Second, because Black women worked outside the home, they were exposed to employers' sexual aggression as white women who worked inside the home were not.

The legal system's denial that Black women experienced sexual abuse by both white and Black men also persisted, although statutes had been made race-neutral. Even if a Black victim's case went to trial—in itself highly unlikely—procedural barriers and prejudice against Black women protected any man accused of rape or attempted rape. The racist rule which facilitated prosecutions of Black offender/white victim attempted rapes by allowing the jury to consider the defendant's race as evidence of his intent, for instance, was not applied where both persons were "of color and there was no evidence of their social standing."[8] That is, the fact that a defendant was Black was considered relevant only to prove intent to rape a white woman; it was not relevant to prove intent to rape a Black woman. By using disparate procedures, the court implicitly makes two assertions. First, Black men do not want to rape Black women with the same intensity or regularity that Black men want to rape white women. Second, Black women do not experience coerced sex in the sense that white women experience it.

These attitudes reflect a set of myths about Black women's supposed promiscuity which were used to excuse white men's sexual abuse of Black women. An example of early twentieth century assumptions about Black women's purported promiscuity was provided by the Florida Supreme Court in 1918. In discussing whether the prior chastity of the victim in a statutory rape case should be presumed subject to defendant's rebuttal or should be an element of the crime which the state must prove, the court explained that:

> What has been said by some of our courts about an unchaste female being a comparatively rare exception is no doubt true where the population is composed largely of the Caucasian race, but we would blind ourselves to actual conditions if we adopted this rule where another race that is largely immoral constitutes an appreciable part of the population.[9]

Cloaking itself in the mantle of legal reasoning, the court states that most young white women are virgins, that most young Black women are not, and that unchaste women are immoral. The traditional law of statutory rape at issue in the above-quoted case provides that women who are not "chaste" cannot be raped. Because of the way the legal system considered chastity, the association of Black women with unchastity meant not only that Black women could not be victims of statutory rape, but also that they would not be recognized as victims of forcible rape.

The criminal justice system continues to take the rape of Black women less seriously than the rape of white women. Studies show that judges generally impose harsher sentences for rape when the victim is white than when the victim is Black. The behavior of white jurors shows a similar bias. A recent study found that sample white jurors imposed significantly lighter sentences on defendants whose victims were Black than on defendants whose victims were white. Black jurors exhibited no such bias.

Evidence concerning police behavior also documents the fact that the claims of Black rape victims are taken less seriously than those of whites. A ... study of Philadelphia police processing decisions concluded that the differential in police decisions to charge for rape "resulted primarily from a lack of confidence in the veracity of Black complainants and a belief in the myth of Black promiscuity."

The thorough denial of Black women's experiences of rape by the legal system is especially shocking in light of the fact that Black women are much more likely to be victims of rape than are white women.[10] Based on data from national surveys of rape victims, "the profile of the most frequent rape victim is a young woman, divorced or separated, Black and poverty stricken."...

CONCLUSION

The legal system's treatment of rape both has furthered racism and has denied the reality of women's sexual subordination. It has disproportionately targeted Black men for punishment and made Black women both particularly vulnerable and particularly without redress. It has denied the reality of women's sexual subordination by creating a social meaning of rape which implies that the only type of sexual abuse is illegal rape and the only form of illegal rape is Black offender/white victim. Because of the interconnectedness of rape and racism, successful work against rape and other sexual coercion must deal with racism. Struggles against rape must acknowledge the differences among women and the different ways that groups other than women are disempowered. In addition, work against rape must go beyond the focus on illegal rape to include all forms of coerced sex, in order to avoid the racist historical legacy surrounding rape and to combat effectively the subordination of women.

NOTES

1. *State v. Petit*, 119 La., 44 So. (1907).
2. *McQuirter v. State*, 36 Ala., 63 So. 2d (1953).
3. *Dorsey v. State*, 108 Ga., 34 S.E. (1899).
4. Out of 343 rapes reported to the Philadelphia police, 3.3% involved Black defendants accused of raping white women; 42% involved complaints of stranger rape; 20.5% involved brutal beatings; 43% involved group rapes.
5. In answer to the question, "Among which racial combination do most rapes occur?" 48% of respondents stated Black males and white females, 3% stated white males and Black females, 16% stated Black males and Black females, 33% stated white males

and white females. Recent victim survey data contradict this prevalent belief; more than four-fifths of illegal rapes reported to researchers were between members of the same race, and white/Black rapes roughly equaled Black/white rapes.

6. Suffolk Superior Court indictment (1980).
7. *George v. State*, 37 Miss. (1859).
8. *Washington v. State*, 38 Ga., 75 S.E. (1912).
9. *Dallas v. State*, 76 Fla., 79 So. (1918).
10. Recent data from random citizen interviews suggest that Black women are much more likely to be victims of illegal rape than are white women.

56

Interpreting and Experiencing Anti-Queer Violence

Race, Class, and Gender Differences among LGBT Hate Crime Victims

DOUG MEYER

Several studies of hate crime victims have documented the ways in which lesbian women and gay men determine that violence is based on their sexuality (Herek et al., 1997; Herek et al., 2002). These studies, however, make no reference to race, class, and gender. In this [reading], I employ an intersectionality framework to explore how lesbian, gay, bisexual, and transgender (LGBT) people determine that violence is based on their sexuality or gender identity. Employing an intersectionality framework reveals how LGBT people's violent experiences differ along the lines of race, class, and gender. Furthermore, an intersectionality approach facilitates our understanding of the ways in which LGBT people interpret and experience hate-motivated violence. To accomplish these research goals, I designed a qualitative research project in which I interviewed 44 people who experienced violence because they were perceived to be lesbian, gay, bisexual, or transgender.

... Studies of hate crime victims have revealed the degree to which hate motivated violence can have traumatic psychological effects. Unfortunately, the focus on the psychological effects of hate-motivated violence has left other research questions unexplored. In particular, studies of hate crime victims have overlooked important sociological research questions. We know little, for example, about the ways in which race, class, gender, and sexuality structure victims' experiences of hate-motivated violence. Intersectionality, a theoretical framework that has been highly influential in other bodies of literature, has remained absent from studies of hate crime victims.

Intersectionality theory denotes the ways in which institutional power structures such as race, class, gender, and sexuality simultaneously structure social relations. It conceptualizes these institutional power structures as distinct but mutually reinforcing systems of oppression. Conceptualizing systems of oppression as distinct prevents researchers from collapsing one system into another.

SOURCE: Meyer, Doug. 2008. "Interpreting and Experiencing Anti-Queer Violence: Race, Class, and Gender Differences among LGBT Hate Crime Victims," *Race, Gender, and Class* 15: 262–282.

Racism, for example, cannot be reduced to gender oppression and heterosexism cannot be understood as a product of patriarchy. Furthermore, because systems of oppression are interlocking, social scientists must account for multiple forms of social inequality to understand patterns of behavior. Thus, they cannot understand gendered patterns of activity without also understanding raced, classed, and sexualized ones....

BRINGING AN INTERSECTIONALITY APPROACH TO HATE CRIME RESEARCH

Although intersectionality has remained absent from studies of hate crime victims, it has been featured in other areas of the literature. Barbara Perry (2001), an eminent hate crime scholar, incorporates elements of intersectionality into her theoretical account of hate crime. She argues that traditional criminological theory has failed to account for hate-motivated violence. According to traditional criminological theory, hate crime occurs because relatively powerless individuals are unable to achieve society's goals through socially acceptable means.... Perry argues that this theoretical framework cannot fully account for hate crime because perpetrators are often relatively privileged members of society. Instead, she argues that a more satisfying explanation must incorporate power relations: hate crime should be understood as a social control mechanism rooted in institutional power structures. She conceptualizes hate crime as an outgrowth of systems of oppression; it is one of the ways in which perpetrators maintain social hierarchies. For instance, racially-motivated violence affirms White privilege, while anti-lesbian and anti-gay violence reinforces the subordination of women and the cultural devaluation of homosexuality (Perry, 2001).

Perry has advanced a persuasive sociological theory of hate crime. She reveals the cultural context in which hate-motivated violence can flourish and she documents the ways in which institutional power structures may lead to hate crime. Similarly, groundbreaking studies of organized hate groups and white supremacist discourse have contributed to intersectionality scholarship. They have revealed the factors leading to women's involvement in racist activism and they have demonstrated how white supremacist discourse reinforces patterns of social inequality. By doing so, they have revealed some of the dynamics that contribute to hate crime. While these studies have produced numerous contributions, they have focused on the causes and the perpetrators of hate-motivated violence more than the victims.

... While most studies of hate crime victims have examined the psychological effects of hate-motivated violence, some social scientists have begun to explore other research questions.... Rather than accepting abstract definitions of hate crime as given, this research approach allows victims to explain their understanding of hate-motivated violence. By doing so, it privileges the voice of victims and it constructs their understanding of hate crime as significant.

Herek and colleagues (Herek et al., 1997; Herek et al., 2002), focusing on anti-lesbian and anti-gay victimization, have revealed some of the ways in which hate crime victims interpret and understand their violent experiences. They have found that lesbian women and gay men often examine their perpetrators' statements to determine that violence is based on their sexuality. In these situations, lesbian women and gay men confront explicit, unambiguous homophobic remarks. In other, less common situations they identify incidents as anti-lesbian or anti-gay by relying on contextual cues. For instance, victims perceive violence as rooted in homophobia when it occurs near a gay-identified location or when it occurs after public displays of affection between same-sex couples.

In this [reading], I employ an intersectionality approach to expand upon studies that have examined how lesbian women and gay men determine that violence is based on their sexuality. As I argue throughout this [reading], systems of oppression affect how LGBT people make this determination. Thus, employing an intersectionality framework improves our understanding of hate crime by revealing some of the ways in which LGBT people's violent experiences differ along the lines of race, class, gender, and sexuality.

METHODS

To examine how victims experience hate-motivated violence, I designed a qualitative research project in which I interviewed 44 people who experienced violence because they were perceived to be lesbian, gay, bisexual, or transgender.

... During the interview, I asked participants to describe their violent experiences and their response to the violence. To understand how they perceived the violence, I asked about their perpetrators' motivations ("As you look back on this incident, why do you think he/she/they used violence?"). I then asked detailed, follow-up questions about their understanding of why the violence had occurred.

Because I wanted to examine the intersection of multiple systems of oppression, it was important to have a diverse sample of LGBT people. Thus, I recruited participants from a wide range of advocacy and service organizations, many of which provide services for LGBT people of color. I interviewed 17 women, 17 men, and 10 transgender people. All of the men identified as gay and 13 of the women identified as lesbian; two women identified as heterosexual and two as bisexual. Eight of the transgender people identified as male-to-female (MTF), one as female-to-male (FTM), and one as intersexed. Participants ranged from 20 to 62 years old; the median age was 41. Twenty-one participants identified as Black, 13 as White, eight as Latino, and two as Asian. In terms of educational background, five participants had dropped out of high school, 14 had a high school diploma, six had taken some college, 16 had a college degree, and three had a postgraduate degree. The interviews lasted from approximately one to three hours....

RESULTS

Determining that Anti-Queer Violence Is Rooted in Multiple Systems of Oppression

One of the ways in which queer people determined that violence was based on their sexuality was by examining what their perpetrators said about gender. When examining the intersection of gender and sexuality, queer people often determined that violence directed against their gender identity was rooted in homophobia. When doing so, they acknowledged societal processes that conflate gender nonconformity with homosexuality. Dorothy, a 49-year-old White lesbian woman, addressed this dynamic when arguing that violence directed against her gender identity was also rooted in homophobia. The violence occurred when she was attacked by three men on the street. One of the men punched her and stole her purse. The men called her a "bitch" and said she had "no business being on the street." Dorothy believed that her gender nonconformity marked her as visibly "out." As Dorothy explained, she was dressed "very aggressively—suit and tie." Although the perpetrators never mentioned homosexuality, she perceived the violence as homophobic: "I wasn't doing anything, but it was obvious that I was a lesbian. That's why they attacked me. They hated gay people."

Dorothy's experience was common among respondents. They were routinely victimized for violating gender norms. In some of these situations, perpetrators referred to the victim's gender nonconformity but not the victim's homosexuality; queer people frequently defined such violence as homophobic. When queer people made this determination, they perceived attempts to punish their gender performance as attempts to regulate their sexuality. For instance, Paul, a 57-year-old White gay man, perceived violence directed against his gender identity as an attempt to punish him for publicly identifying as gay. He was attacked by three strangers on the street. His perpetrators told him "you're not a woman" after pushing and hitting him. Paul thought that he was targeted because of the way he walks—"very feminine," as he described it. Moreover, he saw a relationship between how he performed gender and how his sexuality was perceived: "I think it happened because I'm gay. They didn't like that I'm feminine because it showed that I'm gay."

The examples above are a few among many. They illustrate that queer people sometimes perceived violence as related to their sexuality even when perpetrators did not explicitly address it. When queer people perceived gender-based forms of violence as homophobic, they acknowledged the cultural intersection of gender and sexuality. They recognized that in the United States one's gender display is often understood as indicative of one's sexuality—that is, conformist gender displays are associated with heterosexuality and nonconformist ones are associated with homosexuality. As a result, attempts to punish gender nonconformity could be perceived not only as attempts to enforce gender conformity but also as attempts to restrict homosexuality.

While some queer people highlighted the importance of gender and sexuality in structuring their experiences of violence, others argued that their violent experiences could not be reduced to these two aspects of their identity. These arguments were particularly common among queer people of color. Many queer people of color highlighted the role of racism, as well as homophobia and sexism, in structuring their violent experiences. Kevin, a 62-year-old Black gay man, maintained that his violent experiences could not be separated from his race: "I've experienced violence because I'm black *and* gay. When the police beat me up, they called me a fag ... I would be surprised if they had done the same thing to a White gay guy, though." Here, Kevin argues that violence directed against his sexual identity was also rooted in racism. He highlights the significance of race in structuring forms of anti-queer violence and he suggests that if he were White and queer, he might not experience homophobic violence to the same degree or in the same way.

... Although queer people of color frequently determined that their violent experiences were at least partially rooted in homophobia, they usually thought that violence was based on more than their sexuality. When doing so, they advanced arguments that seemed to borrow from intersectionality. Page, a 45-year-old Latina woman, argued that her violent experiences could not be explained by only a few factors: "It's much more complicated politically than 'I'm a woman so this happened.' Things are just not necessarily about any one category, misogyny or homophobia or whatever." Similarly, Aisha, a 53-year-old Black lesbian woman, argued that her violent experiences could not be reduced to a few aspects of her identity: "I'm a Black lesbian woman who works in a job where mostly men work. Change any of those things and [the violence] would not have gone down in the same way." Highlighting how multiple systems of oppression structured her violent experiences, Aisha argued that homophobic violence can only be fully understood within the context of a racist, male-dominated, and capitalist society.

EXPRESSING UNCERTAINTY: DIFFERENCES ALONG RACIAL LINES

Until this point, I have emphasized situations in which queer people determined that violence was at least partially rooted in homophobia. Many queer people, however, found it impossible to determine whether violence was based on their sexuality. They frequently responded to questions concerning why they thought the violence had occurred with phrases such as "I don't know" or "I'm not sure." In these situations, they expressed uncertainty as to whether violence was based on their sexuality.

Queer people most typically responded with a sense of uncertainty for two reasons: (1) the violence occurred in situations in which the perpetrator insulted many aspects of the victim's identity; or (2) the violence occurred in situations in which the perpetrator said very little about the victim's sexuality. These two situations are, in some sense, opposites. The latter occurred when perpetrators said very little; the former occurred when they said a lot. In both of these situations, however, victims struggled to make sense of their violent experiences.

Queer people of color were more likely than White gay men to express uncertainty as to the cause of their violent experiences. This difference reflects the reasons I have outlined above: queer people of color often faced situations in which many aspects of their identities were attacked and they frequently encountered situations in which their perpetrators did not mention homosexuality. For these reasons, queer people of color often found it more difficult than White gay men to determine whether violence was rooted in homophobia.

When queer people of color experienced violence in which their perpetrators did not mention homosexuality, it was usually intra-racial violence—that is, the victim and the perpetrator were the same race. In contrast, when queer people of color felt as if many aspects of their identity had been attacked, the violence was most typically interracial. Thus, patterns of activity reveal that these two situations were raced—they differed with regard to the racial make-up of the victim and the perpetrator.

Queer people of color had the most difficulty determining whether violence was based on their sexuality or gender identity when their perpetrators were White. In such situations, they often felt as if multiple aspects of their identity had been attacked. Dominique, a 23-year-old Black transgender woman, described this dynamic rather succinctly: "When I'm called a fag or a freak by a White person, I have a hard time telling if they hate me because I'm trans or because I'm Black." White perpetrators often mixed homophobic or transphobic insults with racist ones. This blurring of racist and homophobic insults made it difficult for queer people of color to determine whether violence was based on their sexuality. In these situations, they could not be certain that violence was rooted in homophobia because racism may have played an equal or even more significant role.

... While queer people of color sometimes found it difficult to determine whether interracial violence was based on their sexuality, White gay men almost always argued that their violent experiences were rooted in homophobia. Responses such as "it happened because of my sexuality" or "it happened because I'm gay" were common among White gay men. Even when perpetrators mentioned race, White gay men usually determined that interracial violence was based on their sexuality. For instance, Greg, a 43-year-old White gay man, believed that he was attacked by two Latino men because of his sexuality. His attackers called him a "fag" and told him to "take that White shit somewhere else." He explained his perpetrators' motivations in a rather matter-of-fact way: "Oh, I think it happened because I'm gay. What else could be the reason?" In stark contrast to the uncertainty expressed by some queer people of color, White gay men almost always determined that interracial violence was based on their sexuality.

... Queer people of color were more likely than White gay men to report violence in which their sexuality was not explicitly addressed. Some queer people of color focused on race when explaining why their perpetrators did not mention homosexuality. For instance, Cole, a 33-year-old Black gay man, explained his perpetrators' actions in racial terms:

INTERVIEWER: So, they didn't use any homophobic slurs?

COLE: Slurs? Well, you see, in the Black Community, it's a little different. They don't always say "faggot." They'll say "too sweet mother fucker" or they'll just call you a sissy.... It's more about you being weak than being gay.... For them to call me a "faggot" would have meant that homosexuality exists, so they'd rather just beat me and not say anything.

Here, the intersection of race, gender, and sexuality seems particularly stark. Cole's argument suggests that Black queer people frequently encounter violence in which their perpetrators focus on gender nonconformity rather than homosexuality. As a result, Black queer people may often confront violence in which their perpetrators do not explicitly address homosexuality.

DIFFERENCES AMONG LESBIAN WOMEN OF COLOR AND GAY MEN OF COLOR

While gay men of color sometimes experienced violence in which their perpetrators did not explicitly mention homosexuality, lesbian women of color encountered such situations even more frequently. Gay men of color most typically reported violence in which their perpetrators used homophobic insults such as "homo" or "faggot." In contrast, lesbian women of color reported more violent incidents in which their perpetrators did not use homophobic insults. In some of these situations, their perpetrators used misogynistic insults rather than homophobic ones. For instance, Leslie, a 50-year-old Black lesbian woman, was spat on and called a "bitch" by a man on the street. She thought that the violence might have been rooted in homophobia because it occurred when she was with her girlfriend. However, she found it difficult to determine whether the violence was based on her sexuality, since it also seemed to be rooted in sexism: "I don't know if it happened because I'm lesbian.... It could have happened just because I'm a woman, but it seems like it happened because I'm gay, too. I don't know why they chose me and not [my girlfriend], though."

Lesbian women of color sometimes found it difficult to distinguish between misogynistic and homophobic forms of violence. Judy, a 43-year-old Latina lesbian woman, encountered discourse that was simultaneously misogynistic and homophobic when she was sexually assaulted. The sexual assault occurred at a party when she was 20. A man grabbed her breasts and tore open her t-shirt. When he could not forcibly remove her pants because Judy was holding onto her belt, he told her, "All I want to do is fuck you and I bet you'll come back straight." As he continually tried to remove her belt, he called her a "bitch" and a "whore." Before he could remove her belt, another woman entered the room where the sexual assault had occurred. The two women yelled and threw items at him. Shortly thereafter, he left the party.

Judy's experience illustrates the difficulty of unpacking misogynistic and homophobic forms of violence from one another. It's difficult to determine where the line for one begins and the other ends. Would Judy have been sexually assaulted had she been a heterosexual woman? Was her attacker trying to punish her for what he saw as a deviant sexuality? Did he actually believe that he could make her "come back straight," as he stated? Or was her homosexuality merely the most readily available discourse that he could draw upon to justify his own behavior while simultaneously shifting blame onto her? These questions, it seems to me, may be impossible to answer. It seems unlikely that even the attacker could fully explain all of his unconscious thoughts and feelings at the moment. While these questions may be unanswerable, my research suggests that lesbian women of color often ask themselves such questions as they struggle to make sense of their violent experiences. Indeed, Judy had asked herself many of these questions following the sexual assault. Examining the violence approximately 23 years later, she concluded: "I can't be sure if it occurred because of my sexuality or just because I'm a woman. Both probably played a role."

Lesbian women of color most frequently expressed uncertainty because their perpetrators did not use homophobic insults. Gay men of color, in contrast, more frequently encountered violence in which their perpetrators used many homophobic insults. This dynamic can be explained in part by patterns of victimization: heterosexual men perpetrated most of the anti-queer violence reported by respondents. Given this pattern of victimization, one would expect that male perpetrators would use homophobic insults more frequently against gay men than lesbian women. Using homophobic insults against gay men allows male perpetrators to distance themselves from homosexuality. It allows heterosexual men to construct themselves in opposition to the deviant men—the "fags" or "homos"—whom they attack. Thus, since heterosexual men appear to perpetrate most anti-queer violence, lesbian women of color might encounter fewer homophobic insults than gay men of color. In other words, the gender of the victim and the perpetrator affect the degree to which victims confront homophobic insults.

THE EFFECTS OF SOCIAL CLASS ON WHETHER QUEER PEOPLE EXPRESSED UNCERTAINTY

Social class also affected how queer people determined that violence was based on their sexuality. Middle- and upper-class queer people usually expressed more willingness than low-income queer people to examine whether their violent experiences were rooted in homophobia. Eva, a 46-year-old Black transgender woman, described herself as middle-class, but also said that she had very little money when she experienced transphobic violence several years prior to the interview. She described the effects of social class on her willingness to determine whether violence was based on her gender identity:

> "I didn't want to think if it was a hate crime. I didn't have heat. I didn't have heat! ... How was I supposed to sit around and spend time thinking about whether I had been bashed?"

Eva's experience suggests that working-class and low-income queer people may have more pressing concerns than determining whether violence is rooted in bias. She indicates that poverty hinders the willingness of queer people to determine whether violence is based on their sexuality or gender identity. Indeed, many queer people who were living in poverty began to wonder over the course of the interview whether more of their violent experiences were rooted in homophobia than they had previously thought. Nevada, a 36-year-old White person who identified as intersexed and lived in a homeless shelter at the time of the interview, conveyed this feeling: "I had never thought about all of this as related to my sexuality. Maybe it was now that I think about it." Nevada's response was common among low-income queer people. They often began the interview by describing a violent incident that they thought was rooted in homophobia or transphobia. As the interview progressed, they frequently described more violent incidents. When describing these incidents, they sometimes said that they had not thought about whether their violent experiences were based on their sexuality or gender identity prior to the interview. Conversely, middle- and upper-class queer people seemed to have considered prior to the interview whether their experiences were rooted in homophobia. They frequently responded with phrases such as "I've thought about that before" or "I've thought about this a lot" when explaining whether they thought their violent experiences were based on their sexuality or gender identity. Thus, social class affects the degree to which queer people are willing to determine whether violence is based on their sexuality or gender identity.

DISCUSSION

Previous studies that have documented the ways in which lesbian women and gay men determine that violence is based on their sexuality have made no reference to race, class, and gender (Herek et al., 1997; Herek et al., 2002). As I have argued throughout this [reading], these systems of oppression affect how queer people determine that violence is based on their sexuality. Previous studies, then, have overlooked some of the ways in which lesbian women and gay men make this determination....

Race ... structured how queer people determined that violence was rooted in homophobia. White gay men generally expressed certainty as to the cause of their violent experiences—that is, they usually believed that violence had occurred because of homophobia. Conversely, queer people of color sometimes found it difficult to determine whether violence was based on their sexuality.

They often felt as if multiple aspects of their identity had been attacked and they frequently encountered violence in which their perpetrators said very little about homosexuality. For these reasons, queer people of color were more likely than White gay men to express uncertainty as to the cause of their violent experiences. Thus, the degree to which queer people are willing to determine that violence is based on their sexuality differs along racial lines.

If queer people of color find it more difficult than White gay men to determine whether violence is based on their sexuality, then hate crime statutes may primarily serve to protect the interests of White gay men. Hate crime statutes, which increase criminal sanctions against hate crime perpetrators, benefit victims who are willing to define violence as bias-motivated. Victims who cannot classify violence as bias-motivated will be less likely to report it as a hate crime and, consequently, less likely to have it prosecuted as one. As a result, victims who find it easiest to determine that violence is rooted in bias will benefit disproportionately from hate crime statutes....

... Considering the experiences of lesbian women of color suggests that they may find it particularly difficult to pursue hate crime legislation. As my results suggest, lesbian women of color often confront violence in which their perpetrators do not use homophobic insults. Because perpetrators' hate speech is often used to prosecute hate-motivated violence, hate crime statutes may rarely serve the interests of lesbian women of color.

... Social class further complicates the ways in which victims pursue hate crime statutes. The experiences of working-class and low-income queer people suggest that they may not pursue hate crime statutes because of the financial demands of their lives. Financial anxieties, in other words, make it more difficult for victims to pursue hate crime statutes. As a result, middle- and upper-class victims may benefit disproportionately from hate crime legislation.

Considering the experiences of queer people of color reveals that racism makes possible certain forms of homophobic violence and homophobia makes possible some forms of racist violence. As a result, queer people of color face situations that neither heterosexual people of color nor White LGBT people must confront. Of course, queer people of color are not a monolithic group.... While some queer people of color argued that both racism and homophobia were implicated in anti-LGBT violence, others expressed uncertainty as to the cause of their violent experiences. Although I have tried not to ignore these differences, my primary focus has been to examine the obstacles confronted by queer people of color and to explore how their experiences may differ from the experiences of White gay men....

Because of the exploratory nature of this research project, I have highlighted victims' voices as much as possible.... Hate crime researchers should ... continue to explore how victims' experiences differ along lines of race, class, gender, and sexuality. Indeed, as I have shown throughout this [reading], an intersectionality approach can provide a better understanding of the ways in which hate crime statutes concern the lives of all LGBT people.

REFERENCES

Herek, G. M., Cogan, J. C., & Gillis, J. R. (2002). Victim experiences in hate crimes based on sexual orientation. *Journal of Social Issues*, 58(2): 319–339.

Herek, G. M., Gillis, J. R., & Cogan, J. C. (1999). Psychological sequelae of hate crime victimization among lesbian, gay, and bisexual adults. *Journal of Consulting and Clinical Psychology*, 67(6): 945–951.

Herek, G. M., Gillis, J. R., Cogan, J. C., & Glunt, E. K. (1997). Hate crime victimization among lesbian, gay, and bisexual adults: Prevalence, psychological correlates, and methodological issues. *Journal of Interpersonal Violence*, 12(2): 195–215.

Perry, B. (2001). *In the name of hate: Understanding hate crimes*. New York: Routledge.

PART IV

An Intersectional Framework for Change

From the Local to the Global

MARGARET L. ANDERSEN AND
PATRICIA HILL COLLINS

Upon reaching the end of this book, students often want to know "What can I do?" As the editors of this volume who compile and revise it every couple of years, we know there is not a simple answer to this question. Once people know about the social injustices brought about by race, class, and gender, they may feel overwhelmed by the possibility of changing society. Some may still think that social inequities happen to other people, not to them, and they tune out.

We understand that developing and then acting on an intersectional analysis of race, class, and gender is a lot to expect from our readers, mostly undergraduate students. For some, reading this book will be the first time they have even thought about such things. Others will have experienced some of what is written about here, but perhaps they will not before have thought beyond the particulars of their own lives. We know that developing an inclusive perspective and then deciding what to do about it is a complex process—one without simple solutions or ways of thinking. Thus, we have developed this last section of the book to examine various ways that different people and groups have thought about social change from an intersectional perspective and have, in many different ways, acted to create change in the race, class, and gender systems we have been examining in this book.

The different articles included in this section exhibit change at different levels—some of them local, others global. Together, these articles show what it means to pull together and act on a new way of thinking that recognizes and engages the complexity of race, class, gender, and the other social factors that together make up this complex system we have called a *matrix of domination.*

Is something in your life that you care about so much that it would spur you to work for social justice? Is it your family? Something happening in your school or community? Something that affects your children, your friends, your faith, or your neighborhood? Perhaps for you there is a social issue that you are passionate about—violence against women, climate change and the sustainability of the earth, global poverty and inequality, as examples. Most people think that people who work for social justice must be somehow extraordinary like Martin Luther King Jr., or other heroic figures, but most people who engage in social activism are ordinary, everyday people who decide to take action about something that touches their lives.

Whether social actors are located inside the institutions they wish to change or whether they stand outside its boundaries, the strategies they select reflect the opportunities and constraints of each specific site. Working from within organizations can mean trying to change the institutional policies and practices that overtly discriminate based on race, class, and gender, or it might mean creating new policies and practices that serve people's needs better. Other people and groups work outside formal social institutions to effect social change. The familiar boycotts, picketing, public demonstrations, leafleting, and other direct-action strategies long associated with social movements of all types typically constitute actions taken outside an institution. Although activities such as these can be trivialized in the media, it is important to remember that direct action from outsider locations represents one important way to work for social justice.

The effort to generate a more just society develops in many social contexts—schools, homes, communities, churches, and other locales. Change also occurs at many levels, ranging from personal change and small group-based change to institutional change and large-scale national and global social movements. We have organized the articles in this section to reflect the different spaces where social change occurs, beginning within one's own mind, as illustrated by Jesse A. Steinfeldt and Matthew Clint Steinfeldt's article "Multicultural Training Intervention to Address American Indian Stereotypes." They describe an educational training program developed for counseling students that is intended to educate these students about the impact of stereotypical mascots on Native American people. The training exercise begins by helping the group participants become aware of the pervasiveness and impact of such stereotypes. The authors then present an empathy-building exercise to assist counselors who will be working multicultural and multiracial settings to develop multicultural skills. As you read this article, you might ask yourself in what other settings such an exercise might be fruitful.

Social change often occurs through social movements, but social movements can sometimes be so narrowly focused on a particular issue that they inadvertently exclude some people or groups, thus missing opportunities to connect an issue to the complexities of race, class, and gender relations. This tendency can discourage the participation of people of color in movements that might otherwise serve the needs of multiple groups. You have seen an example of this in the article in Part III by Janani Balasubramanian, in which she criticizes the sustainability movement for being so anchored in White, middle-class experiences. Here, Alfonso Morales ("Growing Food and Justice") analyzes an organization that is part of the sustainable food movement—the Growing Food and Justice for All Initiative (GFJI). This organization deliberately empowered a coalition of groups to expand the food movement to address the needs of low-income, racially and ethnically diverse communities.

Natalie Sokoloff ("The Intersectional Paradigm and Alternative Visions to Stopping Domestic Violence") makes a similar argument in her discussion of how an intersectional framework has expanded organized efforts to stop domestic violence. Violence against women occurs in all social classes and races, and also occurs within lesbian and gay relationships. Sokoloff shows that the most marginalized women—that is poor women, women of color, immigrant women, and lesbians—are not only subjected to violence, but can also be empowered to tackle this issue within diverse communities. Sokoloff argues that an intersectional framework is important to to understanding domestic violence so that activists do not simply focus on dominant group experiences. Developing such an inclusive framework, Sokoloff shows, has enabled poor, immigrant, and racial-ethnic women to have a voice in the understanding of domestic violence. Such a framework analyzes domestic violence not just as individual behavior, but also as stemming from the interlocking social structures that produce violence in the first place. With this analysis in hand, social service and social justice organizations are better able to assist women who are terrorized by domestic violence and to develop social action and social policy that gets to the root cause of such violence. Sokoloff gives numerous examples of community groups and coalitions that have developed such a framework.

Dorothy Roberts and Sujatha Jesudason ("Movement Intersectionality: The Case of Race, Gender, Disability, and Genetic Technologies") provide another example where an intersectional perspective has enabled groups to collaborate when their differing vantage points might otherwise have pulled them apart.

Looking at an organization called Generations Ahead, founded in 2008, Roberts and Jesudason show how two different groups—women concerned about women's reproductive rights and disability rights activists—worked together despite their differences. Focusing on things that connected them, rather than what divided them they were about to form a coalition for change, even though they were unlikely partners. Building such coalitions requires first acknowledging and understanding the different experiences that divide people and then confronting those differences without one group asserting power over another. In the case that Roberts and Jesudason study, women of color concerned about reproductive rights and disability rights advocates worked together through face-to-face discussion. They identified and articulated common values to construct bridging frameworks, thus cultivating a shared advocacy agenda.

These articles focus on issues that are immediate and local in specific places. Many such actions for change are local in this way. Michael Kimmel examines social movements in a different context. He examines two broad-scale movements: The white supremacist movement in the United States and Canada and the Islamic terrorists responsible for attacks on the United States on September 11, 2001. Kimmel gives a perspective on such movements rarely found in the political punditry, situating his analysis squarely in what he calls "global hegemonic masculinity." Kimmel ("Globalization and Its Mal(e)contents") shows how both of these reactionary movements emerge from the perceived threats to hegemonic masculinity. According to Kimmel, gender is one of the central organizing principles of these right-wing social movements. Furthermore, Kimmel shows how the fusion of race, gender, and anti-Semitism in these movements produces violent action. Without understanding this intersectional dynamic of racism, sexism, homophobia, and anti-Semitism, efforts to reduce the violence of such movements are likely to fail.

Working for a more just society requires looking beyond both the borders of the United States as well as beyond what already exists. Such a shift in our thinking moves us in the direction of inclusive thinking that is essential to race, class, and gender studies. In the work of Bandana Purkayastha ("Intersectionality in a Transnational World"), we see ways to think relationally and to see new possibilities that move us in the direction of a broader, social justice vision. More theoretically than the previous article, Purkayastha argues that transnational ties are transforming people's lives, both connecting but also marginalizing some groups. The transnational character of modern life—made possible not only by expanding technologies, but also by changes in the world economy—also increases social control of people through enhanced surveillance and a culture of fear. This transnational climate suggests to Purkayastha that we need new ways of conceptualizing concepts like race, class, gender, ability, age, sexuality, and nation if we are to recognize the increasing global realities of contemporary life.

Whether working for individual empowerment or social activism, we must learn to see beyond what is in order to imagine what is possible. If men and women truly learned to work together, how might economic security be better provided for all? Thinking inclusively about race, class, and gender stimulates this type of vision. All sites of change contain emancipatory possibilities, if only we can learn to imagine them.

57

Multicultural Training Intervention to Address American Indian Stereotypes

JESSE A. STEINFELDT AND MATTHEW CLINT STEINFELDT

The crowd roars as the mascot enters the gym. The student section of the East High School *Catholics* erupts with the anticipated entrance of Father Guido. He is dressed in flowing robes, with golf-ball-sized rosary beads flopping around his neck, and an oversized miter atop his head with the words *CATHOLICS RULE* written on his tall cloth hat. He begins his much anticipated halftime routine by tossing faux-Eucharistic hosts into the crowd, much to the delight of the fans who gobble them up or toss them back and forth to each other. Throughout the routine, his genuflections are accompanied by his pantomimed crucifix consecrations of the crowd. After his flamboyant flipping of holy water into the crowd. Father Guido goes for the money shot—he grabs the incense urn and completes his frenzied blessing of the crowd before suddenly becoming stoic, dropping to a knee in prayer, then rising to scream in unison with the crowd, "Pope Benedict, lead us to victory!" The fans feel honored to receive Father Guido's blessing, in hopes that it will inspire the crowd to cheer diligently for the *Catholics'* second-half surge to victory. Now, given that East High School is predominantly Muslim (and Father Guido is incidentally a student of color who paints his face white to play the role), I wonder how the small handful of Catholic students at East High feel about this portrayal of their people, of their faith. How does it make you feel? How likely is this scenario to occur? I want you to reflect on these questions as we walk together on this journey.

This vignette represents an element of a training intervention designed by the first author to address stereotypes of American Indians in society. The use of American Indians and corresponding imagery for sports mascots, nicknames, and logos is a common societal practice that perpetuates stereotypes of American Indians (Baca, 2004; Davis, 2002; Russel, 2003; Staurowsky, 1999; Williams, 2006)....

SOURCE: Steinfeldt, Jesse A., and Matthew Clint Steinfeldt. 2010. "Multicultural Training Intervention to Address American Indian Stereotypes." *Community Education & Supervision* 51 (March): 17–32.

LOGISTICS OF THE TRAINING INTERVENTION

This article provides a detailed description of the components of a training intervention that is designed for use with counseling students, as well as other audiences in need of critical perspectives of social justice issues....

PROFESSIONAL RESPONSES TO NATIVE-THEMED MASCOTS, NICKNAMES, AND LOGOS

In 2001, the American Counseling Association (ACA) published a resolution condemning the use of American Indian imagery and symbols as mascots by schools and athletic teams (ACA, 2001). Several other prominent organizations (e.g., American Psychological Association [APA], National Association for the Advancement of Colored People [NAACP], and United States Commission on Civil Rights) have also published similar resolutions. According to the APA (2005) resolution, Native-themed mascots should be immediately retired because they undermine the educational experiences of members of all communities, establish an unwelcome and hostile learning environment for American Indian students, have a negative impact on the self-esteem of American Indian children, and undermine the ability of American Indians to portray accurate and respectful images of their culture (APA, 2005). The ACA (2001) resolution specifically encourages its members to work toward the elimination of these stereotypical American Indian images in institutions where they are used.

However, training counselors to be advocates for this change requires both educators and students within counseling programs to first address their own awareness and knowledge of stereotypes perpetuated by race-based mascots....

MULTICULTURAL AWARENESS

The first portion of the training intervention facilitates awareness of attitudes toward Native-themed mascots, nicknames, and logos by providing students with critical perspectives on the issue....

... Students are provided with examples of other groups that could have teams appropriate their name and likeness for use in sports (e.g., Martinsville *Faggots,* Bloomington *Slant Eyes,* Indiana *Wetbacks*, Purdue *Jews*) in an effort to highlight the connection between the actual experience of American Indians and what it might be like for other marginalized groups in society to be exposed to this practice.... With stereotypical pictures of these offensively named team mascots on the screen, the presenter in the following vignette provides commentary on potential matchups of these teams:

> On the college stage Saturday, the Indiana *Wetbacks* take on the Purdue *Jews* in an exciting Big Ten matchup. The *Wetbacks*, who are known for

> their lackadaisical, listless, and sometimes just plain lazy effort on the court, will have to find a way to solve the *Jews*' stingy defense. These *Jews* have a nose for the ball, and they are covetous of victories. The professional matchup on Sunday features the Washington *Redskins* against the Nashville *Nigg* … [stopping mid-sentence] … oops, sorry, my fault. I need to update this slide. You see, Nashville recently caved to the pressures of "political correctness" [said with sarcasm] and changed their team nickname despite tremendous pressure from fans, boosters, and supporters who cited years of vaunted tradition—not to mention the massive merchandising empire—surrounding their beloved team mascot, "Sammy Bo."

After providing the audience with these examples, the presenter asks audience members to reflect on feelings they are having regarding these fictitious group team names. Audience members are informed that although this commentary on the offensively named matchups may seem unrealistic or excessive, it is important to note that the media still uses race-specific language to describe teams with Native-themed mascots.…

In conducting this portion of the training intervention, we are acutely aware of the potential byproduct of offending people from the marginalized groups depicted. Providing stereotypical images (e.g., visuals, corresponding epitaphs) of other marginalized groups effectively illustrates powerful parallels to society's portrayal of American Indians as sport mascots. However, doing so unfortunately exposes those from marginalized groups to the insensitive rhetoric and practices that this training intervention intends to address and eliminate. Thus, the presenter needs to explicitly acknowledge this paradox of perspective-eliciting, both prior to and during the presentation. The presenter of this training intervention should also create space and make him- or herself available afterward to audience members who may feel victimized by a hypothetical portrayal of the marginalized group with which they self-identify.

… To generate a greater understanding of the impact this practice has on the developmental trajectory of children, the presenter of the training intervention provides the audience with a vignette that highlights the perspectives and experience of kids. Borrowing text directly from an article by Baca (2004), the presenter provides audience members with concrete observations of the impact of a sample Native-themed mascot. To facilitate the audience's ability to envision the reality of an American Indian child who attends a school with a Native-themed nickname (i.e., *Brave*), the presenter reads aloud the following vignette. Concurrently, the picture on the screen depicts an American Indian girl standing next to a picture of the school logo (i.e., a severed head of a stereotypic Plains Indian man) and a school banner that reads, "Home of the *Braves*":

> When she gets off the bus at school, she is greeted by this gigantic picture of a "Brave." When she enters the school doors, on the wall she may see a cartooned or caricatured version of an Indian with a big belly, an overexaggerated nose, who is wearing only a loin cloth and a

> headband with a bent feather. When she gets to class, she sees the faux image on textbook covers. When she goes to gym class, she watches her classmates run and bounce balls over the same ubiquitous image painted on the floor. If she goes to a sports event, it is likely that a White student will dress up in some form of American Indian costume and perform fake ritualistic dances to the delight of her peers. These images are omnipresent in the life of this American Indian child, and given the fact that it is a public school, it is all done with the acquiescence of the state. She doesn't see any other race singled out for this caricatured, mocking treatment. She internalizes that her race is treated differently, and her classmates see her as different. This isn't the "I'm interested in learning more about you" or "I want to be your friend" kind of uniqueness. Rather, it is the kind of different that allows others to mock and ridicule freely, thus perpetuating this badge of racial inferiority. This little girl receives this message weekly, daily, hourly.

After reading this vignette, the presenter rhetorically asks students to consider the effects this experience could have on the developmental trajectory of this young girl. Students are then asked to consider the effects this experience may have on non-American Indian children. Again borrowing directly from Baca (2004), the presenter provides the audience with a contrasting example that is intended to illustrate how non-American Indian children are aware that their culture is not caricatured. The presenter comments that for non-American Indian children at this same school,

> Their religious heritage is held with respect such that their iconography would not be used in a secular manner at school. No person of another race would paint their face White and engage in imitations of what they associate with their race. The conclusion of the White child is that his or her culture is superior to this other culture.

The presenter concludes this vignette with a reflective summary statement, "When people are reduced to stereotypes, they are not real. They do not have to be listened to. It is easier to hurt them."

This perspective-facilitating experience intends to raise awareness of how Native-themed mascots, nicknames, and logos allow mainstream society to appropriate American Indian culture while systematically perpetuating an ideology of White supremacy (Pewewardy, 1991). The continued acceptance of race-based mascots, nicknames, and logos demonstrates how schools are constructed as White public spaces (Farnell, 2004). Because "assumptions of Whiteness circulate undetected throughout discussions and debates about the continued use of American Indian imagery" (Staurowsky, 1999, p. 385), society rarely questions the possibility that Native-themed mascots could create a racially hostile educational environment (Baca, 2004). Part of the problem with Native-themed mascots, nicknames, and logos is that their unquestioned acceptance cloaks racism in a seemingly benign disguise (Staurowsky, 2007).

MULTICULTURAL KNOWLEDGE

The intent of the training intervention is to generate awareness among participants by providing examples that facilitate critical thinking and perspective-taking on the issue. The training intervention is also intended to help participants increase their knowledge on this issue by presenting relevant information specific to the nature of race-based mascots, nicknames, and logos. Participants need to acquire specific knowledge of American Indian issues to address misinformation they have been taught about American Indians in schools (Loewen, 2008) and in the media (e.g., newspaper online forums; Steinfeldt et al., 2010)....

One informational aspect of the knowledge component of the training intervention addresses the way that Native-themed mascots, nicknames, and logos perpetuate stereotypes of American Indians (e.g., noble savage, bloodthirsty savage, nonexistent people, one pan-Indian culture; Baca, 2004; King et al., 2002; Staurowsky, 2004). When relegated to mascot status, American Indians are stereotypically seen by mainstream America as people of the past who no longer exist (Staurowsky, 2004). The training intervention addresses this stereotype by presenting a contrast between Native-themed mascots and other human mascots that represent past civilizations (e.g., *Vikings, Spartans, Trojans*). Proponents of Native-themed mascots, nicknames, and logos often cite the argument that if other race-based groups are not offended, then American Indians should also not he offended (Steinfeldt et al., 2010). To debunk this myth and address the shortcomings of this argument, the presenter can answer in following way (without the academic citations inserted for this article):

> Unlike American Indians, Vikings are in fact people of the past. However, because society is filled with images of contemporary Swedes and Norwegians, images of Vikings in sport do not serve as default social representations (e.g., Moscovici, 1988) of contemporary Scandinavians. The National Football League's Minnesota *Vikings* do not represent Scandinavians in the same way that Major League Baseball's Cleveland *Indians* represent American Indians. Unlike Vikings, American Indians are not people of the past. However, the portrayal of American Indians as mascots, without ample societal images to counter the stereotypes that are perpetuated by mascots (e.g., Fryberg et al., 2008), locks American Indians into this stereotypic and past-tense status.

Another piece of specific information presented in the knowledge component of the training intervention is how American Indians are uniquely subjected to this race-based practice. Proponents of Native-themed mascots, nicknames, and logos often cite the argument, "I'm Irish, and I'm not offended by the Notre Dame *Fighting Irish* nickname, so why are American Indians offended?" (King et al., 2002; Steinfeldt et al., 2010). The training intervention addresses the discrepancy underlying the inadequacy of this argument by providing a specific example that highlights aspects of self-identification that differentiate the *Fighting Sioux* from the *Fighting Irish* nickname. The presenter can say,

> If we overlook Notre Dame's use of a mythical creature (i.e., leprechaun) as opposed to an actual human being (i.e., Indian warrior), self-identification is the primary reason to explain why people of Irish descent can purport not to be offended by the *Fighting Irish* nickname. Notre Dame is a Catholic institution with strong historic and current connections to the Irish and Irish American community. As such, people of Irish descent have a vested interest (and pride) in Notre Dame's nickname, and they exercise a degree of control over how the image is presented. In contrast, no American Indians play baseball for the Cleveland *Indians*, American Indians do not profit or gain from this logo, and American Indians do not have control over how their images are used by teams that employ Native-themed mascots (Farnell, 2004; Fenelon, 1999; Staurowsky, 2007).

The presenter of the training intervention can then provide an example that illustrates how people of Irish descent can exert control over their self-identified race-based nickname. A relevant example involves an actual incident that occurred during a college football game between two universities, Stanford and Notre Dame, in 1997 (Stanford Online Report, 1997). The Stanford band performed a half-time routine titled, "These Irish, Why Must They Fight?" This routine included a parody of an Irish potato famine, complete with a band member dressed in a Catholic cardinal uniform. The response from Notre Dame and the Irish American community was swift and overwhelming: They were outraged at how the image of their people was appropriated and used in such a demeaning and offensive manner. As a result, Stanford's band was banned from performing at Notre Dame from 1997 through 2000 (Stanford Online Report, 1997). The relevance of this example is that people of Irish heritage can protect their image, and as a result they have the power to say whether they choose to be offended by their own race-based mascot. Stated more clearly, because Irish and Irish American people control the Image and portrayal (i.e., Notre Dame had the power to ban Stanford's band from enacting this stereotypic presentation of their culture), the assertion of a person of Irish descent should more appropriately say, "I have the power to choose not to be offended by the *Fighting Irish*."

In contrast, American Indians do not have the power to say they are offended.... [T]he training intervention addresses how Native-themed mascots misuse sacred cultural symbols and spiritual practices. The continued use of eagle feathers, dancing, and chanting during mascot performances violates the sanctity of these aspects of Native American culture (Russel, 2003; Staurowsky, 1999). Immediately before reading the East High School *Catholics* vignette, the presenter of the training intervention opens with a 2-minute You Tube video clip of the University of Illinois' former mascot, Chief Illiniwik, dancing his last dance before his forced retirement in 2007. The video clip shows a White student dressed in stereotypic Plains Indian clothing running onto the court, dancing and gyrating to what is presumably intended to represent some kind of American Indian ritual. These two scenarios (i.e., Father Guido and Chief Illiniwik), when juxtaposed, illustrate the ramifications of having a White student dress in a faux-Indian costume to perform fake ritualistic dances that mimic American Indian religious ceremonies. Doing so violates sacred American Indian customs and spiritual practices while simultaneously contributing to mainstream America's ignorance of American Indian culture (Staurowsky, 1999).

Finally, the intent of the knowledge component of the training intervention is to address the way that Native-themed mascots, nicknames, and logos dehumanize American Indians (King, 2004). Visual comparisons are put on the screen that show the similarities between racist images of African Americans (e.g., Lil' Black Sambo) and the Cleveland *Indians* Chief Wahoo logo.... Both caricatured images use exaggerated facial features that bear little human resemblance, yet these images inundate society with bigoted misrepresentations of the depicted group (King. 2004)....

The overarching goal of the knowledge component of the training intervention is to provide counselor trainees with the ability to be critical consumers of information that society perpetuates about American Indians. Much of the daily clinical work conducted with American Indian clients is influenced by societal stereotypes (Duran, 2006). Therefore, counselors need to possess knowledge of the life experiences, cultural heritage, and historical background of clients who are culturally different (Arredondo et al., 1996). As King et al. (2002) noted, "An increase in accurate information about Native Americans is viewed as necessary for the achievement of other goals such as poverty reduction, educational advancements, and securing treaty rights" (p. 392).

MULTICULTURAL SKILLS

In addition to facilitating awareness of attitudes and conveying specific knowledge about the nature of Native-themed mascots, nicknames, and logos, the training intervention focuses on cultivating skills that intend to help audience members begin to conceptualize how they can engage in social justice advocacy at multiple levels. Suggestions include educating other people about Native-themed mascots (professional level), speaking out against the use of Native-themed mascots in schools and colleges (organizational level), and advocating for nonstereotypical media representations of American Indians (societal level; Sue, 2001). However, if social justice skills are to be fully developed, audience members need to understand advocacy experiences not only in abstract form but also in concrete terms (Westheimer & Kahne, 1998). To accomplish these goals, the presenter can inform audience members about the presence of activist discussion groups they could join (e.g., "Say NO to Mascots" on http://www.yahoo.com) that provide information on current events related to Native-themed mascots, nicknames, and logos. Audience members learn about letter-writing campaigns that convey consumer dissatisfaction concerning corporate partnerships with organizations with a Native-themed mascot, nickname, or logo (e.g., FedEx, the naming-rights sponsor of the Washington *Redskins* football stadium). Additionally, audience members are exposed to multimedia presentations (e.g., YouTube video that expresses the perspective of American Indians subjected to racialized mascotery) that are intended to inspire their creativity as they contemplate ways they can develop social justice action plans. These examples highlight a few of the options audience members can explore as they work to cultivate a concrete understanding of social justice skills and strategies, particularly relating to the societal marginalization of American Indians.

In addition to specific strategies for cultivating social justice skills, audience members are presented with theory and research on race-based mascots, nicknames, and logos during the training intervention. These resources are presented in an effort to enhance audience members' theoretical and empirical skills so that they can more effectively evaluate societal portrayals of American Indians....

The content within the training intervention exposes audience members to a variety of empirical research results in order to validate the theoretical and perspective-eliciting information on the deleterious impact of Native-themed mascots, nicknames, and logos. Fryberg et al. (2008) demonstrated that when exposed to images of Native-themed mascots, American Indian students reported higher levels of depressed state self-esteem, lower levels of community worth, and fewer achievement-related possible selves. Fryberg and colleagues concluded that race-based mascots remind American Indians of the narrow view society has of them, which serves to limit the possibilities they see for themselves. Native-themed mascots, nicknames, and logos provide a context that promotes stereotypical representations of American Indians, and contexts that activate stereotypical representations of racial groups are likely to threaten group members' psychological functioning (Fryberg et al., 2008). Subsequent empirical studies on this topic have supported these conclusions by demonstrating that Native-themed nicknames and logos promote stereotyping of other groups (e.g., Asian Americans; Kim-Prieto et al., 2010), provide misinformation about American Indians, and facilitate an environment of racial discrimination and harassment through online forums (Steinfeldt et al., 2010). In sum, audience members can use research, theory, and examples of advocacy strategies to develop the skill component of their multicultural competency as it relates to Native-themed mascots, nicknames, and logos.

At the end of the training intervention, it is important for the presenter to facilitate discussion with audience members so that they can process some of the content they have been exposed to. Defensiveness is part of the process of unlearning stereotypes and misinformation, but this can be at least partially mitigated by the presenter attempting to create an open environment for the members to begin to explore the meanings of the content presented. At the beginning of each training intervention, the presenter should make an explicit acknowledgment such as the following:

> I am not here to tell you what to think. I am not here to tell you how to think. Let me say that again: I am not here to tell you what to think, nor am I here to tell you how to think. I am here to provide you with perspectives so that you can use these perspectives to be critical consumers of your own experience. What you do with these perspectives is

up to you, and that is the challenge facing you after you leave here tonight.

Although this is a serious social justice issue that requires attitudinal and societal change, it is important that audience members be assured that they have a degree of agency in the process of what they decide to do with the information presented in this training intervention....

IMPLICATIONS FOR MULTICULTURAL COUNSELING TRAINING

The overarching goal of the training intervention presented in this article is to provide audience members with perspectives on the issue of American Indian stereotypes and Native-themed mascots so they can become critical consumers of societal portrayals of American Indians. In doing so, we hope that people leave the training intervention experience with the ability to articulate why Native-themed mascots are considered problematic (e.g., inflict psychological harm, perpetuate stereotypes, misuse cultural symbols, deny American Indians control of societal images of themselves, and create a racially hostile educational environment). Many people, even those opposing race-based mascots, often have difficulty articulating reasons for their opposition (Davis, 2002). Thus, counselors-in-training who are exposed to this training intervention can be better equipped to understand how Native-themed mascots can have a negative impact on all members of society, both American Indian and non-American Indian. Knowledge of stereotypes perpetuated by Native-themed mascots can help counselors serve as advocates for change at multiple systemic levels (i.e., professional, organizational, and societal; Sue, 2001) and may help counselors better serve American Indian clients (Sutton & Broken Nose, 2005)....

REFERENCES

American Counseling Association. (2001, November). *Resolution: Opposition to use of stereotypical Native American images as sports symbols and mascots*. Retrieved from http://www.aistm.org/2001aca.htm

American Psychological Association. (2005). *APA resolution recommending the immediate retirement of American Indian mascots, symbols, images, and personalities by schools, colleges, universities, athletic teams, and organizations*. Retrieved from http://www.apa.org/about/governance/council/policy/mascots.pdf

Arredondo, P., Toporek, M. S., Brown, S., Jones, J., Locke, D. C., Sanchez, J., & Stadler, H. (1996). Operationalization of the Multicultural Counseling Competencies. *Journal of Multicultural Counseling and Development, 24*, 42–78.

Baca, L. R. (2004). Native images in schools and the racially hostile environment. *Journal of Sport and Social Issues, 28*, 71–78.

Davis, L. R. (2002). The problem with Native American mascots. *Multicultural Education, 9*, 11–14.

Duran, E. (2006). *Healing the soul wound: Counseling with American Indians and other native peoples* (Multicultural Foundations of Psychology and Counseling Series). New York. NY: Teachers College Press.

Farnell, B. (2004). The fancy dance of racializing discourse. *Journal of Sport and Social Issues, 28*, 30–55.

Fenelon, J. W. (1999). Indian icons in the world series of racism: Institutionalization of the racial symbols of Wahoos and Indians. *Research in Politics and Society, 6*, 25–45.

Fryberg, S. A., Markus, H. R., Oyserman, D., & Stone, J. M. (2008). Of warrior chiefs and Indian princesses: The psychological consequences of American Indian mascots. *Basic and Applied Social Psychology, 30*, 208–218.

Kim-Prieto, C., Goldstein, L. A., Okazaki, S., & Kirschner, B. (2010). Effect of exposure to an American Indian mascot on the tendency to stereotype a different minority group. *Journal of Applied Social Psychology, 40*, 534–553.

King, C. R. (2004). This is not an Indian: Situating claims about Indianness in sporting worlds. *Journal of Sport and Social Issues, 28*, 3–10.

King, C. R., Staurowsky, E. J., Baca, L., Davis, L. R., & Pewewardy, C. (2002). Of polls and prejudice: *Sports Illustrated's* errant "Indian wars." *Journal of Sport & Social Issues, 26*, 381–402.

Loewen, J. (2008). *Lies my teacher told me: Everything your American history textbook got wrong* (2nd ed.). New York, NY: Norton.

Moscovici, S. (1998). The history and actuality of social representations. In U. Flick (Ed.), *The psychology of the social* (pp. 209–247). Cambridge, England: Cambridge University Press.

Pewewardy, C. D. (1991). Native American mascots and imagery: The struggle of unlearning Indian stereotypes. *Journal of Navajo Education, 9*, 19–23.

Russsel, S. (2003). Ethics, alterity, incommensurability, honor. *Ayaangwaamizin: The International Journal of Indigenous Philosophy, 3*, 31–54.

Stanford Online Report. (1997, October 15). *"Tasteless performance" earns ban for band.* Retrieved from http://news-service.stanford.edu/news/1997/october15/irish.html

Staurowsky, E. J. (1999). American Indian imagery and the miseducation of America. *Quest, 51*, 382–392.

Staurowsky, E. J. (2004). Privilege at play: On the legal and social fictions that sustain American Indian sport imagery. *Journal of Sport and Social Issues, 28*, 11–29.

Staurowsky, E. J. (2007). "You know, we are all Indian": Exploring White power and privilege in reactions to the NCAA Native American mascot policy. *Journal of Sport and Social Issues, 31*, 61–76.

Steinfeldt, J. A., Foltz, B. D.. Kaladow, J. K., Carlson, T., Pagano, L., Benton, E., & Steinfeldt, M. C. (2010). Racism in the electronic age: Role of online forums in expressing racial attitudes about American Indians. *Cultural Diversity and Ethnic Minority Psychology, 16*, 362–371.

Sue, D. W. (2001). Multidimensional facets of cultural competence. *The Counseling Psychologist, 29*, 790–821.

Sutton, C. T., & Broken Nose, M. A. (2005). American Indian families: An overview. In M. McGoldrick, J. Giordano, & N. Garcia-Preto (Eds.), *Ethnicity and family therapy* (pp. 43–54). New York, NY: Guilford Press.

Westheimer, J., & Kahne, J. (1998). Education for action: Preparing youth for a participatory democracy. In W. Ayers, J. A. Hunt, & T. Quinn (Eds.), *Teaching for social justice* (pp. 1–20). New York, NY: New Press.

Williams, D. M. (2006). Patriarchy and the "Fighting Sioux": A gendered look at racial college sports nicknames. *Race, Ethnicity, and Education, 9*, 325–340.

58

Growing Food and Justice

Dismantling Racism through Systainable Food Systems

ALFONSO MORALES

... Many Americans, particularly low-income people and people of color, are overweight yet malnourished. They face an overwhelming variety of processed foods, but are unable to procure a well-balanced diet from the liquor stores and mini-marts that dominate their neighborhoods.

These groups are food insecure, but furthermore, they are victims of food injustice.... For the last twenty years there has been a kind of "call and response" that has produced a web of relationships among government, scholars, nonprofit organizations, and foundations all interested in understanding food insecurity and food injustice. Definitions of important concepts like food security have been developed, organizations like the Community Food Security Coalition have grown up, foundations, universities, and government have developed programs to fund food research and practice, and nascent food justice organizations have emerged and are now populating communities around the country.

In this article I describe one of the newest threads in this web of activity, that of the Growing Food and Justice for All Initiative (GFJI), a loose coalition of organizations developed under the auspices of Growing Power, Inc., a food justice organization based in Milwaukee, Wisconsin, with offices in Chicago, Illinois, and a loose coalition of regional affiliates. Food justice organizations borrow from most every strand in the web of interrelated organizations and ideas, but they focus on issues of racial inequality in the food system by incorporating explicit antiracist messages and strategies into their work. This article chronicles in part how GFJI developed in response to the relative absence of people of color in the food system.... I show how food justice organizations have responded to GFJI in different ways and how they are weaving together various threads from the larger web into their own activities and toward their own goals as they develop their own approaches to food justice.

SOURCE: Morales, Alfonso. 2011. "Growing Food and Justice: Dismantling Racism through Systainable Food Systems." Pp. 149–176 in *Cultivating Food Justice: Race, Class, and Sustainability*, edited by Alison Hope Alkon and Julian Agyeman. Cambridge, MA: Massachusetts Institute of Technology Press.

FOOD JUSTICE IN HISTORICAL AND CONTEMPORARY ECONOMIC CONTEXT

… Over the last fifty years major grocery chains have sought suburban locations to accommodate larger stores, more parking spaces, and higher profits (USDA 2009). Eisenhauer (2001) refers to this trend as "supermarket redlining," or the process by which corporations avoid low-profit areas. Consider the impact on food access of such decision making. In 1914 American cities had fifty neighborhood grocery stores per square mile, an average of one for every street corner (Zelchenko 2006). Mayo (1993) documents store design and industry changes that transformed groceries from small neighborhood operations to large chains. Just as some were rediscovering healthy food in the 1960s, grocers began following the migration of the middle class from the city to the suburbs. The Business Enterprise Trust indicated the attitude, "It makes no sense to serve distressed areas when profits in the serene suburbs come so easily" (qtd. in Eisenhauer 2001). For instance, between 1968 and 1984, Hartford, Connecticut, lost eleven out of its thirteen grocery chains, and between 1978 and 1984 Safeway closed more than 600 inner city stores around the country (Eisenhauer 2001).

This mass departure reduced food access for low-income and minority people. Morland's multistate study (2002) found four times as many grocery stores in predominantly white neighborhoods as predominantly black ones, and other studies have noted that inner-city supermarkets have higher prices and a smaller selection of the fresh, wholegrain, nutritious foods (Sloane 2004).… When taken with a general retreat from "hunger" by the USDA … market-driven relocation of groceries to the suburbs left behind the conditions for a public health disaster.

Food, and poor nutrition in particular, is a risk factor in four of the six leading causes of death in the United States—heart disease, stroke, diabetes, and cancer.… We know that race and class inequalities produce insufficient nutrition and increase food-related disease. We know that what people eat and how they eat contributes significantly to mortality, morbidity, and increasing health care costs. By contrast, we know how food relates to good health (Institute of Medicine 2002). And when we think of access to fresh, whole foods we typically think it is dependent on income and on where one lives. Thus, decision making on locating grocery stores created "food deserts," and public health problems, but also germinated a new food coalition: the Community Food Security Coalition.…

FOOD JUSTICE AND THE GROWING FOOD AND JUSTICE FOR ALL INITIATIVE

The effort to reconstruct the foodscape for people of color has augmented the discussion of food security with organizing around the concept of food justice. This idea grew from racial inequality in food access and its accompanying public health problems. In the same way that the civil rights movement grew from racial inequalities in housing, voting, transportation, and the like, new voices are naming the racism in food, but they are not alone. Tom Vilsack expressed these racial inequalities, in his first major speech as the U.S. secretary of agriculture under President Obama. Vilsack wondered aloud what the founder of the USDA, Abraham Lincoln, would find if he walked into the USDA building today and asked, "How are we doing?" "And he'd be told," said Vilsack, "Mr. President, some folks refer to the USDA as the last plantation.' And he'd say, 'What do you mean by that?' 'Well it's got a pretty poor history when it comes to taking care of folks of color. It's discriminated against them in programming and it's made it somewhat more difficult for some people of color to be hired and promoted. It's not a very good history, Mr. President" (Federation of Southern Cooperatives 2009)....

While racially motivated food justice has scant and scattered organizational infrastructure, its current manifestation does have a name and a place of origin.... The GFJI was established under the auspices of Growing Power, Inc., the Milwaukee-based organization founded by MacArthur Award winner Will Allen, Erika Allen's father. Since 1993, Will, Erika, and Growing Power staff have worked with diverse local communities to develop community food systems responsive to the circumstances of people of color.... Will Allen started a two-acre urban farm in a food-insecure Milwaukee community. The produce is sold in the community at affordable prices. But Growing Power is much more than an urban farm: it sponsors national and international workshops on food security, maintains flourishing aquaponics and vermicomposting programs, helps teach leadership skills, and provides on-site training in sustainable food production. The organization embodies the community-based, systemic approach by reconnecting vulnerable populations to healthy food and by developing empowered individuals in economically and socially marginalized communities....

By choosing to focus explicitly on racism and sustainable food systems, GFJI created space for a diverse community to join together and support one another in the eradication of racism and the growth of sustainable food systems. As Pothukuchi points out, "in the 1990s, the community food security concept was devised as a framework for integrating solutions to the problems faced by poor households (such as hunger, limited access to healthy food, and obesity), and those faced by farmers (such as low farm-gate prices, pressures toward consolidation, and competition from overseas)" (2007, 7). By adding the additional concern of racism, GFJI has effectively tightened the connections already implicit in the concept of community food security: racially diverse households and farmers are some of the most at-risk groups in the communities targeted by USDA Community Food Projects. GFJI demonstrates the widespread appeal of an organization that combines the challenging topics of racism, sustainable agriculture, and community food security. But as Will Allen points out, these problems cannot be untangled, and must be tackled simultaneously. He notes, "We are all responsible for dismantling racism and ensuring more sustainable communities, which is impossible without food security."

For its member organizations, GFJI acts as a coordinating body, a source of emotional and spiritual sustenance, and a site for germinating and sharing fresh ideas. Each month one or more "germinators" convene a conference call, on a subject of interest to member organizations. Topics range widely, from the impact of the Obama administration on food-related problems to power sharing; and from developing effective multicultural leadership to sharing strategies for getting food to low-income communities. These monthly calls, which sustain the initiative without having to support an organizational infrastructure, are an important and ongoing source of ideas and support for these organizations....

Often led by people of color, food justice organizations see dismantling racism as part of food security. By taking an explicitly racialized approach, the food justice movement moves away from the colorblind perspective.... The food justice approach aligns movement organizations explicitly with the interests of communities and organizations whose leaders have felt marginalized by white-dominated organizations and communities. By creating a space explicitly intended for the exploration of the particular challenges facing communities of color, the food justice movement has encouraged these communities to get the help and the support that they require to continue their work. The GFJI provides some logistical support, an annual conference, and networking opportunities, but perhaps most important, a sense of community, continuity, and connection among colorful communities working to improve their food security.

By bringing the individual organizations together in a cohesive body, GFJI relates racial social critique to antiracist tools of sustainable agriculture in the service of creating a nationwide network dedicated to implementing food justice strategies. In addition, GFJI provides its members with the important sense of belonging to an organization whose goals are directly aligned with their own: although many sustainable agriculture and CFS (community food security) organizations have fostered the ambitions of people and communities of color over the years, they have not always shared the antiracist agenda. GFJI places racism front and center in the context of food and agriculture, allowing its members to feel confident that they are engaged in a community that shares their ideals and aspirations, and creating an infrastructure through which these organizations can support each other. This is a particularly important point in light of the overwhelming challenges involved in challenging the deeply entrenched and richly funded agribusiness industry, in addition to government agencies with deeply racist histories, such as the USDA. In the first coalition conference—GFJI I—organizations from around the country presented their work across race and the food system, and not surprisingly responded to the agenda in different ways, making clear two things: first, the variety of approaches and practices there are among antiracist food system organizations; and second, how the GFJI fulfilled its purpose by becoming an opportunity for each organization to articulate its unique approach, learn from the nuances others described, and even find that the GFJI might not be what they need....

REFERENCES

Eisenhauer, Elizabeth. 2001. In Poor Health: Supermarket Redlining and Urban Nutrition. *Geojournal* 53:125–133.

Federation of Southern Cooperatives. 2009. Transcript of U.S. Agriculture Secretary Tom Vilsack's speech to Federation of Southern Cooperatives/Land Assistance Fund's Georgia Annual Farmer's Conference, January 21, <http:/www.federationsoutherncoop.com/albany/Vilsackspeech.pdf>.

Institute of Medicine. 2002. *The Future of the Public's Health.* Washington, DC: National Academy Press.

Mayo, James M. 1993. *The American Grocery Store.* Santa Barbara, CA: Greenwood Publishing Group.

Morland, Kimberly, Steve Wing, Ana Diez Roux, and Charles Poole. 2002. Neighborhood Characteristics Associated with the Location of Food Stores and Food Service Places. *American Journal of Preventive Medicine,* 22(1):23–29.

Pothukuchi, Kami. 2007. *Building Community Food Security: Lessons from Community Food Projects,* 1999–2003. Community Food Security Coalition, <http://www.foodsecurity.org/BuildingCommunityHoodSecurity.pdf>.

Sloane, David C. 2004. Bad Meat and Brown Bananas: Building a Legacy of Health by Confronting Health Disparities around Food. *Planners Network* (Winter): 49–50.

United States Department of Agriculture (USDA). 2009. *Access to Affordable and Nutritious Food: Measuring and Understanding Food Deserts and Their Consequences.* Washington, DC: Economic Research Service.

Zelchenko, Peter. 2006. Supermarket Gentrification: Part of the "Dietary Divide." *Progress in Planning* 168(1):11–15.

59

The Intersectional Paradigm and Alternative Visions to Stopping Domestic Violence

What Poor Women, Women of Color, and Immigrant Women Are Teaching Us About Violence in the Family

NATALIE J. SOKOLOFF

This article focuses on how poor women and women of color and their allies ... are working against domestic violence in the United States in communities marginalized by race, class, gender, sexuality, and immigrant status....

In contrast to the earlier feminist approaches, the *intersectional* domestic violence approach challenges gender inequality as *the primary factor* explaining domestic violence: gender inequality is neither the most important nor the only factor that is needed to understand violence against many marginalized women in the home (Crenshaw, 1994). Gender inequality is only part of their marginalized and oppressed status....

In the intersectional domestic violence literature in the U.S., two distinct (and sometimes conflicting) objectives emerge: (1) *giving voice* to battered women from diverse social locations and cultural backgrounds (2) while still focusing on *socially structured inequalities* (e.g., race, gender, class, sexuality) that constrain and shape the lives of battered women, albeit in different ways. The first has been described as a *race/class/gender* (or *multicultural*) perspective, whose focus is on multiple, interlocking oppressions of individuals—i.e., on issues of *difference*; the second has been described as the *structural* perspective requiring analysis and criticism of existing systems of power, privilege and access to resources (see Andersen and Collins, 2001; Mann and Grimes, 2001). It is my position that an intersectional analysis must draw from *both* approaches because women's voices must be situated within the social structure of women's lives....

SOURCE: Sokoloff, Natalie J. 2008. "The Intersectional Paradigm and Alternative Visions to Stopping Domestic Violence." *International Journal of Sociology of the Family* 34 (Autumn): 153-185.

...The major topic of this article ... [i]s to discuss some of the alternative models for paradigms that women marginalized by race, class, gender, sexuality, and immigrant status have begun using to combat violence against women in their homes and communities....

HOW AN INTERSECTIONAL ANALYSIS CAN HELP TO UNDERSTAND MARGINALIZED WOMEN'S EXPERIENCES OF DOMESTIC VIOLENCE

It is still common to hear that domestic violence cuts across all classes, races, and ethnic groups. While this is true, intersectional scholars challenge this uncritical view by arguing that poor women of color are the "most likely to be in both dangerous intimate relationships and dangerous social positions" (Richie, 2000, p. 1136). Richie argues that the anti-domestic violence movement's avoidance of a race, gender and class analysis of violence against women "seriously compromises the transgressive and transformative potential of the antiviolence movement's potential [to] radically critique various forms of social domination" (p. 1135). She concludes, failure to address the multiple oppressions of poor women of color jeopardizes the validity and legitimacy of the anti-domestic violence movement.

One dilemma is the problem of how to report race and class differences in domestic violence prevalence rates, especially in the U.S. where race is such a dominant part of the social agenda and class is typically ignored as a concept.... [T]here is tremendous diversity among women regarding the prevalence, nature, and impact of domestic violence—even within ethnic, racial, religious, socio-economic groups and sexual orientations (Hampton et al., 2005; West, 2005) Several studies indicate that Black women are severely abused (West, 2005) and murdered at significantly higher rates (Della-Giustina, 2005; Hampton et al., 2005; Websdale, 1997; West, 2005) than their representation in the population.

By itself, this information may serve little purpose but to reinforce negative stereotypes about African Americans and the Black community's alleged "culture of violence" in the U.S. That is one reason why Yasmin Jiwani (2001) focuses on structural issues—especially in immigrant of color communities—where culture is all too typically blamed for violence against women in their families. One solution to this problem of representation is to *contextualize* these findings within a *structural* framework—one that looks at socially organized systems of social inequality.... Research literature concludes that (1) Black women are *less* likely than white women to be battered when one controls for income and marital status (see Farmer and Tiefenthaler, 2003), (2) neighborhood context (a high percent of unpartnered parents, unemployment, poverty, and public assistance) is a more effective predictor of domestic violence than race/ethnicity (see Benson et al., 2003), and (3) the relationship between race and intimate partner violence may be spurious—"more likely, race is a proxy measure for neighborhood" (Potter, 2007: 370).

This being said, we must never forget the profound racism that exists in U.S. society, including the effects of living in racially segregated communities. Thus, for example, the degree of poverty is more intense in some African American communities. Whereas 75 percent of poor Blacks live in communities with other poor Blacks—and all its attendant disadvantages, only 25 percent of poor whites live in poor white communities. Instead, poor whites are more likely to live in communities with working class and some middle class white residents, which provide an immeasurable degree of resources available to that community (e.g., see Rusk, 1995). So comparing poor Blacks and poor whites is simply not "comparable." This must be taken into account in working toward lowering levels of domestic violence in Black, white and other communities of color.

Finally, recent research suggests that the level of collective efficacy in a community can be related to domestic violence (Almgren, 2005; Benson et al., 2003; Brown et al., 2005; Lauristen and White, 2001).... [W]omen who live in neighborhoods with greater collective efficacy are more likely to inform others of their abuse, thereby leading to greater levels of support for the women. Thus, it is not only socially structured inequalities but also the community's ability to feel empowered in relation to its many structured obstacles that lead to better opportunities for battered women to escape or challenge abusive situations at home. Lack of good jobs, decent education, livable housing, decent health care and good childcare, etc. can create conditions for communities and its members to not feel up to being able to protect their members, including the women in their homes.

HOW AN INTERSECTIONAL ANALYSIS CHALLENGES STEREOTYPES OF MARGINALIZED COMMUNITIES AND THEIR CULTURES

All too often, whites in the U.S., including feminists, have been quick to allocate blame to non-white cultures (especially Black and immigrant) for domestic violence....

According to Dasgupta (2005), "American mainstream society still likes to believe that woman abuse is limited to minority ethnic communities, lower socio-economic stratification, and individuals with dark skin colors" (212–213). This leads to stereotyping of battered women from "other" cultures. But it also fails to look at the strengths of non-dominant cultures and how they provide protective factors for battered women (Yoshioka and Choi, 2005)....

In terms of the negative stereotyping many women from immigrant and non-white communities experience, Uma Narayan (1997) describes how in the U.S., when we hear that women in India die "by fire" in dowry deaths (a man or some of his family members kill the woman because they are dissatisfied with her dowry to the paternal family line with whom she usually lives), the "culture" tends to be blamed. Thus, many people in the United States call dowry deaths horrendous—which of course they are, but then go on to explain their cause as less "enlightened" attitudes toward women and the "backwardness" of the South Asian culture. However, when women in the U.S. are killed by guns (at the same rate as dowry deaths in India), this is rarely, if ever, said to be due to the culture; rather it is usually blamed on the individual man's unstable personality at best and patriarchy at worst (Narayan, 1997). But the American culture is not said to be "backward" or to blame for the death....

In the U.S., safety for battered women and their children is typically premised on the idea that she will *separate from* or *leave* her abuser. We know that most women *do* leave their batterers: it takes on average 7 attempts before she is ultimately able to leave and be safe; and that the most violent and dangerous time for a woman in a battering relationship is just before, at the point of, or just after leaving (e.g., see Browne, 2003; Campbell et al., 2003). If she chooses to leave the batterer, she needs a strong shelter system or other support especially during the first 3 months, and then the first year. After that, the leave-taking process has a good chance of getting and keeping her out of an intolerable, abusive relationship (Campbell, 2008). But leaving their abusers might mean turning to "outsiders"—police, courts, doctors, domestic violence agencies, etc.—in the mainstream or dominant communities. Thus, while it may be true that she will face violence in her family or community, it is just as true that if she goes outside her home and her community, she may have to face a whole other set of hostilities. Moreover, leaving the batterer may mean leaving her entire community and its supportive aspects—often a not insignificant source of services and/or protections. [E.g., see the experience of an Orthodox Jewish woman in Baltimore who did all the "right" things to leave her abusive husband and whose community provided the kind of support that nurtured her through these terrible times and stigmatized her husband until his abuse stopped (Kay, 2006). Likewise, Holmes (2005) reports that certain Muslim Imams in Philadelphia shun men who abuse their wives].

The stereotyping of marginalized groups of battered women leads to major misunderstandings about what might help the women in their struggles to be free of violence: incarcerating the men or taking away the man's economic contribution to her and her children is counterproductive in many marginalized communities. Arguing that calling the police is the most important thing for them to do often backfires. Thus, as Garfield (2005, 2006) reminds us, when the violence against women movement changed from an advocacy to a state-based criminal justice operation, blame and punishment became the focus, *not social structural change* which is needed for violence against women to stop.

The discrimination in the criminal justice system, as in the larger U.S., is deep and profound. In the Black community, as many as one-third of all young African American men are in prison, jail, on probation or parole in the U.S. (Sokoloff, 2004). In some cities, like Baltimore and Washington D.C., it is over half (Donziger, 1996; Lotke and Zeidenberg, 2005)! Black women who are battered often do not feel the police and the criminal justice system will solve their problems; rather they may just intensify them (Richie, 2005). Native American women have argued that violence against Native women has been part and parcel of the colonization of Native Americans in the U.S.; thus, violence by the state is intimately intertwined with violence by individual Native men against their partners....

Thus, many women of color argue, interpersonal violence and state violence against women and oppressed communities are intertwined. Moreover, many immigrants—both documented and undocumented—often fear the police, whether because of negative experiences in their home countries or because of the arrest and deportation of immigrants here in the U.S. which has only intensified since 9/11 2001.... As with African Americans and Native Americans, given the language, cultural, and structural barriers facing many battered immigrant women, the police, courts, and social services that are available often do not ensure their safety—in fact, may increase their vulnerability (Abraham, 2000; Bui, 2004; Dasgupta, 2005). In short, an intersectional analysis helps us see that the simplistic advice to either "leave" the abusive partner or to "call the police" when she does not have an adequate safety plan is often misguided. This may actually harm rather than help certain groups of women who must deal with violence in their homes as well as in and against themselves and their communities by larger outside forces in the U.S. (Coker, 2005)....

Community Based Models of Social Justice: Community Engagement/Community Accountability

In ethnically, racially, and economically marginalized and immigrant communities of color especially, an important movement has developed to reduce women's reliance on the criminal legal system.... Rather, the goal is to develop a community's capacity to mobilize and organize against, and assume accountability for, changing violence against women in that community. This is important in large part because most batterers are *never* seen by the criminal justice system—even under the best of circumstances. Those arguing for less reliance on the criminal legal system argue for a need to think outside the traditional criminal justice (or criminal legal) and social service models for addressing and eradicating violence. The degree to which these traditional models are incorporated into new paradigms varies considerably, but many see them as "adjuncts" or "back up" at best....

Smith (2005) suggests strategies that must deal simultaneously with the reality of structural as well as state violence, especially state violence within the U.S. criminal justice system. She argues marginalized women too often cannot look to the criminal justice system for safety because state violence is intimately connected to domestic violence. Thus, the "challenge women of color face in combating personal *and* state violence is to develop strategies for ending violence that *do* assure safety for survivors of sexual/domestic violence and *do not* strengthen our oppressive criminal justice apparatus"—a struggle that has been particularly powerful in the African American, Native American, and more recently Latino/a communities (Incite!, 2006). To understand the violence against women in their homes by intimate partners and by the state, one need only look at the ways in which women of color in particular have been brutalized both by their husbands/partners and by the police, prisons, etc. (see Ritchie, 2006; Faith, 2004). It is in large part because of the, at best, inadequate and, at worst, hostile treatment by police and courts in the U.S. that methods of engaging communities beyond the criminal justice system must be found to support poor, minority and immigrant women who are abused in their homes as well as in their communities....

Also, it is important to clarify that while culturally competent services ... are required to create social justice for marginalized battered women in the U.S., one should not be lured into thinking that these social services are adequate *by themselves* without large-scale structural changes. Domestic violence is part of the larger societal systems of violence and inequality (e.g., imperialism, racism, colonialism, patriarchy, etc.) and as such domestic violence must be attacked at its root causes: the socially structured systems of inequality—of race, class, gender, sexual orientation, immigrant status and the like.

In *Safety and Justice for All: Examining the Relationship between the Women's Anti-Violence Movement and the Criminal Legal System* (MsFoundation, 2003)... the MsFoundation elaborates on some of the community responses that are relevant to marginalized communities. Each one of these approaches must be seriously considered and thoroughly researched as well. They include: community squads to intervene with batterers; alternative 911 services that rush community residents to a crisis scene; community groups overseeing children's safety; alternative accountability such as restorative justice; popular education through alternative means like music and street theater; and men teaching men about domestic violence and how to challenge its destructive forces....

Empowering marginalized women to become their own advocates (e.g., see *Building Rhythms of Change*, 2001; Williams and Tibbs, 2002) and to organize against violence against women throughout their communities has been repeated in one version or another as a movement for social change in the African American, Latina, Native American, and immigrant domestic violence literature. Given that these actions are "home grown," the advocacy and organizing are "inherently culturally competent" and do not need to be "imported" from the outside (Dutton et al., 2000). As Alianza (2004) states in its analytic framework:

> We need to develop systems of support for victims/survivors within our communities. Latina survivors need to be recognized as *experts* in meeting these challenges; they must be involved in program design and service delivery at all levels. Programs need to include services that will give survivors better options and opportunities for becoming independent and more able to create relationships and homes free from violence.

Likewise, Mujeres Unidas y Activas, based in San Francisco and Oakland, California, sees itself as different from other domestic violence programs because it is "founded on the concept that immigrant women are uniquely equipped to find solutions to the problems that most directly affect their lives."... Thus battered immigrant women from the Latina community are used as peer mentors, group facilitators, community educators, and organizers (Mujeres, 2006).

Finally, Sista II Sista is a collective of young African American and Latina women, ages 13–25, that began in Bushwick, Brooklyn (NY) in 1995. These young women work in a non-hierarchical structure (using a flower instead of a hierarchy to express its organizational format) and sees as one of its main goals in the Freedom School (founded in 1996) the need to help young women of color see themselves as leaders in their communities. Leadership is encouraged through:

> identifying the issues that are central to their community and learning important ways to fight to shape their community positively around these issues…[like] creating concrete changes (culturally and institutionally) against violence against the women of color in their Bushwick, Brooklyn community (Sista II Sista, n.d.)…

A FINAL NOTE ON THE IMPORTANCE OF MICRO AND MACRO STRUCTURAL AND ECONOMIC CHANGE

Again, it is also important for us to see where we can learn from those in the U.S. and in other countries. At the micro level, the fact that criminal justice responses cannot protect women from re-victimization because they do not systematically address women's underlying economic, social, and political disadvantages is demonstrated by the work of Websdale and Johnson (2005) in rural Kentucky. They show how a more effective way to reduce woman battering is to empower battered women by providing the underlying structural conditions for independent housing, job training and opportunities, affordable childcare, and social services.…

SUMMARY AND CONCLUSION

(1) What might it look life if communities had the resources to explore effective interventions that keep decision-making power within the community, and make it possible for women to stay?

(2) Where might we be if government accountability did not aim its efforts on criminal legal punishment, but instead centralized responsibility for basic needs [i.e., *human, economic, civil, political, and cultural rights of all human beings*] and human dignity and affirmed the human rights of all?

The content of this article, I hope, makes it very clear that we must go far beyond the inner sanctum of the mainstream traditional isolated nuclear family if we are to challenge violence against women in families, especially those living in communities that are marginalized by race, class, gender, ethnicity, immigrant status, and the like. Looking at violence against communities, families within communities, and against the women in those communities is key to understanding violence against women in their homes. And this cannot be done outside a framework of the socially structured conditions of inequality and oppression and movement toward greater structural change and social justice. Both micro- and macro-level analyses and social movements for structural and cultural change (for human, economic, civil, political, cultural, gender, sexual rights and human dignity for all) must be a part of the work that we do to eliminate violence against women in their homes....

... [T]o understand domestic violence generally, and particularly in the most marginalized communities, we must *contextualize* family violence within the larger systems of social, political, racial, gender, economic, and sexual inequality. Nothing less will do if we are to not only study but also work toward *changing* violence against women in the family....

REFERENCES

Abraham, Margaret (2000). *Speaking the Unspeakable: Marital Violence among South Asian Immigrants in the United States.* Piscataway, NJ: Rutgers University.

Alianza Latina Nacional para Erradicar la Violencia Domestica (2004). Nacional Directory of Domestic Violence Programs Offering Services in Spanish. Retrieved on January 5, 2006: http://www.dvalianza.org

Almgren (2005). The Ecological Context of Interpersonal Violence: From Culture to Collective Efficacy. *Journal of Interpersonal Violence*, 20: 218–224.

Andersen, Margaret and Patricia Hill Collins (2001). Introduction. In M. Andersen and P. H. Collins (Eds.), *Race, Class & Gender: An Anthology*, 4E (pp. 1–9). Belmont, CA: Wadsworth.

Benson, Michael, et al. (2003). The Correlation between Race and Domestic Violence is Confounded with Community Context. *Social Problems*, 51: 326–342.

Browne, Angela (2003). Fear and the Perception of Alternatives: Asking "Why Battered Women Don't Leave" Is the Wrong Question. In Barbara Raffel Price and Natalie J. Sokoloff (Eds.). *The Criminal Justice System and Women: Offenders, Prisoners, Victims, and Workers*, 3rd Ed. (pp. 343–359). New York: McGraw-Hill.

Bui, Hoan (2004). *In the Adopted Land: Abused Immigrant Women and the Criminal Justice System.* Westport, CT: Praeger.

Campbell, Jacqueline, et al. (2003). Risk Factors for Femicide in Abusive Relationships: Results of a Multisite Case Control Study. *American Journal of Public Health*, 93(7): 1089–1097.

Campbell, Jacqueline (2008). Danger Assessment: A Tool to Help Identify the Risk of Intimate Partner Homicide and Near Homicide as Part of Routine Mental Health Assessments. Jewish Women International National Training Institute Teleconference. (Online- January 17).

Coker, Donna (2005). Shifting Power for Battered Women: Law, Material Resources, and Poor Women of Color. In Natalie J. Sokoloff with Christina Pratt (Eds.). *Domestic Violence at the Margins: Readings on Race, Class, Gender & Culture*. NJ: Rutgers University, pp. 369–388.

Crenshaw, Kimberle (1994). Mapping the Margins: Intersectionality, Identity, and Politics. In Martha Albeitson (Ed.). *The Public Nature of Private Violence*.

Dasgupta, Shamita Das (2005). Women's Realities: Defining Violence against Women by Immigration, Race and Class. In Natalie J. Sokoloff (Ed.) (with Christina Pratt). *Domestic Violence at the Margins: Readings in Race, Class, Gender & Culture* (pp. 56–70). NJ: Rutgers University.

Della-Giustina, Jo-Ann (2005). *Gender, Race and Class as Predictors of Femicide Rates: A Path Analysis*. Ph.D. Dissertation, John Jay College of Criminal Justice and the Graduate Center, City University of New York.

Donziger, Steven (Ed.) (1996). *The Real War on Crime: The Report of the National Criminal Justice Commission*. NY: HarperPerennial.

Dutton, Mary Ann, Leslye Orloff, and Gail Aguilar Hass (2000). Summer. Characteristics of Help-Seeking Behaviors, Resources and Service Needs of Battered Immigrant Latinas: Legal and Policy Implications. *Georgetown Journal of Poverty Law and Policy*, 7(2): 245–305.

Faith, Karlene (2004). Progressive Rhetoric, Regressive Policies: Canadian Prisons for Women. In Barbara Raffel Price and Natalie J. Sokoloff (Eds.). *The Criminal Justice System and Women: Offenders, Prisoners, Victims, and Workers*, 3rd Ed., pp. 281–288.

Farmer, Amy and Jill Tiefenthaler (2003). Explaining the Recent Decline in Domestic Violence. *Contemporary Economic Policy*, 21: 158–I72.

Garfield, Gail (2005). *Knowing What We Know: African American Women's Experiences of Violence and Violation*. Rutgers University.

Garfield, Gail (2006). Does the Violence Against Women Act Provide Justice for Battered Women? Unpub. Ms. John Jay College of Criminal Justice. New York City.

Hampton, Robert, et al. (2005). Domestic Violence in African American Communities. In Natalie J. Sokoloff (Ed.) (with Christina Pratt). *Domestic Violence at the Margins: Readings in Race, Class, Gender & Culture* (pp. 127–141). NJ: Rutgers University.

Holmes, Kristen (2005). Muslims in Philly Shun Men Who Abuse Wives. At www.CentreDaily.com (July 23).

Incite!/Critical Resistance Statement (2005). Gender Violence and the Prison Industrial Complex: Interpersonal and State Violence against Women of Color. In Natalie J. Sokoloff (Ed.) (with Christina Pratt). *Domestic Violence at the Margins: Readings in Race, Class, Gender & Culture* (pp. 102–114). NJ: Rutgers University.

Incite! Women of Color Against Violence (2006). *Color of Violence: The Incite! Anthology*. Boston: South End.

Jiwani, Yasmin (2001). *Intersecting Inequalities: Immigrant Women of Colour, Violence & Health Care*. At http://www.harbour.sfu.ca/freda/articles/hlth.htm

Kay, Liz (2006). A Woman's Plea for Closure: Orthodox Jewish Community Rallies Against Husband Who Denied a Religious Divorce. *The (Baltimore) Sun*, September 19, pp. IB + 2B.

Lauristen and White (2001). Putting Violence in Its Place: The Influence of Race, Ethnicity, Gender and Place on the Risk for Violence. *Criminology and Public Policy*, 1(1): 37–59.

Lotke, Jason and Eric Zeidenberg (2005). *Tipping Point: Maryland's Overuse of Incarceration and the Impact on Public Safety*. At: http://www.justicepolicy.org/images/upload/05-03_REP_MDTippingPoint_AC-MD.pdf

Mann, Susan and Michael Grimes (2001). Common Grounds: Marxism and Race, Gender and Class Analysis. *Race, Gender, Class*, 8 (2).

Mujeres Unidas y Activas. Retrieved on April 10, 2006. At: http://www.mujeresunidasnetwork.com/ncn/index.php?option=com_content&task=vie

Narayan, Uma (1997). *Dislocating Cultures Identities, Traditions, and Third World Feminisms*. NY: Routledge.

Potter, Hillary (2007). Reaction Essay: The Need for A Multi-faceted Response to Intimate Partner Abuse Perpetrated by African-Americans. *Criminology and Public Policy*, 6(2): 367–376.

Richie, Beth (2005). A Black Feminist Reflection on the Antiviolence Movement. In Natalie J. Sokoloff (Ed.) (with Christina Pratt). *Domestic Violence at the Margins: Readings in Race, Class, Gender & Culture* (pp. 50–55). NJ: Rutgers University.

Ritchie, Andrea (2006). Law Enforcement Violence against Women of Color. In Incite! The Color of Violence. *Color of Violence: The Incite! Anthology*. Boston: South End, pp. 138–156.

Rusk, David (1995). *Baltimore Unbound: A Strategy for Regional Renewal*. Balto.: Johns Hopkins University.

Sista II Sista-Social Justice Wiki. Got to http://socialjustice.ccnmtl.columbia.edu/index.php/Sista_II_Sista

Smith, Andrea (2005). Looking to the Future: Domestic Violence, Women of Color, the State and Social Change. In Natalie J. Sokoloff with Christina Pratt (Eds.), *Domestic Violence at the Margins: Readings in Race, Class, Gender, & Culture* (pp. 416–434). NJ: Rutgers University.

Sokoloff, Natalie J. (2004). Impact of the Prison Industrial Complex on African American Women *Souls: A Critical Journal of Black Politics, Culture, and Society*, 5(4): 31–46.

Websdale, Neil (1997). *Understanding Domestic Homicide*. Boston: Northeastern University.

West, Carolyn (2005). The "Political Gag Order" Has Been Lifted: Violence in Ethnically Diverse Families. In Natalie J. Sokoloff with Christina Pratt (Eds.). *Domestic Violence at the Margins: Readings in Race, Class, Gender & Culture* (pp. 157–173). NJ: Rutgers University.

Williams, Oliver J. and Carolyn Tibbs (2002). *Community Insights on Domestic Violence among African Americans: Conversations about Domestic Violence and Other Issues Affecting Their Community* (San Francisco and Oakland). Office of Justice Programs, Dept. of Justice, Office on Violence Against Women.

Yoshioka, Marianne and Deborah Choi (2005). Culture and Interpersonal Violence Research: Paradigm Shift to Create a Full Continuum of Domestic Violence Services. *Journal of Interpersonal Violence*, 20(4): 513–519.

60

Movement Intersectionality

The Case of Race, Gender, Disability, and Genetic Technologies

DOROTHY ROBERTS AND SUJATHA JESUDASON

INTRODUCTION

Intersectional analysis does not apply only to the ways identity categories or systems of power intersect in individuals' lives. Nor must an intersectional approach focus solely on differences within or between identity-based groups. It can also be a powerful tool to build more effective alliances between movements to make them more effective at organizing for social change. Using intersectionality for cross movement mobilization reveals that, contrary to criticism for being divisive, attention to intersecting identities has the potential to create solidarity and cohesion. In this article, we elaborate this argument with a case study of the intersection of race, gender, and disability in genetic technologies as well as in organizing to promote a social justice approach to the use of these technologies. We show how organizing based on an intersectional analysis can help forge alliances between reproductive justice, racial justice, women's rights, and disability rights activists to develop strategies to address reproductive genetic technologies. We use the work of Generations Ahead to illuminate how intersectionality applied at the movement-building level can identify genuine common ground, create authentic alliances, and more effectively advocate for shared policy priorities.

Founded in 2008, Generations Ahead is a social justice organization that brings diverse communities together to expand the public debate on genetic technologies and promote policies that protect human rights and affirm a shared humanity. Dorothy Roberts is one of the founding board members of Generations Ahead, and Sujatha Jesudason is the Executive Director.

Since its inception, Generations Ahead has utilized an intersectional analysis approach to its social justice organizing on reproductive genetics. Throughout 2008–2010, the organization conducted a series of meetings among reproductive

SOURCE: Roberts, Dorothy, and Sujatha Jesudason. 2013. "Movement Intersectionality." *Du Bois Review* (Fall): 313–328.

justice, women's rights, and disability rights advocates to develop a shared analysis of genetic technologies across movements with the goals of creating common ground and advancing coordinated solutions and strategies. This cross-movement relationship- and analysis-building effort laid the foundation for successfully resisting historical divisions between reproductive rights, racial justice, and disability rights issues in several important campaigns. In examining the ways in which the theory and practice of intersectionality are used here we hope to demonstrate the kinds of new alliances that now become possible—alliances that can be both more inclusive and effective in the long term.

FROM DIFFERENCE TO RADICAL RELATEDNESS

In her classic article, "Demarginalizing the Intersection of Race and Sex," Kimberlé Crenshaw (1989) focused on Black women to show that the "single-axis" framework of discrimination analysis not only ignores the way in which identities intersect in people's lives, but also erases the experiences of some people. As a result, she argued, "[b]lack women are sometimes excluded from feminist theory and antiracist policy discourse because both are predicated on a discrete set of experiences that often does not accurately reflect the interaction of race and gender" (p. 140). The intersectional framework revealed that Black women suffer the combined effects of racism and sexism and therefore have experiences that are different from those of both White women and Black men, experiences which were neglected by dominant antidiscrimination doctrine (Crenshaw 1989). Extending from the example of Black women, an intersectional perspective enables us to analyze how structures of privilege and disadvantage, such as gender, race, and class, interact in the lives of all people, depending on their particular identities and social positions.... Furthermore, intersectionality analyzes the ways in which these structures of power inextricably connect with and shape each other to create a system of interlocking oppressions, which Patricia Hill Collins (2000) termed a "matrix of domination" (p. 18).

The value of intersectional analysis, however, is not confined to understanding individual experiences or the ways systems of power intersect in individuals' lives. Over the last two decades, feminist scholars have discussed and debated the potential applications of intersectionality. As a "framework of analysis" or "analytic paradigm," intersectionality has been applied to theory, empirical research, and political activism....

By highlighting the differences in experiences among women, it might seem that an intersectional approach would make coalition building harder. Some scholars have criticized its attention to identity categories for hindering both intra- and cross-movement mobilization by splintering groups, such as women, into smaller categories, and accentuating the significance of separate identities (Brown 1997). As Andrea Canaan (1983) observed in *This Bridge Called My Back,* the singular focus on identity can lead us to "close off avenues of communication and vision so that individual and communal trust, responsibility, loving and knowing are impossible" (p. 236).

Yet intersectionality presents an exciting paradox: attending to categorical differences *enhances* the potential to build coalitions between movements and makes them more effective at organizing for social change.

How can illuminating differences build solidarity? First, it is only by acknowledging the lived experiences and power differentials that keep us apart that we can effectively grapple with the "matrix of domination" and develop strategies to eliminate power inequities. This is not a matter of *transcending* differences. To the contrary, activists interested in coalition building must confront their differences openly and honestly. "Our goal is not to use differences to separate us from others, but neither is it to gloss over them," writes Gloria Anzaldua and AnaLouise Keaton (2002, p. 3). Intersectionality avoids the trap of downplaying differences to reach a false universalism and superficial consensus—a ploy that always benefits the most privileged within the group and erases the needs, interests, and perspectives of others. An intersectional approach should not create "homogenous 'safe spaces' " where we are cordoned off from others according to our separate identities (Cole 2008, p. 443). Rather, it can force us into a risky place of radical self-reflection, willingness to relinquish privilege, engagement with others, and movement toward change.

Second, once differences are acknowledged, an intersectional framework enables discussion among groups that illuminates their similarities and common values.... Commonality is not the same thing as sameness. Searching for and creating commonalities among people with differing identities through active engagement with each other is one of intersectionality's most important methodologies not only for feminist theorizing but also for political activism.

Third, analysis of our commonalities reveals ways in which structures of oppression are related and therefore highlights the notion that our struggles are linked. Despite our distinct social positions, we discover that "we are all in the same boat" (Morales 1983, p. 93). Not only does intersectionality apply to everyone in the sense that all human beings live within the matrix of power inequities, but also that the specific intersections of multiple oppressions affect each and every one of us.

Of course, these intersecting systems affect individuals differently, depending on the specific context and their specific political positions. This is why engagement between groups with differing perspectives is critical to understanding the dynamics of inequality and to organizing for social change. Rather than erasing our identities for the sake of coalition, we learn from each other's perspective to understand how systems of privilege and disadvantage operate together and, therefore, to be better equipped to dismantle them....

Far from building walls around identity categories, then, intersectionality forces us to break through these categories to examine how they are related to each other and how they make certain identities invisible. This shift from seeing our differences to seeing our relatedness requires that we understand identity categories in terms of matrices of power that are connected rather than solely as features of individuals that separate us (Cole 2008; Dhamoon 2011).... "While analytically we must carefully examine the structures that differentiate us, politically we must fight the segmentation of oppression into categories such as 'racial issues,' 'feminist issues,' and 'class issues,'" writes Bonnie Thornton Dill (1983, p. 148). Indeed, our radical *interrelatedness* is equally as important as our differences. To us, the radical potential for intersectionality lies in moving beyond its recognition of difference to build political coalitions based on the recognition of connections among systems of oppression as well as on a shared vision of social justice. The process of grappling with differences, discovering and creating commonalities, and revealing interactive mechanisms of oppression itself provides a model for alternative social relationships.

AN INTERSECTIONAL ANALYSIS OF RACE, GENDER, DISABILITY, AND REPRODUCTIVE GENETIC TECHNOLOGIES

... Reproductive justice is a prime example of applying an intersectional frame to both political theorizing and political action. Women of color developed a reproductive justice theory and movement to challenge the barriers to their reproductive freedom stemming from sex, race, and class inequalities (Nelson 2003; Roberts 1997, 2004; Silliman et al., 2004). Reproductive justice addresses the inadequacies of the dominant reproductive rights discourse espoused by organizations led by White women that was based on the concept of choice and on the experiences of the most privileged women. Thus, women of color contributed to the understanding of and advocacy for reproductive freedom by recognizing the intersection of race, class, and gender in the social control of women's bodies.

What if we complicated the matrix even more by including disability as an identity and political category in theorizing and organizing by women of color? Far from being a marginal social division because it affects fewer people, disability helps to shape reproductive and genetic technologies and policies that affect everyone.... Like intersectionality's central claim that "representations of gender that are 'race-less' are not by that fact alone more universal than those that are race-specific" (Crenshaw 2011, p. 224), representations of race and gender that neglect disability are no more universal than those that are based solely on able bodies.... It was only when we engaged with disability rights activists that we began to grapple with their perspectives on reproductive politics and changed our own perspectives in concrete ways.

Just as the dominant conception of discrimination imposed by courts erases Black women, organizing for social change along certain categories can obscure the importance of other perspectives and opportunities for building coalitions to achieve common social justice goals. Disability rights discourse largely has failed to encompass racism, and anti-racism discourse largely has failed to encompass disability. The disability rights and civil rights movements are often compared as two separate struggles that run parallel to each other, rather than struggles that have constituents and issues in common, ... even as both people of color and people with disabilities share a similar experience of marginalization and "othering" and even though there are people of color with disabilities (Pokempner and Roberts, 2001).

Race, gender, and disability do not simply intersect in the identities of women of color with disabilities, however. Rather, racism, sexism, and ableism work together in reproductive politics to maintain a reproductive hierarchy and enlist support for policies that perpetuate it (Roberts 2009, 2011). In her past work, Roberts (1997) has contrasted policies that punish poor women of color for bearing children with advanced technologies that assist mainly middle- and upper-class White women not only to have genetically-related children, but to also have children with preferred genetic traits. While welfare reform laws aim to deter women receiving public assistance from having even one additional healthy baby, largely unregulated fertility clinics regularly implant privileged women with multiple embryos, knowing the high risk multiple births pose for premature delivery and low birth weight that requires a fortune in publicly-supported hospital care. Rather than place these policies in opposition, however, Roberts argued in "Privatization and Punishment in the New Age of Reprogenetics" (2005) and "Race, Gender, and Genetic Technologies: A New Reproductive Dystopia?" (2009) that they are tied together. Policies supporting both population control programs and genetic selection technologies reinforce biological explanations for social problems and place reproductive duties on women that privatize remedies for illness and social inequities.

Advances in reproduction-assisting technologies that create embryos in a laboratory have converged with advances in genetic testing to produce increasingly sophisticated methods to select for preferred genetic traits, and de-select for disability. Liberal notions of reproductive choice obscure the potential for genetic selection technologies to intensify both discrimination against disabled people and the regulation of women's childbearing decisions. These technologies stem from a medical model that attributes problems caused by the social inequities of disability to each individual's genetic make up and that holds individuals, rather than the public, responsible for fixing these inequities. Disability rights activists have pointed out that prenatal and pre-implantation genetic diagnosis reinforce the view that "disability itself, not societal discrimination against people with disabilities, is the problem to be solved" (Parens and Asch, 1999, p. s13). This medicalized approach to disability assumes that difficulties experienced by disabled people are caused by physiological limitations that prevent them from functioning normally in society, rather than the physical and social limitation enforced by society on individuals with disabilities (Saxton 2007). Although disabilities cause various degrees of impairment, the main hardship experienced by most people with disabilities stems from pervasive discrimination and the unwillingness to accept and embrace differing needs to function fully in society.

Locating the problem inside the disabled body rather than in the social oppression of disabled people leads to the elimination of these bodies becoming the chief solution to impairment. By selecting out disabling traits, these technologies can divert attention away from social arrangements, government policies, and cultural norms that help to define disability and make having disabled children undesirable (Wendell 1996). Genetic selection is also discriminatory in that it reduces individual children to certain genetic traits that by themselves are deemed sufficient reasons to terminate an otherwise wanted pregnancy or discard an embryo that might otherwise have been implanted (Asch 2007).

The expectation of genetic self-regulation may fall especially harshly on Black and Latina women, who are stereotypically defined as hyperfertile and lacking the capacity for self-control (Gutierrez 2008; Roberts 1997). In an ironic twist, it may be poor women of color, not affluent White women, who are most compelled to use prenatal genetic screening technologies. This paradox is revealed only by a political analysis that examines the interlocking systems of inequity based on gender, race, and disability that work together to support policies that rely on women's management of genetic risk rather than social change. This intersectional analysis also reveals that reproductive justice, women's rights, and disability rights activists share a common interest in challenging unjust reprogenetics policies and in forging an alternative vision of social welfare.

THE DYNAMICS OF INTER- AND INTRA-MOVEMENT MOBILIZATION ROOTED IN AN INTERSECTIONAL FRAMEWORK

... An intersectional approach provides a method for overcoming ... barriers to collaboration and even using differences between identity categories and causes as a tool for more effective strategizing and action....

In the last several years, as scholars and activists, working with the staff and board of Generations Ahead, we have used an intersectional framework as an integral part of our organizing work. Intersectionality has been an essential tool in shaping the mission, vision, and work of the organization, in deepening our understanding of the social and ethical implications of genetic technologies, and in building unlikely partners to advance a social justice agenda....

At the heart of Generations Ahead's method of cross-movement alliance-building are three main elements: honestly and openly discussing in face-to-face conversations key areas of conflict among movements; articulating common values upon which bridging frameworks could be constructed; and cultivating a shared advocacy agenda, followed by joint strategizing and collective action, to address specific issues. These elements put in practice the key theoretical insight of an intersectional analysis ... —that uncovering how dominant discourses and systems marginalize certain groups in intersecting ways and at specific sites can be a basis for solidarity. By acknowledging differences, not transcending them, activists can more effectively grapple with the "matrix of domination" because an intersectional analysis ultimately reveals how structures of oppression are related and therefore our struggles are linked. To be successful, this process required building trust by learning about each other's movements and concretely demonstrating solidarity for each other's issues, for example, by co-sponsoring and attending each other's events (Generations Ahead 2009).

Based on this model, Generations Ahead organized a series of meetings among reproductive rights and justice, women of color and Indigenous women, and disability rights advocates to dig deeper into the areas of tension between movements and to develop a shared analysis of genetic technologies across movements, with the goals of creating common ground and advancing coordinated solutions and strategies....

In order to openly and honestly identify the distinctive ways in which reproductive and genetic technologies affected different constituencies, the participants were asked to divide themselves up into self-identified constituency groups. It was clearly acknowledged that participants were not being asked to privilege or prioritize any one identity over others, but rather that they were being asked to share the unique and distinguishing perspectives of different constituencies. The twenty-one participants divided up into the following groups: Indigenous women, Asian women, women of African descent, women (of color) with disabilities, and Latinas living in the United States. Queer identified people agreed to raise their specific concerns within all of the other groups. Each group's members then spent time identifying the particular benefits and concerns genetic technologies raised for their group, and the values that they wanted to see integrated into any advocacy on this issue.

Rather than starting the discussion about the benefits and risks of genetic technologies based on a universal and generic human being, these constituency groups were able to do several interesting things simultaneously. First, when asked to consider these technologies from the standpoint of their identity-specific perspective, these issues became more relevant for all participants. None of the participants were users of these technologies, and, up until that moment, most felt that they were not relevant to their lives and social justice advocacy. But once they were able to connect what felt like an abstract, futuristic, and privileged issue to their lives and communities, their investment in the issue shifted. Most participants were now able to reflect on and attach genetic technologies to issues that they deeply cared about: sex selection and son preference for Asian women; genetic determinism and eugenics for women of African descent; prenatal disability de-selection for women with disabilities; blood quantum and tribal identity for Indigenous women; and family formation and fertility for Latinas. By the end of the discussion, all participants were able to understand the issues raised by genetic technologies as an extension of their existing social justice commitments and concerns (Generations Ahead 2008).

Second, the participants were able to make these linkages as a part of a larger, shared "matrix of domination," rather than as a hierarchical analysis of oppression. Because everybody was able to speak to the intersections with their lived experiences, and since all identities were equally valued, the discussion quickly and easily transitioned to shared struggles and solidarity, rather than a debate over who was more or less oppressed or privileged in the development and use of genetic technologies. Shared concerns were quickly visible in the similar histories of reproductive oppression and genetic determinism, and the ways in which biology, bodies, and reproduction have been historically categorized, regulated, stigmatized, and controlled for some groups.

In addition, participants in each group discussed other intersecting identities that clearly cut across all groups, such as immigration status, class, sexual orientation, gender identities, and age. Acknowledging these other intersections prevented any one individual or one group to claim the "most oppressed" or "most victimized" identity. It meant that everyone in the room enjoyed privilege in at least one, if not more, of their identities. Since no one in the room could be either pure victim or pure oppressor, participants were more willing and comfortable acknowledging their own privilege and less attached to any presumed victim status. This led to, as Gloria Anzaldua (2002) noted, more thorough self-reflection and openness to learning from and engaging with others. Everybody felt like they belonged together because of, not in spite of, their differences.

And finally, owing to the sense of "we are all in this together" and newly recognized links between genetic technologies and their existing social justice commitments, the whole group was able to identify and articulate a shared set of values and perspectives that they wanted to promote in any analysis of the social implications of genetic technologies. They pinpointed values that they felt were important to help guide work in this area, values such as: start with an intersectional analysis of power and inequities at the center of any analysis of benefits and risks, include community in identifying solutions, and make sure to address the underlying factors that cause unequal outcomes and don't just blame it on the technology per se (Generations Ahead 2008).

Participants then worked together to develop a condensed list of shared values that everyone could take back to their organizations and continue to use to inform any shared or individual advocacy in this area. The group collectively affirmed values such as: put human welfare, not profit, at the center of the use of these technologies; recognize that individuals, families, and communities are socially, culturally, and politically determined, not solely biologically or genetically; include those most impacted by these technologies to be a part of the decision-making about their use; and acknowledge the intersectionality of diverse lived experiences and advocate for long-term, holistic solutions (Generations Ahead 2008).

This convening laid the groundwork for future, more challenging conversations and collaborations between reproductive justice and disability rights leaders. The lessons and praxis of using an intersectional approach were then applied to a series of five roundtable conversations between two groups that have a long history of tension, mistrust, and aversion to working together—reproductive rights and disability rights advocates. These roundtable discussions started with the most difficult area of disagreement between these two movements—their differing approaches to genetic testing technologies and abortion....

These discussions were started with an open acknowledgment of this third rail of disagreement, and recognition that there was a mutual history of hurt and fear, where each movement felt that the other did not appreciate its perspective or deep concerns about the other movement's perspective....

As a result of their engagement over conflicts and common values, the advocates were able to agree on a shared alternative paradigm for addressing genetic technologies based on "long-term, comprehensive, intersectional policies that create structural changes in social inequality" (Generations Ahead 2009, p. 6).

Instead of these two groups being at loggerheads over whether to regulate abortion and prenatal screening to prevent the de-selection of people with disabilities or allow unfettered reproductive freedom that could lead to the eugenic elimination of disability, participants were able to define a set of shared values. These include:

- Reproductive autonomy should include support for people making the choice to have children, including children with disabilities, and support to raise their children with dignity.
- All women who choose to parent should be valued as parents and all children should be valued as human beings, including children with disabilities.
- Policy advocacy should focus on providing social and material supports to women, families, and communities, not on when life begins, whose life is more valued, or who can be a parent.
- Both movements should broaden their agendas to fight to improve the social, political, physical, and economic contexts within which women and people with disabilities make decisions about their lives. The focus should be on changing society, not on individual decision-making (Generations Ahead 2009, p. 2).

Through these shared values all participants were able to affirm women's self-determination and the value of people with disabilities, so that one was not pitted against the other. And they were able to include an analysis that encompassed concerns about race, class, immigration, and sexual orientation....

CONCLUSION

As the work of Generations Ahead illustrates, the radical potential for intersectionality lies in moving beyond its acknowledgement of categorical differences to build political coalitions based on the recognition of connections among systems of oppression as well as on a shared vision of social justice.... In the process we have learned several important lessons for how to "do" intersectionality in organizing and advocacy.

First, a good process for radical relationship- and alliance-building requires forthrightly acknowledging the multiple intersecting lived experiences of all participants. Radical alliances can only be built on the basis of being honest about differences and disagreements. This honesty is what creates the potential for new solidarities based on shared but different experiences. Second, trust must be developed through the process. Alliance building is a progressive, developmental process where trust is built through repeated contact, connection, conversation, and collective action. Identifying multiple and intersecting interests is crucial to creating repeated opportunities for collaboration. The third lesson is related to a willingness on the part of all participants to change their perspectives and politics. An intersectional framework is a critical tool for disrupting oppressed-oppressor binaries, and opening up the possibilities for discovering values and experiences in common. And the final lesson is to keep the focus on shared values. While scholars and advocates for social change might disagree on general strategy and tactics, they can more easily agree on shared values that can form the basis for a common vision, as well as for joint action on specific campaigns. Here again, an intersectional approach is useful in deconstructing disagreements and reconstructing similar experiences and hopes....

In acknowledging that all of us have multiple identities and by including all of those identities in the organizing process, intersectionality in practice can be a powerful tool for grappling with differences and uncovering shared values and bridging frameworks. This process provides a basis for collective action and a model for alternative social relationships rooted in our common humanity. Instead of separating groups, as some have argued, using an intersectional framework can create new and authentic alliances even among historically oppositional groups that can lead to more inclusive, focused, and effective efforts for social change. Intersectionality as a theory and practice for social change can, and should, be used as a critical tool in struggles for social justice that seek to include us all.

REFERENCES

Anzaldua, Gloria and AnaLouise Keating (Eds.) (2002). *This Bridge We Call Home: Radical Visions for Transformation*. New York: Routledge.

Asch, Adrienne (2007). Why I Haven't Changed by Mind about Prenatal Diagnosis: Reflections and Refinements. In Erik Parens and Adrienne Asch (Eds.), *Prenatal Testing and Disability Rights*, pp. 234–258. Washington, DC: Georgetown University Press.

Brown, Wendy (1997). The Impossibility of Women's Studies. *Difference: A Journal of Feminist Cultural Studies*, 9(3): 79–101.

Canaan, Andrea (1983). Brownness. In Cherrie Moraga and Gloria Anzaldua (Eds.), *This Bridge Called My Back: Writings by Radical Women of Color*, pp. 232–237. New York: Kitchen Table, Women of Color Press.

Cole, Elizabeth (2008). Coalitions as a Model for Intersectionality: From Practice to Theory. *Sex Roles*, 59: 443–453.

Crenshaw, Kimberlé (1989). Demarginalizing the Intersection of Race and Sex: A Black Feminist Critique of Antidiscrimination Doctrine, Feminist Theory and Antiracist Politics. *University of Chicago Legal Forum*, pp. 139–167.

Crenshaw, Kimberlé (2011). Postscript. In Helma Lutz, Maria Teresa Herrera Vivar, and Linda Supik (Eds.), *Framing Intersectionality: Debates on a Multi-Faceted Concept in Gender Studies*, pp. 221–233. Surrey, England: Ashgate Publishing Limited.

Dhamoon, Rita Kaur (2011). Considerations on Mainstreaming Intersectionality. *Political Research Quarterly*, 64(1): 230–243.

Generations Ahead (2008). A Reproductive Justice Analysis of Genetic Technologies: Report on A National Convening of Women of Color and Indigenous Women. <http://www.generations-ahead.org/files-for-download/articles/GenAheadReport_ReproductiveJustice.pdf> (accessed May 17, 2012).

Generations Ahead (2009). Bridging the Divide: Disability Rights and Reproductive Rights and Justice Advocates Discussing Genetic Technologies, convened by Generations Ahead 2007–2008. <http://www.generations-ahead.org/files-for-download/articles/GenAheadReport_BridgingTheDivide.pdf> (accessed May 21, 2012).

Generations Ahead (2010). Robert Edwards, Virginia Ironside, and the Unnecessary Opposition of Rights. <http://www.generations-ahead.org/resources/the-unnccessary-opposition-of-rights> (accessed May 17, 2012).

Gutierrez, Elena R. (2008). *Fertile Matters: The Politics of Mexican Origin Women's Reproduction*. Austin, TX: University of Texas Press.

Hill Collins, Patricia (2000). *Black Feminist Thought: Knowledge, Consciousness and the Politics of Empowerment*, 2ed. New York: Routledge.

Morales, Rosario (1983). We're All in The Same Boat. In Cherrie Moraga and Gloria Anzaldua (Eds.), *This Bridge Called My Back: Writings by Radical Women of Color*, pp. 91–93. New York: Kitchen Table, Women of Color Press.

Nelson, Jennifer (2003). *Women of Color and the Reproductive Rights Movement*. New York: NYU Press.

Parens, Erik and Adrienne Asch (1999). The Disability Rights Critique of Prenatal Genetic Testing: Reflections and Recommendations. *The Hastings Center Report*, 29(5): s1–s22. The Hastings Center.

Pokempner, Jennifer and Dorothy E. Roberts (2001). Poverty, Welfare Reform, and the Meaning of Disability. *Ohio State Law Journal*, 62: 425–463.

The Prenatally and Postnatally Diagnosed Conditions Awareness Act Fact Sheet. <http://www.generations-ahead.org/files-for-download/success-stories/InfoSheetBrownbackKennedy Legislation_final.pdf> (accessed May 17, 2012).

Roberts, Dorothy (1997). *Killing the Black Body: Race, Reproduction, and The Meaning of Liberty*. New York: Pantheon.

Roberts, Dorothy E. (2004). Women of Color and the Reproductive Rights Movement. *Journal of the History of Sexuality*, 13: 535–539.

Roberts, Dorothy E. (2005). Privatization and Punishment in the New Age of Reprogenetics. *Emory Law Journal*, 54: 1343–1360.

Roberts, Dorothy (2009). Race, Gender, and Genetic Technologies: A New Reproductive Dystopia? *Signs*, 34: 783–804.

Roberts, Dorothy (2011). *Fatal Invention: How Science, Politics, and Big Business Re-create Race in the Twenty-first Century*. New York: The New Press.

Saxton, Marsha (2007). Why Members of the Disability Community Oppose Prenatal Diagnosis and Selective Abortion. In Erik Parens and Adrienne Asch (Eds.), *Prenatal Testing and Disability Rights*, pp. 147–164. Washington, DC: Georgetown University Press.

Silliman, Jael, Marlene Gerber Fried, Loretta Ross, and Elena R. Guttierez (2004). *Undivided Rights; Women of Color Organize for Reproductive Justice*. Boston, MA: South End Press.

Thornton Dill, Bonnie (1983). Race, Class, and Gender: Prospects for an All-Inclusive Sisterhood. *Feminist Studies*, 9(1): 131–150.

Wendell, Susan (1996). *The Rejected Body: Feminist Philosophical Reflections on Disability*. New York: Routledge.

61

Globalization and Its (Mal)econtents

The Gendered Moral and Political Economy of Terrorism

MICHAEL S. KIMMEL

... Globalization changes masculinities—reshaping the arena in which national and local masculinities are articulated, and transforming the shape of men's lives. Globalization disrupts and reconfigures traditional, neocolonial or other national, regional or local economic, political and cultural arrangements, and thus transforms local articulations of both domestic and public patriarchy (see Connell, 1998). Globalization includes the gradual proletarianization of local peasantries, as market criteria replace subsistence and survival. Small local craft producers, small farmers and independent peasants traditionally stake their notions of masculinity in ownership of land and economic autonomy in their work; these are increasingly transferred upwards in the class hierarchy and outwards to transnational corporations. Proletarianization also leads to massive labor migrations—typically migrations of male workers—who leave their homes and populate migrant enclaves, squatter camps and labor camps.

In addition, the institutional arrangements of global society are equally gendered. The marketplace, multinational corporations and transnational geopolitical institutions (World Court, United Nations, European Union) and their attendant ideological principles (economic rationality, liberal individualism) express a gendered logic. As a result, the impact of global economic and political restructuring is greater on women. At the national and global level, the world gender order privileges men in a variety of ways, such as unequal wages, unequal labor force participation, unequal structures of ownership and control of property, unequal control over one's body, as well as cultural and sexual privileges....

... The processes of globalization and the emergence of a global hegemonic masculinity have the ironic effect of increasingly "gendering" local, regional and national resistance to incorporation into the global arena as subordinated entities. Religious fundamentalism and ethnic nationalism use local cultural symbols to express regional resistance to incorporation.... However, these religious and ethnic expressions are often manifest as gender revolts, and include a virulent resurgence of domestic patriarchy (as in the militant misogyny of Iran or Afghanistan); the problematization of global masculinities or neighboring masculinities (as in the former Yugoslavia); and the overt symbolic efforts to claim a distinct "manhood" along religious or ethnic lines to which others do not have access and which will restore manhood to the formerly privileged (white militias in the U.S. and skinhead racists in Europe).

Thus gender becomes one of the chief organizing principles of local, regional and national resistance to globalization, whether expressed in religious or secular, ethnic or national terms. These processes involve flattening or eliminating local or regional distinctions, cultural homogenization as citizens and social heterogenization as new ethnic groups move to new countries in labor migration efforts. Movements thus tap racialist and nativist sentiments at the same time as they can tap local and regional protectionism and isolationism. They become gendered as oppositional movements and also tap into a vague masculine resentment of economic displacement loss of autonomy and collapse of domestic patriarchy that accompany further integration into the global economy. Efforts to reclaim economic autonomy, to reassert political control and revive traditional domestic dominance thus take on the veneer of restoring manhood....

Here I discuss white supremacists and Aryan youth in both the U.S. and Scandinavia, and compare them briefly with the terrorists of Al Qaeda who were responsible for the attack on the U.S. on 11 September 2001. All these groups, I argue, use a variety of ideological and political resources to re-establish and reassert domestic and public patriarchies. All deploy "masculinity" as a form of symbolic capital, an ideological resource, (1) to understand and explicate their plight; (2) as a rhetorical device to problematize the identities of those against whom they believe themselves fighting; and (3) as a recruitment device to entice other, similarly situated young men to join them. These movements look backward, nostalgically, to a time when they—native-born white men and Muslim men in a pre-global era—were able to assume the places in society to which they believed themselves entitled. They seek to restore that unquestioned entitlement, both in the domestic sphere and in the public sphere. They are movements not of revolution, but of restoration....

RIGHT-WING MILITIAS: RACISM, SEXISM, AND ANTI-SEMITISM AS MASCULINE REASSERTION

In an illustration in an 1987 edition of *WAR*, the magazine of the White Aryan Resistance, a working-class white man, in hard hat and flak jacket, stands proudly before a suspension bridge while a jet plane soars overhead. "White Men Built This nation!!" reads the text, "White Men Are This nation!!!" Here is a moment of fusion of racial and gendered discourses, when both race and gender are made visible. "This nation," we now understand, "is" neither white women, nor non-white.

The White Aryan Resistance that produced this illustration is situated on a continuum of the far right that runs from older organizations such as the John Birch Society, Ku Klux Klan and the American Nazi Party, to Holocaust deniers, neo-Nazi or racist skinheads, White Power groups like Posse Comitatus and White

Aryan Resistance, and radical militias, like the Wisconsin Militia or the Militia of Montana. The Southern Poverty Law Center cites 676 active hate groups in the U.S. including 109 Klan centers, 209 neo-Nazi groups, 43 racist skinheads groups and 124 neo-Confederate groups, and more than 400 U.S.-based websites (*Intelligence Report*, Spring 2002).

These groups are composed of young white men, the sons of independent farmers and small shopkeepers.... Buffeted by the global political and economic forces that have produced global hegemonic masculinities, they have responded to the erosion of public patriarchy (displacement in the political arena) and domestic patriarchy (their wives now work away from the farm) with a renewal of their sense of masculine entitlement to restore patriarchy in both arenas. That patriarchal power has been both surrendered by white men—their fathers—and stolen from them by a federal government controlled and staffed by legions of the newly enfranchised minorities, women and immigrants, all in service to the omnipotent Jews who control international economic and political life. Downwardly mobile rural white men—those who lost the family farms and those who expected to take them over—are squeezed between the omnivorous jaws of capital concentration and a federal bureaucracy which is at best indifferent to their plight, and at worse, facilitates their further demise. What they want, says one, is to "take back what is rightfully ours" (cited in Dobratz and Shanks-Meile, 2001: 10).

In many respects, the militias' ideology reflects the ideologies of other fringe groups on the far right, from whom they typically recruit, especially racism, homophobia, nativism, sexism and anti-Semitism. These discourses of hate provide an explanation for the feelings of entitlement thwarted, fixing the blame squarely on "others" who the state must now serve at the expense of white men. The unifying theme of these discourses, which have traditionally formed the rhetorical package Richard Hoft-sadter labeled "paranoid polities," is gender. Specifically, it is by framing state policies as emasculating and problematizing the masculinity of these various "others" that rural white militia members seek to restore their own masculinity....

White supremacists see themselves as squeezed between global capital and an emasculated state that supports voracious global profiteering. NAFTA took away American jobs; the "Burger King economy" leaves no room at the top, so "many youngsters see themselves as being forced to compete with nonwhites for the available minimum wage, service economy jobs that have replaced their parents" unionized industry opportunities' (Coplon, 1989: 84).

Of course, these are the same men whose ardent patriotism fueled their support of American involvement in Vietnam and the Gulf War. It is through a gendered rhetoric of masculinity that this contradiction between loving America and hating its government, loving capitalism and hating its corporate iterations is resolved. First, like others on the far right, militia members believe that the state has been captured by evil, even Satanic forces; the original virtue of the American political regime deeply and irretrievably corrupted. Environmental regulations, state policies dictated by urban and northern interests, the Internal Revenue Service, are the outcomes of a state now utterly controlled by feminists, environmentalists, blacks and Jews.

In their foreboding futuristic vision, communalism, feminism, multiculturalism, homosexuality and Christian-bashing are all tied together, part and parcel of the New World Order. Multicultural textbooks, women in government and legalized abortion can individually be taken as signs of the impending New World Order. Increased opportunities for women can only lead to the oppression of men. Tex Marrs proclaims, "In the New Order, woman is finally on top. She is not a mere equal. *She is Goddess*" (Marrs, 1993: 28). In fact, she has ceased to be a "real" woman—the feminist now represents the confusion of gender boundaries and the demasculinization of men, symbolizing a future where men are not allowed to be real men....

... [I]n the logic of militias and other white supremacist organizations, gay men are both promiscuously carnal and sexually voracious and effete fops who do to men what men should only have done to them by women. Black men are both violent hyper-sexual beasts, possessed of an "irresponsible sexuality," seeking white women to rape (*WAR* 8(2), 1989:11) and less than fully manly, "weak, stupid, lazy" (*NS Mobilizer,* cited in Ferber, 1998: 81). Blacks are primal nature—untamed, cannibalistic, uncontrolled, but also stupid lazy—and whites are the driving force of civilization....

Most interesting is the portrait of the Jew. On the one hand, the Jew is a greedy, cunning, conniving, omnivorous predator; on the other, the Jew is small, beady-eyed and incapable of masculine virtue. By asserting the hyper-masculine power of the Jew, the far right can support capitalism as a system while decrying the actions of capitalists and their corporations. According to militia logic, it is not the capitalist corporations that have turned the government against them, but the international cartel of Jewish bankers and financiers, media moguls and intellectuals who have already taken over the U.S. state and turned it into ZOG (Zionist Occupied Government). The U.S. is called the "Jewnited States" and Jews are blamed for orchestrating the demise of the once-proud Aryan man.

In white supremacist ideology, the Jew is the archetype villain, both hyper-masculine—greedy, omnivorous, sexually predatory, capable of the destruction of the Aryan way life—and hypo-masculine, small, effete, homosexual, pernicious, weasely....

In the militia cosmology, Jews are both hyper-masculine and hypo-masculine. Hyper-masculinity is expressed in the Jewish domination of the world's media and financial institutions, and especially Hollywood. They are sexually omnivorous, but calling them "rabid, sex-perverted" is not a compliment. *The Thunderbolt* claims that 90 percent of pornographers are Jewish. At the same time, Jewish men are seen as wimpish, small, nerdy and utterly unmasculine—likely, in fact, to be homosexual.

In lieu of their brawn power, Jewish men have harnessed their brain power in their quest for world domination. Jews are seen as the masterminds behind the other social groups who are seen as dispossessing rural American men of their birthright. And toward that end, they have coopted blacks, women and gays and brainwashed cowardly white men to do their bidding....

Since Jews are incapable of acting like real men—strong, hardy, virtuous manual workers and farmers—a central axiom of the international Jewish conspiracy for world domination is their plan to "feminize White men and to masculinize White women" (*Racial Loyalty,* No. 72, 1991: 3).

Embedded in this slander is a critique of white American manhood as soft, feminized, weakened—indeed, emasculated. Article after article decries "the whimpering collapse of the blond male," as if white men have surrendered to the plot (cited in Ferber, 1998: 127)....

WHITE SUPREMACISTS IN SCANDINAVIA

While significantly fewer in number than their American counterparts, white supremacists in the Nordic countries have also made a significant impact on those normally tolerant social democracies. Norwegian groups such as Bootboys, NUNS 88, the Norsk Arisk Ungdomsfron (NAUF), Varg and the Vikings; the Green Jacket Movement (Gronjakkerne) in Denmark; and the Vitt Ariskt Motstand (VAM, or White Aryan Resistance), Kreatrivistens Kyrka (Church of the Creator, COTC) and Riksfronten (National Front) in Sweden have exerted an impact beyond their modest numbers. Norwegian groups number a few hundred, while Swedish groups may barely top 1000 adherents, and perhaps double that number in supporters and general sympathizers.

Their opposition seems to come precisely from the relative prosperity of their homelands, a prosperity that has made the Nordic countries attractive to ethnic immigrants from the economic South. Most come from lower middle-class families; their fathers are painters, carpenters, tilers, bricklayers, road maintenance workers. Some come from small family farms. Several fathers own one-person businesses, are small capitalists or self-employed tradesmen (Fangen, 1999b: 36)....

Like the American white supremacists, Scandinavian Aryans understand their plight in terms of masculine entitlement which is eroded by state immigration policies, international Zionist power and globalization. All desire a return to a racially and ethnically homogeneous society, seeing themselves, as one put it, as a "front against alienation, and the mixing of cultures" (Fangen, 1998a: 214).

Anti-gay sentiments also unite these white supremacists. "Words are no use; only action will help in the fight against homosexuals," says a Swedish magazine, *Siege.* "With violence and terror as our weapons we must beat back the wave of homosexual terror and stinking perversion whose stench is washing over our country" (cited in Bjorgo, 1997: 127). And almost all have embraced anti-Semitism casting the Jews as the culprits for immigration and homosexuality....

The anti-Semitism, however, has also inhibited alliances across the various national groups in Scandinavia. Danish and Norwegian Aryans.

Masculinity figures heavily in their rhetoric and their recruitment. Young recruits are routinely savagely beaten in a "baptism of fire". One Norwegian racist recounted in court how his friends had dared him to blow up a store owned by a Pakistani in Brumunddal. He said he felt a lot of pressure, that they were making fun of him, and he wanted to prove to them that he was a man after all. After he blew up the shop, he said, the others slapped his back and cheered him. Finally, he felt accepted (Fangen, 1999b: 92)....

Like their American counterparts, Scandinavian white supremacists also exhibit the other side of what Connell calls "protest masculinity"—a combination of stereotypical male norms with often untraditional attitudes respecting women. All these Nordic groups experience significant support from young women, since the males campaign against prostitution, abortion and pornography—all of which are seen as degrading women (see Durham, 1997). On the other hand, many of these women soon become disaffected when they feel mistreated by their brethren, "unjustly subordinated," or just seen as "mattresses" (in Fangen, 1998a)....

THE RESTORATION OF ISLAMIC MASCULINITY AMONG THE TERRORISTS OF SEPTEMBER 11

Although it is still too soon, and too little is known to develop as full a portrait of the terrorists of A1 Qaeda and the Taliban regime in Afghanistan, certain common features warrant brief comment. For one thing, the class origins of the A1 Qaeda terrorists appear to be similar to these other groups. Virtually all the young men were under 25, well educated, lower middle class, downwardly mobile (see also Kristof, 2002a).

Other terrorist groups in the Middle East appear to have appealed to similar young men, although they were also organized by theology professors—whose professions were also threatened by continued secularization and westernization. The Taliban, itself formed in 1994 by disaffected religious students, seems to have drawn from a less fortunate class. Taliban soldiers were uneducated, and recruits were drawn often from refugee camps in Pakistan, where they had been exposed to the relative affluence of the West through aid organizations and television (see Marsden, 2002: 70).

Several of the leaders of A1 Qaeda are wealthy. Ayman al-Zawahiri, the 50-year-old doctor who was the closest advisor to Osama bin Laden, is from a fashionable suburb of Cairo; his father was dean of the pharmacy school at the university there. Bin Laden was a multi-millionaire. By contrast, many of the hijackers were engineering students, for whom job opportunities had been dwindling dramatically. (From the minimal information I have gleaned, about one-quarter of the hijackers had studied engineering.)...

Most of these Islamic radical organizations developed similar political analyses. All were opposed to globalization and the spread of western values; all opposed what they perceived as corrupt regimes in several Arab states (notably Saudi Arabia and Egypt), which were mere puppets of U.S. domination. Central to their political ideology is the recovery of manhood from the devastatingly emasculating politics of globalization....

This fusion of anti-globalization politics, convoluted Islamic theology and virulent misogyny has been the subject of much speculation. Viewing these through a gender lens, though, enables us to understand the connections better. The collapse of public patriarchal entitlement led to a virulent and violent reassertion of domestic patriarchal power. "This is the class that is most hostile to women," said the scholar Fouad Ajami (cited in Crossette, 2001: 1). But why? Journalist Barbara Ehrenreich (2001) explains that while "males have lost their traditional status as farmers and bread-winners, women have been entering the market economy and gaining the marginal independence conferred even by a paltry wage." As a result, "the man who can no longer make a living, who has to depend on his wife's earnings, can watch Hollywood sexpots on pirated videos and begin to think the world has been turned upside down."

Taliban policies were designed to both remasculinize men and to refeminize women. "The rigidity of the Taliban gender policies could be seen as a desperate attempt to keep out that other world, and to protect Afghan women from influences that could weaken the society from within" (Marsden, 2002: 99). Thus, not only were policies of the Afghani republic that made female education compulsory immediately abandoned, but women were prohibited from appearing in public unescorted by men, from revealing any part of their body, or from going to school or holding a job. Men were required to grow their beards, in accordance with religious images of Mohammed—but also, I believe, because wearing beards has always been associated with men's response to women's increased equality in the public sphere. Beards especially symbolically reaffirm biological natural differences between women and men, even as they are collapsing in the public sphere. Such policies removed women as competitors and also shored up masculinity, since they enabled men to triumph over the humiliations of globalization, and as well to triumph over their own savage, predatory and violently sexual urges that would be unleashed in the presence of uncovered women.

Perhaps this can be best seen paradigmatically in the story of Mohammed Atta, apparently the mastermind of the entire operation and the pilot of the first plane to crash into the World Trade Center Tower. The youngest child of an ambitious lawyer father and pampering mother, Atta grew up a shy and polite boy. "He was so gentle," his father said. "I used to tell him 'Toughen up, boy!'" (cited in *The New York Times Magazine,* 7 October). Atta spent his youth in a relatively shoddy Cairo neighborhood. Both his sisters are professionals—one is a professor, the other a doctor. Atta decided to become an engineer, but his "degree meant little in a country where thousands of college graduates were unable to find good jobs."... His father had told him he "needed to hear the word 'doctor' in front of his name. We told him your sisters are doctors and their husbands are doctors and you are the man of the family." After he failed to find employment in Egypt, he went to Hamburg, Germany, to study to become an architect. He was "meticulous, disciplined and highly intelligent," yet an "ordinary student, a quiet friendly guy who was totally focused on his studies" according to another student in Hamburg.

But his ambitions were constantly thwarted. His only hope for a good job in Egypt was to be hired by an international firm. He applied and was constantly rejected. He found work as a draftsman—highly humiliating for someone with engineering and architectural credentials and an imperious and demanding father—for a German firm involved with razing lower-income Cairo neighborhoods to provide more scenic vistas for luxury tourist hotels. Defeated, humiliated, emasculated, a disappointment to his father and a failed rival to his sisters, Atta retreated into increasingly militant Islamic theology. By the time he assumed the controls of American Airlines flight 11, he evinced a gendered hysteria about women. In the message he left in his abandoned rental car, he made clear what really mattered to him in the end. "I don't want pregnant women or a person who is not clean to come and say good-bye to me," he wrote. "I don't want women to go to my funeral or later to my grave" (on CNN, 2 October 2001).

MASCULINE ENTITLEMENT AND THE FUTURE OF TERRORISM

Of course such fantasies are the fevered imagination of hysteria; Atta's body was without doubt instantly incinerated, and no funeral would be likely. But the terrors of emasculation experienced by the lower middle classes all over the world will no doubt continue to resound for these young men whose world seems to have been turned upside down, their entitlements snatched from them, their rightful position in their world suddenly up for grabs. And they may continue to articulate with a seething resentment against women, "outsiders," or any other "others" perceived as stealing their rightful place at the table.

The common origins and common complaints of the terrorists of September 11 and their American counterparts were not lost on American white supremacists (see also Kristof, 2002b). In their response to the events of September 11, American Aryans said they admired the terrorists' courage, and took the opportunity to chastise their own compatriots. Bill Roper of the National Alliance publicly wished his members had as much "testicular fortitude' (Intelligence Report, Winter 2001). 'It's a disgrace that in a population of at least 150 million White/Aryan Americans, we provide so few that are willing to do the same," bemoaned Rocky Suhayda, Nazi Party chairman from Eastpointe, Michigan. "A bunch of towel head/sand niggers put our great White Movement to shame" (cited in Ridgeway, 2001: 41). It is from that gendered shame that mass murderers are made.

BIBLIOGRAPHY

Bjorgo, Tore (1997) *Racist and Right-Wing Violence in Scandinavia: Patterns, Perpetrators, and Responses.* Leiden: University of Leiden.

Cornell, R. W. (1998) "Masculinities and Globalization," *Men and Masculinities* 1(1).

Coplon, J. (1989) "The Roots of Skinhead Violence: Dim Economic Prospects for Young Men," *Utne Reader* May/June.

Crossette, B. (2001) "Living in a World Without Women," *The New York Times* 4 October.

Dobratz, B. and Shanks-Meile, S. (2001) *The White Separatist Movement in the United States: White Power! White Pride!* Baltimore, MD: Johns Hopkins University Press.

Durham, Martin (1997) "Women and the Extreme Right: A Comment," *Terrorism and Political Violence* 9: 165–8.

Ehrenreich, B. (2001) "Veiled Threat," *The Los Angeles Times* 4 November.

Fangen, Katrine (1998a) "Living Out our Ethnic Instincts: Ideological Beliefs among Rightist Activists in Norway," in Jeffrey Kaplan and Tore Bjorgo (eds.) *Nation and Race: The Developing Euro-American Racist Subculture*, pp. 202–30. Boston, MA: Northeastern University Press.

Fangen, Katrine. (1999a) "Pride and Power: A Sociological Interpretation of the Norwegian Radical Nationalist Underground Movement," PhD dissertation, Department of Sociology and Human Geography, University of Oslo.

Fangen, Katrine (1999b) "Death Mask of Masculinity," in Soren Ervo (ed.) *Images of Masculinities: Moulding Masculinities*. London: Ashgate.

Ferber, A. L. (1998) *White Man Falling: Race, Gender and White Supremacy*. Lanham, MD: Rowman and Littlefield.

Kristof, N. (2002a) "What Does and Doesn't Fuel Terrorism," *The International Herald Tribune* 8 May: 13.

Kristof, N. (2002b) "All-American Osamas," *The New York Times* 7 June: A-27.

Marrs, T. (1993) *Big Sister is Watching You: Hillary Clinton and the White House Feminists Who Now Control America—And Tell the President What To Do*. Austin, TX: Living Truth Publishers.

Marsden, P. (2002) *The Taliban: War and Religion in Afghanistan*. London: Zed Books.

Ridgeway, J. (2001) "Osama's New Recruits," *The Village Voice* 6 November.

62

Intersectionality in a Transnational World

BANDANA PURKAYASTHA

...SOCIAL LIVES IN TRANSNATIONAL SPACES

Over the past decade, a rapidly growing literature has described how individuals and groups maintain connections across countries so that social lives are constructed, not only in single countries, but in transnational spaces. Transnational spaces are composed of tangible geographic spaces that exist across multiple nation-states *and* virtual spaces. With improvements in personal and media communication and travel technology, the ability to move money easily across the globe, and the marketing and ease of consuming "cultural" products—including fashions, cosmetics, music, foods, and art—have made it easier for many groups to create lives that extend far beyond the boundaries of single nation-states. We now know about first-generation immigrant "transnational villagers" who build lives in more than one country by traveling back and forth regularly, organizing family lives across countries, and remitting and investing money, as well as engaging in politics in "homelands".... We also know about post-immigrant generations who actively maintain links with their parents' homelands; ... cyber migrants who work for Northern employers but are geographically based in the South; ... and participants in web-based communities, some of whom seek community, while others try on less essentialist, choice-driven, multiple, fragmented, and hybrid identities on the web and thus dilute the consequences of gendering, racialization, class, and other social hierarchies to which they are subjected in their tangible lives.... As a rapidly growing number of people are tied to transnational spaces—that is, they build lives that combine intersecting local, regional, national, and transnational spaces—single nation-states no longer wholly contain their lives.

At the same time, nation-states have responded in a variety of ways to control social lives in transnational spaces. For instance, the literature on immigration shows how nation-states are creating gendered categories of "overseas citizens"

SOURCE: Purkayastha, Bandana. 2012. "Intersectionality in a Transnational World," *Gender & Society* 26: 55–66.

in order to attract remittances from migrants ... or to draw on the expertise or lobbying power of people settled in other countries (Purkayastha 2009). Equally important, nation-states have attempted to expand their ability to control people across transnational spaces; ideologies, interactions, and institutions that have sustained raced/gendered/classed and other hierarchies *within nations* have expanded in new ways across nations.... For most of the twentieth century, nation-states maintained separate apparatus for controlling groups within nations (e.g., police, prisons) from the apparatus used to dominate and control groups/states outside nations (e.g., the military, foreign intelligence agencies, facilities to house prisoners of wars). Now, these tools of control are increasingly blurred within nations; for instance, policies such as the PATRIOT Act and organizations such as Homeland Security have blurred the distinction between foreign surveillance and national surveillance in the United States. Security agreements *across* nations have created transnational security regions, where profiles developed in one powerful country are likely to be rapidly disseminated and acted on in other countries within the transnational security regime (Purkayastha 2009; Vertovec 2001). A *suspect* in a terrorism case in Scotland or Spain can, almost immediately, be arrested in Australia or the United States.... The profile of a "turbaned terrorist" or the suspicions against "Muslim-looking" people have generated contemporary racial profiles so that Sikh men—who wear turbans to comply with their religious tenets—and a range of people of Middle Eastern and South Asian origin are profiled as potential security threats. They are searched more stringently at airports, subject to extra questioning at national borders, subject to surveillance for communicating with people in "enemy countries," frequently visiting these countries, or sending money to "suspect" organizations through institutional arrangements as they travel through security regime (see Iwata and Purkayastha 2011; Purkayastha 2009). These new global security arrangements intersect with other processes for controlling racially marked populations within nations.

Overall, then, transnational spaces are composed of tangible and virtual social spaces that exist through and beyond single nation-states. Individuals, groups, corporations, and nation-states continue to expand their purview into such spaces.... At the same time, those who cannot access transnational spaces—for a variety of reasons, including the digital divide and stringent government control over travel and internet access—are marginalized in new ways within this expanded context. Contemporary discussions of marginalization and privilege have to take these new developments into account.

INTERSECTIONALITY IN TRANSNATIONAL SPACES

... [I]t is not always clear when and how we are to conceptualize "race" within the intersectionality matrix if we study transnational social lives.

I will begin with a simple example that focuses on women of color. A Ugandan Black immigrant and a Ugandan Indian immigrant—whose family lived for many generations in Uganda before being forcibly evicted by Idi Amin—are both racially marginalized, though in different ways, in the United States. While both share the effects of gendered/racialized migration policies that would prohibit or slow the process through which they might form families in the United Slates, their experiences differ in other ways. The Ugandan Black migrant is likely to experience the gamut of racisms experienced by African Americans, while the Indian Ugandan is likely to experience the racisms faced by Muslims and "Muslim-looking" people in the United States, and they may share other structural discriminations experienced by Asian Americans (Narayan and Purkayastha 2009). These similarities and differences are consistent with racist ideologies, interactions, and institutional arrangements in the United States. But if both return to their home country Uganda, they would encounter a different set of privileges and marginalization in this Black-majority country; the Black Ugandan migrant is advantaged here (though the other intersecting factors would together shape her exact social location). If both visit or temporarily live in India, the Indian-origin Uganda-born person may experience the privileges associated with the dominant group in the country. However, if she is a Muslim or a low-caste Hindu, she might experience a different set of social hierarchies. Similarly, Japanese-origin people from Brazil who returned to Japan, or Japanese-origin Americans who were forced to return to Japan, encountered different sets of social hierarchies.... A broadly similar argument could be made about the relative position of Blacks and Indians under the different historical circumstances, for instance, during the apartheid regime and after apartheid in South Africa (see Govinden 2008).

There are variations of who is part of the privileged majority versus the marginalized minority *within* a country, and these hierarchies do not always fit the white-yellow/brown-Black hierarchy extant in Western Europe and North America. Thus concepts such as "women of color"—which act as an effective framework for indicating the social location of these women in Western Europe and North America, and continuing global hierarchies *between* countries in the global North and South—do not work as well if we wish to track the array of the axes of power and domination within countries *along with* existing global-level hierarchies. Yet considering these multiple levels is important if intersectionality is to retain its explanatory power in an increasingly transnational world where within-country *and* between-country structures shape people's experiences.

The possibility of forming community on virtual spaces and using the web to maintain meaningful connections with people in other countries also emphasizes the need to consider transnational spaces. A South African Black female immigrant in the United States who is able to maintain active connections with her friends, family, and political networks in her home country (via phone, email, and a variety of web-based media) may be able to minimize some of the toll of racism she experiences by making her South African relationships most salient in her life. The Indian-origin post-immigrant-generation American who regularly participates in a religio-social Hindu online community and visits India regularly is also able to position herself as a member of the majority group in India (and the Indian diaspora) even as she experiences the deleterious effects of structural racism in the United States. In other words, people who can access transnational social spaces attempt to balance their lack of privilege in one country (their raced/classed/gendered/ability/sexual/age/nationality status in one nation-state) by actively seeking out privilege and power in another place and/or in virtual spaces.

... Being able to build transnational lives—the ability of groups to live within and beyond single nation-states—suggests that it is quite possible for groups to be part of *the racial majority and minority simultaneously* (Purkayastha 2010). Indeed, in places where caste and religious or ethnic hierarchies—with their own set of ideologies, interactions, and institutional structures—are more salient, we should consider the relative importance of these axes of domination within those countries (and the extent to which these structure transnational social lives) as we use intersectional frameworks.

I do not intend to suggest that we stop considering racial hierarchies. Along with variations in who makes up the racial majority or minority *within* different countries, hierarchies among nation-states continue to promote Western hierarchies of race and whiteness, yellowness, and Blackness across the world in ways that are broadly similar to the period of colonialism (see, e.g., Gilroy 1989; Kim 2008; Kim-Puri 2007; Nandy 2006; Sardar et al. 1993). Such racial hierarchies are maintained through ideologies, actions, and institutional arrangements associated with political and economic control. As Evelyn Nakano Glenn and her colleagues have documented, color-based hierarchies continue to structure people's lives in many countries around the globe, especially as "fairness as beauty" is marketed to places where the majority of the people are nonwhite (Glenn 2008)....

INTERSECTIONALITY AND A NOTE ON RELIGION

... It is important to think further about our assumptions of religion within the framework of intersectionality. As I mentioned earlier in this essay, religion (i.e., the idea that certain religions promote tendencies toward violence and terrorism) is being used to create racial profiles within nations and across nations....[U]ndocumented immigrants are not the only ones who have been targeted; government agencies and organized groups have targeted "Muslims" irrespective of the state of documentation.... Such surveillance extends to transnational spaces—for example, communications in virtual spaces can now be monitored by several countries for security purposes; people originating from several countries of the world now need transit visas to touch down in airports in Western Europe—so that it seems important to consider these religion-based forms of racialization at national and transnational levels. As Evelyn Nakano Glenn described in her work *Unequal Freedom* (2002), the construction of the category "American" is predicated on the ability to create and sustain hierarchies between Christians and non-Christians, native and foreign, and white and nonwhite.... A similar process is underway again in our times. We have understood that culture and phenotype form the bases of racism structures. Now, we have to pay attention to the ways in which processes of marking religions, that is, marking phenotypes, cultures, and nationalities, acts in the service of racism.

LAST REMARKS

... Understanding and attending to the complexities of transnationalism—composed of structures within, between, and across nation-states, and virtual spaces—alerts us to look for other axes of domination and the limits of using "women of color" concepts, as we use them now, to look across *and* within nation-states to understand the impact of transnationalism. My examples here were focused on those who can access transnational spaces. A focus on transnational intersectionality should alert us to the position of those who are unable to afford access to technology to build virtual communities, to participate in a medium because they are not proficient in English, which has become the dominant language in virtual spaces or to build transnational social lives because of active government surveillance and control of their lives or because they are too poor and isolated to access transnational tangible and virtual spaces....

REFERENCES

Gilroy, Paul. 1989. *There ain't no Black in the Union Jack*. Chicago: University of Chicago Press.

Glenn, Evelyn Nakano. 2002. *Unequal freedom: How race and gender shape American citizenship*. Cambridge, MA: Harvard University Press.

Glenn, Evelyn Nakano. 2008. *Shades of citizenship*. Stanford, CA: Stanford University Press.

Govinden, Devarakhsnam. 2008. *Sister outsiders: The representation of identity and difference in selected writings by South African Indian women*. Pretoria: University of South Africa Press.

Iwata, Miho, and Bandana Purkayastha. 2011. Cultural human rights. In *Human rights in our backyard: Social justice and resistance in the U.S.*, edited by William Armaline, Davita Glasberg, and Bandana Purkayastha. Philadelphia: University of Pennsylvania Press.

Kim, Nadia. 2008. *Imperial citizens: Koreans and race from Seoul to LA*. Stanford, CA: Stanford University Press.

Kim-Puri, H.-J. 2007. Conceptualizing gender/sexuality/state/nation: An introduction. *Gender & Society* 19:137–59.

Nandy, Ashis. 2006. *The intimate enemy: Loss and recovery of self under colonialism*. New Delhi, India: Oxford University Press.

Narayan, Anjana, and Bandana Purkayastha. 2009. *Living our religions: South Asian Hindu and Muslim women narrate their experiences*, Stirling, VA: Kumarian Press.

Purkayastha, Bandana. 2009. Another word of experience? South Asian diasporic groups and the transnational context. *Journal of South Asian Diasporas* 1:85–99.

Purkayastha, Bandana. 2010. Interrogating intersectionality: Contemporary globalization and racialized gendering in the lives of highly educated South Asian Americans and their children. *Journal of Intercultural Studies* 31:29–47.

Sardar, Ziauddin, Ashis Nandy, Merryl Wyn Davies, and Claude Alvares. 1993. *The blinded eye: 500 years of Christopher Columbus*. New York: Apex Press; Goa, India: The Other India Press.

Vertovec, Steven. 2001. Transnational challenges to the "new" multiculturalism. www.transcomm.ox.ae.uk/working papers.

Index

Note: Page numbers followed by "f" or "t" indicate figures and tables, respectively.

N